Generating Electricity in a Carbon-Constrained World

Edited by

Fereidoon P. Sioshansi

Menlo Energy Economics

AMSTERDAM • BOSTON • HEIDELBERG • LONDON
NEW YORK • OXFORD • PARIS • SAN DIEGO
SAN FRANCISCO • SINGAPORE • SYDNEY • TOKYO

ELSEVIER Academic Press is an imprint of Elsevier

Academic Press is an imprint of Elsevier
30 Corporate Drive, Suite 400, Burlington, MA 01803, USA
525 B Street, Suite 1900, San Diego, California 92101-4495, USA
84 Theobald's Road, London WC1X 8RR, UK

Library of Congress Cataloging-in-Publication Data
Application submitted

British Library Cataloguing-in-Publication Data
A catalogue record for this book is available from the British Library.

ISBN: 978-1-85617-655-2

For information on all Academic Press publications
visit our Web site at www.elsevierdirect.com

Printed in the United States of America
09 10 11 9 8 7 6 5 4 3 2 1

Contents

This book is dedicated to the memory of
João Lizardo Rodrigues Hermes De Araujo
(1941–2008)

Lizardo in front, left, editor and his wife are behind him, Professor Reinhard Hass, TUV front right.
Photo taken June 2006 in Potsdam, Germany.

Lizardo, as he was known to his friends and colleagues, was born on May 26, 1941, in Manaus in Amazonas, Brazil. In 1965 he married Hildete Pereira de Melo, with whom he had three children: Diogo, Pedro, and Rodrigo.

Lizardo earned a B.S. in electronic engineering from the Technological Institute of Aeronautics (ITA) in 1964 and a Ph.D. in applied mathematics from the University of Toulouse in France in 1968.

Lizardo dedicated 43 years to teaching and research, including supervising 26 master's and doctoral theses by his students. His scientific career began at the Federal University of Paraíba in 1965, followed by a long association with the Graduate School in Engineering (COPPE) at the Federal University of Rio de Janeiro, starting in 1970. At COPPE, he participated in the creation of the Program of Systems Engineering and of the Interdisciplinary Area of Energy. From 2004 to 2008 he was general director of the Electric Energy Research Centre (CEPEL) of the Brazilian Electrical Centrals (ELETROBRÁS).

He was a visiting scholar at the Imperial College in London, the Lawrence Berkeley National Laboratory in Berkeley, California, and the Science Policy Research Unit (SPRU) at the University of Sussex in the United Kingdom.

During his distinguished career, Lizardo published numerous articles, book chapters, and books on energy, optimization, mathematical modelling, and regulation. He contributed two chapters on the Brazilian electricity sector to two prior books by the editor of this volume, published in 2006 and 2008. At the time of his death, Lizardo was in the midst of completing a third chapter for the present volume.

Lizardo was a man of integrity and wisdom who blessed all who knew him as family, friend, teacher, and colleague.

Foreword

Electricity has become such a fundamental part of our lives that we take it for granted. Despite this fact, generating power has faced many challenges as it developed and now finds itself at the center of another storm: how to prevent massive disruption to the world's climate. In facing this challenge, electricity curiously finds itself with a Janus's head, looking both ways: simultaneously under pressure as the world's biggest source of greenhouse gas emissions, yet widely touted as the solution to other, even more intractable sources of CO_2 from transport and even heat. Whether electricity can really both decarbonize and expand hinges on one main question: whether a carbon-constrained world can effectively foster low-carbon electricity generation at scale.

With a new U.S. administration at last committed to getting serious about climate change and global negotiations gathering pace toward a grand deal on new commitments after the current provisions of the Kyoto Protocol expire, this book presents a timely overview and analysis of this question. In its sheer breadth and scope it represents a hugely valuable and ambitious undertaking.

The book is structured neatly in three broad parts. Part 1 sets the context and sketches the different policy instruments potentially in play—the emerging experience and some of the new frontiers. Part 2 assesses the technical options for decarbonizing power generation—a wide sweep of options assessed from standpoints around the world. Part 3 looks at the initiatives, plans, and prospects in five diverse regions of the world before homing in on some of the key initiatives and instruments in the United States. In adopting this approach, the book neatly frames the way that the combined challenges of policy and technological choice are played out in many diverse locations—with lessons for all.

I recently had the experience of editing a volume focused just on the prospects for decarbonizing the power sector in the United Kingdom.[1] It proved a much bigger task than we had bargained for, though highly relevant, as the U.K. Climate Change Committee, established to advise the U.K. government on future carbon targets, subsequently recommended an ambitious set of goals based on underlying analysis that placed the power sector at the absolute heart of implementation. To attempt such analysis at a global level

---

[1] *A low-carbon electricity system for the UK: technology, economics and policy*, Cambridge University Press, May 2008.

is a daunting but wonderful undertaking, and Perry Sioshansi and all the contributing authors should be congratulated for a tremendous compendium of insights and experience from around the world. If you want a timely, diverse, and global perspective on one of the great challenges of our time—read this book.

Michael Grubb
November 2008

Preface

Reducing the Carbon Footprint: A Multidimensional Problem

Wolfgang Pfaffenberger

Decarbonization of energy supply is among the key issues facing policymakers in the years ahead. And it is a daunting task, given the enormity and immediacy of the problem as described in the 2008 edition of the *World Energy Outlook,* released by the International Energy Agency (IEA) in November. To address the problem requires careful consideration and balance among multiple dimensions, technical, economic, social, and political. In this preface I give a short summary of these dimensions and their implications for successful approaches to address the problem.

The technical dimension

To get an idea about the possibilities of reducing carbon emissions, it is useful to take a look at the factors determining the level of emission. The following formula contains six elements, highlighted beneath with reflections on their relevance for decarbonizing electricity generation—the topic of the present volume.

$$CO_2 = CO_2 \ / \ C * C \ / \ PE * PE \ / \ FE * FE \ / \ UE * UE \ / \ GDP * GDP^1$$
$$\quad\quad (1) \quad\quad (2) \quad\quad (3) \quad\quad (4) \quad\quad (5) \quad\quad (6)$$

C = carbon, PE = primary energy consumption, FE = final energy consumption, UE = useful energy consumption (after losses), GDP = gross domestic product

---

1. *CO_2 emission per unit of carbon.* At present this factor is equal to 1, meaning that for one unit of carbon burned, one unit of CO_2 will be released.

---

[1]Hensing, Pfaffenberger, and Ströbele (1998; S. 183).

Theoretically there is the possibility of reducing this factor by using an appropriate "end-of-the-pipe" technology. Such technologies are currently in an experimental phase, and only future research and development will teach us whether it will become a technologically viable and economically feasible option in the future that can contribute considerably to decarbonizing the electricity generation industry.

2. *Carbon content of primary energy.* The fuel mix available for electricity generation determines the amount of CO_2 emitted per unit of electricity. In this context, fuel switching provides an important potential source for reducing carbon emissions. The availability of low-carbon or carbon-free fuels is partly determined by natural conditions (e.g., potential for hydropower, availability of wind resources sites, and many other factors) and partly by the political framework and policies (e.g., availability of nuclear power generation, among others).

3. *Primary energy needed per unit of electricity or efficiency of energy transformation from primary energy into electricity.* Carbon emissions are reduced if fuels are used more efficiently in the process of generating electricity. The average efficiency of fossil-fuel plants in many countries is far below the efficiency level available by present best technology. So, modernization of existing suboptimal plants offers significant opportunities for reducing carbon emissions.

   For thermal plants, the efficiency can be further improved if it is possible to use some of the by-product heat for heating and cooling processes in a combined heat and power (CHP) process, also called *cogeneration.* For this to be effective, an integrated approach for producing heat and electricity is required. For some time, there has been the vision for distributed generation allowing such an integrated approach. Increasing the future share of CHP is an important determinant of CO_2 emissions from the electricity sector. How much of this potential will be utilized in the future depends on many currently unknown variables, including availability of low-cost, durable CHP generators, network management for a high share of distributed generation, compatibility of CHP with a higher share of renewables from intermittent sources, and other factors.

4. *Electricity needed per unit of useful energy demanded (distribution and end-use efficiency).* It is a well-known fact that electricity consumers use electricity to meet specific energy service needs such as lighting, heating, or air conditioning (AC). The amount of electricity needed to produce one unit of AC service depends on the losses of electricity in the generation, transmission, and distribution system *and* in the AC device that delivers the final service. The losses in distribution in modern systems are relatively low and can be controlled by the network operators, but the loss in the end-use appliances depends on the quality of the appliances, and there

may be a trade-off between the efficiency of the appliance and its price. If consumers are nearsighted[2] and select appliances with the lowest front-end costs, they will end up with higher electricity consumption over the life of the devices. This represents a lost opportunity for society and a market failure requiring some sort of policy intervention.[3]

5. *Useful energy needed per unit of GDP.* Following with the energy services example, the amount of AC service needed to maintain a comfortable temperature in a building is dependent on the quality of the building, notably the amount of insulation. The better the building is protected against external temperature fluctuations, the lower will be the demand for AC services. In addition, the level of required AC services depends on the habits of the building habitants. Both of these factors are outside the control of the electricity generation industry but can be influenced through education, better monitoring and thermostats, and other means.

Modernization of capital stock and change of habits and lifestyles are important factors for decarbonization of electricity generation. As the chapters of this book make clear, we cannot rely entirely on changes in the supply side of the equation to reduce the industry's carbon footprint. Changes in the demand side as well as changes in energy consumption habits and—perhaps more profoundly—lifestyle changes could ultimately be needed to address the carbon problem.

6. *The level of economic activity.* The level of economic activity clearly influences the level of energy use and carbon emissions. The relationship between energy use and economic output, however, is not linear. Higher economic growth, for example, may lead to substitution of more efficient capital stock, with positive consequences for carbon emissions. Many studies have documented that robust economic growth may be sustained with frugal energy use. The relationship between aggregate energy use and economic output plays a significant role, especially for rapidly developing economies such as China and India.

The economic dimension

Looking at the various factors influencing carbon emissions to find an efficient path for reducing carbon emissions, one might expect marginal cost of reducing CO_2 to be about the same for all alternative options. Many studies,

---

[2]Stoft (2008, S. 9) mentions nearsightedness of consumers as one of the important problems for CO_2 reduction.
[3]Chapter 8, by Prindle et al., in this volume covers some of these market failures and how to address them.

however, have shown[4] that the cost of reducing CO_2 is considerably different between sectors and for applications within each sector. To find the most economically efficient path, it is important to seek and pursue opportunities for carbon reduction that have the lowest costs among all the sectors of the economy.

In this context, it is not economically efficient, for example, to pursue high-cost but low-carbon opportunities in the electricity generation sector if electricity conservation can produce the same results at lower costs. By the same token, if low-cost opportunities exist in the transportation sector, these must be pursued before higher-cost opportunities in electricity generation are captured. To reach an efficient equilibrium, an integrated approach is necessary across all sectors—on a global scale, since the carbon problem is global in scope.

The social dimension

Economists are mainly interested in rational allocation of resources. In society, however, people have an implicit understanding of the free-riding potential available in a complex and partly transparent social system. Therefore, aspects of fairness and distribution are of high importance, and often there is a strong tension between the perceived or expected distributional impact of changes and their economic rationale. The pure economic solution is therefore often not socially acceptable. Reducing the carbon footprint is not free and will have significant cost impacts on many components of the economy. The transformation to a decarbonized economy is a secular task. It is quite likely, therefore, that the political system representing the social dimension and the economic sectors facing the changes will be in conflict. In such a situation, "irrational" solutions are often the only solutions available. The current European debate on climate policy delivers many examples of irrational decisions that are politically expedient.[5] A recent publication by Stoft[6] gives several interesting ideas on how to shape more rational policies.

The political dimension

The problem's political dimension is further complicated by the different time horizons affecting climate change (decades to centuries); the time horizon of assets in the electricity generation business (decades); and the time horizon

---

[4]For example, see Vattenfall (2007).
[5]See Weimann (2008) for more details.
[6]Stoft (2008).

of consumers, voters, and politicians (typically months to years). Decision making in these various segments necessarily follows different patterns with different discount rates and different priorities. To address these significant disparities, yet again, an integrated approach is needed. In my view, this could be the most difficult challenge facing us in trying to address the carbon problem.

I am convinced that a market solution would be best for electricity-producing companies as well as society as a whole. For such a market solution to be successful, we need a clearly defined set of long-term goals for CO_2 reduction that would give investors a clear framework for decision making and for energy-intensive consumers a level of cost transparency and certainty that currently does not exist.

The likelihood of international agreements that would help stabilize the concentration of carbon in the atmosphere is at best uncertain at present. If nations or groups of nations such as the United States, the European Union, and others follow different targets and strategies defined by themselves, a large free-rider potential is created and issues of international competitive advantages and disadvantages will arise. An implicit condition for a successful global climate policy is international cooperation.

REFERENCES

[1] Hensing I, Pfaffenberger W, Ströbele W. Energiewirtschaft. München: Oldenbourg; 1998.
[2] IEA. World Energy Outlook. Paris; 2008.
[3] Stoft S. Carbonomics: How to Fix the Climate and Charge It to OPEC. Diamond Press: Nantucket, Mass; 2008.
[4] Vattenfall. Global Mapping of Greenhouse Gas Abatement Opportunities up to 2030. Stockholm; 2007.
[5] Weimann J. Die Klimapolitik katastrophe. Marburg: Metropolis; 2008.

About the Contributors

Parviz M. Adib is director with APX Inc., a leading platform provider for energy and environmental markets, including carbon commodities. He is involved in consulting services covering power and environmental markets. His responsibilities include analyses of markets and development of successful strategies, including development of renewable energy credits and carbon offsets.

Dr. Adib's main research interests include the development of environmental commodity markets, energy and public policy, and market oversight. Prior to joining APX, Dr. Adib worked at the Public Utility Commission of Texas in various capacities, including the director of the Market Oversight Division, where he performed the duties of the ERCOT Market Monitor. He is also the former chairman of the Energy Intermarket Surveillance Group.

Dr. Adib has a Ph.D. in Economics from the University of Texas at Austin, where he taught graduate and undergraduate courses.

Erica Allis is currently a consultant for the United Nations Environment Programme, Division of Technology, Industry, and Economics (UNEP DTIE), where she provides technical assistance and policy recommendations in Cleaner Production and Energy Efficiency (CPEE) initiatives in developing economies. In this position she works to promote sustainable production and consumption patterns among industry professionals and consumers.

Ms. Allis's research interests include water and energy policy and conflict resolution methods in resource-scarce regions, where competing demands have the potential to escalate into conflict.

Ms. Allis holds a bachelor of science in geology from the University of Arkansas and a master's of public policy from the Lyndon B. Johnson School of Public Affairs at the University of Texas.

Geoff Bertram is senior lecturer in economics at Victoria University of Wellington, New Zealand. Prior to joining the faculty at Victoria University, he was employed as a research officer at the Institute of Economics and Statistics at Oxford University.

Dr. Bertram's research interests include the economics of utility regulation and industry restructuring, with particular reference to energy and infrastructure sectors. He has also published widely on the economics of small island states and the economic histories of New Zealand and Peru.

Dr. Bertram holds a B.A. Honors degree from Victoria University and a D. Phil. in economics from Oxford University. He has published over 50 peer-reviewed papers and books.

Jean-Paul Bouttes is executive vice president of Corporate Strategy, Prospective and International Affairs at Electricité de France (EDF). His responsibilities and main interests include issues related to the economics of the electricity sector and related public policies such as security of supply, climate change, competition, and regulation.

He joined EDF in 1982 and has held various positions in the Corporate Strategy Division, Economics Department, and Industrial Strategy for Generation and Trading. Mr. Bouttes was also professor of economics at Ecole Polytechnique in Paris.

Mr. Bouttes is a graduate of Ecole Polytechnique, Paris, and Ecole Nationale de la Statistique et de l'Administration Economique.

Marilyn Brown is a professor of energy policy in the School of Public Policy at the Georgia Institute of Technology. Her research has focused on the impacts of policies and programs aimed at accelerating the development and deployment of sustainable energy technologies. She has led several energy technology and policy scenario studies and is a national leader in the analysis and interpretation of energy futures in the United States. In this capacity, she has testified before Congress and state legislatures.

Prior to her present position she was at Oak Ridge National Laboratory, where she led programs in energy efficiency, renewable energy, engineering science, and technology. She remains a Distinguished Visiting Scientist at ORNL. Dr. Brown serves on the boards of directors of the American Council for an Energy-Efficient Economy and the Alliance to Save Energy and is a member of the National Commission on Energy Policy and the National Academies Board of Engineering and Environmental Systems. Her latest edited book, *Energy and American Society: Thirteen Myths,* was published in 2007 by Springer.

Dr. Brown has a Ph.D. in geography from the Ohio State University and a master's degree in resource planning from the University of Massachusetts.

Qimin Chai is a Ph.D. candidate of management science and engineering in the School of Public Policy and Management at Tsinghua University. He is an assistant researcher in the Institute of Energy and Environmental Economics and in the China Automotive Energy Research Center at the Tsinghua University Low Carbon Energy Laboratory. He is also a visiting scholar in the Precourt Institute at Stanford University.

Mr. Chai's main research interests include energy and environmental economics, climate change, and low-carbon energy policy.

Mr. Chai has a B.A. in engineering from Zhejiang University.

Emile J. L. Chappin is a Ph.D. researcher at the Faculty of Technology, Policy, and Management of Delft University of Technology. His research is on modeling and transition management of energy infrastructures.

Mr. Chappin's main research interests include modeling, simulation and gaming of complex systems, data analysis, visualization and interpretation, and life-cycle analysis.

He holds bachelor's and master's of science degrees in systems engineering, policy analysis, and management from Delft University of Technology.

Douglas Clover is a Ph.D. candidate at the School of Geography Environment and Earth Sciences, Victoria University of Wellington, where he is studying the environmental and energy implications that might arise from the increased use of electric vehicles.

Mr. Clover has worked extensively in energy and transport policy, focusing on the role of market and government measures that promote environmental sustainability. Before returning to study, he was the leader of the Parliamentary Commissioner for the Environment's energy and transport research team.

He has bachelor's degrees in zoology and forestry science and an M.Sc. in resource management, all from the University of Canterbury.

Nigel Cornwall is managing director of Cornwall Energy Associates, where he is engaged in U.K. market design, policy, and regulatory issues for regulators and market participants. He also edits and publishes *Energy Spectrum*, a weekly publication covering the United Kingdom's electricity markets.

He was intimately involved in the restructuring of the electricity market in Britain in the 1980s and has worked on a variety of market reforms elsewhere. More recently, he has focused on design and implementation of the new electricity trading arrangements, NETA, in Britain, and he was one of the main negotiators for electricity suppliers during the new market redesign process. Mr. Cornwall is active in industry governance processes in both the electricity and gas markets in the United Kingdom and is a regular commentator on the low-carbon agenda and its market impact.

Mr. Cornwall has a B.A. in modern history from Pembroke College, Oxford, and an M.Sc. in politics from Birkbeck College, London.

François Dassa is head of Corporate International Affairs within the Prospective and International Affairs Division of Electricité de France (EDF). His experience includes work at EDF's Corporate Strategy Division and in regulatory affairs. He cofounded EDF Italia, the commercial branch of EDF in Italy, where he acted as head of Key Accounts and deputy director for Strategic Affairs and Energy Procurement. He was also appointed visiting professor of economics at the Paris Dauphine University.

Mr. Dassa is a graduate of Ecole des Mines de Paris.

Paul Denholm is a senior energy researcher at the National Renewable Energy Laboratory (NREL), where he examines systems integration of renewable electricity generation sources such as wind and solar.

Dr. Denholm's main research interests include examining the technical, economic, and environmental benefits and impacts of large-scale deployment of renewable electricity generation, including the limits of intermittency and quantifying the need for enabling technologies such as energy storage, plug-in hybrid electric vehicles, and long-distance transmission.

Dr. Denholm holds a B.S. in physics from James Madison University, an M.S. in instrumentation physics from the University of Utah, and a Ph.D. in environmental studies and energy analysis from the University of Wisconsin-Madison.

Laurens de Vries is assistant professor at the Faculty of Technology, Policy, and Management of Delft University of Technology. He performs research and teaches in the field of electricity market design, analyzing the mutual relationships between the physical infrastructure and its economic organization and regulation. He focuses on long-term issues such as generation adequacy and restructuring power sectors in OECD as well as in developing countries.

Dr. De Vries has an M.Sc. in mechanical engineering from Delft University of Technology, holds an M.A. in environmental studies from the Evergreen State College, and obtained his Ph.D. from the Faculty of Technology, Policy, and Management of Delft University of Technology.

Easan Drury is a research associate at the National Renewable Energy Laboratory (NREL), where he works as a solar energy analyst in the Energy Forecasting and Modeling group, examining the technical and economic implications of large-scale renewable energy deployment.

Dr. Drury's main research interests include modeling the U.S. photovoltaics market by quantifying the impacts of PV costs, electricity rate structures, and state and federal incentives on PV adoption.

Dr. Drury has a B.A. in physics from the University of California, Berkeley, and an M.S. and Ph.D. in engineering sciences from Harvard University.

Gerard P. J. Dijkema is associate professor of Energy and Industry, Faculty of Technology, Policy, and Management, TU Delft. His specialization is systems innovation for sustainability. He serves as the chairman of Dutch price for innovation and sustainability and is an elected member of the General Council of the Delfland Water Authority.

His main areas of interest include understanding, development, and transition of energy infrastructure as large-scale sociotechnical systems, effectively and timely transformation of industry networks, and water and energy infrastructure systems. To this end, much of his research involves model-based decision support to help stakeholders develop sustainable innovations, policies, and strategies.

Dr. Dijkema graduated as a chemical engineer from Twente University of Technology and holds a Ph.D. from Delft University of Technology.

Tina Fawcett is a senior researcher based at the Environmental Change Institute within Oxford University's Centre for the Environment. Dr. Fawcett's main research interests are personal energy use and carbon dioxide emissions, U.K. and international energy policy, household energy efficiency, and personal carbon allowances.

Dr. Fawcett has a Ph.D. in energy policy from University College, London. She is coauthor of a book on climate change, energy use, and personal carbon allowances, *How We Can Save the Planet*, which has been published in both U.K. and U.S. editions.

Neilton Fidelis Silva is a professor at the Federal Institute of Education, Science and Technology of Rio Grande do Norte (IFE-RN), Brazil, and a researcher at COPPE, the Post-Graduate Engineering School of the Federal University of Rio de Janeiro (UFRJ).

He is currently working for the Presidential Office of the Brazilian Government as assistant to the Executive Secretary of the Brazilian Forum on Climate Change.

His research focuses on energy planning, renewable energy, and climate change.

Professor Silva's qualifications include a D.Sc. from the Federal University of Rio de Janeiro and a B.Sc. and M.Sc. in electrical engineering from the Federal University of Rio Grande do Norte.

Peter Fraser is manager of Wholesale Power Policy at the Ontario Energy Board. His responsibilities include electricity price regulation, oversight of wholesale electricity markets, regulatory aspects of electric reliability, and the review of power system plans.

Mr. Fraser's main professional interests are in electricity policy and markets. From 1998–2004, Mr. Fraser worked at the International Energy Agency (IEA) in Paris as senior electricity policy advisor. Prior to working for the IEA, he spent nearly 9 years as energy policy advisor at the Ontario Ministry of Energy.

Mr. Fraser holds master's degrees in physics from Queen's University, Kingston, and in environmental studies from York University, Toronto, and a B.Sc. in physics from the University of Toronto.

Rob Graber is vice president of EnergyPath Corporation, a consultancy advising clients on financial and investment strategies in energy industries. Mr. Graber has over 30 years of experience in energy industries and over 20 years at GE Nuclear Energy, where he directed financial and marketing strategy in GE's nuclear services business in San Jose, California.

Mr. Graber's current focus is on the applications of advanced methodologies to value power plants, particularly nuclear power plants, for decision makers in global power markets. Mr. Graber also has a research interest in the history of nuclear power's political, social, economic, and financial currents and the renewed interest in nuclear power throughout the world. He maintains an extensive database on global nuclear power performance, costs, and investments.

He has an advanced degree in nuclear engineering and received an MBA in economics with honors from the Stern School of Business at New York University.

Michael Grubb is chief economist at the U.K. Carbon Trust and chairman of the international research organization Climate Strategies. He is also a part-time senior research associate at Cambridge University and a visiting professor at Imperial College, London. He was recently appointed to the U.K. Climate Change Committee, established under the U.K. Climate Change Bill to advise the government on future carbon budgets and to report to Parliament on their implementation.

Professor Grubb is author of seven books and numerous journal publications and reports and most recently edited a book examining the economic and policy dimensions of developing a low-carbon electricity system in the United Kingdom. He has held advisory positions with governments, companies, and international studies on climate change and energy policy and has been a lead author for several IPCC reports on mitigation, including the IPCC Fourth Assessment Report. He is editor-in-chief of the journal *Climate Policy* and is on the editorial board of *Energy Policy* and *Environmental Science and Policy*. His most recent works have been directing Climate Strategies research on carbon market instruments, covering the evolution, competitiveness, and design dimensions of the EU ETS and assessing the international carbon mechanisms of the Kyoto Protocol.

He has a degree from Cambridge University and a doctoral thesis from the Cavendish Laboratory on the integration and analysis of intermittent electricity sources.

Maureen Hand is a senior engineer at the National Renewable Energy Laboratory (NREL) in the National Wind Technology Center. She recently participated in a collaborative effort among public and private organizations to investigate the potential for wind energy to provide 20 percent of future U.S. electricity needs. Her role was to coordinate the analytical effort to support the study.

Prior to this project, Dr. Hand investigated active control designs for utility-scale wind turbines. She also contributed to experimental investigation of wind turbine aerodynamics, both in field test conditions as well as the NASA National Full-Scale Aerodynamics Complex $80' \times 120'$ wind tunnel.

Dr. Hand has a B.S. in mechanical engineering from the University of Wyoming and holds an M.S. and a Ph.D. in mechanical engineering from the University of Colorado.

Frank Harris is senior environmental economist at Southern California Edison. He is responsible for evaluating environmental and energy legislation and regulation, providing policy recommendations to SCE management and developing SCE submittals to California regulatory agencies. In this capacity, he represents SCE before the California Air Resources Board, the California Public Utilities Commission, and the California Energy Commission.

Dr. Harris's main research interests include environmental policy and energy economics. Prior to his current position with SCE, he was an adjunct professor of economics at the University of California at Irvine.

Dr. Harris holds B.A. degrees in economics and finance from the University of Texas at Dallas. He holds a Ph.D. in economics from the University of California at Irvine, where he specialized in energy and health policy and public choice.

Udi Helman is a principal market economist with the California Independent System Operator (CAISO). His current work is focused on market design and policy analysis, including policies to achieve California targets for greenhouse gas emissions reductions and renewable portfolio standards.

Prior to joining CAISO he worked at the Federal Energy Regulatory Commission (FERC) on electricity market design issues, including wholesale energy, ancillary service and installed capacity markets, transmission usage pricing, and transmission property rights. He has published extensively, including contributing two book chapters to prior books edited by F. P. Sioshansi in 2006 and 2008.

He has B.A. and M.A. degrees from the University of Toronto and a Ph.D. from Johns Hopkins University in applied economics and systems analysis.

Roy Hrab is policy advisor in the Regulatory Policy and Compliance Branch of the Ontario Energy Board, where he develops policy options and recommendations related to electricity operations, conducts research on current and emerging issues, and undertakes strategic regulatory planning.

Prior to joining the Board, he was engaged in public policy and economic research projects at the Institute for Competitiveness and Prosperity and the Government of Ontario's Panel on the Role of Government.

He holds B.A. and M.A. degrees in economics from the University of Toronto.

Frede Hvelplund is professor in energy planning at the Department of Development and Planning, Aalborg University. He is a member of an interdisciplinary research group for sustainable energy with interest in public regulation, sustainable energy, political economy, and social anthropology.

His current research is focused on technologies that can facilitate transition from fossil fuel–based to sustainable energy systems. He is author and coauthor of several books and articles on the interrelationships between sustainable energy systems and public regulation.

Dr. Hvelplund's educational background is in economy and social anthropology, and he has a Dr.Techn. degree from Aalborg University.

Klaus S. Lackner is the Ewing-Worzel Professor of Geophysics in the Department of Earth and Environmental Engineering and a member of the Earth Institute at Columbia University. He is the director of the Lenfest Center for Sustainable Energy and the chair of the Department of Earth and Environmental Engineering in the School of Engineering and Applied Sciences. He is a member of GRT, a company that hopes to develop a commercially viable device to capture CO_2 directly from the atmosphere.

Professor Lackner's career includes work at the Stanford Linear Accelerator Center and Los Alamos National Laboratory. His current interests are focused on developing environmentally acceptable technologies for the use of fossil fuels, including carbon capture and sequestration, zero-emission coal plants, carbon electrochemistry, and the study of large-scale energy infrastructures.

Dr. Lackner received his Ph.D. in 1978 in theoretical physics from the University of Heidelberg and held postdoctoral positions at the California Institute of Technology.

Joanna Lewis is an assistant professor of science, technology, and international affairs at Georgetown University's Walsh School of Foreign Service.

Her research focuses on renewable energy industry and policy development in China, mechanisms for low-carbon technology transfer in the developing world, and expanding options for multilateral engagement in a post-2012 international climate change agreement. Dr. Lewis also serves as the primary technical advisor for the Asia Society's Initiative for U.S.-China Cooperation on Energy and Climate and as an international advisor to the Energy Foundation China Sustainable Energy Program in Beijing. She has previously worked at the Pew Center on Global Climate Change and Lawrence Berkeley National Laboratory and was a visiting scholar at Tsinghua University in Beijing.

Professor Lewis holds a master's degree and a Ph.D. in energy and resources from the University of California, Berkeley, and a bachelor's degree in environmental science and policy from Duke University.

Robert M. Margolis is a senior energy analyst in the Washington, DC, office of the National Renewable Energy Laboratory (NREL), where he leads analysis of markets and policies related to solar photovoltaics.

His main research interests include energy technology and policy; research, development, and demonstration policy; and energy-economic-environmental modeling. Previously, he was a member of the research faculty at Carnegie Mellon University and a research fellow at Harvard University.

He holds a B.S. in electrical engineering from the University of Rochester, an M.S. in technology and policy from the Massachusetts Institute of Technology, and a Ph.D. in science, technology, and environmental policy from Princeton University.

Mark Mehos is the principal program manager for Concentrating Solar Power (CSP) at the National Renewable Energy Laboratory (NREL), where he supervises the development of low-cost, high-performance, and high-reliability systems that use concentrated sunlight to generate power.

He has led the High Temperature Solar Thermal Team at NREL since 1998 and has managed the Concentrating Solar Power Program since 2001. He has participated on and conducted analysis for several high-profile task forces on solar energy and is currently the leader for the International Energy Agency's SolarPACES Solar Thermal Electric Power Systems task.

Mr. Mehos holds a B.S. in mechanical engineering from the University of Colorado and an M.S. in mechanical engineering from the University of California at Berkeley.

Niels I. Meyer is Emeritus Professor of Physics at the Technical University of Denmark.

Dr. Meyer's main research interests include policies for sustainable energy development, with focus on renewable energy systems. Dr. Meyer is past president of the Danish Academy of Technical Sciences and past member of the Danish Energy Supervisory Commission.

Dr. Meyer has a Ph.D. and a Dr.Sc. in physics from the Technical University of Denmark.

Bruce G. Miller is the associate director of Penn State's EMS Energy Institute. He leads fossil and biomass fuel utilization, emissions control, and direct coal liquefaction research, development, and demonstration projects.

Mr. Miller has been involved with energy RD&D since 1981, with an emphasis on combustion and gasification systems, advanced fuel characterization, emissions characterization and control, hardware development and testing, behavior of inorganic constituents in utilization systems, and coal liquefaction. He has worked with coal-based systems his entire career and is a leading biomass utilization expert. His publications include two books, *Coal Energy Systems* (2005) and *Combustion Engineering Issues for Solid Fuel Systems* (2008), and numerous journal articles and technical reports.

Mr. Miller received his B.S. and M.S. in chemical engineering from the University of North Dakota.

Alan Moran is director of the Deregulation Unit at the Institute of Public Affairs (IPA). His work covers regulatory issues concerning energy, water, housing, and infrastructure.

He has published widely on energy and regulation, including the Australian chapter in *Electricity Market Reform: An International Perspective*, edited by Sioshansi and Pfaffenberger, in 2006. In 2007 he published *Regulation of Infrastructure*, coauthored by Warren Pengilley.

His work on housing includes a book, *The Tragedy of Planning: Losing the Great Australian Dream* (2006).

He has degrees in economics, including a Ph.D. from the University of Liverpool in the United Kingdom.

Reiner Musier is vice president with APX Inc., a leading platform provider for energy and environmental markets, including carbon commodities. He is responsible for the company's market strategy and represents APX in discussions with regulators, policymakers, environmental groups, corporations, and the trade press.

Dr. Musier's area of specialization is in technology and software for environmental and energy markets. Prior to his present position, he was vice president with the Siemens Corporation division providing solutions for energy and commodities trading, risk management, market operations, and simulation.

Dr. Musier holds Ph.D. and M.S. degrees from the Massachusetts Institute of Technology and a B.S. from Northwestern University.

Ah-Hyung Alissa Park is the Lenfest Junior Professor in Applied Climate Science in the Department of Earth and Environmental Engineering at Columbia University and the associate director of the Lenfest Center for Sustainable Energy of the Earth Institute.

Professor Park's main research interests include carbon sequestration, CO_2 capture using nanoparticle ionic materials, coal-to-liquid and waste-to-liquid technologies, and particle technology. Prior to her current position, she was a postdoctoral researcher at the Ohio State University and led a research group developing a coal-to-jet-fuel technology.

Professor Park has a B.A.Sc. and an M.A.Sc. in chemical engineering from the University of British Columbia in Canada and holds a Ph.D. in chemical engineering from the Ohio State University.

Wolfgang Pfaffenberger is adjunct professor in economics (European Utility Management) at Jacobs University, Bremen, and is professor emeritus in economics at the University of Oldenburg. He was director of Bremer Energie Institut from 1997 to 2006 and president of the German branch of IAEE (GEE) from 1995 to 1999.

His main interest is energy policy and the reform of natural gas and electricity markets. He has published extensively in this area.

He has degrees in economics, including a Ph.D. from Freie Universität Berlin.

William Prindle is a vice president with ICF International, a global energy-environment consultancy. He helps lead the firm's energy efficiency work for government and business clients, including major support work for U.S. EPA's ENERGY STAR® and related efficiency programs as well as utility efficiency program development. He also supports the firm's corporate energy management, carbon management, and sustainability advisory services.

He was previously policy director at the American Council for an Energy-Efficient Economy (ACEEE), where he led research and advocacy work on energy and climate policy for federal and state governments. Prior to that, he directed buildings and utilities programs for the Alliance to Save Energy and previously was a management consultant.

Mr. Prindle received his B.A. in psychology from Swarthmore College and an M.S. in energy management and policy from the University of Pennsylvania.

Luiz Pinguelli Rosa is director of COPPE, the Institute of Post-Graduate Studies and Research in Engineering at Federal University of Rio de Janeiro, and professor of the Energy Planning Program and Secretary General of the Brazilian Forum on Climate Change. Formerly he was president of Centrais Elétricas Brasileiras SA (ELETROBRAS). He served on the Intergovernmental Panel on Climate Change from 1998 to 2001 and is a member of the Brazilian Academy of Sciences.

His current research interests include energy planning, energy technology, and climate change. Through his long professional career, he has been engaged in theoretical physics, nuclear engineering, and studies of the contribution of the energy sector to the greenhouse effect, including measurements of greenhouse gas emissions from hydro reservoirs.

Professor Rosa has an M.Sc. in nuclear engineering from Federal University of Rio de Janeiro and a D.Sc. in physics from Catholic University of Rio de Janeiro.

Geoffrey Rothwell is senior lecturer and director of the Honors Programs in the Department of Economics, Stanford University, and associate director of Stanford's Public Policy Program. Since 2001, he has been the chief economist of the Economic Modeling Working Group of the Generation IV International Forum through the U.S. Department of Energy, Office of Nuclear Energy.

Dr. Rothwell has written extensively on the economics of nuclear power and electricity regulation, including *Electricity Economics: Regulation and Deregulation,* with Tomas Gomez, published in 2003. His current research focuses on building econometric-cost engineering models of the nuclear fuel cycle—in particular, uranium enrichment, nuclear fuel fabrication, reprocessing, and geologic disposal.

He received both his M.A. in jurisprudence and social policy and his Ph.D. in economics from the University of California, Berkeley.

Harry Singh is a vice president with RBS Sempra Commodities in Stamford, Connecticut, where he works on power and environmental market issues across North America.

Prior to his current position, he was a senior advisor at the Federal Energy Regulatory Commission (FERC), where he worked in the Office of Energy Market Regulation on policy issues in electric power markets and with the Office of Enforcement on several investigations. He worked on the design of long-term transmission rights, Order 890, the Competition NOPR leading to Order 719, Order 679 on Market Based Rates, a report to Congress on the state of electric competition and market redesign in California. He held various positions at PG&E Corporation and its regulated and unregulated subsidiaries.

He holds a Ph.D. in electrical engineering from the University of Wisconsin-Madison.

Fereidoon Sioshansi is president of Menlo Energy Economics, a consulting firm based in San Francisco serving the energy sector. Dr. Sioshansi's professional experience includes working at Southern California Edison Company (SCE), the Electric Power Research Institute (EPRI), National Economic Research Associates (NERA), and, most recently, Global Energy Decisions (GED), now called Ventyx. He is the editor and publisher of *EEnergy Informer*, is on the Editorial Advisory Board of *The Electricity Journal,* and serves on the editorial board of *Utilities Policy*.

Dr. Sioshansi's interests include climate change and sustainability, energy efficiency, renewable energy technologies, regulatory policy, corporate strategy, and integrated resource planning. His two recent edited books, *Electricity Market Reform: An International Perspective,* with W. Pfaffenberger, and *Competitive Electricity Markets: Design, Implementation, Performance,* were published in 2006 and 2008, respectively.

He has degrees in engineering and economics, including an M.S. and a Ph.D. in economics from Purdue University.

Paul Sotkiewicz is senior economist in the Market Services Division at the PJM Interconnection, LLC, where he provides analysis and advice on market design and performance with particular attention to demand response mechanisms, intermittent and renewable resource integration, market power mitigation strategy, and the potential effects of climate-change policy on PJM's markets.

Prior to joining PJM, Dr. Sotkiewicz was the director of Energy Studies at the Public Utility Research Center (PURC), University of Florida, involved in executive education and outreach programs in regulatory policy and strategy. Prior to that he was at the Federal Energy Regulatory Commission (FERC), where he conducted research, analysis, and advice on market design issues related to ISO/RTO markets. His research and publications cover electricity market design; tariff and rate design related to the deployment of distributed generation, energy efficiency, and demand response; and the effect of different regulatory policies or industry conditions on the cost-effectiveness of cap-and-trade air pollution programs.

Dr. Sotkiewicz received a B.A. in history and economics from the University of Florida, an M.A. in economics from the University of Minnesota, and a Ph.D. in economics from the University of Minnesota in 2003.

Benjamin K. Sovacool is a research fellow in the Energy Governance Program at the Centre on Asia and Globalization at the National University of Singapore. He is also an adjunct assistant professor at the Virginia Polytechnic Institute and State University, where he has taught for the Government and International Affairs Program and the Department of History.

Dr. Sovacool recently completed work on a grant from the U.S. National Science Foundation's Electric Power Networks Efficiency and Security Program, investigating the impediments to distributed and renewable power systems. He has also worked closely with the Virginia Center for Coal and Energy Research, the New York State Energy Research and Development Authority, the Oak Ridge National Laboratory, and the U.S. Department of Energy's Climate Change Technology Program. His recent edited book, *Energy and American Society: Thirteen Myths,* was published in 2007; *The Dirty Energy Dilemma: What's Blocking Clean Power in the United States* was published in 2008.

He holds a Ph.D. in science and technology studies from the Virginia Polytechnic Institute and State University, an M.A. in rhetoric from Wayne State University, and a B.A. in philosophy from John Carroll University.

Gary Stern is the director of Market Strategy and Resource Planning for Southern California Edison Company (SCE), where he manages resource planning, capacity and energy market design, and monitoring the wholesale electricity market in California.

Dr. Stern's main research interests include greenhouse gas legislation focusing on cap-and-trade market design, electricity market design, simultaneous optimal auctions, resource adequacy, and capacity markets. His recent endeavors have included working on carbon cap-and-trade market design, implementation of an increased renewable portfolio standard for California, and electrification of transportation as a carbon-reducing option. He is also involved in the design and development of a capacity market for California.

Dr. Stern earned a Ph.D. in economics, an M.A. in economics, and a B.A. in mathematics, all from the University of California at San Diego.

Jean-Michel Trochet is EDF senior economist within the Prospective and International Affairs Division of Electricité de France (EDF). He is a member of the Steering Committee of the Sustainable Development Chair at Ecole Polytechnique and General Secretary of the French Affiliate of the International Association for Energy Economics (IAEE).

Prior to his current position, he worked as an economist in EDF in the Corporate Strategy Division, the Industrial Strategy for Generation and Trading Division, and the General Economic Studies Department. He has taught at Institut d'Economie Publique in Marseille (IDEP) and Centre d'Etudes et Programmes Economiques in Paris.

Mr. Trochet is a graduate of Ecole des Mines de Paris.

Kenneth H. Williamson is a consultant in geothermal energy. He spent much of his career with Unocal Corporation, the world's largest developer of geothermal resources, where he was involved in the exploration and development of geothermal resources in the United States, Europe, Latin America, and Southeast Asia in both geoscientific and engineering roles. As general manager of Geothermal Technology and Services, he was responsible for worldwide exploration, technology development, and technical support.

Prior to joining Unocal in 1981, Dr. Williamson was a principal scientific officer in the British Geological Survey, where he carried out geothermal exploration and research in the Caribbean, Latin America, South Pacific, and Africa.

Dr. Williamson has a B.Sc. in physics from the University of Aberdeen and a Ph.D. in geophysics from Imperial College, London.

Ryan Wiser is a staff scientist at Lawrence Berkeley National Laboratory. He leads research in the planning, design, and evaluation of renewable energy policies and on the costs, benefits, and market potential of renewable electricity sources.

Dr. Wiser's recent analytic work has included studies on the economics of wind power; the treatment of renewable energy in integrated resource planning; the impact of state-level renewables portfolio standards; trends in solar costs; and the risk mitigation value of renewable electricity. Dr. Wiser regularly advises state and federal agencies in the design and evaluation of renewable energy policies, is an advisor to the Energy Foundation's China Sustainable Energy Program, and is on the Corporate Advisory Board of Mineral Acquisition Partners. Prior to his employment at Berkeley Lab, Dr. Wiser worked for Hansen, McOuat, and Hamrin, Inc., the Bechtel Corporation, and the AES Corporation.

Dr. Wiser received a B.S. in civil engineering from Stanford University and holds an M.S. and a Ph.D. in energy and resources from the University of California, Berkeley.

Jay Zarnikau is president of Frontier Associates, where he provides consulting assistance to utilities, retail electric providers, large industrial energy consumers, and retail trade associations in the design and evaluation of energy-efficiency programs, retail market strategies, electricity pricing, demand forecasting, and energy policy.

Dr. Zarnikau formerly served as a vice president at Planergy, manager at the University of Texas at Austin Center for Energy Studies, and a division director at the Public Utility Commission of Texas. His publications include articles in *The Energy Journal, Resource and Energy Economics, Energy Economics, IEEE Transactions on Power Systems, Energy,* and *The Electricity Journal.*

Dr. Zarnikau has a Ph.D. degree in economics from the University of Texas at Austin, where he also teaches statistics as a part-time visiting professor.

Xiliang Zhang is professor of energy economics and executive director of Institute of Energy and Environmental Economics, Tsinghua University. He serves as the secretary general of the New Energy Committee of China Energy Research Society and associate editor of the international journal *Energy for Sustainable Development.*

Dr. Zhang has conduced research on sustainable energy technology innovation and diffusion, markets, and policies for China. He played a key role in drafting China's Renewable Energy Law, sponsored by the Environmental Protection and Resource Conservation Committee of the National People's Congress. He is the coprincipal investigator of two key research projects funded by the National Science Foundation of China: *China's Energy-Related CO_2 Control Technology and Policy* and *Modelling Energy Development and Utilization Strategy in Western China.* He is also leading a team to carry out research into the *China Roadmap for Renewable Energy Technology Development* at Tsinghua University. Dr. Zhang is a lead author for "Energy Supply" of the *4th IPCC Assessment Report.*

Dr. Zhang received his Ph.D. in management science and engineering at Tsinghua University in 1997.

Introduction

Carbon Constrained: The Future of Electricity Generation

Fereidoon P. Sioshansi

Menlo Energy Economics

Abstract

The electric power sector—like the energy sector in general—has entered a new phase in its evolution, one in which emissions of greenhouse gases can no longer be assumed to be costless. This fact will gradually and profoundly change electricity generation—and utilization—in the coming decades. This volume of collected chapters is focused on examining why these changes are necessary, how they may be implemented, and what might be their implications for the generation, cost, and consumption of electricity.

▌ Historical context

Human welfare and energy use are strongly correlated. Access to affordable energy improves our quality of life and increases productivity. Electricity, the most convenient, versatile, and (at the point of consumption) cleanest form of energy, plays a crucial role in sustaining today's modern economies. As countries develop and mature, their relative dependence on electricity increases. This phenomenon means that more advanced economies typically convert a higher percentage of their primary energy to generating and consuming electricity.

Looking through human history, two gradual transitions are observed. With carbon as a new constraint, a third transition is taking shape.

■ The first period, beginning from the dawn of human civilization to the Industrial Revolution, can be characterized by agrarian societies using relatively little energy, virtually all from renewable sources, and a negligible per capita carbon footprint.

Doi: 10.1016/B978-1-85617-655-2.00035-3.

■ The second period, beginning from around the 1800s to the beginning of the Information Age, can be characterized by significantly increased reliance on cheap and abundant fossil fuels and virtual abandonment of renewable resources, with the exception of hydroelectricity. During this period, per capita carbon emissions increased dramatically, especially among developed countries—a trend that is now sweeping across rapidly developing countries.

■ The third period, covering the recent past and going forward, can be characterized by increasing concerns about climate change and a concentrated effort to rely more heavily on renewable resources, low-carbon or noncarbon fuels, and efficient utilization of energy.[1]

The predominant challenge since the Industrial Revolution has been to find and exploit large quantities of fossil fuels to feed the growing needs of the global economy. With rapidly improving technology and by taking advantage of ever-increasing economies of scale, the unit cost of energy—measured in dollars per barrel of oil, cents per lumen of lighting, or cents/kWh of electricity—fell continuously for decades. The rise of the unit cost of energy, exemplified by increases in oil prices, is a relatively recent phenomenon.

Some experts believe that with the continued growth in world population (Figure 1) and growing per capita energy consumption, we are entering a new phase, one characterized by diminishing marginal returns. If this proves to be the case, natural resources in general, and energy resources in particular, will cost more.

The carbon problem in context

Since the dawn of the Industrial Age, large quantities of greenhouse gases (GHGs) have been and continue to be released into the atmosphere. These emissions may be characterized as *natural*—say, from the decay of organic material— or anthropogenic—say, from burning coal to generate electricity. Carbon dioxide (CO_2) is the principal but not the only GHG, often used as the proxy in discussing global climate issues, and this convention is used in this volume.

There is no disagreement that burning of fossil fuels, whether from the transportation sector, for power generation, or otherwise, releases large

---

[1] In his seminal book *The Third Wave*, Alvin Toffler described human civilization as having gone through three distinct stages of evolution: from primitive agrarian societies to industrialization to the contemporary dawn of the Information Age, corresponding to the three phases of energy use outlined here.

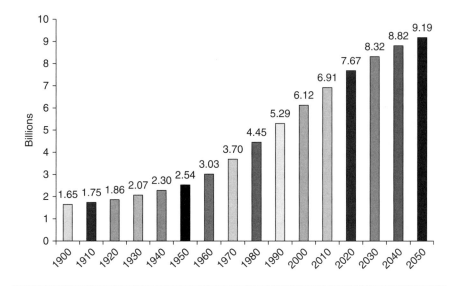

Figure 1 World population, 1900–2050.
Source: The United Nations.

quantities of GHG into the atmosphere. Likewise, there is no disagreement that the concentration of CO_2 in the atmosphere has been dramatically rising in the recent past relative to its historical levels[2] (Figure 2).

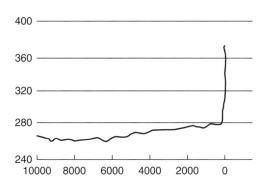

Figure 2 Concentration of CO_2 in the atmosphere in parts per million, years before 2005.
Source: EIA.

[2]To put it in perspective, the preindustrial concentration of CO_2 in the atmosphere is estimated to have been around 280 parts per million (ppm); the current level is approximately 385 ppm and rising.

Until recently, GHG emissions and ambient CO_2 concentrations were not considered a significant issue, certainly not a serious constraint on energy markets or economic growth. But over the past two decades a growing body of scientific evidence has been accumulating that indicates that increased accumulation of GHG in the atmosphere is responsible for gradual increases in mean global temperature, and that could result in unpredictable and potentially catastrophic consequences.

The debate on the science of climate change, however, is complex and controversial. It is *not* the focus of the current volume. This volume adheres to the scientific consensus that is in favor of limits to global growth of GHG emissions in the near future, followed by longer-term absolute reductions in anthropogenic emissions.[3]

The 1997 Kyoto Protocols accepted the provisional assessment of the United Nations Intergovernmental Panel on Climate Change (IPCC) by establishing international guidelines to reduce GHG emissions. Discussions are currently focused on reaching a post-Kyoto agreement involving broader international cooperation. The leaders of the G8, the eight largest global economies, at their summit in July 2008 in Japan agreed to reduce GHG emissions by 50 percent by 2050.[4]

Alternative targets and dates have been proposed and are being debated in political, scientific, and environmental forums.[5] Former U.S. vice president Al Gore, for example, has proposed that the United States should set a goal to generate *all* its electricity from zero-carbon energy sources within a decade.[6] German Chancellor Angela Merkel has proposed that global per capita carbon emissions should be halved.[7] The new U.S. administration has indicated its intention to lead international efforts to curb global carbon emissions.

In the meantime, a number of large and small jurisdictions have set caps and adopted future reduction targets on GHG emissions. The previous New Zealand government, for example, banned the construction of new plants using fossil fuels for power generation for at least a decade and set a goal of

---

[3]Scientists have proposed that atmospheric concentrations of CO_2 in the atmosphere should not exceed a range between 450–550 ppm by 2050 and preferably should be reduced after that.

[4]The G8 leaders agreed to "consider and adopt … the goal of achieving at least 50% reduction of global emissions by 2050, recognizing that this global challenge can only be met by a global response." The developing countries, however, did not agree to cut their emissions in half by 2050, stating that the G8 must cut their emissions by 25–40 percent by 2020 and 80–95 percent by 2050 from 1990 levels before they would consider reducing theirs.

[5]For example, see Steven Stoft's *Carbonomics: How to Fix the Climate and Charge It to OPEC*, 2008.

[6]"Gore seeks 100% green energy," *Financial Times*, July 18, 2008.

[7]For further details, refer to Chapter 4, by Fawcett et al., in this volume.

meeting 90 percent of the nation's electricity needs from renewable resources.[8] The state of California passed a law to reduce its 2020 GHG emissions to 1990 levels.[9] The Province of Ontario in Canada has decided to phase out its existing coal-fired plants by 2014.[10] The city of Boulder, Colorado, has set a goal to reduce its carbon footprint by 22 percent from its 2006 levels. There are numerous other examples from around the world.[11]

These efforts, though helpful—and certainly moving in the right direction—are nevertheless piecemeal, uncoordinated, and not necessarily efficient, economically speaking. Addressing the global climate challenge, as the term implies, will ultimately require global agreements with broader, more coordinated and efficient policies. Yet the experience of these early adapters will provide useful lessons that could form the basis of broader future global policies.

Objectives of the book

This volume purposely avoids the debate about *whether, when, and how much* governments should seek to reduce GHG emissions, by what means or at what cost. Those debates, though critically important, rest *outside* the scope of the present book.

The starting point for this book is to *assume* that a decision—or a number of decisions—will be made in the near future that would impose physical limits on GHG emissions, principally CO_2, and/or would impose an explicit or implicit cost on such emissions. The former might be in the form of a local, regional, or global cap, as has already been proposed in some jurisdictions. The latter might be in the form of a carbon-trading scheme with a cap-and-trade mechanism or possibly some form of direct carbon tax.

The decisions may be localized, as in the case of the city of Boulder in Colorado; statewide or provincial, as in the case of California; countrywide or regional, as in the case of the Regional Greenhouse Gas Initiative (RGGI) in the northeastern United States or the emission trading scheme (ETS) in the European Union; or global, as in the case of a post-Kyoto treaty sponsored by the UN, potentially covering both developed and developing countries.

Although the ultimate form and scope of such decisions are critical, it is fair to assume that the electric power sector will be affected and should begin to

---

[8]For further details, refer to Chapter 14, by Bertram and Clover, in this volume.
[9]For further details, refer to Chapter 18, by Stern and Harris, in this volume.
[10]For further details, refer to Chapter 13, by Hrab and Fraser, in this volume.
[11]Sovacool and Brown (Chapter 5 in this volume) discuss the relative merits of centralized versus localized efforts to curb emissions.

internalize the future costs of carbon emissions.[12] In this sense, the timing and the specifics of carbon reduction mandates are not as important as the high probability that they will be in place in the near future. This makes it prudent for the industry to prepare, plan, and respond.

This book is focused on an examination of the effect of such GHG emission reduction schemes and, more important, on how the development of such policies can be shaped to achieve the maximum benefits at minimum costs. Moreover, this book is focused primarily, but not exclusively, on the issues pertinent to the electric power sector.

It is recognized that anthropogenic emissions are *not* the sole source of GHG emissions and that emissions from the power sector are only a part of anthropogenic emissions, with the transportation and the industrial sectors as the other major sources. It is, however, also acknowledged that the power sector is a major GHG contributor and is likely to be a relatively convenient sector to be targeted, given its concentration among a number of large emitters and the fact that it is heavily regulated in most parts of the world.

On a positive note, the power sector is blessed in the sense that it can use many forms of primary energy—coal, oil, natural gas, uranium, and renewables—to generate electricity, a versatility not necessarily available in other sectors.[13] This versatility also suggests that the electricity sector could be a potentially significant contributor to reducing GHG emissions in other sectors—for example, through electrification of transportation.[14]

The organization of this book

The present volume consists of 19 chapters organized into three complementary parts:

- The first part describes the enormity of the carbon challenge facing humankind and various proposed schemes to control and reduce GHG emissions.

---

[12]There are indications that this is already happening. In February 2008, three major U.S. banks—Citi Group, JPMorgan Chase, and Morgan Stanley—announced new carbon principles in assessing the risks of investing in plants that emit GHGs, particularly coal-fired power plants.

[13]Although flexibility exists in all sectors to some degree. For example, transportation fuels could be supplemented by biofuels, or cars can run on natural gas, electricity, or hydrogen.

[14]The auto industry, for example, is showing interest in plug-in electric hybrids to reduce transportation emissions associated with cars. This, in turn, puts additional demand on the electric power sector to meet the increased load generated by electric vehicles.

■ The second part describes alternative solutions to the carbon challenge—namely, various options to decarbonize electricity generation and to make more efficient use of what is generated.

■ The third part provides case studies of various proposals from around the world to address the carbon challenge.

A brief outline of these chapters follows.

PART ONE: THE CARBON CHALLENGE

In **Chapter 1, Alan Moran** sets the context and the goal posts for the chapters of the book that follow. He points out that to address concerns about climate change, *significant* reductions in GHG emissions, ranging from 20 percent below to half of prevailing levels, will be necessary. These goals would require per capita global emissions to be reduced to 3.4 and 2 tonnes of carbon dioxide equivalent per capita, respectively. To put things in perspective, U.S. per capita emissions are currently 20 tonnes; the OECD average is 11.5; and China already exceeds 3.4.

Developed countries appear willing to commit to serious emission cutbacks, but these are not adequate to achieve the necessary targets while developing countries currently have no significant reduction plans. Moran points out that fast-growing developing countries, which currently have relatively low levels of per capita energy use and GHG emissions, are likely to remain dependent on lowest cost fossil fuels—principally carbon-rich coal—and this presents a difficult obstacle to stabilizing global emissions as these economies continue to grow.

Moran underscores the reality that meeting the targets that are necessary to avert climate change, while maintaining high standards of living in developed countries *and* allowing for improvements in standards of living in developing countries, will take extraordinary resourcefulness and ingenuity and may require different lifestyles.

In **Chapter 2, Emile Chappin, Gerard Dijkema**, and **Laurens de Vries** describe the relative merits and drawbacks of the two principal carbon policies in use today: emission trading and carbon taxation.

Using a quantitative model that reflects today's technology, EU policy, and economic decision making in the power generation sector, the authors compare the impact of these carbon policies on CO_2 emission, technology, and fuel choice as well as electricity prices. Under both policies, CO_2 emissions continue to increase for 10–15 years due to the long life cycle of power plants. Carbon taxation leads to earlier investment in CO_2 abatement and a better balance between capital and operating costs, and in lower electricity prices. Maintaining proper taxation or a tight cap, both policies result in dramatic CO_2 emission reductions in 20–40 years.

In **Chapter 3**, **Reiner Musier** and **Parviz Adib** describe current efforts to develop verifiable and transparent markets for trading carbon allowances in what promises to become the world's largest commodity market. The value of tradable permits in environmental markets will be measured in trillions of dollars, with major national and international players, particularly financial institutions, as active participants.

The authors argue that the design of such markets should be based on sound economic principles with a robust supporting infrastructure that allows trading, verification, reporting, and compliance with national or global requirements. They outline a number of success factors essential to effective operation of such markets, including requirements for effective management and regulatory oversight.

In **Chapter 4**, **Tina Fawcett**, **Frede Hvelplund**, and **Niels I. Meyer** discuss a novel approach to addressing the carbon challenge by making it personal through a Personal Carbon Allowance (PCA) scheme. Just as commodities such as food and petrol are rationed at times of scarcity, carbon, in principle, can also be rationed on a per capita basis. The idea that consumers are ultimately responsible for an economy's overall carbon emissions and must be empowered to ration their environmental impact is gaining support in some circles.

The authors review the fundamentals of PCA and analyze its merits and shortcomings. The United Kingdom and Denmark are used as case studies because the energy situation and the institutional setup are quite different between the two. The authors conclude that PCA is an idea worthy of serious further consideration.

In **Chapter 5**, **Benjamin K. Sovacool** and **Marilyn A. Brown** explore the topic of appropriate scale in addressing the carbon challenge and the pros and cons of taking initiatives at individual, community, state, regional, national, or international levels. Currently, in the absence of a broad international coalition that includes all major developed and developing countries and with a lack of consensus about how to get everyone on board, various countries, regions, states, and even some cities have adopted GHG reduction goals and targets. But since global climate change is global in nature, it begs to be addressed on a broader level.

The authors point out that *what* we regulate is just as important as *how* we regulate it. The chapter discusses the costs and benefits of addressing climate change at both local and global scales and concludes by offering a proposal that attempts to blend the two.

PART TWO: SOLUTIONS

In **Chapter 6, Klaus Lackner, Alissa Park**, and **Bruce Miller** note that the world is roughly 300 Gigatons of carbon away from the frequently cited stabilization goal of 450 ppm of CO_2 in the atmosphere.[15] With current emissions of around 8 Gigatons of carbon per year, it will take about 40 years to reach that level. Hence, every coal-fired power plant built today represents locking in a 70-year stream of emissions. For a Gigawatt power plant, this amounts to roughly half a Gigaton of CO_2, which indicates the urgency of addressing the problem.

The authors point out that it is not the use of fossil fuels *per se* that needs to be curtailed but rather the emission of the CO_2 associated with their use. They argue that carbon capture and storage (CCS) offers a solution, provided it is applied on a large scale and relatively soon.

Initially, carbon dioxide capture was considered the difficult part of the problem, but the current thinking has come to realize that developing the necessary storage capacity poses the more serious challenge. The authors argue that together, the various storage options offer an immediate pathway to stopping CO_2 emissions from the fossil-fuel sector and that in the long term several of these options appear to have the capacity to store vast amounts of CO_2 from continued reliance on fossil fuels.

In **Chapter 7, Geoffrey Rothwell** and **Rob Graber** describe the potential role and contribution of virtually carbon-free nuclear energy, currently enjoying a renaissance in the United States. In China, Russia, India, Korea, and a number of other countries, nuclear construction is expected to be expanded after years of relative dormancy. Despite its considerable promise, nuclear energy is not risk free, is capital intensive, and is opposed by many observers based on safety and lack of long-term waste disposal. Yet many believe that nuclear energy must play a key role in any effort to address global climate change.

Although nuclear power is promoted for its role in curtailing GHG emissions, there may be limitations on its ability to achieve these lofty expectations. The reality is that nuclear plants' high capital costs could limit the number that can reasonably be built. Perhaps even more important, the high-growth emerging economies are building coal plants at a rate that far outpaces nuclear power. Unless a global determined effort is made to not only duplicate the nuclear construction pace seen in the 1980s but to increase it substantially, it is likely that too many coal plants will be built for nuclear to overcome. Nuclear can be part of the solution, but it will require the equivalent of a modern-day Marshall Plan.

---

[15]For example, the often-cited *2007 Stern Report*.

In **Chapter 8**, **William Prindle**, **Jay Zarnikau**, and **Erica Allis** describe the important role that energy efficiency can play in reducing the growth in carbon emissions and reducing the cost of climate policies. Many energy and climate policy analyses fail to account for efficiency potential and assume increased energy consumption based on historical trends. The authors challenge this notion by summarizing the numerous studies that have demonstrated that using energy efficiently holds enormous potential and that tapping this potential is an essential element of any serious effort to combat climate change.

Despite efficiency's enormous potential, a number of formidable market and policy barriers continue to block realization of that full potential. For example, carbon cap-and-trade programs typically do not allow end-use efficiency to be used in tradable credits for carbon markets. The principal agent problem and transaction cost barriers chronically inhibit efficiency investments in major end-use markets. Utility regulatory policies typically discourage utilities from investing in customer usage reduction. The authors describe these barriers and discuss how they can be overcome through effective programs and policies.

In **Chapter 9**, **Ryan Wiser** and **Maureen Hand** describe the potential contribution of wind power in combating the inexorable rise of carbon emissions. They point out that the global wind power market has been growing at a phenomenal pace in recent years, driven by favorable policies toward renewable energy and the improving economics of wind projects. But how much more wind power can be expected, how soon, and at what ultimate cost?

Drawing on a number of studies, the authors conclude that wind power is poised for significant additional scale-up, with the potential to contribute substantially to reducing the carbon footprint of the electricity sector. The technology has matured, and costs, in good resource regimes, are comparable to fossil-fuel generation. The authors contend that the global wind resource is vast and that, though accessing this potential is not costless or lacking in barriers, wind power can be developed at scale in the near to medium term at what appears to be an acceptable cost.

In **Chapter 10**, **Paul Denholm**, **Easan Drury**, **Robert Margolis**, and **Mark Mehos** point out that solar energy is the planet's largest source of renewable energy. Yet up to now "direct" solar energy technologies have been a nearly insignificant contributor to the world's electrical energy supply. Many questions remain regarding the prospects for cost-competitive solar energy on a large scale.

Examining recent trends in solar markets and costs, the authors conclude that direct solar electric technologies may emerge from their current role as "niche" technologies and become mainstream generation technologies. Solar photovoltaics (PVs), for example, have seen continued improvements in efficiency and costs in proven technologies now being deployed at scale, while concentrating solar power (CSP) has emerged as a promising source of utility scale generation, utilizing low-cost thermal storage that creates a highly reliable and dispatchable source of solar-generated electricity.

In **Chapter 11, Kenneth Williamson** covers the growth of geothermal energy worldwide and describes advances in technology that could transform geothermal energy into a significant renewable energy option. Historically, geothermal projects have depended on finding and exploiting naturally occurring reservoirs of hot water or steam, and that has limited their growth potential.

For the past 30 years, experiments to create artificial reservoirs in hot rock formations have been conducted with limited success, but recent developments have been encouraging. The author argues that this baseload, carbon-free, renewable power source may be on the verge of a major technological breakthrough, with significant implications.

In **Chapter 12**, the late **Lizardo Araujo, Luiz Pinguelli Rosa**, and **Neilton Fidelis da Silva** describe the additional potential contribution of the oldest form of renewable energy known to man: water. The authors point out that hydroelectricity has the potential to play a critical role in global carbon management, especially in developing countries where significant untapped resources remain.

The authors consider the issue of greenhouse gas emissions from hydro reservoirs and show that these emissions are significantly lower than those from thermal power plants generating the same amount of energy. They describe lessons learned from mistakes of the past that caused inadvertent environmental and ecological damage in building large reservoirs; these remain among the most significant challenges in developing additional hydro schemes.

PART THREE: CASE STUDIES

In **Chapter 13, Peter Fraser** and **Roy Hrab** describe efforts by the Province of Ontario in Canada to phase out all coal-fired generation, which currently accounts for roughly 20 percent of generation mix, by 2014. The authors describe how this mandate will be implemented and how much it could cost. Can the experience of Ontario serve as a model for other countries that aspire to reduce their own carbon footprints?

The authors conclude that Ontario's experience illustrates that the elimination of a large quantity of coal-fired generation from an electricity network is feasible. However, the implementation of the government's strategy involves overcoming many financial and technical challenges. Furthermore, the ultimate economic and environmental impacts of the policy will not be observable for several years. Moreover, the ultimate success of the Ontario scheme may be difficult to judge because of the interconnected nature of power systems, where reduced production in one jurisdiction can be replaced by increased production elsewhere.

In **Chapter 14, Geoff Bertram** and **Doug Clover** describe the implications of a 2008 parliamentary decision in New Zealand to ban the construction of new fossil-fired baseload generation and to aim for a target of 90 percent renewables in electricity by 2025.[16]

The authors claim that meeting the 90 percent renewable target is feasible in light of the country's resource endowment and past history, although the current market structure and the dominance of large generators may have foreclosed a number of possible ways forward.

The authors conclude that the policymakers have failed to create an open, competitive environment for new entry by distributed generation or real-time demand response; in these areas New Zealand has yet to learn key lessons from other countries. Nevertheless, the authors believe that the newly legislated emissions trading scheme, in combination with falling costs of key renewable technologies, will drive New Zealand's electricity sector toward renewables, even without market reform.

In **Chapter 15, Nigel Cornwall** describes the significant challenges faced by the U.K. electric power sector, which is facing a cliff face of new investment while falling behind on a number of ambitious government-imposed as well as European carbon reduction targets. Whether and how the United Kingdom can address significant coal retirement, an aging fleet of nuclear power plants, and, maybe, growing demand while increasing renewables and other low-carbon technologies may be a measure of how others can arrive at an affordable, sustainable low-carbon electricity future.

Early indications are that progress against renewable targets is off-course and that existing low-carbon measures have flattered to deceive. More progress has been made through energy efficiency programs as a result of obligations placed on energy suppliers. Significant further support measures are in the pipeline, but with new nuclear builds at least 12 years away, the overall message is that energy and environmental policy in the United Kingdom remains a work in progress.

In **Chapter 16, Jean-Michel Trochet, Jean-Paul Bouttes,** and **Francois Dassa** describe how France and the EDF group are responding to EU policies on

---

[16]The construction ban was revoked following a change of government at the end of 2008, but this revocation does not affect the chapter's overall conclusions.

climate change. The authors do this by providing an analysis of climate regulation at the European level, given the present state of regional integration of electricity and CO_2 markets.

The authors reach two key conclusions. First, investing in available CO_2-free electricity technologies can provide the means to achieve ambitious climate targets within the coming decades, and cap-and-trade as well as other public policy instruments should be appropriately designed to send the right signals for the right investments. Second, relevant public policy instruments depend on the maturity of technologies and policy objectives. Cap-and-trade alone cannot shape R&D policy or deal with security of supply issues in the electricity sector.

In **Chapter 17**, **Joanna Lewis**, **Zhang Xiliang**, and **Qimin Chai** focus on China's power sector, the largest single source of carbon dioxide emissions in the world. The authors explore the current state and options for low-carbon power generation in the context of China's current energy policy priorities, changing politics of climate change, and the outlook for future technology developments.

The authors explain how diversification away from coal and toward a range of low-carbon power sources will need to be at the core of any climate change mitigation strategy that moves China to a lower-carbon development pathway, yet there is currently little incentive for China to pursue low-carbon options that bring added costs and have little benefit other than their low-carbon characteristic, such as carbon capture and storage. In the near term, the authors conclude, China must focus on promoting domestic policy strategies that can create economic benefit and reduce pollution while having the additional benefit of climate change mitigation, including energy efficiency and renewable energy.

In **Chapter 18**, **Frank Harris** and **Gary Stern** provide an overview of efforts by one of the largest investor-owned utilities in the United States to address climate change. Facing multiple mandates from the state of California to integrate renewable energy, pursue energy efficiency, and contribute to an aggressive target to reduce statewide GHG emissions to 1990 levels by 2020, Southern California Edison Company (SCE) is in a race against time. How well major utilities such as SCE are able to respond to such regulatory mandates and what the cost implications of their strategies may be are being followed with keen interest across the United States and around the world.

In **Chapter 19**, **Udi Helman**, **Harry Singh**, and **Paul Sotkiewicz** describe how U.S. independent systems operators (ISOs) and regional transmission operators (RTOs) can facilitate the GHG emission reduction schemes now mandated by individual states or regions as well as by any future federal legislation. ISOs and RTOs are centralized operators of regional power systems and integrated markets for spot energy, ancillary services, and typically also capacity. They also conduct transmission planning and oversee interconnection of system resources.

The authors argue that these organized markets embody a number of important features that can support the integration of renewable energy resources and climate policy objectives, including transparent spot pricing reflecting congestion and losses and the use of efficient methods for scheduling and dispatch of resources, including generation, demand response, storage, and transmission. In tandem with either a carbon tax or a cap-and-trade system and effective mechanisms for resource adequacy and regional grid planning, organized markets can deliver the least-cost combination of existing resources while providing critical price signals for optimal investments under GHG emissions constraints.

In the book's **Epilogue, Fereidoon Sioshansi** argues that although the electricity sector faces a challenging transition to a lower carbon future, there are opportunities to become a part of the solution. He also shares two *personal* surprises and one insight not necessarily shared by the contributing authors to this book.

The Carbon
Challenge

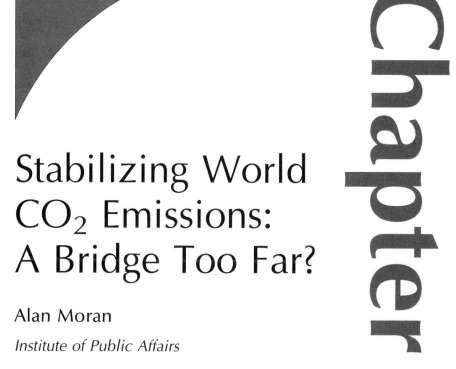

Part 1

Stabilizing World CO₂ Emissions: A Bridge Too Far?

Alan Moran

Institute of Public Affairs

Abstract

Proposals have been made to reduce greenhouse gas emissions by 20 to 50 percent of prevailing levels to address concerns about climate change. These goals would require emissions to be reduced to 3.4 and 2 tonnes of carbon dioxide per capita, respectively, with corresponding reductions in other greenhouse gases. U.S. emissions are currently 20 tonnes per capita, and the Organization for Economic Cooperation and Development (OECD) countries as a whole are at 11.5 tonnes; China already exceeds 3.4 tonnes per capita. Even in those developed economies that are not predominantly reliant on fossil fuel for electricity generation, emissions far exceed the targets under discussion. If emission reduction targets are

Doi: 10.1016/B978-1-85617-655-2.00001-8.

to be met alongside higher living standards, solving this problem will require unprecedented ingenuity.

1.1 Introduction

Contrary to many assertions, including those of the Stern report on the economics of climate change,[1] the costs of reducing emissions of carbon dioxide (CO_2) and other greenhouse gases will be considerable.

Though urging greater reductions in emissions, many countries have lowered their own emission reduction bars; Australia, for example, counts reductions in land clearing as contributing to its goal. Almost all OECD countries, however, have incurred considerable costs in subsidies to renewables and other measures involving regulating use of energy and energy-using goods. In spite of these reductions and notwithstanding that the first level of cuts is likely to be the easiest, few signatories to the Kyoto Convention will meet the obligations to which they agreed. The OECD group as a whole in 2005 had *increased* emissions by 20 percent over 1990 levels.

Far more draconian emission reductions are required than agreed to at Kyoto for the period to 2012 if the world is to see a reduction in the concentration of CO_2 and other gases said to be responsible for climate change. This would require vigorous action by all countries, including developing countries, the emissions of which have now surpassed those of the developed world. On average, developing countries' currently have only one quarter of developed countries' per capita emissions.

Developed countries have been reducing their emissions relative to their GDP levels, partly by increasingly outsourcing energy-intensive manufactured goods to developing countries, including oil- and gas-rich countries. This relocation of production, which the Intergovernmental Panel on Climate Change (IPCC) claims has only had a minor effect, does not, of course, bring a global reduction.

At Bali in 2007, the lowest figure discussed for a reduction in emission levels was 20 percent. This means bringing global emissions down from their 2004 levels of 29 gigatonnes (Gt) of CO_2 equivalent to 23 Gt. Under a business-as-usual scenario, emissions in 2030 would grow to 43 Gt. The July 2008 G8 summit agreed on a target of 50 percent reduction in global emissions by 2050 but specified neither a base date nor any intermediate targets.

---

[1]Stern Review. Economics of Climate Change. 2006. www.hm-treasury.gov.uk/stern_review_final_report.htm

In per capita terms, adjusted for population growth, 23 Gt translates to some 2.7 tonnes down from the present 4 tonnes.[2] Currently the United States represents around 20 tonnes, and even China is now approaching 4 tonnes.

Countries are likely to seek modifications of a starting point of equal amounts per capita, even if agreement is reached on the necessity for action. Developed countries might argue for emissions to be set to reflect units of gross domestic product (GDP). Many will claim special circumstances, as Australia did in negotiating a higher target in the Kyoto Convention. Developing countries might argue for higher allocations than developed countries based on their historically lower cumulative emission levels.[3]

Some detailed, though perhaps fanciful, emission reductions have been "scenarioed" by the IPCC, which has estimated global savings of 9–17 Gt of CO$_2$-equivalent (CO$_2$-e) from a tax of US\$20 per tonne and 12–26 Gt from a US\$50 tax. Savings of such magnitudes have certainly not emerged from the measures already in place in the European Union (EU) and elsewhere, as United Nations Development Program (UNDP) data illustrates in Table 1.1.

Moreover, even if approached, the possible savings need to be weighed against business-as-usual emissions of 43 Gt in 2030.

All forecasts such as those of the IPCC incorporate unproven technological breakthroughs to reduce costs of nonfossil fuels and often include a radical substitution of energy by other goods and services. There is no reliable information on which to base the forecast outcomes of measures that force a markedly lower use of fossil fuel–based energy—hence the IPCC's use of the term *storylines and scenarios*. Improved energy efficiencies have been a permanent feature of economic growth, and such efficiencies will doubtless continue to emerge. Indeed, if energy prices increase relative to those of goods and services in general, we are likely to see lower energy-to-GDP ratios. However, even with a universal adoption of nuclear energy for baseload power or breakthroughs such as cheap carbon capture and storage, unprecedented technology developments and/or much increased fossil fuel prices would be required to

---

[2]As Fawcett, Hvelplund, and Meyer point out in Chapter 4, German Chancellor Angela Merkel was calling for a halving of global emission levels to 2 tonnes per capita, though the global financial crisis has shifted her priorities toward industry assistance.

[3]In this respect, a spokesman for the Bangladesh high commission in London said this in response to the news that U.K. aid for climate change was to be in the form of loans: "The climate situation has not been created by us. The money should come spontaneously from rich countries and not be a loan" (www.guardian.co.uk/environment/2008/may/16/climatechange .internationalaidanddevelopment). China, in its October 2008 white paper, "China's Policies and Actions for Addressing Climate Change," argued that cumulative emission levels were the appropriate measure.

Table 1.1 Per capita CO_2 emissions of selected countries, 1990 and 2004 (tonnes)

	1990	2004
High-income OECD	12	13.2
France	6.4	6.0
Spain	5.5	7.6
United Kingdom	10.0	9.8
Italy	6.9	7.8
Sweden	5.8	5.9
Switzerland	6.2	5.4
Japan	8.7	9.9
United States	19.3	20.6
Canada	15.0	20.0

Source: UNDP, Human Development Report, *2007 (http://hdr.undp.org/en/media/HDR_20072008_EN_Complete.pdf).*

bring about lower levels of emissions of the magnitudes sought while retaining current living standards.[4]

▌ 1.2 The magnitude of the task

1.2.1 THE STERN REPORT

The Stern Report sought reductions in global emissions of CO_2 by 80 percent of current levels by 2050. Stern argued that the economic cost will be a total of 1 percent of world GDP, "which poses little threat to standards of living given that the economic output in the OECD countries is likely to rise by over 200 percent and in developing countries by more than 400 percent" during this period (p. 239).

Stern's forecasts have attracted wide-ranging opinions. Dasgupta,[5] on the assumption of a constant population and no technological change, has written, "Suppose the social rate of return on investment is 4% a year. *It is an easy*

---

[4]The nuclear issues are further addressed in Chapter 7 by Rothwell and Graber; clean coal is examined in Chapter 6 by Lackner et al.

[5]Sir Partha Dasgupta. Comments on the Stern Review's Economics of Climate Change. University of Cambridge; November 11, 2006. Available from: www.econ.cam.ac.uk/faculty/dasgupta/STERN.pdf

calculation to show that the current generation in that model economy ought to save a full 97.5% of its aggregate output for the future!" Nordhaus,[6] who nevertheless believes that action should be taken to mitigate climate change, also argues that Stern's discount rates are too high, saying, "The Review's unambiguous conclusions about the need for extreme immediate action will not survive the substitution of assumptions that are consistent with today's marketplace real interest rates and savings rates." On the other hand, Kenneth Arrow[7] is broadly supportive of the Stern Review, as are reviews of the Stern Report by economists such as Robert Solow, Amartya Sen, and Joseph Stiglitz.[8]

Among the assessments that have been highly critical of his findings has been that of Australia's Productivity Commission (PC).[9] The PC noted that Stern assumes higher temperature increases than the IPCC as a result of CO$_2$ and other emissions. The PC also noted that the Stern Report has no adaptation assumptions, lengthy and spurious suggestions about the cost of health, and, in general, "draws heavily on studies that have a more pessimistic view on climate change and its impacts and gives little attention to more optimistic views." The PC was also critical of the low discount rate Stern uses (1.4 percent), which allows it to come to far higher costs than any other study, and argued that Stern "erred in its failure to present a range of results for different discount rates."

The real economic task is demonstrated by the modest outcomes of changed energy policies that Stern cites. Among these is the relatively minor emission reductions achieved in the EU under its Kyoto commitments. This is in spite of regulations on energy use and subsidies to renewables as well as a carbon-use restraint program based on cap and trade. The capped emission trade has a tax equivalent that has ranged between €32 and €0.08 per tonne of CO$_2$ and was around €14 per tonne in May 2009. This, incidentally, would be enough to increase the Australian wholesale price of electricity by 60 percent. In the absence of a full set of guidelines, Australian carbon credits were trading below the EU price.

Recognizing that any CO$_2$ reduction would need to address the use of fossil fuels, Stern's report saw a form of carbon tax as the key feature of any policy to limit greenhouse gas emission. He estimated that a tax would be required at

---

[6]Nordhaus W. The Stern Review on the Economics of Climate Change. May 2007. Available from: http://nordhaus.econ.yale.edu/stern_050307.pdf

[7]Arrow KJ. Global Climate Change: A Challenge to Policy. Economist's Voice 2007;4(3). Available from: www.bepress.com/ev/vol4/iss3/art2/

[8]www.cambridge.org/catalogue/catalogue.asp?isbn=9780521700801

[9]The Stern Review: as assessment of its methodology, Staff Working Paper, Productivity Commission, January 2008.

an initial level of US$100 per tonne of CO_2, which would increase Australian wholesale electricity prices two and a half fold, and envisaged new technologies cutting in by 2030. These new technologies, Stern envisaged, would drive down the required tax level to around US$35 per tonne.

In addition to these tax effects, the Stern Report estimates also include other measures such as a continuation of existing energy-efficiency taxes and programs. They also incorporate a considerable emphasis on energy saving at the production end. Moreover, they are posited on a major contribution from voluntary energy savings, partly stimulated by education programs, drawing upon what the economist Lionel Robbins famously referred to as "that very scarce commodity, human love."[10]

1.2.2 Sources of Carbon Emissions

IPCC data[11] has sought to identify the share of the various emission sources. CO_2 in fuel is responsible for over half, with methane and CO_2 from deforestation comprising most of the rest. Figure 1.1 illustrates the data.

Within the energy sector, coal and natural gas comprise 46 percent, with only nuclear, solar, and other renewables, at less than 13 percent, genuinely free of emissions. Biomass is potentially largely free of emissions and is renewable but currently it mostly comprises obsolete energy supplies such as animal dung. Table 1.2 shows the UNDP estimates as shares of 2005 energy use.

Emissions themselves are mainly derived from industrial and consumer uses, with about 30 percent coming from agriculture and forestry. Figure 1.2 shows the source of emissions by usage.

There are myriad combinations of ways to reduce emissions. The measure that presents the easiest to envisage with current technologies involves a shift to nuclear of all the coal and 80 percent of the gas plants used for energy production. With those inputs supplying about 75 percent of CO_2 from fossil use, that would mean a reduction of some 40 percent of CO_2 emissions compared to the present supply profile.

As discussed in the following sections, stabilization of emissions would require the high-income OECD countries to reduce their per capita emissions

---

[10]The University of Aston's Professor Julia King has argued for a plan to "educate" children so that they can help shape parents' purchases of cars and other commodities to be "green." Professor King was appointed by the then Chancellor of the Exchequer, Gordon Brown, to lead the King Review to examine the vehicle and fuel technologies to help reduce carbon emissions from road transport. The interim analytical report (Part 1) was published on October 9, 2007. Personal communication, Paul Biggs Birmingham University.

[11]IPCC. Fourth Assessment Report, Working Group III Report, Mitigation of Climate Change.

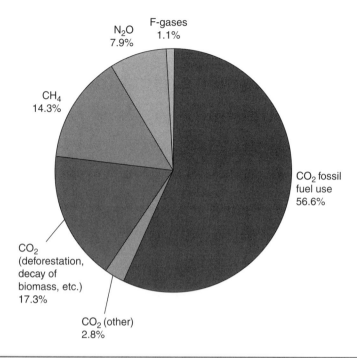

Figure 1.1   Global shares of gases in CO_2 equivalents.
Source: Technical Summary, WG3 IPCC, *Fourth Assessment Report*, 2007
(www.ipcc.ch/pdf/assessment-report/ar4/wg3/ar4-wg3-ts.pdf, p. 28).

Table 1.2   Primary energy supplies, 2005

	Total Primary Energy Supply (Tonne of Oil Equivalent)	Coal (%)	Oil (%)	Natural Gas (%)	Hydro, Solar, Wind, and Geothermal (%)	Biomass and Waste (%)	Nuclear (%)
World	11,433.90	25.3	35.0	20.7	2.6	10.0	6.3

Source: UNDP, Human Development Report, 2007.

by about 70 percent and the rest of the world to show no increase. Hence even far-reaching changes, as envisaged in a nuclear substitution scenario—and a renewables substitution would be similar—would still be insufficient to allow emission stabilization with current energy usage levels.[12]

---

[12]See Chapter 7 on nuclear energy in this book for further details.

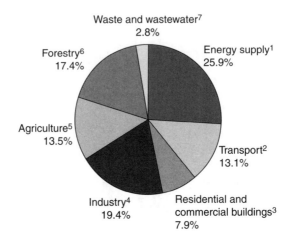

Figure 1.2 Sources of emissions.
1. Excluding refineries, coke ovens, etc., which are included in industry.
2. Including international transport (bunkers), excluding fisheries.
3. Including traditional biomass use.
4. Including refineries, coke ovens, etc.
5. Including agricultural waste burning and savannah burning (non-CO_2). CO_2 emissions and/or removals from agricultural soils are not estimated in this database.
6. Data includes CO_2 emissions from deforestation, CO_2 emissions from decay (decomposition) of above-ground biomass that remains after logging and deforestation, and CO_2 from peat fires and decay of drained peat soils.
7. Includes landfill CH_4, wastewater CH_4 and N_2O, and CO_2 from waste incineration (fossil carbon only).

Source: Technical Summary, WG3 IPCC, *Fourth Assessment Report*, 2007 (www.ipcc.ch/pdf/assessment-report/ar4/wg3/ar4-wg3-ts.pdf).

Shifting electricity generation from coal to nuclear power would have a significant but estimatable price tag representing the premium cost of nuclear power over fossil fueled plants. Though major technological breakthroughs cannot, of course, be ruled out, a series of minor breakthroughs is more likely. Among the latter, one that has attracted considerable attention involves sequestering CO_2 in cement, which the U.S. Environmental Protection Agency (EPA) classes as the third largest source of greenhouse gas pollution.[13] Major possible breakthroughs include work by Atlantic Richfield's ArcTech

---

[13]Carbon sequestering in cities, Calera cement, and maybe Vinod Khosla first trillionaire in 2020. Available from: http://nextbigfuture.com/2008/07/carbon-sequestering-in-cities-calera.html

into using termites to convert coal into methane and humic acid, thereby largely eliminating its CO_2.[14]

1.2.3 THE INITIAL STEPS TAKEN BY THE DEVELOPED COUNTRIES

All developed countries have incurred considerable costs in subsidizing and regulating in favor of high-cost energy sources with low emissions. In spite of this cost and the fact that the early gains are likely to be the easiest because they tap into the fabled "low-hanging fruit," most major signatories will fail to meet their Kyoto obligations.

Although individual European Union countries will meet their targets—Germany because of unification, the United Kingdom because of the shift from coal-powered electricity generation to gas—the EU as a whole is currently falling short, and this is likely to be amplified in the 2008–2012 period, over which commitment results will be measured.

Canada, which has often been in the vanguard of countries urging increased action, is among those falling furthest from the goal to which it agreed.[15]

Australia, which claimed to be only 4.5 percent above 1990 levels in 2005, will, if the economy continues to grow, be some 14 percent above 1990 levels for the Kyoto yardstick average of 2008–2012. Compared to its highly generous Kyoto target of 108 percent of 1990 levels, Australia would be over 30 percent above its 1990 levels were it not to measure its emissions on the basis of the creative "Australia clause" in Article 3.7. That clause permits countries for which land-use change and forestry are a net source of greenhouse gas emissions to count changes to these as part of their measures of net emissions. Norway has also benefited from this inclusion of clearing credits and, as a result, will meet its target.

Table 1.3 is drawn from the latest United Nations Framework Convention report and indicates levels of achievement compared to the 2008–2012 targets expressed as the emissions in excess of or below the 1990 base level. The latest data is for 2005 and the levels are expressed on two bases: with and without counting land use changes as a result of policy toward clearing land for cultivation. Only the EU taken as a whole is close to the targets in the form they were originally agreed.

---

[14]Margonelli L. Gut Reactions. Available from: www.theatlantic.com/doc/200809/termites/5

[15]As described elsewhere in this book, the province of Ontario plans to phase out its coal by 2014.

Table 1.3 Kyoto commitments and achievements over 1990 baselines

	2008–2012 Target (%)	2005 Actual	
		Inc. Clearing (%)	Exc. Clearing (%)
Australia	8	4.5	25.6
Canada	−6	54.2	25.3
European Union	−8	−4.0	−1.5
Japan	−6	7.1	6.9
New Zealand	0	22.7	24.7
Norway	1	−23.1	8.8
United States	−7	16.3	16.3

Source: UNFCC, "National greenhouse gas inventory data for the period 1990-2005", Note by the secretariat, December 2007, Bali. *(http://unfccc.int/resource/docs/2007/sbi/eng/30.pdf).*

▌ 1.3 Achieving global emission reductions

1.3.1 THE GLOBAL SETTING FOR AN EMISSIONS REDUCTION SCHEME

Difficulties that developed countries have experienced thus far in reducing greenhouse gas (GHG) emission levels are likely to be amplified for many developing countries, should they agree to join a post-Kyoto covenant. And if they do not agree to make reductions in their emission levels, it is very difficult to see carbon stabilization occurring without a series of breakthroughs in nonfossil fuel technologies or carbon capture and storage. At the G8 summit in July 2008, developing countries, including China, India, South Africa, and Brazil, rejected the notion of joining in a 50 percent reduction by 2050, indicating that the developed world would need a far more substantial reduction in its emissions if they were to agree to any reductions of their own.

Fast-growing developing countries currently have relatively low per capita levels of carbon emissions. However, their economic growth is highly dependent on consumption of fossil fuels, presenting seemingly insuperable obstacles to stabilizing world CO_2 emissions.

In 2004, global GHG emissions (in CO_2 equivalents) were 28,790 million tonnes. Just over 10 percent of these were from the former Soviet bloc, with the rest split fairly evenly between the OECD countries and the developing world. Emissions from OECD countries grew at 1.3 percent per annum between 1990 and 2004. Those of the developing countries, however, saw annual growth at 5.7 percent during the same period, while the former Soviet bloc's emissions fell by 1.7 percent per annum.

By 2008, developing countries' emissions exceeded those of the OECD countries. The faster growth in emissions within developing countries will increasingly dilute any actions taken by the developed OECD nations,[16] the only group seriously considering abatement measures at present. The dilution is further amplified if abatement in the OECD is achieved by the established trend of smelting and other energy-intensive activities being relocated to developing countries. The IPCC report tended to downplay this leakage issue, arguing, "Estimates of carbon leakage rates for action under Kyoto range from 5 to 20% as a result of a loss of price competitiveness, but they remain very uncertain."[17] Given the globalized nature of production and the incentives and necessities of businesses to relocate to venues where even modest cost savings are available, the IPCC's range for carbon leakage is likely to be too conservative.

Table 1.4 shows the output of aluminum production, probably the most energy intensive of the main industrial products, by major country. A summary of the recent trends in emissions is shown in Table 1.5.

The UN *Human Development Report* illustrates considerable diversity between countries' emission levels per unit of purchasing power parity adjusted for GDP (in terms of kt of CO_2 per million dollars). In 2004, the poorest countries, such as Angola and Congo, emitted less than 100 tonnes of CO_2 per million dollars of GDP. Oil-rich countries had far higher emission levels (e.g., Kuwait was 1810; the UAR, 1570; and Iran, 930). Similarly, many former Soviet bloc countries had high emission levels (e.g., Kazakhstan, 2070; Uzbekistan, 3070; Russia, 1170).

India and China have different progressions to their current emission levels, perhaps reflecting their different development paths, with China focused on manufacturing and Indian development more closely associated with the service industries. Both countries' carbon intensity to GDP dropped between 1990 and 2004—India's from 480 tonnes per million dollars to 440 and China's from 1300 to 700.[18]

In spite of the rapid growth in developing country emissions, their per capita CO_2 emissions remain considerably below those of the OECD countries. In 2004, OECD emissions averaged 11.5 tonnes, the United States and Canada were at 20, and Australia, the United Kingdom, and France at 16, 10, and

---

[16]For a discussion of the role of China, refer to Chapter 17 by Lewis et al., in this book.

[17]www.ipcc.ch/pdf/assessment-report/ar4/wg3/ar4-wg3-chapter11.pdf, p. 622.

[18]Chapter 17 by Lewis et al. discusses the trends for China.

Table 1.4 World aluminum production 2006 (tonnes)

	World	33,410,000
1	People's Republic of China	5,896,000
2	Russia	4,102,000
3	United States	3,493,000
4	Canada	3,117,000
5	Australia	1,945,000
6	Brazil	1,674,000
7	Norway	1,384,000
8	India	1,183,000
9	Bahrain	872,000
10	United Arab Emirates	861,000
11	South Africa	855,000
12	Iceland	721,000
13	Germany	679,000
14	Venezuela	640,000
15	Mozambique	530,000
16	Tajikistan	520,000
17	Spain	399,000
18	France	394,000
19	United Kingdom	366,000
20	New Zealand	330,000
21	Netherlands	313,000
22	Argentina	272,000
23	Romania	270,000
24	Egypt	245,000
25	Iran	240,000
26	Indonesia	225,000
27	Ghana	200,000
28	Italy	198,000
29	Nigeria	193,000
30	Greece	165,000
31	Slovakia	158,000
32	Montenegro	120,000
33	Slovenia	117,000
34	Ukraine	113,000
35	Bosnia and Herzegovina	107,000
36	Sweden	102,000
37	Cameroon	96,000
38	Mexico	75,000

(*Continued*)

Table 1.4 (*Continued*)

39	Turkey	65,000
40	Poland	50,000
41	Switzerland	44,000
42	Azerbaijan	35,000
43	Hungary	28,000
44	Japan	18,000

Source: Wikipedia; http://wapedia.mobi/en/List_of_countries_by_aluminium_production

Table 1.5 Growth of CO$_2$-equivalent emissions by region, 1990 and 2004 (m tonnes)

	1990	2004	Annual Increase (%)
OECD	11,205	13,319	1.3
Former Soviet bloc	4182	3168	−1.7
Developing countries	6833	12,303	5.7
Total	22,220	28,790	2.1

Source: UNDP, Human Development Report, 2007–2008.

6 tonnes, respectively. Per capita emissions in developing countries averaged 2.4 tonnes. Table 1.6 summarizes the position of selected countries and country groupings in 2004.

There have been suggestions that the developing countries should be brought into an emission reduction scheme by granting them tradable emission rights. This offers ostensibly attractive outcomes all around. Developing countries would be given rights that would be surplus to their requirements, rather similar to when post-Communist countries in the former Soviet bloc were brought within the system. Those countries' adoption of capitalist production and pricing methods had encouraged conservation of resources and meant that their previous emission levels were far higher than their reformed economies required. Granting them their existing levels of emissions and allowing them to trade the surplus amounts handed them windfall gains.

The treatment of the former Soviet bloc countries in this way was crucial to getting their agreement to the Kyoto Convention and, in turn, to the Convention receiving the global support necessary for it to come into force as an international treaty. But at the same time, this scheme vastly expanded the

Table 1.6 Carbon intensity of energy emissions, selected countries, 2004

		CO_2 Emissions	
	Per Capita	Per Unit of GDP (kt of CO_2 per Million 2000 PPP US$)	
Selected countries			
Angola	0.7	0.29	
Democratic Republic of Congo	—	0.06	
Kuwait	37.1	1.81	
United Arab Republic	34.1	1.57	
Iran	6.4	0.93	
Uzbekistan	5.3	3.07	
Kazakhstan	13.3	2.07	
India	1.2	0.44	
China	3.8	0.70	
Australia	16.2	0.58	
United States	20.6	0.56	
Canada	20.0	0.69	
United Kingdom	9.8	0.34	
France	6.0	0.23	
Aggregate areas			
Least developed countries	0.02	0.017	
East Asia and Pacific	3.5	0.63	
Former Soviet bloc	7.9	0.97	
High-income OECD	13.2	0.45	
World	4.5	0.55	

Source: UNDP, Human Development Report, *2007–2008.*

quantities of permitted emissions by activating "sleeper" emission rights. In this way it somewhat undermined the basic intent of the protocol.

The far greater magnitude of developing country emissions, their less wasteful use of energy, and their future need for much higher levels of energy use make it impossible to adopt a similar approach. This would be even more difficult if developing countries claimed that they should receive credits for their previously low level of emissions. Figure 1.3 illustrates the overwhelming importance of the developed world in past levels of emissions.

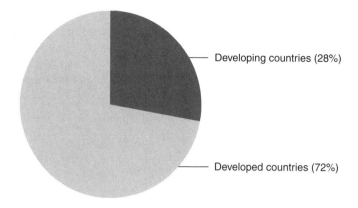

Developing countries (28%)

Developed countries (72%)

Figure 1.3   Contributions to atmospheric concentrations of greenhouse gases, 1850–2002.
Note: This figure shows the relative contribution of developed and developing countries to increases in concentrations from CO_2 emissions from fossil fuels and cement manufacture over the period 1850–2002.
Source: *Carbon Pollution Reduction Scheme*, Australian green paper, July 2008, derived from Kevin Baumert, Timothy Herzog, and Jonathan Pershing, 2005, "Navigating the Numbers: Greenhouse Gas Data and International Climate Policy," World Resources Institute.

An alternative approach to the carrot of incentives to developing country participation is the stick of penalties for nonparticipation. For example, Australia's Garnaut Report cites, with apparent approval, the suggestion of the economist Joseph Stiglitz that a tariff be placed on goods for recalcitrant countries that are not playing the game. Garnaut also notes that the head of the World Trade Organization (WTO), Pascal Lamy, supports such penalties as a "distant second best solution."[19]

Though measures such as WTO tariffs on carbon contents of goods may be a background threat to be used to encourage a "voluntary" solution, should this not emerge, such countervailing duty measures would prove extremely difficult to devise. They would entail a careful estimate of the fossil fuel content of every good and service, an estimate that would clearly be highly variable between products and over time. In the face of sharp disagreements, it is not difficult to see an attempt to require such compliance as bringing about the end of the present rules under which the global trading system operates.

---

[19]Garnaut R. Draft climate change review. Available from: www.garnautreport.org.au/draft.htm, p. 324; July 2008. Border tarrifs on carbon intensive products from countries that fail to take action comparable to the US also featured in the Waxman-Markey Bill.

Set against the case for developing countries to receive more generous credits, Posner and Sunstein[20] argue that developed countries' previous growth also brought benefits to developing countries. Posner and Sunstein also point to further difficulties that might emerge in determining a fair allocation of costs. These include the different levels of benefits said to accrue from taking action; Russia, for example, would be likely to obtain gains from warming, and China, the United States, and Japan are forecast to incur relatively low losses.

Further practical issues in administrating CO_2 reduction programs can be seen by examining measures that are already in place to allow signatories' obligations to be acquitted in developing countries through the Clean Development Mechanism (CDM). There has been a rapid increase in applications to use this approach. However, Wara and Victor[21] identify at least three problems with CDM-type schemes:

- Many offsets have made use of the refrigerant HCF-23 and created perverse incentives whereby the manufacture of the gas becomes a sideline to the credits it can earn.
- Many offsets are claimed for projects that would have proceeded anyway.
- Verification of projects' emission savings is unreliable.

These matters aside, the CDM would require the adoption of as yet unknown fundamental technological developments to achieve stabilization at 2004 levels of 28,790 million tonnes under any practicable or fair apportionment of the emission levels. If the trajectory were global, stabilization by 2030, with OECD countries reducing their emission levels by 20 percent and the former Soviet bloc holding their emissions constant, would require developing countries to limit their increases in emissions to 15 Gt (by 22 percent), as illustrated in Table 1.7.

Though superficially generous to the developing countries, the 22 percent increase is a massive reduction compared with business-as-usual growth levels. Compared with the 15 billion tonnes of CO_2 equivalent projected under this scenario, business-as-usual levels, based on previous growth rates, would see developing countries emitting over 37 billion tonnes in 2030.

Moreover, because of their population growth, limiting developing countries' emission levels to 15 billion tonnes of CO_2 equivalent would result in their emissions per head actually *falling*. Developing countries in 2030 are

[20]Posner AE, Sunstein CR. Global Warming and Social Justice. Regulation 2008; (Spring).

[21]Wara MW, Victor DG. A Realistic Policy on International Carbon Credits. Working Paper 74, Program on Energy and Sustainable Development, Stanford University, April 2008.

Table 1.7 Emission stabilization scenario, 2004 and 2030 (million tonnes of CO$_2$ equivalent)

	2004	2030	2030 bau[22]
OECD	13,319	10,655	18,350
Former Soviet bloc	3168	3168	3168
Developing countries	12,303	14,967	36,671
Total	28,790	28,790	58,188

Source: Derived from UNDP, Human Development Report, 2007–2008.

estimated to have a population at 7.2 billion,[23] and under the scenario in Table 1.7 their per capita emissions would fall from 2.4 tonnes to 2.3 tonnes. This is one fifth of the OECD 2004 per capita average of 11.5 tonnes and only a quarter of the OECD average in 2030 (7.9 tonnes), once a 20 percent reduction and population growth are incorporated.

Many other scenarios can be examined. At one extreme, if developing countries were to maintain their 1990–2004 levels of increase and the former Soviet bloc's emission levels remained constant, this would leave virtually no emissions for the OECD group in 2030. Even under that scenario, the developing countries' 2030 per capita emissions would be less than one third the current OECD level.

Perhaps the most readily supported basis of allocating emissions would be on an equal per capita basis for all countries.[24] With a 20 percent reduction, this would require emissions per capita to be limited to 3.4 tonnes. Such a task would be Herculean for those countries now in the over 15 tonne per capita category. Chancellor Merkel's proposal of halving present global emission levels would further magnify the difficulties involved.

OECD countries' current per capita emissions and the percentage reduction necessary to bring these to a world average of 3.4 tonnes are illustrated in Figure 1.4. The 3.4 tonnes per capita level is equivalent to an average overall 20 percent reduction on 2004 levels. The 2004 emissions are shown by the bars measured against the left-hand axis. The reductions for the selected countries are illustrated by the blue line measured on the right-hand scale.

---

[22]Based on projecting emissions per head forward at the compound average rates, 1990–2004, for OECD (1.24 percent) and developing countries (4.29 percent), with the former Soviet bloc constant.

[23]Derived from www.prb.org/Datafinder/Geography/MultiCompare.aspx? variables = 6,7,4&regions = 1,2,3

[24]Refer to Chapter 4 by Fawcett et al. in this book.

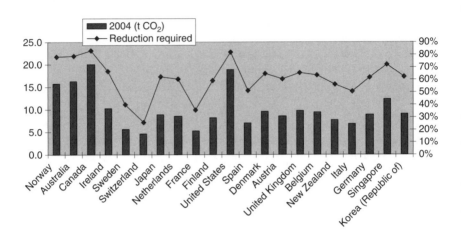

Figure 1.4 OECD countries' per capita emission reduction requirements (with aggregate emissions 20 percent below 2004 levels).
Source: Derived from UNDP, *Human Development Report*, 2007–2008.

Ominously for those seeking lower emissions, China's per capita emission levels have already been noted as indicative of the magnitude of a stabilization task. At 3.8 tonnes of CO_2 equivalent per capita in 2004, China already exceeded the 3.4 tonne global average target in spite of massive reductions in its energy intensity.

Any debate over emission reductions in practice is likely to follow the negotiated course that characterized the Kyoto Protocol's agreement. Countries, at least those intending to abide by obligations they agreed to, will seek to point to special circumstances that make them unable to follow a general standard, the most obvious of which is equal emissions per capita. Many countries will argue that they are producing goods entailing high carbon emissions for consumption in other countries; others will argue that their geography requires intrinsically higher emission levels, that special features of their location or their agricultural profile mean that they require higher-than-average emission credits or that they are natural sinks for CO_2 and should obtain some recognition for this, and so on.

The developed countries are likely to call for a weighting to be given to emission levels per dollar of GDP, which rewards them for their higher levels of carbon "efficiency" in producing each unit of GDP.

By contrast, developing countries are likely to require above-average quotas in recognition of their previously low levels of emissions, summarized in Figure 1.3. Figure 1.5 illustrates this idea in greater detail, showing that by 2030, OECD countries with about 15 percent of the world's population will cumulatively have produced 54 percent of the world's emissions since 1990, whereas developing countries with 85 percent of the population will have accounted for under 20 percent.

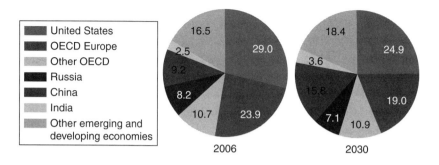

Figure 1.5   Stock of CO_2 emissions beginning in 1990 (percent of world stock). Source: IMF, *World Economic Outlook,* Chapter 4: "Climate Change and the Global Economy," 2008 (www.imf.org/external/pubs/ft/weo/2008/01/index.htm).

1.4 Paths to emission reductions

Barring some unforeseen technology breakthroughs, if developing countries were somehow forced to hold their emissions at their present levels, they would be unable to close the gap with the developed world's living standards. If developed countries were required to reduce their emissions to the current world average of around 4 tonnes per capita, this could only be possible with (1) a fundamental shift to low-carbon economies, (2) a markedly reduced living standard, or (3) a drastically different lifestyle.

As previously discussed, a radical nuclearization of electricity generation would produce less than a 40 percent reduction in CO_2 equivalents. It would also be a "one-shot" reduction requiring additional emission reductions or substitutions out of energy if living standards were to be allowed to increase. Even a massive conversion to nuclear would come at considerable cost in the abandonment of wealth in such assets as coal and in the diversion of capital from more productive venues.

There is, of course, the prospect of new technologies emerging. Draconian cuts in emission levels would require taxes or prices on emission levels that would certainly stimulate the discovery of these as well as energy-use economies. But the necessary technological breakthroughs are, as yet, commercially unproven.

Organizations promoting actions are more sanguine. The IPCC and others have modeled many scenarios that purport to show the task, one version of which Table 1.8 exemplifies.

Many individual country estimates go into great detail in specifying the areas in which energy savings are to be made. A recent Australian study not

Table 1.8 Areas of estimated emission reductions

Chapter of Report	Estimate	Sector-Based ("bottom-up") Potential by 2030 (GtCO₂-eq/yr)				Economy-Wide Model ("top-down") Snapshot of Mitigation by 2030 (GtCO₂-eq/yr)	
		Downstream (indirect) Allocation of Electricity Savings to End-Use Sectors		Point-of-Emissions Allocation (Emission Savings from End-Use Electricity Savings Allocated to Energy Supply Sector)			
		Low	High	Low	High	Low	High
		"Low-Cost" Emission Reductions: Carbon Price <20 US$/tCO₂-eq					
4	Energy supply	1.2	2.4	4.4	6.4	3.9	9.7
5	Transport	1.3	2.1	1.3	2.1	0.1	1.6
6	Buildings	4.9	6.1	1.9	2.3	0.3	1.1
7	Industry	0.7	1.5	0.5	1.3	1.2	3.2
8	Agriculture	0.3	2.4	0.3	2.4	0.6	1.2
9	Forestry	0.6	1.9	0.6	1.9	0.2	0.8
10	Waste	0.3	0.8	0.3	0.8	0.7	0.9
11	Total	9.3	17.1	9.1	17.9	8.7	17.9

"Medium-Cost" Emission Reductions: Carbon Price <50 US$/tCO$_2$-eq

4	Energy supply	2.2	4.2	5.6	8.4	6.7	12.4
5	Transport	1.5	2.3	1.5	2.3	0.5	1.9
6	Buildings	4.9	6.1	1.9	2.3	0.4	1.3
7	Industry	2.2	4.7	1.6	4.5	2.2	4.3
8	Agriculture	1.4	3.9	1.4	3.9	0.8	1.4
9	Forestry	1.0	3.2	1.0	3.2	0.2	0.8
10	Waste	0.4	1.0	0.4	1.0	0.8	1
11	Total	13.3	25.7	13.2	25.8	13.7	22.6

"High-Cost" Emission Reductions: Carbon Price <100 US$/tCO$_2$-eq

4	Energy supply	2.4	4.7	6.3	9.3	8.7	14.5
5	Transport	1.6	2.5	1.6	2.5	0.8	2.5
6	Buildings	5.4	6.7	2.3	2.9	0.6	1,5
7	Industry	2.5	5.5	1.7	4.7	3	5
8	Agriculture	2.3	6.4	2.3	6.4	0.9	1.5
9	Forestry	1.3	4.2	1.3	4.2	0.2	0.8
10	Waste	0.4	1.0	0.4	1	0.9	1.1
11	Total	15.8	31.1	15.8	31.1	16.8	26.2

Source: www.ipcc.ch/pdf/assessment-report/ar4/wg3/ar4-wg3-chapter11.pdf, p. 636.

only made such estimates for domestic appliances but also identified usages that extended to such minutiae as waterbeds.[25]

Table 1.9, also reproduced from the IPCC, estimates the sorts of reductions expected from various effective tax rates. It illustrates two of the 40 "storylines and scenarios," suggesting that even with no tax, 5–14 percent reduction in emissions will take place, that a tax of $50 per tonne of CO_2 would increase that to 13–52 percent, and a tax of $100 would increase it to 16–63 percent.

Though some quite extraordinary detail about the expected reductions is offered in Tables 1.8 and 1.9, the numbers should not be interpreted as providing anything beyond conjecture. We do not have the information to be able to model behavior in response to price (or regulation) with the level of detail offered by the IPCC work. Historical price data that is available to allow estimates to be made is of two sorts. The first is associated with relatively small changes in price. The second relates to infrequent large changes that have previously occurred in the context of readily available alternative supply sources—for example, with dramatic oil price increases where fossil fuel substitution was possible and where energy-intensive activities could shift to

Table 1.9 Global economic mitigation potential in 2030, estimated from bottom-up studies[26]

Carbon Price (US$/ tCO₂-eq)	Economic Potential (GtCO₂-eq/ yr)	Reduction Relative to SRES A1 B (68 GtCO₂-eq/yr) (%)	Reduction Relative to SRES B2 (49 GtCO₂-eq/yr) (%)
0	5–7	7–10	10–14
20	9–17	14–25	19–35
50	13–26	20–38	27–52
100	16–31	23–46	32–63

Source: www.ipcc.ch/pdf/assessment-report/ar4/wg3/ar4-wg3-ts.pdf, p. 77.

[25]Government of Victoria, Department of Premier and Cabinet, Understanding the Potential to Reduce Victoria's Greenhouse Gas Emissions, prepared by the NOUS Groups and SKM, December 2007.

[26]The estimates refer to SRES A1 B, a high economic growth scenario with a median forecast of technology development introducing reduced emissions, and SRES B2, with lower economic growth and diminished material intensity of the GDP and a relatively rapid innovation and take-up of resource-efficient technologies. The 2030 CO_2 equivalent of compounding the 1990–2004 growth rates of the OECD and developing countries (with the former Soviet bloc constant at 2004 levels) is 58 Gt.

locations where energy prices remained low. Neither of these experiences offers sound bases for forecasting the kinds of outcomes required to reduce CO$_2$ emissions by the quantities sought.

1.5 Conclusion

1.5.1 ADDRESSING THE ISSUES

There have been many suggested targets for emission reductions. The Stern Report sought an 80 percent reduction from present levels, and Professor Stern has reiterated such calls.[27] Others have called for stabilization at 2004 levels.

Any of the targets would require sweeping technological innovations to allow the substitution from fossil fuels or living standards, which, at least on a global scale, would be vastly lower than those that are being anticipated.

It is not difficult to reduce consumption of goods and services for which alternatives are available. Thus the world's steel industry adjusted to the oil price hikes in the 1970s by almost eliminating its use of petroleum products and substituting coal. It is even easier to envisage economizing on other products such as beef when substitutes are available.

It would be far less easy to envisage reducing consumption of broadly defined goods such as food or housing to a degree approaching 80 percent. In the case of both products, such economizing would be possible (for food by reducing the consumption of grain-fed animals and other low-efficiency calorie converters). But meeting the proposed GHG targets is likely to cause a considerable loss of consumer satisfaction and a sacrifice in living standards.

Presumably those calling for radical reductions in carbon emissions, especially when they are also largely rejecting nuclear power, consider that reductions in carbon emissions have penalties akin to reducing a subcomponent of a major demand category such as food or shelter. They consider that man's ingenuity, given adequate incentives, will discover low-cost alternatives and means of satisfying demand in ways that use a different mix of inputs and outputs.

Such notions appear to be highly optimistic. Already we have seen prices of low carbon-emitting fuels increase markedly—before falling at the end of 2008, gas prices more than trebled over recent years—in response to an aversion to coal use in Europe and North America.

Using the food analogy, if we were told that the consumption of fish would need to be reduced by 80 percent, this would require a very large tax on fish, but the availability of substitutes is such that the aggregate loss of economic

---

[27]www.theaustralian.news.com.au/story/0,25197,23627804-11949,00.html

welfare would be small. Contrast that with the outcome if the reduction were to be food in general. At issue is whether carbon emissions are so pervasive in the production of energy that taxing them so that they are reduced by 80 percent would have an effect closer to the analogy of food or of fish.

Even for targets involving less than an 80 percent reduction in emissions, considerable restructuring and costs would be required. For stabilization at 2004 levels, a wholesale replacement of coal and gas by nuclear for electricity generation would be capable of achieving some but by no means all of the required reductions.

Some indication of the practicality of emission levels that are achievable from this approach is illustrated by low-carbon economies such as France, Switzerland, and Sweden. In all three cases nuclear power dominates, with hydro playing important roles in Sweden and Switzerland, and all three produce less than 6 tonnes of CO_2 equivalent per capita. Even so, this remains a far cry from the 2.0–3.4 tonnes of CO_2 equivalent per capita that is necessary for stabilization at 2004 levels.

Adoption of nuclear power presents the only low-ish cost option to move the world substantially toward stabilizing emissions; however, many of those pressing most strongly for emission reductions are also tenaciously opposed to this form of electricity generation; astonishingly, the first Australian Garnaut Report[28] did not even mention it.[29]

Though more attainable than some approaches, even an extensive replacement of fossil fuels with nuclear power is a task of considerable magnitude. For many countries that are resource rich in fossil fuels, the journey will mean a huge sacrifice of living standards compared to those that would otherwise prevail. Some would see levels of wealth loss that would be difficult to compensate.

The ubiquity of carbon in our lives offers the medium by which controls can be all encompassing. This was strikingly observed at the 2020 Summit called by Australia's Prime Minister Kevin Rudd in April 2008, just a few months after his election. The Prime Minister told that summit of 1000 of Australia's selected leaders and thinkers that climate change "overarches all." And the leader of the environment panel at the summit called for "robust institutions

---

[28]www.garnautreview.org.au/CA25734E0016A131/WebObj/GarnautClimateChangeReviewInterim Report-Feb08/$File/Garnaut%20Climate%20Change%20Review%20Interim%20Report%20-%20Feb%2008.pdf

[29]The final Garnaut Report (September 2008) does discuss nuclear power but dismisses it as a possible source of energy for Australia because of capital cost increases, better possibilities that Garnaut speculates will become available in carbon capture and storage, and adverse public opinion (www.garnautreview.org.au/CA25734E0016A131/pages/draft-report).

to support" a climate change agenda encompassing "government expenditure, tax, regulation and investment."

Even so, Australia demonstrates considerable policy confusion, not least because measures to suppress domestic CO_2 emissions are accompanied by continued enthusiasm for coal exports, which comprise 23 percent of the country's total exports.

The task of bringing a stabilization of world emissions is difficult to understate. Australia, since the April 2008 Summit, has issued two reports by Professor Garnaut, modeling exercises, and a green and white paper mapping out its agenda for a comprehensive Carbon Pollution Reduction Program involving a tradable rights system, largely based on auctions, and additional regulatory measures. The objective, a reduction of emissions by 60 percent by 2050, would still leave Australia at double the level required for stabilization at a uniform per capita emission level.

Garnaut's draft report, *Targets and Trajectories*,[30] argued that Australia would suffer an 8 percent loss of GDP by 2100 under business as usual (four times the global costs posited by Stern) and recommended a tax of $20 per tonne of CO_2-e in 2010, rising to $30 in 2020. This was projected to bring a 10 percent reduction in emissions by 2020, rather less than the amount inferred by the government's green paper issued in July 2008.[31] The Garnaut report also contemplated halving the Australian 2020 reduction if China and other major emitters failed to participate, a response that begs serious questions, since it would surely be an empty gesture for Australia to impose on itself emission reduction costs if the world at large did not do so.

The modeling by the Australian Treasury,[32] which was issued separately but also formed the basis of the modeling results of Garnaut, used a price of A$20 per tonne of CO_2 (2005 dollars) escalating over time and estimated the cost to the economy of stabilization would be only 2 years' loss of growth by 2050, or 4.7 percent of GDP.

Global stabilization is modeled at 550 parts per million of CO_2-e and is on the basis that all nations phase in targets (China, for example, by 2015).

Although Garnaut suggested that force, international sanctions, and perhaps a new international trading regime might be necessary to ensure that countries reduce their emissions, the Australian Treasury has taken a controversial, almost unworldly, view that, "Where emission pricing is gradually introduced across the world, countries that defer action face higher

---

[30]Ross Garnaut Climate Change Review, Targets and Trajectories, draft report, September 2008.
[31]Department of Climate Change. Carbon Pollution Reduction Scheme, green paper. Available from www.climatechange.gov.au; July 2008.
[32]Australia's Low Pollution Future. Available from: www.treasury.gov.au/lowpollutionfuture/

long-term costs, because global investment is redirected to countries that act early. Australia therefore benefits from being an early mover in a multi-stage world."

The modeling forecasts a great deal of trading. It has Australia, in the central Treasury scenario (called *Garnaut 10*), buying 293 MT CO_2-e, 40 percent of its needs in 2050, at US$91 per tonne (in 2005 prices), or over US $26 billion per annum. Such numbers are largely determined by assumptions as to how cheap it is for countries to mitigate emissions relative to each other. The United States and China are said to find easy opportunities to do this, whereas countries such as Australia, the EU nations, and the OPEC countries are in the world market buying the permits; the former Soviet bloc is spending over $200 billion a year buying credits.

As with all modeling, that of the Australian Treasury is dependent on the assumptions employed. Critical in this respect are the following:

■ Responsiveness of energy demand to higher prices

■ Substitution of energy for other goods and services due to higher prices

■ Technological developments of noncarbon energy sources and the abilities to sequester carbon cheaply

Irrespective of the richness of the input data and the complexity of the interrelationships assembled, the answers are driven by assumptions on these matters that, for the period involved and the magnitude of changes required, are little more than educated guesses. In this respect it is useful to consider how 1960 forecast scenarios might have coped with imagining a world 50 years hence. They would have had to envisage the Internet, mass air travel, mobile phones, the collapse of Communism, and the rise of India and China. No forecasts could have picked these changes.

1.5.2 CONCLUSIONS

■ The sort of GHG reduction target that the scientists believe is necessary to avert climate change involves stabilization of CO_2-e emissions at their present level of around 550 parts per million (ppm). Many argue that a trajectory to 450 ppm is necessary. Politicians in developed countries have in general endorsed the need for emphatic action to address emissions.

■ Under a business-as-usual scenario, emissions are likely to be more than double 2004 levels by 2030 and to continue rising thereafter. In July 2008 the G8 called for a 50 percent reduction in emissions by 2050. Meeting such targets in the timeframes proposed is challenging and requires significant resources and resourcefulness.

■ The following chapters of this book examine the difficulties and explore ways in which the emission reduction targets might be approached.

Carbon Policies: Do They Deliver in the Long Run?

Emile J. L. Chappin, M.Sc., Gerard P. J. Dijkema, Ph.D., and Laurens J. de Vries, Ph.D.

Faculty of Technology, Policy, and Management Delft University of Technology

Abstract

Carbon taxation and emissions trading are policy instruments for achieving significant CO_2 emission reduction by inducing a shift in technology and fuel choice. Simulations with a quantitative agent-based model of a competitive electricity generation sector show that under both policies, CO_2 emissions increase for 10–15 years due to the long life cycle of power plants. Dramatic reductions materialize after 20–40 years, when a tight cap or sufficient tax level is maintained. When taxes are set equivalent to trading prices, taxation induces earlier investment in CO_2 abatement, a better balance between capital and operating costs, and lower long-run electricity prices.

▌ 2.1 Introduction

Currently, electric power production is largely based on the combustion of fossil fuels, predominantly coal and natural gas, except in countries with abundant hydropower. This inevitably leads to the emission of CO_2, since carbon capture and storage and renewable energy sources are not feasible or available yet on a large scale.[1] Global climate change can be considered a "Tragedy of the Commons" for which no effective global coordination, regulation, or enforcement has yet been developed [19]. This has not happened for a variety of reasons. First, CO_2 is a global pollutant, not a regional pollutant such as SO_2 or NO_x, which implies that the regulation of local emissions needs to be coordinated worldwide. Second, fossil fuels have become the lifeblood of industrialized economies: Reducing or replacing their consumption is difficult and costly. Although the cost of abatement is high, doing nothing will eventually be much more expensive (cf. [30]), and the growing consensus that CO_2 emissions need to be stabilized and then reduced in the course of this century has led to much interest in achieving cost-efficient emission reduction through incentive-based instruments rather than command-and-control regulation.

Incentive-based policy instruments such as the European emissions trading system (ETS) and carbon taxation (CT) use market signals to influence decision making and behavior [14]. A market in which emission rights can be traded is expected to yield an economically optimal distribution of emissions among polluters. It remains to be seen, however, whether it creates sufficient investment incentives for electricity producers, because the price of emission rights is volatile and the time horizon of the ETS is limited. A carbon tax represents a more stable price signal, but it is difficult if not impossible to establish *ex ante* which tax level would be required to achieve the desired emission reduction.

In this chapter, we compare these two policy instruments, addressing the following question: What are the effects of taxes and emissions trading upon CO_2 emissions, electricity prices, and the technology portfolio for electricity generation and CO_2 abatement? We address this question by developing and using an agent-based model of a competitive electricity production sector in which noncoordinated decisions are made within a common framework of an electricity market, either with no carbon policy, with an ETS, or with a CT.

---

[1]See Chapter 6, by Lackner et al., in this book for a detailed discussion of carbon capture and sequestration.

In this chapter, first the technology and policy options for CO_2 emission reduction are summarized and the impact of both instruments explored. Second, these insights are translated into an agent-based model. Its structure and approach are described, the scenarios and assumptions that are used for comparing the policy instruments are given, and the agents' behavior and technology options are introduced. Third, the simulation results are presented and interpreted for a large variety of exogenous conditions. Finally the conclusions are summarized.

2.2 Options for CO_2 emission reduction

While the European CO_2 emissions trading scheme (ETS) is the largest in the world, similar systems have been established in at least six states in the United States, and several large companies have implemented internal trading schemes. Carbon taxes have been implemented in Scandinavia. Table 2.1 presents an overview of the main types of carbon policies.

"Economic theory tells us that if cost and benefit functions are known with certainty, then a price based policy (such as a tax) and a quantitative policy (such as tradable permits) are equivalent from an efficiency point of view" [20], page 141. However, one of the key issues in climate policy is that cost

Table 2.1 Characterization of carbon policies

Policy Instrument	Price	Volume of Emissions	Allocation of Emission Rights	Implemented in Practice
Carbon taxation		Set by government	Not limited	Can shift between sectors	Yes
Emissions trading	Cap and trade	Market based	Capped[2]	Grandfathering/ auction[3]	Yes
	Performance standard rate	Market based	Not limited	Benchmarking and performance	No
Command and control		No price	Regulated per source	By government, per source	Only for other pollutants

[2]In the ETS, the total amount of rights granted is capped to reach a certain emission target. This cap has been divided between member states. As of January 2008, inter-member states trade is possible. Member states also can increase the volume of rights via the Clean Development Mechanism (CDM).

[3]An alternative strategy is to ration carbon allowances per capita. Chapter 4, by Meyer et al., in this book discusses this alternative.

and benefit functions are uncertain. Weitzman argued that given uncertainty, the slope of the supply and demand functions should determine the choice [35]. Grubb and Newberry summarize this argument and apply it to CO_2 policy [18]. They conclude that in principle taxes are superior, but they observe practical obstacles such as political acceptability. An advantage of a tax is that it creates less investment risk than emissions trading, because there is no market and thus no price volatility. A risk to investors is, however, that the tax level could be reduced during the economic life of the investments.

In Europe, electricity generation accounts for one third of CO_2 emissions [9, 10]. The success of an emissions trading scheme therefore depends in significant part on the reduction of emissions from the power sector. Will these reductions materialize via operational adjustment or investment? Let us briefly analyze carbon policies and their effects.

Three types of effects of incentive-based carbon policy instruments can be discerned. The first effect is that the pricing of CO_2 leads to higher energy prices, which in turn leads to a reduction of demand and supply substitution. In the short term, the price elasticity of electricity demand is notoriously low, but in the long term higher prices will cause consumers and industry to invest in less energy-intensive equipment. "We need only look back to the oil price shocks of the 1970s to see how well the price mechanism works. Higher fossil fuel prices dampen total energy consumption" [23]. The lower energy intensity of the European economies compared to North American economies provides evidence of the impact of structurally higher end-user prices, which are largely due to higher taxes.[4] However, given the fundamental importance of electricity in our society, the potential for demand reduction alone is limited, compared to the CO_2 emission reduction needed.

The second effect of carbon policies is that CO_2-intensive electricity production becomes less attractive. Higher fossil fuel prices also make fossil fuels less attractive relative to other supply-side alternatives. Hence, carbon taxes create incentives to switch away from carbon-intensive fuels. However, at the level of an individual power plant the options for fuel switching are limited because the technical designs differ too much to make economically attractive a switch from coal to, for instance, natural gas in an existing installation. A single option that is economically feasible is to cofire biomass in a coal-fired power plant, to a maximum of 15 percent fuel input. At the sector level, fuel switching takes place through changes in the merit order: plants switching from base load operation to peak load operation, and vice versa. With any merit order change, the fuel diet of the sector changes.

---

[4]For an overview of electricity price elasticity, see, for instance, Lijesen (2007).

The third effect of carbon policies is to induce investment in CO_2 abatement. Investment options are retrofitting existing installations or extending them with carbon capture and sequestration (CCS) or investing in new, more efficient facilities and carbon-free technologies such as wind power.

Changing the merit order and cofiring will, at best, reduce CO_2 emission by 10–15 percent. Over time, investment decisions will tend toward less CO_2-intense technologies, reducing the average CO_2 intensity of the electricity generation portfolio. The dynamics of process innovation in mature capital-intensive industries are characterized by high risks and long time spans (cf. [13]). The main impact of carbon policies therefore must be achieved through the investment decisions of electricity producers.

Electric power generation is a capital-intensive industry and assets have life cycles of decades. The capital cost of a full-scale, state-of-the-art coal-fired power plant in the EU is around 1000–1200 €/kW, which means more than €1 billion for a 1040 MW plant such as currently planned by E.On. A coal gasification plant costs another 600–800 €/kW more. Investment levels for wind parks or biomass firing are similar. These generation technologies are proven and commercially available, but under which conditions will carbon pricing cause power companies to invest in these low-carbon technologies?

Carbon taxation provides a clear price signal by increasing the variable costs of fossil fuel–based electricity production [22]. It is a classic Pigouvian tax, the ideal level of which should be equal to the marginal social damage [27]. The positive cost of CO_2 emissions provides a monetary incentive for reducing emissions ([28], page 2). An issue with a carbon tax is that the total emissions volume is not constrained. A tax is expected to shift the portfolio balance from coal to more natural gas and perhaps renewables as well as carbon capture and sequestration (CCS). Such a shift is the aggregate result of many separate investment decisions regarding the choice of energy source, electricity generation technology, plant scale, and CO_2 abatement technology. A possible second-order effect of a carbon tax is that it reduces the demand for coal and increases the demand for alternatives such as natural gas, which could cause coal to become relatively cheaper, partly undoing the effect of the tax. At which level of fuel prices, volumes, and CO_2 emission level the market would stabilize is difficult to predict, because they not only depend on the fuel markets' dynamics but also on the availability and price of alternatives such as CCS and renewable energy sources. This is one of the reasons that the effect of a tax on the CO_2 emission level is difficult to estimate *ex ante*.

This would not be a problem if we knew the optimal tax level; then, by definition, the resulting emission level would also be socially optimal. However, a fundamental problem with a Pigouvian tax is that we do not have a reliable measure for the social damage, so it is impossible to establish *ex ante* the correct level of the tax [1]. As Grubb and Newberry argue, we do not know which tax level would reduce CO_2 emissions sufficiently to stabilize the atmospheric concentration at a certain level.[5] A possible solution is to start with a relatively low tax and adjust it over time in response to observed emission reductions. If a firm commitment is made that the tax will not be lowered during the life span of existing investments in less carbon-intense power generation or CO_2 abatement, this would provide significant certainty to investors regarding the minimum level of return on their investment. This way, investment risk can be limited while preserving policy flexibility.

Emissions trading relies on a price signal for internalizing a negative external effect of production [16]. A major argument for tradable emission rights is that "the invisible hand" of the market would lead to least-cost emission reduction [15, 29, 32, 33]. Both within a sector and between sectors, transactions will occur until a CO_2 price develops whereby total emissions equate to the emissions cap and no emitter will invest in further emission reduction. "There is a broad consensus that the costs of abatement of global climate change can be reduced efficiently through the assignment of quota rights and through international trade in these rights" [24]. Box 2.1 presents an overview of the experience with the European ETS.

The main difference between trading and taxation can be summarized as follows: With trading, the total quantity of CO_2 emissions is set but the price is unknown and volatile. Under taxation, the price of CO_2 is set, whereas the volume of emissions is not.

▍ 2.3 An electricity market model with carbon policy

2.3.1 Description of the Model

The electric power production sector may be considered as a large-scale socio-technical system in which a variety of stakeholders (agents) interact with each other and with the physical infrastructure for the production and transport of electricity [6, 8, 26]. Though the technical infrastructure is governed by the rules of nature, the social network is governed by informal and formal social rules and regulations. The combined system is complex and exhibits chaotic behavior. The long life cycles of power plants and electricity networks cause strong path dependence in the development of this system. Consequently, quantitative static equilibrium analyses, as are common in economics, provide

---

[5]Stern (2006) argues that this level should be around 500 ppm.

Box 2.1   EXPERIENCE WITH THE EUROPEAN EMISSIONS TRADING SCHEME

In January 2005, the European emissions trading scheme (ETS) was implemented [4].[6] In the ETS at least 90 percent of emission rights are grandfathered: They are allocated to emitters for free, in volumes based on past emissions. This has led to a highly politicized process in which companies, industrial sectors, and European countries vie for emission allowances in order to minimize the financial consequences of the CO_2 cap. Overallocation of allowances was the consequence. Initially, market parties did not know about this overallocation, but when in April 2006 the European Commission communicated that they had issued too many emission rights, the price collapsed to nearly zero [10]. Between €7 and €8 billion in emission rights value vaporized overnight.

The grandfathering of emission rights also led to substantial windfall profits for power producers. They passed the marginal costs of CO_2 on to the consumers (in perfect accordance with economic theory), which they had largely obtained at zero cost. In addition, with respect to emission reduction, the low-hanging fruit could still be picked at no or limited cost. To solve this problem, in the second phase of the ETS, between 2013 and 2020, all emission rights for the power sector will be auctioned.

In the first phase of the ETS (2005–2007), the prices of tradable CO_2 emission credits were highly volatile. In retrospect, this was due to the limited time horizon of this phase, the highly politicized process for determining the emission cap, uncertainties regarding the cost and availability of abatement options, the mismatch between the actual and forecast demand for emission rights, and the inelasticity of the supply of emission rights. Using the first phase as a learning period, the European Commission proposed improvements to the ETS. The most important change is to set a predictable cap that is to be reduced by 1.7 percent each year to achieve a 20 percent reduction between 2013 and 2020. The Commission also made it clear that ETS will continue beyond 2020 and at least become more stringent. Meanwhile, an extensive program to develop and demonstrate CCS is being developed. Funding of R&D on innovative energy technologies has been increased, and regulation and research to reduce energy consumption is back on the agenda. As in any market, a certain amount of price volatility remains inevitable, but both the design of the ETS and its context are improved to reduce uncertainty.

---

[6]An elaborate discussion of the EU ETS is given in the Chapter 3, by Musier and Adib, in this book: for an overview of the results of Phases 1 and 2 and a discussion of the proposed changes for Phase 3, see, for example, the report by Carbon Trust (2008).

only limited insight into the long-term impact of policy interventions such as a carbon tax or emissions cap.

Therefore, a quantitative *agent-based model* (ABM) was developed to simulate the evolution of the structure and performance of a hypothetical electricity market in the next 50 years using insights from microeconomics, market design, agent theory, process system engineering, and complex system theory [6, 8, 26]. An ABM represents a set of interacting "agents" with certain properties who live in an external world on which they have no influence—a modeling paradigm that matches the electric power production sector, where independent power producers, governments, and consumers can be considered agents that compete and interact via markets. Each agent has a set of goals, a working memory, a social memory, and a set of rules of social engagement.

The model reflects the real-world situation of six independent electricity producers that have different generation portfolios and that make different decisions regarding the operation of their generators, investment, and decommissioning. A schematic overview of the ABM is presented in Figure 2.1. The model contains

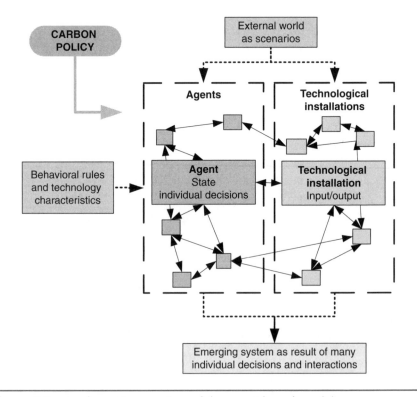

Figure 2.1 A schematic overview of the agent-based model.
Source: Adapted from Chappin et al., 2008b [7].

two subsystems: agents and installations. The external world is represented by exogenous scenarios. The agents in the model, the power producers, need to nego-tiate contracts for feedstock, the sales of electricity, and, in the case of emissions trading, emission rights. In the longer term, the agents need to choose when to invest, how much capacity to build, and what type of power generation technology to select. Agents interact through negotiated contracts and organized exchanges, and the physical flows and their constraints and characteristics are modeled. The characteristics of the modeled system are emergent: The generation portfolio and merit order, fuel choice, abatement options, and electricity and CO_2 prices and emissions emerge as a result of the decisions of the agents. The model has been run for three cases: no carbon policy, ETS, or CT.

The electricity demand profile consists of 10 steps per year that reflect a typ-ical load-duration curve to reflect the different emissions levels, costs, and operating hours of the various power plants. Markets for CO_2 rights, power, and fuels are modeled as exchanges in which 100 percent of the product is traded every time step. The time step of the model is 1 year and the simulations span a horizon of 50 years.

The main policy variable of the ETS is the emissions cap. In the model the cap is set to reflect the likely design of Phase 3 of the EU ETS in which the CO_2 cap is reduced every 5 years by 3 Mton for a market the size of the Netherlands. With an initial cap of 50 Mton, a 50 percent reduction is achieved in little more than 40 years. Another important policy variable is how many emission rights can be obtained through the Clean Development Mechanism (CDM).[7] This is set to 5 Mton/year over the entire simulated time period.

The only CT policy variable to be set is the tax level. To allow a fair com-parison between ETS and CT, the tax level in our model has been calibrated to the average CO_2 price that emerges in the simulated emission market. The ini-tial tax level equates to 20 €/ton, which reflects the current CO_2 price under the ETS. With time, the tax level increases to 80 €/ton (see Figure 2.5).

2.3.2 Scenarios and Assumptions

The electricity producers—the agents—operate in a dynamic world that is represented as exogenous trends: time series of fuel prices, electricity demand, and carbon policy parameters (emission caps or tax levels). We

---

[7]Under pressure of the industry, the Dutch government acquires additional emission rights through the Clean Development Mechanism. In the Dutch ETS allocation plan, it was announced that the government reserved €600 million for this purpose—the equivalent of 20 Mton CO_2 rights. Source: Ministry of VROM and SenterNovem, 2005.

Table 2.2 Exogenous parameters: Scenario and carbon policy settings

Domain	Parameters	Initial Value	Trend
Fuel markets	Natural gas price	0.61 €/Nm^3 [8]	+2%/year
	Coal price	103.3 €/ton[9]	+2%/year
	Uranium price	17 €/kg[10]	+1%/year
	Biofuel price	120 €/ton	+1.5%/year
Power market	Electricity demand	140 TWh/year	+2%/year
Emissions trading	Cap	50 Mton CO_2/year	−3 Mton/5 years
Carbon taxation	Taxation level	20 €/ton	Rising from 20 to 80 €/ton, with the average equal to the average CO_2 price in emissions trading

assume that the electricity producers have no market power, neither in fuel markets nor in the electricity or CO_2 markets. In Table 2.2 an overview of the scenarios and carbon policy parameters, values, and trends used is provided.

The fuel prices in the simulation start at October 2008 market levels and develop as depicted in Figure 2.2. The figure presents the *average* fuel prices used. In individual runs, fuel prices vary randomly around these averages.

The rationale for these choices is as follows:

■ *Natural gas* is and remains relatively expensive because it is a clean fuel, the conversion efficiency (MWh produced per GJ fuel) is high (55–60 percent for new plants), the capital costs of natural gas plants are relatively low,

---

[8]World average gas price in 1984–2007 (BP, 2008).
[9]World average coal price in June 2008 (GlobalCoal, 2008).
[10]World average uranium price in June 2008 (UxConsultingCompany, 2008).

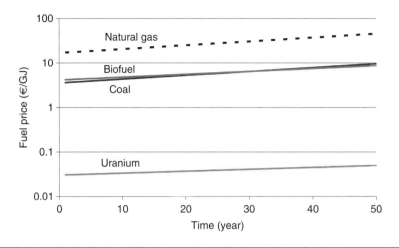

Figure 2.2 Average development of fuel prices.

and natural gas can be used for home heating and combined heat and power generation, also in small facilities. With increasing demand, the production of existing fields leveling off, and a limited amount of new production under way, an increasingly tight supply/demand balance is expected for the coming decades, which leads to continuously increasing prices.

- *Coal* has a much lower price per energy unit than natural gas because it is a polluting fuel that can be used only in large power plants or gasification units at relatively high investment costs, whereas the conversion efficiency (MWh produced per GJ fuel) is relatively low (40–45 percent). World coal resources suffice for over 400 years of present consumption, or even 2500 years of present consumption if all known coal deposits are developed. Therefore the marginal cost of coal production will only gradually increase and average prices are expected to rise only moderately.

- *Biomass* for use in power generation is expected to be traded at a somewhat higher price than coal because, although biomass can be fired in similar installations as coal, it is a more desirable product because we assume that it does not lead to net CO_2 emissions.[11] On the other hand, biomass demand

---

[11]Currently, the net CO_2 emissions associated with biomass production are heavily debated. Some biomass sources appear to have a negative CO_2 impact: the emissions associated with the production chain exceed the emission avoided. The high-level Cramer committee concluded in its advice to the Dutch government that 30–70 percent of the direct CO_2 emission from burning biofuel is compensated for in the biological cycle (Cramer Commission, 2006).

is limited by the higher handling costs, the more expensive installations, and the fact that it is converted at a lower efficiency (35–40 percent). We assume that biomass production can keep pace with demand, so price reflects cost rather than scarcity. The possibility of switching from biomass to coal is an effective cap on the biomass trading price.

■ *Uranium* costs per GJ are assumed to remain near their current low levels.

The following assumptions underlie the models:

1. Fuel is always available. There is an unlimited supply of biomass and natural gas.

2. Fuel prices are exogenous and reflect the relative scarcity of fuels. The modeled system is too small to impact world fuel prices.

3. Biomass is assumed to be 100 percent carbon-neutral.[11] In our model, biomass represents the general characteristics of renewable energy: carbon-free, but more expensive.

4. The main characteristics of Phase 3 of the EU ETS (2013 and beyond) are included: 100 percent of emission rights are auctioned and the cap will decrease over time.

5. The effect of intersector emissions trading is assumed to be negligible compared to intrasector trade.

6. Innovation is limited to learning; available technologies gradually improve in terms of cost and performance, but entirely new technologies do not become available in the model.

7. The generation portfolio, size of the market, CO_2 cap, the number of players, and the attitude toward nuclear power reflect the current (2008) Dutch power sector.

8. All costs and prices are in constant 2008 euros. Electricity prices are wholesale prices; taxes and network fees are not included.

2.3.3 POWER GENERATION TECHNOLOGIES

In the model, power plants are characterized by their fuel type, costs, technical life span, and fuel usage (conversion efficiency). The model includes an extensive set of state-of-the-art power generation technologies as well as technologies that are expected to be commercially available within 10 years' time, most notably CCS. In Figure 2.3 the carbon intensity, currently and in

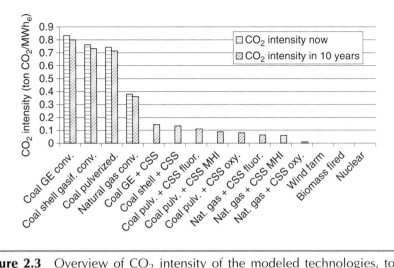

Figure 2.3 Overview of CO_2 intensity of the modeled technologies, today and in 10 years.

10 years, is summarized for the various technologies. The data in Table 2.3 on coal and gas plants—with and without CO_2 capture—is taken from Davidson [12]; the other data sources are cited in Chappin [5]. The effect of learning and incremental innovation is included by gradually increasing the efficiency and reducing the investment costs of new facilities. Carbon capture and storage options are only available after the first 10 years of the simulated period.

2.3.4 AGENT DEFINITION AND BEHAVIOR

The key agents in the model are the power-producing companies. Their tactical decisions consist of offering their output to the power market. They bid their power based on marginal costs, which includes the cost of CO_2 emissions in the cases with a carbon policy. Thus markets are simulated in which power producers negotiate the electricity, fuel, and CO_2 prices. Their strategic decisions cover investment in and decommissioning of power plants. Each agent's decision process is as follows:

1. Decide per power plant whether it should be dismantled. The decision to dismantle is taken when the technical lifetime of a power plant has expired (after 20 years for wind farms, 30 years for gas and coal plants, and 40 years for nuclear) or if the plant caused continuous operational loss for over 5–9 years.

Table 2.3 Power plant characterization

Power Plant Type	Efficiency (%)	Efficiency Modifier (%)	Investment (€/MW)	Investment Modifier (%)	Fixed Operating Cost (€/MWh)
Coal, pulverized	44	0.4	1,144,715	1.0	7
Coal, pulverized, and CSS fluor	35	0.5	1,608,943	1.0	7
Coal, pulverized, and CSS MHI	35	0.5	1,660,976	1.0	7
Coal, pulverized, and CSS oxy	35	0.5	1,792,683	1.0	12
Coal, shell gasification conversion	43	0.4	1,311,382	1.0	12
Coal, shell and CSS	35	0.5	1,791,870	1.0	12
Coal, GE conversion	38	0.4	1,169,919	1.0	9
Coal, GE and CSS	32	0.5	1,475,610	1.0	13
Natural gas conversion	56	0.4	405,691	0.5	2
Natural gas + CSS fluor	47	0.5	706,504	0.5	4
Natural gas and CSS MHI	50	0.5	721,138	0.5	4
Natural gas, CSS oxy	45	0.5	1,245,528	0.5	6
Biomass	35	0.4	1,250,000	1.0	4
Wind	35	—	1,150,000	2.0	3
Nuclear	—	—	2,000,000	0.0	5

2. Estimate whether there is a need for new generation capacity in 3 years. The estimate of the demand for capacity in 3 years is based on an extrapolation of the electricity demand trend of the past 3 years. Capacity expansion decisions take into account investments and decommissioning already announced by competitors. Continuous operational losses will cause unannounced decommissioning; thus agents' planning is not perfect and investment cycles can occur. Limited overinvestment is modeled to dampen those investment cycles.

3. If Step 2 results in an investment decision, the agent needs to select a technology for its new plant. Its decision is based on the life-cycle cost per MWh_e produced. The life-cycle CO_2 cost is based on current CO_2 taxation levels or, under emissions trading, the three-year average CO_2 auction price. The total life-cycle cost must be recovered by electricity income or else the investment is cancelled. In the latter case, another agent will get the opportunity to invest. The order in which the agents make their investment decisions varies randomly. In addition to financial aspects, an agent's conservativeness, aversion to nuclear power, and risk attitude affect its decisions. Despite the large weight of financial considerations, these individual style aspects have an effect, especially when financial differences between options are small. Conservativeness is modeled as "preferring more of the same"; risk attitude translates to different responses to historic variance of CO_2 and electricity prices.

In the case of emissions trading, each year electricity-producing agents complete the following actions concerning the operation of their power plants:

1. *Purchase emission rights in the annual auction.* The auction bids are based on the "willingness to pay" per installation, which is determined as the expected electricity price less the marginal costs of each unit, divided by the CO_2 intensity. The bid volume equals the expected electricity sales volume times the CO_2 intensity of the power plants that are expected to be in merit.

2. *Offer electricity to the market (which is modeled as a power pool).* Each plant's capacity is offered at variable generation cost (fuel cost, variable operating and maintenance cost, and CO_2 cost). The CO_2 costs of a generator equal the CO_2 price times its CO_2 intensity. In case insufficient CO_2 rights have been obtained, CO_2 cost is equal to the penalty for noncompliance.[12]

3. *Acquire the required amounts of fuel from the world market,* which are calculated from the actual production and fuel usage.

4. *Bank surplus CO_2 rights or pay the penalty in case there is a shortage of CO_2 rights.* Surpluses and shortages are calculated from the actual production levels and the volume of emission rights owned by the agent.

A difficulty with this procedure is that the CO_2 and electricity markets are mutually dependent. Though there is only one CO_2 price per year, the use of

---

[12]When the CO_2 price exceeds the penalty level, agents will rationally choose to pay the penalty rather than purchase more CO_2 credits. Consequently, this penalty level functions as a price cap for the CO_2 market.

a load-duration function with 10 steps means that 10 different electricity prices are developed for each year. Therefore we need to model arbitrage between these 10 periods. Because the demand for CO_2 credits is different at every step in the load-duration curve, we had to develop an iterative process in which arbitrage between the demand for CO_2 in these markets takes place in such a way that total annual demand for CO_2 satisfies the emissions cap and a single annual CO_2 price develops. We adopted the following procedure: Since the outcome of the CO_2 market is input to the power market and vice versa, Steps 1 and 2 are computed via an iteration that is complete when stable prices have been established for the entire year. In each simulation interval, we start with the prices of the previous year. In each iteration, first the CO_2 auction is cleared (Step 1), which results in a CO_2 price. This price is then used to calculate power market offers, and for each of the 10 sections of the load-duration curve this market is also cleared (Step 2). This new clearing price for electricity is fed into the bids for the CO_2 auction as the expected price of electricity (Step 1), and so on. Upon completion of this iteration, emissions trading (Step 1) has effectively been completed.

Under carbon taxation, Step 1 is skipped; in Step 2, the CO_2 cost is the carbon tax times the CO_2 intensity. In this case, the CO_2 price is exogenously determined. The electricity market bids simply incorporate this price. No iteration is necessary. Step 4 is replaced by paying carbon tax to the government. In case there is *no carbon policy*, the calculation procedure consists of Steps 2 and 3, with a CO_2 price equal to zero.

2.4 Simulation results

Where the model design and the technology representation are generic, results are presented for a CO_2 market that is modeled after the European ETS and for generation portfolio and market data that reflect the Dutch power sector. In reality and in the model, carbon policy is only one of several factors that affect emissions. The evolution of the system is also determined by: (1) the scenarios (exogenous factors such as fuel prices and electricity demand), (2) the system's components and properties, (3) and the starting conditions. To provide a good representation of the possible development of the system over 50 years, we present the aggregated results of 60 simulation runs in which the scenario parameters were varied evenly across the entire scenario space and the initial set of power plants is randomly distributed among the agents.

2.4.1 AVERAGE TOTAL CO_2 EMISSIONS

Figure 2.4 shows what carbon policies deliver in the long run. Emissions are lowest under the carbon tax. In the long run, emissions trading generally leads to emissions close to the cap. This might not come as a surprise, but in some

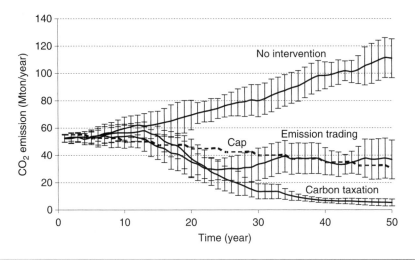

Figure 2.4   Average CO_2 emission levels for three carbon policies.

simulation runs the cap is not met at all. In these cases abatement investments are made too late, given their long lead time and the fact that the cap continues to decrease. High CO_2 prices result. Despite the spread in outcomes (indicated by the error bars in Figure 2.4), the difference between the trajectories caused by the three carbon policies is statistically significant.

The results reflect the tremendous inertia in capital-intensive energy systems. Without intervention, emissions continue to rise indefinitely and neither carbon policy guarantees a rapid decrease of emissions. To the contrary, emissions increase in the first 10–15 years in all scenarios, due to the system's inertia: Even at high CO_2 prices, it is not attractive to replace relatively new power plants, even if they emit much CO_2.

2.4.2 ELECTRICITY PRICES

The pressure that carbon policies put on the power generation system is reflected in the electricity prices (Figure 2.5), since power companies ultimately pass on their CO_2 cost to consumers. The prices shown are outcomes of the simulated negotiation between the six operating companies and simulated demand. Three important observations can be made:

■ The three carbon policies cause significant structural differences in the electricity prices.

■ Under emissions trading, CO_2 prices are highly volatile for the first three to four decades.

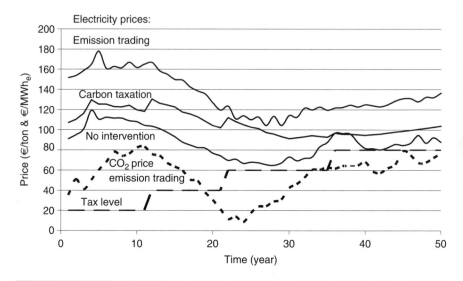

Figure 2.5 Electricity and CO_2 prices, averaged over all runs, under various carbon policies.

■ Under emissions trading, the CO_2 price is strongly correlated with the electricity price, whereas the correlation between a carbon tax and electricity prices is much weaker.

Electricity prices drop during the first two decades, which appears to be counterintuitive when an ETS or CT is introduced. In all three cases, however, the power plant portfolio at the start of the simulation is not economically optimal for existing market conditions. With time, generators invest, adjusting their portfolio and lowering their marginal cost of electricity production.

In the case without intervention, coal becomes increasingly dominant because it is more attractive. Innovation leads to further cost reductions. Toward the end of the modeled period, electricity prices begin to rise again due to the assumption that fuel prices will gradually increase.

In the case of an emissions trading scheme, both the price of emission rights and the CO_2 emissions remain high for the first 15 years, which leads to extremely high electricity prices. These can be explained by the inertia of the generation portfolio and risk aversion. Inertia results from the economic rationale for keeping existing power plants and the lead time for building new ones. Power producers exhibit risk aversion toward the capital-intensive investments required for CO_2 abatement due to CO_2 price volatility. The high prices lead to an abatement overshoot in most runs, which causes a CO_2 price collapse in the third decade. This discourages further abatement measures, and emissions creep back to the cap and stabilize.

Under emissions trading, the CO_2 price is volatile (see Figure 2.5). It contributes to an already high investment risk. The consequences for abatement efforts are a delay of investments and a bias toward less capital-intensive abatement technologies, many of which are more costly in the long run. A carbon tax does not have the disadvantage of volatility and thus minimizes the price risk of abatement measures, provided there is no regulatory uncertainty about the tax level—the risk of later governments backtracking on earlier taxation decisions. However, there is also regulatory uncertainty with emissions trading, since later governments could decide to loosen the cap. Regulatory uncertainty increases investment risk under both policies.

The impact of carbon taxation on the electricity prices is relatively small. The tax starts at a fairly low level of 20 €/ton. When the tax level rises, investment in abatement reduces the CO_2 intensity of electricity generation, which reduces the impact of the tax on electricity prices. Clearly, one cannot simply add the cost of CO_2 under the two carbon policies to the electricity prices under no intervention. The price is determined by CO_2 price and the CO_2 intensity of the portfolio, which evolves differently under each policy option (see Figure 2.6).

2.4.3 CO_2 INTENSITY

Given the continuous rise in electricity demand, CO_2 emissions can only be reduced significantly by changing the generation portfolio, that is, by shutting down existing facilities and by investing in new ones. Figure 2.6 presents the

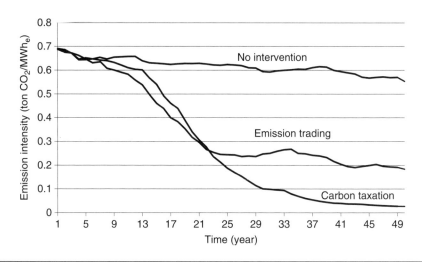

Figure 2.6 Average CO_2 intensity of capacity and supply under the various carbon policies.

CO_2 intensities of the carbon policies, averaged over the runs. This figure shows that the absolute emission levels shown in Figure 2.4 are achieved via a dramatic reduction of the CO_2 intensity of the generation portfolio. Without intervention, CO_2 emissions rise, but the CO_2 intensity is relatively stable; natural gas is replaced by coal while its fuel efficiency increases through innovation.

The impact of CO_2 prices on the variable cost of installations may change the merit order of generation. At higher CO_2 prices, CO_2-intensive installations may move from base load to peak load. Under all scenarios, including no intervention, a merit order shift takes place from CO_2-intensive toward CO_2-extensive base load facilities.

2.4.4 GENERATION PORTFOLIO DEVELOPMENT

The simulation results clearly show that various carbon reduction policies profoundly affect the generation portfolio. Without a carbon policy, the economics favor coal, which replaces natural gas, nuclear, and biomass. Under emissions trading, the generation portfolio becomes more diverse. Coal without CCS remains important, but its share stabilizes. Coal with CCS emerges in the second decade and replaces natural gas because of the declining cost of CCS and the increasing price of natural gas. The introduction of carbon-free biomass is the second largest source of emissions reductions. An increasing carbon tax prompts an almost complete switch to carbon-free electricity generation in the long run. Coal with CCS first replaces natural gas capacity and later coal without CCS, with biomass taking a far greater share than it does under emissions trading. Traditional coal is phased out. The volumes of wind energy are stable and small under all three policy instruments. Nuclear energy does not appear in any of the cases due to its substantially higher cost (see Table 2.3). In Figure 2.7 the evolution of the average portfolio of technologies is displayed.

The portfolios in Figure 2.7 are diverse, whereas one would expect economic rationality to lead to a single preferred technology. However, other factors also affect investment decisions. Especially when the costs of options do not differ much, secondary criteria can be decisive. These include greenness (measured in CO_2 intensity) and conservativeness (a preference for proven technologies, which is measured by the current adoption level of the technologies). The adoption of wind—in continuous but limited amounts—offers a typical illustration. In the simulations there is one agent who is relatively conservative and green. Since wind often is close to the cheapest option, it is sometimes adopted by this agent. And in some runs, early adoption combined

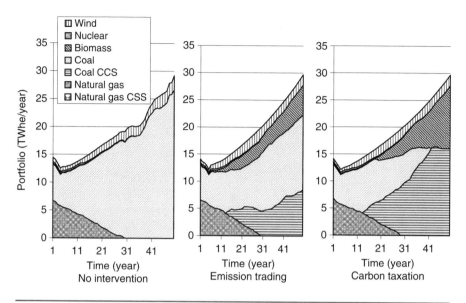

Figure 2.7 Average generation portfolio evolution for the three scenarios.

with the agent's conservativeness will cause it to adopt wind again. Via the same mechanism, wind gradually may disappear in other runs. When the runs are averaged, this produces a gradual and limited share of wind in the portfolio.

2.4.5 SENSITIVITY TO THE ASSUMPTIONS

However complex a model might be, it remains a simplification of reality. The results are influenced by the following types of assumptions:

- The way the carbon policies are modeled
- The assumptions regarding the model's inputs: the (relative) prices of natural gas, coal, biomass and uranium, the set of available generation technologies, and the demand for electricity
- The structure of the energy market that was modeled
- Assumptions regarding investment behavior and the way prices are formed in the market

A cap-and-trade scheme is more complicated than a carbon tax because it involves more design variables. Choices need to be made about the method of allocating the emission rights (auctions are theoretically superior but not always politically favored), about credit issuing and continuous registration, banking and borrowing credits, and whether to issue negative credits to CO_2 sinks. In both systems, emissions must be monitored and verified, the scope of the system (which sectors and countries to include) must be decided on as well as where to place the obligation to obtain credits or pay tax (at the consumer, the power producer, or further upstream), and so on.

A difficult choice is how to model fuel prices, because structural changes (such as China's economic emergence) can create lasting price effects. We assumed prices to be exogenously determined. This assumption holds for a small system, such as a single country or state, but if carbon policies are widely implemented, this could decrease the demand for carbon-intensive energy sources worldwide, making them cheaper and hence economically more attractive and reducing the effectiveness of the reviewed carbon policies.

We did not assume any technological revolutions. The existing technologies, including CCS, would continue to be available and gradually improve in terms of cost and performance. We assumed that a technology's maturity determines its pace of improvement, with learning being exogenous to the market. In reality, adoption and improvements reinforce each other, so technological learning is endogenous. A second technical issue is that most existing coal plants are not suitable for running peak load (in case high carbon prices cause them to shift their position in the merit order). This could lead to block bidding and reduced flexibility of the power system. Future technologies, such as coal gasification, will probably be more flexible.[13]

The abatement options differ by country. In most countries, only a limited amount of CO_2-free generation options, such as hydropower or geothermal energy, are available. In these countries, the options that were reviewed in this chapter are the main ones. But there are exceptions, such as New Zealand, Canada, Brazil, and Norway, since they have an abundance of hydro.

Electricity demand is modeled exogenously, without price elasticity. One may assume that there is, in reality, some price elasticity, which would dampen price swings. Perhaps price elasticity will be improved through applications that make use of the digital electricity meters that are beginning to be installed across the world. Finally, it might not be a correct assumption that electricity demand will grow perennially; perhaps there is a saturation point, or conservation efforts could outweigh natural demand growth.

Society's acceptance of a carbon policy might be affected by the way in which the revenues are spent. Stoft favors returning revenues (both from a tax or an auction of emission allowances) to people on a per capita basis [31]. This

---

[13]See, for instance, Chapter 11, by Williamson, in this book.

avoids a net income transfer from consumers to government while maintaining the incentive to reduce emissions. The revenues may also be returned to the affected industry sector to maintain an international competitive position. Other options are to use the revenues to finance CCS infrastructure or support R&D or to let them flow to the treasury. This question of political acceptability and allocation of the revenues, however, is outside the scope of this chapter.

The market is modeled with a limited number of generating companies, which is realistic, but they act as perfect competitors, which is not realistic. Oligopolistic behavior is likely to be observed in electricity markets, given the regional nature of the product, and may lead to different investment behavior. Oligopolistic rents may offset investment risks, allowing companies to invest more proactively in an emissions market than the model suggests, but it is uncertain whether they will choose to do so.

2.5 Conclusion

Taxation and CO_2 emissions trading schemes should yield similar results, in theory. In this chapter we analyzed, for a hypothetical electricity sector, the effects of both instruments under realistic circumstances, such as policy uncertainty, risk aversion by investors, and long construction lead times.

Both carbon policies are effective in reducing CO_2 emissions in the long run, provided that the tax or cap level is set at an ambitious level. In the first 10–15 years, CO_2 emissions from power generation continue to increase under all three policies (no intervention, CT, or ETS). Operational adjustments, which both CT and ETS can be expected to invoke in the short term, do not have sufficient potential. A substantial change in the generation portfolio is needed to obtain the policy goals for emission reduction. Under emissions trading, natural gas is replaced by coal with CCS and biomass. Conventional coal retains a certain share. An increasing tax leads to a complete phasing out of natural gas and conventional coal, leading to a portfolio with almost only coal with CCS and biomass. No new nuclear capacity is developed under any of the three policies. In the absence of intervention, absolute emission levels grow dramatically (50 percent), even though the CO_2 intensity of electricity generation is stable due to technological improvements.

A key result is that given a certain CO_2 cost to producers—whether it is due to a tax or the price of CO_2 emission rights—carbon taxation leads to lower electricity prices than does emissions trading. The explanation is the difference in investment risk: A tax is predictable, whereas CO_2 prices are volatile. This uncertainty leads to an investment cycle under emissions trading that is absent under carbon taxation. High CO_2 prices frequently occur when the CO_2 intensity of electricity generation is high. This cyclical behavior is a significant disadvantage of emissions trading. In contrast, under taxation, high tax levels occur only in the second half of the simulated period. At that time, they do not cause large income transfers, because the CO_2 intensity is already low, so the impact on the electricity price is limited. Predictability is a key advantage of taxation, which allows investors to minimize cost over a longer time horizon. Given the capital intensiveness of many of the abatement options, this leads to substantially lower overall cost as well as lower emissions in the long term. This confirms the ideas of Grubb and Newberry [18].

Both trading and taxation are instruments that create current pain while yielding significant results in the future. When these policies are kept in place for decades, their long-term impact is significant. From the modeling exercise, however, we also conclude that for both instruments to have an effect, affordable and competitive low-CO_2 electricity generation options must become available on a large scale. In our simulations, options included were CCS, nuclear and renewables biomass, and wind. In practice, other technologies, such as solar power, may also be part of the solution. Though it cannot be concluded that the very portfolio shifts that were observed in the model are the most likely to occur in practice, it is safe to conclude that carbon policies do deliver in the long run.

REFERENCES

[1] Bimonte S. An algorithm for optimal Pigouvian taxes without benefits data. Environmental and Resource Economics 1999;13(1):1–11.

[2] BP. Statistical Review of World Energy. June 2008.

[3] Carbon Trust. Cutting Carbon in Europe: The 2020 plans and the future of the EU ETS, Carbon Trust; 2008.

[4] CEC. Directive 2003/87/EC of the European Parliament and of the Council of 13 October 2003 establishing a scheme for greenhouse gas emission allowance trading within the Community and amending Council Directive 96/61/EC. Official Journal of the European Union 2003;275:32–46.

[5] Chappin EJL. Carbon Dioxide Emission Trade Impact on Power Generation Portfolio, Agent-based Modelling to Elucidate Influences of Emission Trading on Investments in Dutch Electricity Generation. Delft: Delft University of Technology; 2006.

[6] Chappin EJL, Dijkema GPJ. On the Design of System Transitions: Is Transition Management in the Energy Domain Feasible? IEEE IEMC: International Engineering Management Conference, Estoril, Portugal, IEEE; 2008a.

[7] Chappin EJL, Dijkema GPJ. On the Impact of CO_2 Emission-Trading on Power Generation Emissions. Technological Forecasting & Social Change 10.1016/j.techfore.2008.08.004: 10.1016/j.techfore.2008.08.004; 2008b.

[8] Chappin EJL, Dijkema GPJ. Towards the Assessment of Policy Impacts on System Transitions in Energy. 31st IAEE International Conference, Bridging Energy Supply and Demand: Logistics, Competition and Environment, Istanbul, Turkey: IAEE; 2008c.

[9] Cozijnsen J. Website over emissiehandel. Retrieved December 19, 2005, from www. emissierechten.nl; 2005.

[10] Cozijnsen J, Weijer H, editors. Dossier CO_2-Emissiehandel. Stromen Dossiers. Den Haag: Sdu Uitgevers; 2005.

[11] Cramer Commission. Criteria voor duurzame biomassa productie - Eindrapport van de projectgroep "Duurzame productie van biomassa," Task Force Energietransitie; 2006.

[12] Davidson J. Performance and costs of power plants with capture and storage of CO_2. Energy 2007;32:1163–76.

[13] Dijkema GPJ. Process System Innovation by Design. Delft: Delft University of Technology; 2004.

[14] Egenhofer C. The compatibility of the Kyoto Mechanisms with traditional environmental Instruments. In: Carraro C, Egenhofer C, editors. Firms, Governments and Climate Policy: Incentive-Based Policies for Long-Term Climate Change. Cheltenham: Edward Elgar; 2003.

[15] Ehrhart KM, Hoppe C, Schleich J, Seifert S. Strategic aspects of CO_2-emissions trading: Theoretical concepts and empirical findings. Energy and Environment 2003;14(5):579–98.

[16] Ekins P, Barker T. Carbon taxes and carbon emissions trading. Journal of Economic Surveys 2001;15(3):325–76.

[17] GlobalCoal. GlobalCoal news: RB Index Coal, Spot prices (USD/tonne) for coal to be delivered in the 3 calendar months following publication date. Retrieved July 17, 2008, from http://sog.globalcoal.com/default.asp; 2008.

[18] Grubb M, Newberry D. 2007; Pricing carbon for electricity generation: National and International dimensions. Cambridge Working Papers in Economics 0751.

[19] Hardin G. The tragedy of the commons. Science 1968;(162):1243–8.

[20] Hovi J, Holtsmark B. Cap-and-trade or carbon taxes? The feasibility of enforcement and the effects of non-compliance. International Environmental Agreements: Politics, Law and Economics 2006;6(2):137–55.

[21] Lijesen MG. The real-time price elasticity of electricity. Energy Economics 2007;29(2):249–58.

[22] Lowe R. Defining and meeting the carbon constraints of the 21st century. Building Research and Information 2000;28(3):159–75.

[23] Manne AS, Richels RG. The EC proposal for combining carbon and energy taxes: the implications for future CO_2 emissions. Energy Policy 1993;21(1):5–12.

[24] Manne AS, Stephan G. Global climate change and the equity-efficiency puzzle. Energy 2005;30(14):2525–36.

[25] Ministry of VROM and SenterNovem. Allocation plan for CO_2 emission allowances 2005–2007, Dutch National allocation plan regarding the allocation of greenhouse gas emission allowances to companies. SenterNovem; 2005.

[26] Nikolic I, Chappin EJL, Davis C, Dijkema GPJ. On the development of agent-based models for infrastructure evolution. International Conference on Infrastructure Systems: Building Networks for a Brighter Future, Rotterdam, the Netherlands: NGInfra Foundation; 2008.

[27] Pigou AC. A Study in Public Finance. London: Macmillan; 1947.

[28] Pizer WA. Choosing price or quantity controls for greenhouse gases. Resources for the Future: Climate Issues Brief 1999;99(17).

[29] Smith A. An Inquiry into the Nature and Causes of the Wealth of Nations; 1776.

[30] Stern N. The Economics of Climate Change: The Stern Review. Cambridge: Cambridge University Press; 2006.

[31] Stoft S. Carbonomics: How to fix the climate and charge it to OPEC; 2008.

[32] Svendsen GT. The idea of global CO_2 trade. European Environment 1999;9(6):232–7.

[33] Svendsen GT, Vesterdal M. How to design greenhouse gas trading in the EU? Energy Policy 2003;31(14):1531–9.

[34] UxConsultingCompany. UxC Nuclear Fuel Price Indicators. Retrieved July 18, 2008, from www.uxc.com/review/uxc_Prices.aspx; 2008.

[35] Weitzman M. Prices vs. quantities. Review of Economic Studies 1974;41(4):477–91.

Emerging Carbon Markets and Fundamentals of Tradable Permits

Reiner Musier and Parviz Adib

APX, Inc., Santa Clara, California, U.S.A.

Abstract

Environmental commodities represent the largest and fastest growing new commodity market in North America and the world, with a value measured in trillions of dollars. They capture much of the value that society places on safeguarding the environment and energy independence and are essential to achieving environmental objectives. The increasing involvement of major national and international players, particularly financial institutions, is a clear indication of the importance of environmental commodities. Therefore, the design of compliance and voluntary markets should be based on sound economic principles, and market trust should be established through a national and global technology infrastructure that supports effective markets and their regulatory oversight.

3.1 Introduction

The impact of climate change is one of the largest global problems facing us today. After several years of debate, the adoption of the United Nations Framework Convention on Climate Change (UNFCCC) in 1992 was an important step forward. After intense negotiations, the Kyoto Protocol[1] was adopted in December 1997; its detailed rules for implementation were adopted in Marrakesh in 2001 and entered into force in February 2005.[2] Significantly, the United States, one of the world's largest emitting nations, failed to ratify the agreement.[3] More recently, at the G8 summit in Hokkaido, Japan, in July 2008, the G8 leaders agreed to adopt the goal of achieving a 50 percent reduction in global emissions of greenhouse gases by 2050.[4]

The publication of the 2007 report by the Intergovernmental Panel on Climate Change [19] and the Stern Report [32] have further emphasized the need for action. Like the Europeans and Japanese, policymakers in the United States are now considering options to reduce greenhouse gas emissions.[5] Chief among these options is a cap-and-trade system similar to the one in use in Europe.

This chapter focuses on environmental market mechanisms and the potential market size for environmental commodities. A brief discussion of the cap-and-trade approach appears in Section 3.3, followed by a discussion of the emerging markets for tradable permits in Section 3.4. Issues of market oversight and the technology needed to support the operation of these markets are presented in Sections 3.5 and 3.6, followed by the chapter's conclusions.

3.2 Carbon market size

Up to now the size of the global greenhouse gas or carbon market has been limited due to lack of broad global participation. However, the recent decision by the G8 and expected progress at Copenhagen at the end of 2009 promise to significantly expand the size of the carbon market.

---

[1]The Kyoto Protocol is posted at http://unfccc.int/resource/docs/convkp/kpeng.html
[2]More details are posted at http://unfccc.int/kyoto_protocol/items/2830.php
[3]In contrast, several U.S. states, such as California and several states in the Northeast, demonstrated leadership in forming their own environmental action plans without waiting for the federal government to take the first step.
[4]For a summary of all agreements, see www.g8summit.go.jp/eng/doc/doc080709_10_en.html
[5]For an assessment of the impact of some of these policies on the power industry, see Ford (2008).

The recent reports by the International Energy Agency [17, 18] include projections of the amount of energy-related carbon emission reductions needed to achieve 50 percent reduction by 2050 compared to 2005. Under a business-as-usual scenario, the amount of global energy-related CO_2 emissions could reach 62 gigatons (Gt) by 2050 compared to 27 Gt in 2005. To meet such a goal, significant steps,[6] including reliance on alternative energy technology, must be taken to reduce carbon emissions to below 14 Gt (Figure 3.1).

The worldwide environmental commodity markets consist of both voluntary markets and mandatory or compliance markets established by government regulations, which are currently many times larger than the voluntary markets. Trade in carbon permits is one of the driving factors in the development of significant wind and other renewable resources throughout Europe [29]. Renewable resources are planned to play a major role in emission reductions.[7] Such actions will have significant impacts on the number of environmental certificates and tradable permits to be offered in the European Union Emission

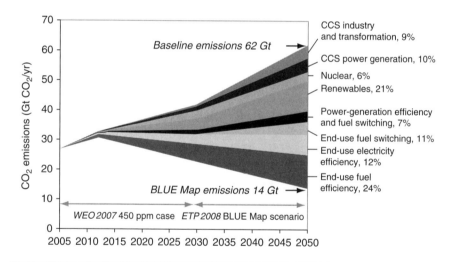

Figure 3.1   Reducing global CO_2 emissions by 50 percent by 2050.
Note: Baseline scenario refers to business as usual. BLUE scenario requires extensive emission reductions. *WEO 2007* and *ETP 2008* refer to the *IEA World Energy Outlook 2007* and *Energy Technology Perspectives 2008* reports, respectively. Source: IEA [17], Slide 5. This graph originally appeared in *IEA World Energy Outlook 2007*.

[6]Chapter 1, by Moran, in this book outlines the myriad of issues involved. See Pew Center (2005) for insights and opportunities regarding European Emission Trading Scheme.
[7]For information regarding European countries, see Massy et al. (2008) and Hass et al. (2008).

Trading Scheme (EU ETS).[8] Similarly, 28 states plus the U.S. District of Columbia have established Renewable Portfolio Standards (RPSs) to achieve goals ranging from a few percent to more than 25 percent of their power use.[9] Texas is leading such programs with 8135 MW of installed wind generation by December 2008, expected to pass 9000 MW by 2010.[10] The Renewable Energy Credit (REC), resulting from the production of electricity from renewable resources, has played a major role in increasing liquidity in the environmental commodity markets to meet increasing demand in both regulated (mandatory) and voluntary markets.

Worldwide transaction volumes for carbon emission reductions in both regulated and voluntary markets have been increasing dramatically in the past few years due to the active response of European countries (Table 3.1).

The volume of activity in the voluntary carbon markets, both exchange traded and over the counter, is projected to increase from 65 metric tons of CO_2 equivalent ($MtCO_2e$) in 2007 to between 428 $MtCO_2e$ and 1400 $MtCO_2e$ in 2020 [9]. The growth in the regulated markets is projected to be significant given the large impact that should be expected by the upcoming passage of a U.S. federal carbon regulation in the near future.

The value of U.S. allowances and offsets could be as much as $300 billion annually[11] and the value of traded allowances and offsets could be orders of magnitude greater than transactions in the sulfur dioxide (SO_2) and nitrogen oxides (NO_X) markets. The recent commitments by major industrialized countries to reduce carbon emission by at least 50 percent by 2050, and similarly the change in the U.S. administration, are expected to have a sizable impact on the scale of environmental commodities. By all accounts, the environmental markets are poised to become the largest new commodity markets in the world in the next 20 years.[12,13]

---

[8]For more information regarding the EU ETS and its evaluation, see Grubb and Neuhoff (2006) and other works at www.electricitypolicy.org.uk/TSEC/2/prog1.html

[9]For state-by-state information, see the Pew Center on Global Climate Change at www .pewclimate.org/what_s_being_done/in_the_states/rps.cfm. See Hass et al. (2008) for similar information regarding European countries.

[10]See ISO/RTO Council (2007) for summary discussion of the steps taken by various U.S. electric markets to encourage more renewable resources and Musier (2006) for Renewable Energy Credit Markets in U.S. For more information on the worldwide potential for wind resources, see Chapter 9, by Wiser and Hand, in this book.

[11]Press release by Senator Dianne Feinstein's office, "Senator Feinstein Introduces Measure to Establish Federal Oversight for New Carbon Markets," December 6, 2007.

[12]See an interview with James E. Newsome, president and CEO of the New York Mercantile Exchange, by Martin Rosenberg (2007), in which Newsome predicts that "It [environmental commodity contracts] can dwarf oil."

[13]See an interview with Paul Ezekiel, the Credit Suisse's Carbon Emission Emissary, by David Moss (2007), in which Ezekiel predicts that "... with a potential global market that some have valued in the trillions, the sky, in this case, might not be the limit."

Table 3.1 Worldwide transaction volumes and values in carbon markets, 2006 and 2007

	Volume (MtCO$_2$e)		Value (US$million)	
Markets	2006	2007	2006	2007
Voluntary OTC market	14.3	42.1	58.5	258.4
CCX	10.3	22.9	38.3	72.4
Total voluntary markets	**24.6**	**65.0**	**96.7**	**330.8**
EU ETS	1104	2061	24,436	50,097
Primary CDM	537	551	6887	6887
Secondary CDM	25	240	8384	8384
Joint implementation	16	41	141	495
New South Wales	20	25	225	224
Total regulated markets	**1702**	**2918**	**40,072**	**66,087**
Total global market	**1727**	**2983**	**40,169**	**66,147**

Source: Ecosystems Marketplace and New Carbon Finance [9], page 6; and World Bank [37], page 1.

In a recent study published by Point Carbon, the global carbon market is projected to reach about €2 trillion ($3.1 trillion) and account for a total transaction volume at 38 Gt (CO$_2$e) per year by 2020.[14] The United States is projected to account for 67 percent of such a global carbon market.[15]

3.3 Cap-and-trade schemes[16]

Although there are alternative mechanisms,[17] such as carbon taxes or fees on emissions, cap-and-trade schemes have emerged as an effective tool for addressing GHG emissions. In the United States, the most well-known cap-and-trade success story is acid rain. In the early 1990s, high levels of SO$_2$ emissions resulted in acid rain problems in the eastern United States. Rather than addressing the problem through a

---

[14]Point Carbon. Carbon Prices "set to triple by 2020" reaching to €79 per MT of carbon. Available at: www.pointcarbon.com/news/1.983583; October 10, 2008.

[15]For a news release distributed by Reuters, see: May 22, 2008, article at www.reuters.com/article/pressRelease/idUS187544+22-May-2008+BW20080522

[16]The authors wish to acknowledge the substantial contribution of Clare Briedenich to this section.

[17]See Stern (2008, pp. 23–26) for a brief evaluation of alternative policy instruments. For an assessment of the quantitative impacts of each alternative policy, see Chapter 2, by Chappin, Dijkema, and Vries, in this book.

tax on emissions or traditional command-and-control regulation, Title IV of the 1990 Clean Air Act established a cap-and-trade program to reduce these emissions.[18] The actual costs of the program, around $1.0–1.4 billion per year, were far below early projections of $3–25 billion per year.[19,20] The drastic cost savings in early years enabled many firms to reduce emissions more quickly than required by law.[21] Figure 3.2 demonstrates environmental achievements by this program.

A more recent application of a cap-and-trade mechanism is the European Union Environmental Trading Scheme (EU ETS). In contrast to the U.S. experience, the initial phase of EU ETS was not successful in producing the expected level of emission reduction.[22] The EU has since improved its mechanism and tightened its cap for Phase II (Kyoto implementation).[23,24] However, some policymakers and stakeholders in the United States and elsewhere question the efficacy of a cap-and-trade system to reduce GHG.[25]

There is still significant debate on the merits of alternative emission reduction schemes and some experts argue that cap and trade cannot produce socially optimal results in addressing global climate change [28].[26] However, most experts agree that, given the certainty in the emission reductions and the economic flexibility favored by industry, a cap-and-trade mechanism tends to be more politically attractive than a tax-based approach and more likely to

---

[18]The program, which began in 1995, established a goal of reducing SO_2 emissions to 50 percent of 1980 levels by 2000 and maintaining these levels thereafter.

[19]See Ellerman et al. (2003).

[20]Environmental Defense. The Cap and Trade Success Story. Available at: www.edf.org/page.cfm?tagID = 1085

[21]EPA ACID Rain Site: www.epa.gov/airmarkets/progsregs/arp/basic.html#phases. See World Resources Institute (2005) for similar program dealing with the U.S. NO_x Budget Program and Environmental Protection Agency (2002) for lessons in environmental markets and innovation.

[22]For a review of lessons learned from the European experience, see Betz and Sato (2006) and Carbon Trust (2007).

[23]See the decision of the European Commission establishing guidelines for the monitoring and reporting of greenhouse gas emissions pursuant to Directive 2003/87/EC of the European Parliament and of the Council, January 29, 2004 (European Commission 2003).

[24]For more on European carbon regulation, see the Forward piece by Grubb in this book as well as Chapter 15, by Cornwall, on the United Kingdom and Chapter 16, by Jean-Michel Trochet, Jean-Paul Bouttes and Francois Dassa, on France. See Reilly and Paltsev (2005) for further analysis of the European Emission Trading Scheme.

[25]See Chapter 16 in this book by Jean-Michel Trochet, Jean-Paul Bouttes, and Francois Dassa, on environmental policies in France. Also see Chapter 15, by Nigel Cornwall, on carbon challenges in the United Kingdom.

[26]The Carbon Tax Center lists six fundamental reasons to demonstrate that carbon tax is superior to cap and trade in achieving environmental goals. For details, see www.carbontax.org/issues/carbon-taxes-vs-cap-and-trade/. Similarly, the American Enterprise Institute has published an online essay titled "Climate Change: Caps vs. Taxes," in which an attempt is made to demonstrate that a carbon tax is the superior option. For details, see www.aei.org/include/pub_print.asp?pubID = 26286

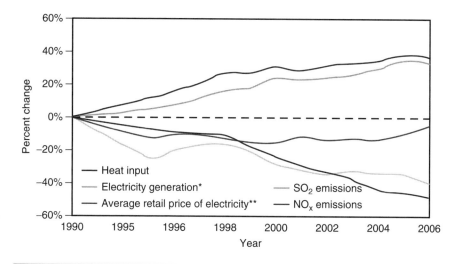

Figure 3.2 Trends in electricity generation, fossil energy use, prices, and emissions from the U.S. electric power industry, 1990–2006.
Source: See EPA [36], Figure 1, p. 7.

garner the support needed for adoption. This section argues that, if designed properly, cap and trade can achieve environmental goals while using market mechanisms to find the lowest costs.

3.3.1 DESIRABLE FEATURES OF CAP AND TRADE

The most desirable features of cap and trade include these[27]:

■ *Compliance flexibility results in lower compliance costs.* Under a cap-and-trade system, regulated firms have two compliance options. They can either reduce emissions to the level of their allowance holdings, through whatever technology or process improvements are available, or they can purchase allowances from others who have reduced their emissions at lower costs, to cover additional emissions. Because of such flexibility, the cap-and-trade system utilizes the lowest cost emission reduction opportunities first.[28]

---

[27]For a detailed review of economic arguments for a carbon tax versus cap and trade, see Jacoby and Ellerman (2004). Also, see Ford (2008) for a brief discussion of alternative methods to allocate allowances.

[28]This was exactly the function of the United Nations' Clean Development Mechanism (CDM) programs. The CDM allows developed countries to partly meet their Kyoto targets by financing more cost-effective carbon emission reduction projects in developing countries. For more information, see UNFCCC information on CDM/JI posted at http://unfccc.int/kyoto_protocol/mechanisms/items/1673.php

■ *Allows targeting emission reduction.* A cap-and-trade system sets the maximum level of emissions as a guaranteed and legally enforceable cap but lets the market determine the allowance price, and hence the cost.[29]

■ *Spurring technology investment and development.* Addressing climate change will require reliance on new technologies well beyond what could be economically and technically achievable under the existing alternative energy sources. Some of the technologies, such as carbon capture and sequestration or carbon capture storage (CCS), are still under development. Consequently, a cap-and-trade system provides an important incentive for development of these new technologies by providing a "carbon price signal" that will enable firms to capture the value of these new technologies.

3.3.2 Success Factors for Designing Cap and Trade

Historical experience with cap-and-trade systems in both the United States and internationally has provided adequate evidence of what should be included in designing a successful system, including the following[30]:

■ Reporting and verification are fundamental for meeting the cap, requiring rigorous methods and procedures for measuring, reporting, and verifying facility emissions.

■ Reliable tracking of allowances, offset credits, and market activity must be in place to ensure certainty regarding the ownership, generate confidence in the market system, and help regulated corporations and entities plan for and achieve compliance by managing their carbon allowance and offset portfolio at both the firm and facility levels.

■ Meaningful penalties must be in place and enforced to make the atmosphere whole (e.g., result in the required emission reductions) and must be set at a level that is higher than the market price of allowances to encourage compliance.[31]

■ Transparency of market conditions is important to ensure timely access to up-to-date information on market conditions and allow market

---

[29]Pizer (1999) acknowledges that the two policies are different when we take into consideration the uncertainties associated with costs of emission reduction technologies. However, he adds that the two policies can result in the same outcomes if the costs of emission reductions are known.

[30]For a review of lessons learned from the European experience, see Betz and Sato (2006) and Carbon Trust (2007).

[31]See Stranlund (2007) for discussion of significant financial penalty in the U.S. acid rain program.

participants to forecast allowance demand and to plan for compliance while allowing regulators to make more informed decisions.

■ A liquid allowance market providing compliance opportunities with inter-temporal compliance flexibility, such as banking and borrowing, to increase the availability of allowances in a given period and moderate price volatility. Further, features that reduce the transaction costs of trading—such as trading platforms, market intermediaries, and effective trading rules—as well as the frequency of auctions will also tend to increase the liquidity of the market.

■ Market oversight to enhance confidence through the establishment of an oversight body with adequate resources and authority to respond to attempts at market manipulation or to intervene in the event of market failure.

3.4 New environmental markets

To reduce carbon emissions, solutions must be found to reduce both the *energy intensity*, through reliance on less energy per unit of gross domestic product, and the *carbon intensity*, through more effective management of the amount of carbon emissions per unit of energy consumed. Market mechanisms establishing environmental commodity markets will be a key element in addressing both energy and carbon intensities.[32]

3.4.1 COMPLIANCE AND VOLUNTARY MARKETS

There are two broad categories of environmental markets: namely, *compliance markets*, also called *mandatory markets*, and *voluntary markets*. Governments have led the development of the environmental compliance markets, which mandate performance, set industry compliance rules, conduct enforcement, and impose penalties. In contrast, voluntary markets have been driven by corporations that place a high value on social responsibility or a positive environmental brand image or that want to gain valuable experience and recognition through early actions.

The sellers of the environmental credits are generally renewable energy generators or project owners for carbon emissions reduction or energy efficiency projects. However, a key difference between markets is in the nature of the

---

[32]No attempt is made in this chapter to discuss the importance of emissions trading in addressing global warming and climate change. For such discussion, see Stern (2007 and 2008).

buyers. In compliance markets, buyers are typically manufacturing sector companies, utilities, or other capped sectors, obligated to comply with regulatory requirements or a renewable portfolio standard. In contrast, buyers in voluntary markets are corporations or individuals seeking to offset their carbon footprints and/or support renewable energy. In addition, there may also be companies who are taking voluntary actions to anticipate or shape future mandatory programs or hoping to get credit for their early actions under future regulatory programs, as has occurred in Canada and is currently advocated in the United States.[33]

3.4.2 TRADABLE CERTIFICATES

The following four broad categories of environmental commodities are either developed or under development:

- *Carbon allowances or permits.* This is a widely used product in the European market. In the United States, with California's passage of carbon legislation, many states are active and very likely to follow suit.[34] And with recent national election results, federal action on carbon is increasingly likely. Experts today believe the question is *when*, not *whether*, a compliance market for carbon will become a reality, alongside today's voluntary carbon market. Most experts believe regulators will first focus on the power and manufacturing industries, and government is expected to rely on a market-based mechanism to issue carbon allowances or permits to meet environmental goals.

- *Renewable energy credits or certificates (RECs).* The European countries have been very active in increasing the share of renewable resources in their energy use, resulting in a healthy trade in RECs. Similarly, the mandatory and voluntary REC markets in the United States are predicted to double in the next 3 years. The National Renewable Energy Laboratory (NREL) market report[35] indicates that states will continue to adopt Renewable Portfolio Standards as a means to promote investment in renewable

---

[33]Canadian Government Launches Credit for Early Action Program. http://ecoaction.gc.ca/news-nouvelles/20080629-eng.cfm?rss. Another example: the 1999 NARUC resolution posted at www.naruc.org/Resolutions/Resolution%20on%20Early%20Action%20Credits%20for%20Greenhouse%20Gas.pdf
[34]For more details on California, see Chapter 18, by Stern and Harris, in this book. See CARB (2007) and CARB (2008) for actions by the California Air Resources Board.
[35]See Holt and Bird (2005).

energy, such as wind, solar, geothermal, biomass, and other renewable resources.

■ *Energy efficiency and conservation credits or certificates (EECs).* The market for energy efficiency (EE) and conservation credits is just now emerging, with a huge potential for monetizing energy efficiency savings. Some experts believe that the market potential for energy efficiency and conservation credits is even larger than the REC markets because EE is applicable across all states and industries and the MWh savings potential is much larger at relatively lower costs.[36]

■ *Carbon offsets.* This category of environmental product is very broad and typically represents any verifiable actions with measurable outcome that could result in reduction of harmful environmental emissions. In addition to covering the preceding categories, carbon offset can also account for credits, resulting in reliance on cleaner technologies that emit fewer pollutants into the air. The United Nations Clean Development Mechanism (CDM) and European Joint Implementation (JI) programs fall within this category.[37]

Table 3.2 shows varying measures used to define the tradable units for the new environmental commodities.

Table 3.2 Environmental commodity units of measurement

Environmental Commodity	Typical Unit of Measurement
Carbon allowance or permit	Represents permission to emit 1 metric ton of CO_2 equivalents
Renewable energy credit	Represents 1 megawatt hour of power generated by a renewable resource (Green Tags)
Energy efficiency certificate	Represents 1 megawatt hour of power conserved or load reduced (White Tags®)
CO_2-equivalent carbon emission reduction credit	Represents a 1 metric ton reduction in CO_2 equivalents emitted

---

[36]For more potentials regarding the impacts of energy efficiency programs in addressing greenhouse gas emissions, see Hamrin and Sharick (2007).

[37]For more information, see UNFCCC information on CDM/JI programs, posted at http://unfccc.int/kyoto_protocol/mechanisms/items/1673.php

Finally, various gases, including carbon monoxide, methane, and others, contribute to global warming. The environmental impact of these gases can be converted to an equivalent CO_2 basis (CO_2e), a helpful common measure to assess the overall environmental impact, using standard and generally agreed-upon formulas. These credits are sometimes called *carbon credits, carbon offsets, verified emission reductions* (VERs), *emission reduction units* (ERUs), *voluntary carbon units* (VCUs), *carbon reduction tons* (CRTs), or *certified emission reductions* (CERs).

3.4.3 THE COMMODITY LIFE CYCLE

To have confidence in environmental markets, corporations, regulators, market participants, and the public need the ability to quickly and accurately obtain answers to important questions regarding environmental commodities, such as:

- What are the permits (credits, certificates, allowances) supporting the compliance obligation?
- Where did they come from?
- What type of activity created them?
- Who verified them? When?
- What are their serial numbers?
- For what compliance obligation or time period are they valid?
- Where have they been transacted?
- Have they been retired?
- Have position limits or trading transaction limits (if any) been observed for all participants?
- What was the price information?

Registry and tracking system deployments can provide answers to such questions and help ensure that the environmental certificates (offsets, credits, or emission allowances) are verified, unique, and valid:

- *Verified.* The environmental certificate must be measured, calculated, allocated, or auctioned and then tracked and recorded. In addition, the data related to the environmental certificate must be created and entered into the database via established procedures, with all necessary attestations and related documentation.

■ *Unique.* The environmental certificates (allowances or offsets) must be given a unique serial number and tracked from inception to retirement for the lifetime of the program to ensure that such certificates are not double counted and/or sold multiple times or used to fulfill multiple regulatory requirements.

■ *Valid.* If the environmental certificate is being purchased to fulfill a compliance requirement, then it must comply fully with the required rules of the sector or time period for which it will be applied. This element is particularly important in the complex multisector approach to environmental markets. Various sectors may potentially have different rules regarding allowance allocation, auction, and use of offsets, banking, or borrowing that must be managed.

Major components of such a trusted and reliable tracking mechanism with adequate market transparency for environmental commodities are presented in Figure 3.3.

Registry systems play an important role in the oversight, transparency, and integrity of certificates, ensuring the public trust and integrity of these new environmental markets. Market participants may use the tracking and registry

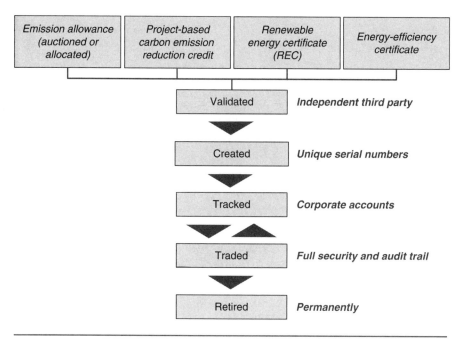

Figure 3.3 Trusted and reliable tracking mechanism.

system of record to confirm information regarding their purchases of allowances and offsets, empowering themselves to help thwart attempts to manipulate or defraud the marketplace, and help interested parties resolve deeper questions related to the origin and pedigree of a certificate. Finally, registry systems can provide the basis for information required for environmental oversight.

3.4.4 TRANSACTION MECHANISMS

To meet the environmental goals and reduce the threat of climate change, more environmental commodities or certificates are expected to be exchange traded in the future, compared with current practice that is dominated by direct transactions between buyers and sellers.

The discussion of a trading system for tradable permits in a cap-and-trade approach should center on the orderly transfer of the environmental certificate or credit between buyers and sellers and their retirement to meet compliance obligations. Most typically, buyers will have a regulatory obligation or public benefit purpose to fulfill, but buyers can also include intermediaries—marketers, brokers, or financial institutions—of these certificates or credits. As experience in Europe particularly demonstrates, as the size of the market grows, the role of brokers, banks, and exchanges can also be expected to grow. In U.S. environmental markets, the primary trading system mechanisms include:

- *Transactions within divisions of a large company.* Companies may choose to transact between subsidiaries or divisions before going to outside markets.

- *Direct transactions between companies.* Companies may choose to buy and sell environmental commodities directly with each other or perform swaps without any reliance on third parties.

- *Brokered transactions.* Brokers, banks, marketers, or other institutions may match buyers and sellers of an environmental commodity for a minimal service charge.

- *Electronic bulletin board transactions.* This service is provided to facilitate the matching of buyers and sellers. If a match is made, the buyer and seller contact each other to complete the transaction.

- *Private buying network transactions.* Similar to an electronic bulletin board, this mechanism might be managed by a financial institution, bank, broker, or marketer on behalf of its clients.

- *Exchange traded transactions.* The exchanges match buyers and sellers through either a floor-driven or electronic bid/ask mechanism.

Experience in environmental markets has shown that regulators need not specify the exact mechanism by which the environmental commodity is bought and sold. The act of creating obligations and the scarcity of the environmental commodity by itself will drive market behavior, as will the evolution of the appropriate trading approach and platform, as observed under the acid rain program. However, regulators do play an important role in shaping the development of market mechanisms through their influence in fundamental areas by establishing policies that enhance market liquidity, price transparency, and tracking mechanisms and reporting of essential market information.

3.4.5 MARKET DYNAMICS

The EU has in place a system to handle tradable permits under the EU ETS for emission allowances and the UN's CDM/JI programs for offset credits in the developing world. While independent steps are taken in various U.S. regions, the landscape of the U.S. environmental markets over the next 5 years will be more complex and fragmented than any commodity market in U.S. history. Further complexity is expected, depending on decisions as to which greenhouse gases will be regulated, what industry sectors should be covered, and where emissions should be regulated [25].

Dealing with such complexities is a major challenge for corporations in the coming years. In addition, corporations will need to find answers for many basic questions: How many RECs, energy efficiency certificates, carbon offsets, and allowances do we own? What are they worth? Are we short or long? What is our exposure in out-years? How is this impacted with acquisitions or divestitures? How will this affect our environmental financial balance sheet? This provides a business opportunity for banks and financial institutions in helping corporate clients answer these questions related to their environmental balance sheets. Financial institutions can also develop derivative instruments and assist in managing the risk and opportunities created by the jurisdictional complexity across the state, regional, and federal markets. Advanced registry technology can support financial institutions and corporations in the management of these new environmental commodities.

3.5 Market oversight

Environmental markets are likely to be tightly regulated and have strict oversight requirements. In part, these requirements are being driven by more general scrutiny of energy trading in the wake of the Enron and other scandals in the California power market.[38] But there is also added pressure, mainly due to excesses in the subprime mortgage and credit markets that resulted in the recent worldwide financial crisis, to ensure that emerging carbon markets are transparent and operating in the public interest. Legislators and regulators across the United States are sensitive to the concern that certain segments or entities might manipulate or profit excessively in the new environmental markets. Hence, the advanced tracking systems and registries will play essential roles in operating these emerging markets as intended to achieve policy objectives.[39]

3.5.1 CURRENT STATE OF PLAY ON MARKET OVERSIGHT

Actions by the U.S. Congress point to an increasing trend toward more oversight of emerging greenhouse gas emissions markets:

■ In September 2007, Senator Carl Levin of Michigan introduced legislation that would increase regulation of energy trading markets. Senator Levin's bill, "The Close the Enron Loophole Act," would require reporting of transactions and price information in exempt commercial markets (ECMs).[40]

■ Senator Dianne Feinstein of California has introduced a measure to establish federal oversight for new carbon emissions trading markets, designed to prevent fraud and manipulation in greenhouse gas credit markets. The bill establishes a maximum $1 million fine and 10 years in jail for each offense and clarifies that the Commodities Future Trading Commission (CFTC) maintains its exclusive jurisdiction over futures markets, including carbon dioxide futures markets.[41,42]

■ The Lieberman-Warner climate bill establishes a Carbon Market Efficiency Board that would handle a wide range of responsibilities related to the

---

[38]For detailed discussion of the essentials of market oversight for the power market, see Adib and Hurlbut (2008).

[39]For specifics of market oversight regarding environmental commodities markets, see APX (2007b) and APX (2008).

[40]For information on Senator Levin's bill, see www.senate.gov/~levin/newsroom/release.cfm? id=283461

[41]Information on Senator Feinstein's Emission Allowance Market Transparency Act (S.2423) is posted at www.govtrack.us/congress/bill.xpd?bill=s110-2423

[42]For information on the Energy Policy Act of 2005 and the Federal Energy Regulatory Commission's (FERC's) authority to oversee the energy and emission markets, see www.doi.gov/iepa/EnergyPolicyActof2005.pdf

functioning of the allowance market and the impact of allowance prices on the economy. The Board is also required to report to Congress quarterly on the status of the emission allowance market; its economic impact on regions and consumers; prices, incidents, and effects of any market fraud or manipulation; recommendations to relieve any excessive costs to the economy; and to make its reports available on the Internet.[43]

■ Voluntary markets are also receiving more scrutiny. Congressman Ed Markey has held hearings to determine whether the Federal Trade Commission (FTC) should intervene to set standards and to ensure that voluntary offsets are real and valid.[44]

The broad use of existing registry technology represents the surest way to ensure broad consumer protection and the most efficient regulatory oversight and substantiation of environmental claims at the lowest cost to society. It would be prudent to use extensions of these systems for the management of corporate compliance claims for new GHG markets, where they are certificate-based and enhance market oversight.

3.5.2 MARKET TRANSPARENCY

The new models of emissions market oversight for emerging GHG markets require greater transparency due to their significant financial importance. Furthermore, an important feature of the GHG market is the new concept of an emission offset, whereby certain emissions reduction activities are qualified to be used for compliance purposes in lieu of emissions allowances.[45] The complexity associated with the future greenhouse gas emissions markets will require greater transparency and must be subject to greater oversight. This is because:

■ The Federal Energy Regulatory Commission (FERC), Commodity Futures Trading Commission (CFTC), or Securities and Exchange Commission (SEC) may need more frequent and timely data on environmental credits and transactions to better understand the physical market for carbon

---

[43]Lieberman-Warner Climate Security Act information is posted at http://thomas.loc.gov/cgi-bin/bdquery/z?d110:s.02191.

[44]Information on the FTC process can be found at www.ftc.gov/opa/2008/02/greenguides.shtm

[45]Reforestation, landfill gas methane reduction, new agricultural practices, and certain types of renewable energy are typical examples of offset projects that should be verified, certified, tracked, and reported.

emission allowances and offsets and how it interacts with wholesale power markets.

■ The cost impact of a new carbon regime on the price of electricity will be significant and may even approach the fuel cost for power generation. This will drive a desire for oversight by electricity regulatory authorities and others. Having a system with the maximum integrity is important to build confidence in what will be the most important mechanism used to address climate change.[46]

■ Better and more frequent market data on emissions, allowances, and offsets will remove an important degree of uncertainty for market players, improve market efficiency, and ensure that the type of collapse in market prices that occurred after the first year of the EU ETS will be less likely. In addition, it would make clear to the public the societal cost of programs and compliance.

Given that mandatory policies regarding climate change will inevitably lead to higher energy prices, the public will demand more proactive market oversight of the emissions markets that are transparent and free from even the appearance of collusion or manipulation.

3.5.3 Registry Systems and Implementation

Technology systems for the expected GHG emission mechanism and allowance registries should meet regulatory reporting requirements for all transactions, handle complexity factors (such as offsets), reflect significant resulting economic impact, and present an adequate and transparent market reporting requirement to reflect price and transaction information in a more frequent and timely manner. These factors are all crucial for the success of an effective carbon market operation in which financial and market needs play as important a role as the need for regulatory compliance.

The current state-of-the-art registry technology provides proven, large-scale information systems that meet the compliance infrastructure that environmental regulators need, at the same time managing the offset and allowance information with the security and completeness needed by participants and market regulators. This is likely to include both electricity sector and commodities/securities regulators. Current registry systems with full audit trail and capability to handle cross-jurisdictional markets have a more robust platform and offer a further advantage over existing data systems.

---

[46]The letter by Representative Ed Markey to the Federal Trade Commission was issued on July 18, 2007, and is available at www.house.gov/apps/list/press/global_warming/July18offsetFTC.shtml

Finally, the scale of market activity for the new environmental commodities will require an information technology infrastructure, security, backup systems, geographically redundant data centers, help desk services, Web-based access and reporting, and a robustness commensurate with the needs of a large global financial market to be expected to play a major role in the operation of new environmental commodities. This essential aspect is often overlooked in regulatory discussions but is an important factor in the success of any market operation and its oversight.

3.5.4 IMPLEMENTATION OF OVERSIGHT MECHANISMS

Because the full pedigree and life cycle of an environmental certificate is tracked and known, a tracking and registry system can act as a single system of record and provide the basis of information required for multiple levels of oversight—environmental compliance oversight and commodities/securities market oversight—as well as providing information to corporations, environmental interest groups, and the general public. Based on experience in deploying market systems over the past decade, these levels of oversight can be achieved using a single system of record in a way that is not burdensome to any stakeholder. Such an oversight system could be available through a highly secure, rapid, large-scale, automated market system and database technology that can allow varying degrees of access to different segments of data as appropriate for each stakeholder. Figure 3.4 shows a desired registry and tracking system for administrative oversight and market transparency.

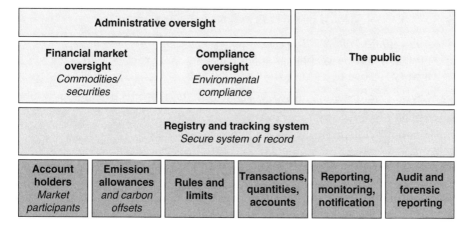

Figure 3.4   Registry and tracking system for administrative oversight and market transparency.

While legislation and rules regarding oversight are still being hammered out, certain elements under discussion can be readily handled with existing registry technology and could be implemented in any combination, as specified by policy. Specifically:

- Tracking of emission allowances, offset credits, or permits by an individual or corporation can be supported. Such information can be private or public and aggregated at any level required for oversight and reporting.
- Quantities and types of allowances and offset credits can be reported and tracked.
- Position and trading limitations for all market participants or certain types of market participants can be enforced through current systems and monitored and reported.
- Aggregated market prices as well as price history for every certificate, from auction or allocation through intermediate transactions to final purchase and retirement, can be tracked and reported.
- Other transaction information, including quantities sold, bought, transferred, banked to future compliance periods, or borrowed from future allocations, can be tracked, monitored, and reported.
- Limits on the number or volume of transactions can be tracked, reported, and monitored.
- Transaction information can be aggregated to monitor and identify market trends and possible abnormal market behavior.
- If any thresholds or limits for transaction, price, quantity, or overall market activity are exceeded, processes for immediate notification and actions can be implemented in the registry and tracking system.
- Full support for "forensic reporting" and an audit trail are available to an oversight authority that wants immediate access to a full transaction history for one or more market participants.
- Levels of information and disclosure ensure that an effective market can be supported.

3.6 Market registries and tracking systems

Because registries track an environmental certificate and all associated data across its full life cycle, from creation to retirement, the use of a registry system enables full oversight for regulators as well as complete data for corporations needing to demonstrate compliance. A registry system also provides a level playing field for all market participants, large and small. Five U.S. regions have currently deployed power generation information and certificate tracking systems. These systems already track certificate volumes and are enabled to track and store transaction price information. The regional systems are deployed in markets operated within the Electric Reliability Council of Texas (ERCOT) (Texas RECs), New England Power Pool (NEPOOL GIS), Pennsylvania-New Jersey-Maryland (PJM) (PJM GATS), the Western States (WREGIS), and the Upper Midwest States (M-RETS). Today most of these systems track all transactions and volumes for Renewable Energy Certificates and Energy Efficiency Certificates.[47] Tracking systems ensure that rules are transparently implemented so that every participant knows what he or she has to do to participate in the market and comply with the rules.

Market registries will also need to integrate with the systems used by governments to distribute and allocate emission allowances (permits), such as auction systems. One example of such an auction system is the quarterly auction implemented by the Regional Greenhouse Gas Initiative (RGGI), a power sector cap-and-trade market implemented across 10 U.S. northeastern states and for which the first compliance year will be 2009.

3.6.1 FUNCTIONAL REQUIREMENTS OF MARKET REGISTRIES

U.S. market experience across jurisdictions in deploying transparent and reliable environmental market systems indicates that it is never too early to start thinking about the critical infrastructure that will underlie any such system. Such deployments are strikingly similar, with the following common elements across certificate-based market systems:

- Creation of unique serialized certificates that are traded and transferred to meet stakeholder regulatory compliance obligations
- Tracking of ownership of certificates
- Transactions of certificates and trading between parties[48]
- Integration with financial markets, including brokerage, exchange, and auction mechanisms

---

[47]An additional mechanism is established through the creation of the North American Renewable Registry to track other areas not covered by the five listed regional systems. For more details, see www.apx.com/news/pr-APX-Launches-North-American-Renewables-Registry.asp

[48]The tracking systems are not exchange or brokering systems; rather, they record and fully track transfers of ownership of certificates and enable such transfers.

■ Management of certificate balances by account holders

■ Potential banking of certificates for use in future compliance periods or borrowing

■ Full audit trail and transaction history for all certificates, from origination to retirement, including the full pedigree and origin of each certificate

■ Web-based access with full transparency for stakeholders and regulators

■ A level playing field for all participants, including equal access for large and small firms across all sectors

■ A full and clear implementation of the rules according to the proscribed guidelines for access and use of the system

■ A clear workflow process so that all market participants know what they have to do to participate

■ Reporting mechanisms for participants and state agencies to verify compliance and enable market oversight

■ A secure market system, with redundant data centers and geographic redundancy for secure 24/7 operations

3.6.2 TECHNOLOGY PLATFORMS

An often overlooked topic in the discussion of environmental markets and tradable permits is the technology infrastructure needed to successfully operate a market system. Lessons could be learned from the European ETS system, in which the technology infrastructure needs were overlooked without accurately projecting the level of activities in the upcoming years. As explained by James Atkins,[49] "The heart of the problem is the use of a system which is not suitable for fast, high volume trading which has become the norm in the EU ETS. A design was accepted which appears inappropriate for the ambitions of the scheme. ..." To design an adequate and effective technological infrastructure, the following essential elements and attributes should be addressed:

■ *Security.* A multiple data center configuration and Web hosting services are essential, combined with the distributed capabilities of such a system. Fully redundant secondary data centers are typically called for, not just a

---

[49]James Atkins is the Chairman of Vertis Environmental Finance. His article is available at http://thebustard.blogspot.com/2008/09/taking-responsibility-in-eu-ets.html

disaster recovery site. Data center operations that consider environmental impacts and are geographically redundant are strongly recommended. Secondary data centers typically operate in a warm-standby mode, capable of taking over the primary role within a short period of time. Additionally, they provide a quality assurance staging area for testing new enhancements in the production environment without interrupting the system operation.

■ *Scalability.* System architecture must be highly scalable so that growing numbers of users can be handled by the addition of Web servers without any interruption in production or the need of software changes. For example, the WREGIS renewable energy market platform for Western states has been successfully tested to support more than 10,000 private account holders, with 600 concurrent users. Additionally, the PJM GATS system creates over 700 million certificates per year. The NEPOOL GIS database serving the New England renewable energy market is currently 600 Mbytes in size and stores the entire system activity since April 2002 (user login, transaction records, and over 700 million certificates).[50]

■ *Service-oriented architecture.* The use of service-oriented architectures (SOAs) is a design approach that allows for autonomy in implementation and provides for implementation of distributed, loosely coupled components that help to localize, and therefore minimize, the impact of change. Well-defined public service methods will be necessary to fulfill interoperability requirements between the wide and diverse variety of market participants and systems. The essence of this approach is that individual services (or software object classes) collaborate and cooperate, orchestrated by business process or systems events. SOAs can be implemented in any programming language on any operating system; object-oriented languages such as Java and C# are commonly used in most new system developments and are expected to be the preferred languages for new environmental market applications.

■ *Software and hardware platforms.* Today's commodity (meaning standard, commercially available) hardware, database management systems, and industry-standard middleware products are sufficiently robust and mature to alleviate the need for proprietary components and bespoke development. They offer the prerequisite reliability, resiliency to operational failure, and security for environmental and financial markets.

Platform construction will inevitably be an integration exercise in which environmental market-specific components are integrated with the current and much larger range of financial and commodities markets software components.

---

[50]This information is based on the authors' communication with software developers at APX, Inc.

With advances in hardware, the platforms required for environmental markets can be readily built from commodity hardware components such as racks of standard processors, advanced memory components, and multiple redundant storage area networks, with failover to one or more sites using dedicated, secure, high-bandwidth communications paths.

Commercially available Unix-derived operating systems, possibly open source, and standard relational database management systems such as Oracle or DB2, will be required for transaction processing and for data warehouses for reporting and analysis. The high data volumes in measurement data streamed from physical meters, trading desks, and wholesale and retail market systems will require the specialized data management capabilities typically found in Complex Event Processing (CEP) engines from specialist suppliers.[51] The middleware that facilitates the collaboration among services will also be "off the shelf" products.[52]

Master Data Management (MDM), which includes essential market participant information, will be an important solution component both internally for the service orchestration and externally for managing the reference data used to facilitate interoperability. Here again, this capability is found in most SOA middleware.

Applications would be provided using a software-as-a-service (SaaS) delivery model with a user interface typically rendered in an industry-standard browser using Adobe, Google, or Microsoft technology. In such a delivery model, vendors handle all necessary software upgrades, including revisions necessary to address changes in laws and regulations, resulting in further elimination of uncertainties facing regulatory agencies and corporations.

■ *The important role of standards.* Because the information technology infrastructure will rely so heavily on integration, it is important to discuss the role of standards. Both business standards and technology standards are fundamental to the smooth operation of the environmental market as a whole and the interoperability between the market participants.

Business standards covering master agreements, securities templates and contracts, credit rating, environmental reporting, reference data, and credit event protocols will be required. Technology standards governing data exchange message formats and communications protocols will also be required. Any internal (local) standards that occur within market

---

[51]Some of the examples are Coral8, Aleri, or Streambase or the more generic offerings bundled with the middleware from IBM, TIBCO, and Oracle.
[52]Some examples are IBM Websphere, Oracle (BEA) WebLogic, or TIBCO Enterprise Service Bus, with adapters for financial market integration such as those provided by iWay.

applications should be aligned with external standards to the fullest extent possible using extensions and/or transformations to map between the two as required. Some of these "standards" are really operational procedures: credit event protocols in the event of offset project default, bankruptcy of an actor, or default on payment by a counterparty to a trade, for example. These procedures should be modeled on their financial market equivalents to the extent possible.

Technology standards run the gamut from communications protocols, such as Transmission Control Protocol/Internet Protocol (TCP/IP) at the network platform layer, to the data exchange standards used by cooperating services. Data exchange standards for environmental attributes have been defined by the UNFCCC and underpin the Community Independent Transaction Log (CITL) for interoperability in the EU ETS, the International Transaction Log (ITL) allowing interoperability with the UNFCCC CDM, and the registries cooperating in the life-cycle management of voluntary offsets accredited by the Voluntary Carbon Standard.

Data exchange with participants in the financial markets will inevitably use the messaging standards defined for pre-trade, trade, and post-trade clearing and settlement in financial markets. Each financial market participant organization will likely also have its own rules defining the semantics of these messages, and each implements a specific version of the standard; these variations will need to be considered carefully for interoperability.

Finally, environmental market reporting will undoubtedly become a contributor to broader financial reporting and financial performance management as environmental markets develop.[53]

The attributes listed here are essential to establishing full confidence in the operation of such a system and assurance that the system can be fully trusted to work reliably and efficiently to meet the high expectations of regulators and financial institutions.[54]

3.6.3 ENVIRONMENTAL REGISTRIES

Although an effective and comprehensive environmental registry is important to meet regulators' expectations, it is also essential for other stakeholders, particularly various industrial and financial corporations with an obligation to comply

---

[53]The authors thank Michael Bonaventura of Noumenal Inc. for his contributions in this section.
[54]For more detailed explanation of beneficial features of an environmental market registry, see APX (2007a).

with regulations. Not-for-profit and public interest groups can also enhance their abilities to effectively pursue their environmental objectives. For corporations, the benefits of such registries can go well beyond regulatory compliance and can easily provide worldwide business opportunities and the ability to substantiate voluntary marketing claims [26]. Additional benefits include:

- Having a "one-stop shop" for environmental market data and customized reports

- Comparing their own environmental commodity standing and market position (quantity and price) against benchmarks (statistical averages), reflecting all corporations within the existing databases and markets

- Conducting economic evaluation of various emission reduction alternatives to identify, evaluate, and prioritize economic and practical environmental options that will maintain or enhance their competitive position

- Identifying opportunities to enhance and optimize their competitive position via environmental commodities markets.

3.7 Conclusion

Global climate change is now widely recognized as a serious problem, and economies around the world are challenged to identify the best approaches to cope with climate change while managing unintended consequences. Although the debate continues regarding the best economic approach, it is highly practical to achieve policy objectives using a market-based cap-and-trade mechanism as an environmental regulatory approach, if properly designed. This chapter outlines a number of success factors essential in addressing both emission reductions and resulting markets for tradable permits, as well as some of the technological requirements for effective management and regulatory oversight of such markets. Also, a number of larger conclusions, which are important to corporations impacted by these regulations, are derived:

- New environmental regulations are complex and multijurisdictional, sometimes even within one country, so corporations will need to take compliance cost into consideration and obtain infrastructure to manage this complexity.

- Effective regulatory oversight of such complex markets requires infrastructure capable of tracking prices, volumes, positions, trading limits, intermediate transactions, exchange transactions, forensic reporting, and full audit trail capabilities for every transaction for the lifetime of the program, as well as the ability to investigate the possibility of fraud or manipulation in the marketplace via queries and reports of historical information.

■ Markets and participants will need a trustworthy and reliable technology infrastructure to manage their obligations and to track credits and transactions with high integrity in the business processes.

■ Environmental market registries with desirable features can enhance market operations, facilitate transactions, integrate with financial markets, support corporate environmental asset management, and improve market efficiency and transparency.

■ Participants will need extensive support and information technology infrastructure in the new environmental markets to handle massive volumes of trade and transactions, with the flexibility to adapt to structural changes in these emerging markets and the ability to integrate with existing financial and transaction systems through integrated technology and standards.

REFERENCES

[1] Adib P, Hurlbut D. Market Power and Market Monitoring. 2008. In: Sioshansi FP, editor. Competitive Electricity Markets: Design, Implementation and Performance. Elsevier.

[2] APX. Creating a Trusted Environmental Commodity: A Guide to APX Environmental Market Depository and Its Pivotal Role in How Environmental Commodities Are Created, Verified, and Managed. Santa Clara, CA; 2007.

[3] APX. Market Oversight of Emissions Markets: The Essential Role of Tracking and Registry Systems. APX Inc., Santa Clara, CA; 2007.

[4] APX. Comments to the Federal Trade Commission Workshop, January 9, 2008, Washington, DC. Santa Clara, CA: APX, Inc.; January 24 2008.

[5] Betz, Regina, and Misato Sato. Emission Trading: Lessons Learnt from the 1st Phase of the EU ETS and Prospects for the 2nd Phase, Climate Policy 2006;6(5):351–9. Available at www.electricitypolicy.org.uk/TSEC/2/betz.pdf

[6] CARB (California Air Resources Board Market Advisory Committee). Recommendation for Designing a Greenhouse Gas Cap-and-Trade System for California. Draft for Public Review June 2007.

[7] CARB. Climate Change Draft Scoping Plan: A Framework for Change. Draft for Public Review June 2008.

[8] Carbon Trust. EU ETS Phase II allocation: implications and lessons. United Kingdom; May 2007.

[9] Ecosystem Marketplace and New Carbon Finance. Forging a Frontier: State of the Voluntary Carbon Markets 2008. May 8, 2008.

[10] Ellerman AD, Joskow PL, Harrison Jr D. Emission Trading in the U.S.: Experience, Lessons, and Considerations for Greenhouse Gas. Arlington, VA: Pew Center on Global Climate Change; May 2003.

[11] European Commission. Directive 2003/87/EC Establishing a Scheme for Greenhouse Emission Allowance Trading within the Community and Amending Council Directive 96/61/EC. Brussels: European Commission; 2003.

[12] Ford A. Global Climate Changes and the Electric Power Industry. In: Sioshansi FP, editor. Competitive Electricity Markets: Design, Implementation and Performance. Elsevier; 2008.

[13] Grubb M, Neuhoff K. Allocation and Competitiveness in the EU Emissions Trading Scheme: Policy Overview. Research Article, Climate Policy 2006;6:7–30.

[14] Hamrin J, Vine E, Sharick A. The Potential for Energy Savings Certificates (ESC) as a Major Tool in Greenhouse Gas Reduction Programs, prepared for the Henry P. Kendall Foundation. Boston, Center for Resource Solutions; May 24, 2007.

[15] Hass R, Meyer NI, Held A, Finon D, Lorenzoni A, Wiser R, et al. Promoting Electricity from Renewable Energy Sources: Lessons Learned from the EU, U.S., and Japan. In: Sioshansi FP, editor. Competitive Electricity Markets: Design, Implementation and Performance. Elsevier; 2008.

[16] Holt E, Bird L. Emerging Markets for Renewable Energy Certificates: Opportunities and Challenges. Technical Report, NREL/TP-620-37388 Golden, Colorado: National Renewable Energy Laboratory; January 2005.

[17] International Energy Agency. Energy Technology Perspectives: Scenarios and Strategies to 2050, A Luncheon Presentation in Support of the G8 Plan of Action, Tokyo, Japan; June 6, 2008. Paris, France: International Energy Agency; 2008.

[18] International Energy Agency. IEA Work for the G8-2008 Messages: Report to the G8 Summit, A Report Provided in Support of the G8 Plan of Action. Hokkaido, Japan; July 2008. Paris, France: International Energy Agency; 2008.

[19] Intergovernmental Panel on Climate Change. Climate Change 2007: Synthesis Report, Summary for Policymakers, Fourth Assessment Report, United Nation. Available at: www .ipcc.ch/pdf/assessment-report/ar4/syr/ar4_syr.pdf; 2007.

[20] ISO/RTO Council. Increasing Demand Response and Renewable Energy Resources: How ISOs and RTOs Are Helping Meet Important Public Policy Objectives, Summary Presentation, October. Also available at: www.isorto.org/atf/cf/%7B5B4E85C6-7EAC-40A0-8DC3-003829518EBD%7D/IRC_Demand_Renewables_Glossy.pdf; 2007.

[21] Jacoby HD, Ellerman AD. The Safety Valve and Climate Policy. Energy Policy 2004;32:481–91.

[22] Massy J, Knight S, Moller T, McGovern M, O'Brian H, Dodd J, et al. An Acceptable Basis for Negotiation. Wind Power Monthly 2008; February: 57–61.

[23] Moss D. Seeing Green. An Interview with Paul Ezekiel of Credit Suisse's Carbon Emission Emissary, TraderDaily.com, September. Also available at: www.traderdaily.com/magazine/article/9612.html; 2007.

[24] Musier R. US Mandatory REC Markets: An Established Environmental Infrastructure. North American Windpower 2006; November.

[25] Musier R, Shults R. Managing the Mosaic. Environmental Finance 2007; April.

[26] Musier R, Melby J. The Age of Substantiation. Environmental Finance 2008; September.

[27] Pew Center. The European Union Emission Trading Scheme (EU-ETS): Insight and Opportunities. Alexandria, VA: Pew Center; 2005.

[28] Pizer W. Choosing Price or Quantity Controls for Greenhouse Gases, Climate Issues. Brief No. 17. Washington, DC: Resources for the Future; July 1999.

[29] Rajgor G. Carbon Trade to Derive Renewables Growth. Wind Power Monthly 2008; February: 58.

[30] Reilly J, Paltsev S. An Analysis of the European Emission Trading Scheme. Report No. 127. MIT Joint Program on the Science and Policy of Global Change, October 2005.

[31] Rosenberg M. The Future of Futures: Look for Boom Times in Cap and Trade. An Interview published in *Energy Biz* by James E. Newsome, the president and chief executive officer of the New York Mercantile Exchange, Online, September/October; 2007.

[32] Stern N. The Economics of Climate Change: The Stern Review. Cambridge, UK: Cambridge University Press; 2007.

[33] Stern N. The Economics of Climate Change. The American Economic Review, Papers and Proceedings of the 120th Annual Meeting 2008;98(2):1–37.

[34] Stranlund JK. The Regulatory Choice of Noncompliance in Emissions Trading Programs. Environmental and Resource Economics 2007;38(1):99–117.

[35] U.S. Environmental Protection Agency. An Evaluation of the South Coast Air Quality Management District's Regional Clean Air Incentives Market: Lessons in Environmental Markets and Innovation. Washington, DC: Environmental Protection Agency; November 2002.

[36] U.S. Environmental Protection Agency. Acid Rain and Related Programs: 2006 Progress Report. Report No. EPA-430-R-07-011. Washington, DC: Environmental Protections Agency; 2006.

[37] World Bank. State and Trends of the Carbon Market 2008. Washington, DC: World Bank Institute; May 2008.

[38] World Resources Institute. (Aulisi, Farrell, Pershing, VanDeveer). Greenhouse Gas Emission Trading in U.S. States: Observations and Lessons from the OTC NO_X Budget Program. White Paper. Washington, DC: World Resource Institute; January 2005.

Making It Personal: Per Capita Carbon Allowances

Tina Fawcett,* Frede Hvelplund,** and
Niels I. Meyer***

*Oxford University, UK
**Aalborg University, Denmark
***Technical University of Denmark, Denmark

Abstract

*Consequences of global climate change are appearing faster than estimated
by the Intergovernmental Panel on Climate Change (IPCC). This highlights
the importance of introducing new, efficient schemes for mitigation of global
warming. One such scheme is Personal Carbon Allowances (PCA), whereby
individuals are allotted a tradable ration of CO_2 emissions per year. This chapter
reviews the fundamentals of PCA and analyzes its merits and problems.
The United Kingdom and Denmark have been chosen as case studies because*

the energy situation and the institutional setup are quite different between the two countries. As we conclude, PCA is an idea worthy of serious further consideration.

4.1 Introduction

Recent results on the melting of polar ice [21, 27] illustrate that the consequences of global climate change are happening at a faster pace than anticipated by the latest IPCC reports [24]. This knowledge has not been reflected, however, in the international political negotiations on mitigation of global warming. Despite the increasing urgency for serious action, the UN Climate Conference on Bali (COP 13) in December 2007 and the G8 summit in Japan in July 2008 failed to establish an agreement on efficient strategies and substantial commitments for the period after 2012.

Prior to the Bali talks, German Chancellor Angela Merkel had suggested a way ahead. In a public speech in New York in September 2007 she stated that international agreement on a principle of equal emission rights for all human beings in the world would be essential to include the developing world in commitments for emission targets [29]. She further argued that present climate models, together with official prognoses for population growth, will lead to an emission cap between 1 and 2 tons of CO_2 per person per year by 2050 (i.e., 10–20 percent of typical per capita EU emissions today). Her message, however, did not include detailed proposals for meeting this target.[1] This chapter takes as starting points both the principle of equal emissions rights and the need for massive emissions cuts in developed nations.

Given this background, there is an urgent need for investigating new and efficient means for reducing emission of greenhouse gases (GHGs). Several schemes are presented in other chapters of this volume. To simplify the discussion, this chapter focuses on CO_2 as the dominant GHG arising from combustion of fossil fuels. The chapter covers all uses of energy within households as well as the energy used for personal transport, rather than electricity alone.

Before introducing the concept of Personal Carbon Allowances (PCA), which is the focus of this chapter, we briefly discuss the success of energy policy to date in reducing carbon dioxide emissions. The traditional methods for reduction of CO_2 emissions include norms (e.g., building codes) and economic means (e.g., taxes on fossil fuels, subsidies for energy conservation and renewable energy sources). So far, these tools have not prevented the general increase of CO_2 emissions in the world. The quantitative effects of the

---

[1]Chapter 1, by Alan Moran, in this volume illustrates some of the challenges in meeting this target.

methods, either alone or in combination, are rather uncertain, and the desired targets for energy reduction are often not attained.

In addition, the EU has some years of experience with caps (quotas) for industrial emissions of CO_2, the EU Emissions Trading Scheme (EU ETS). Emissions from agriculture, service sector, transportation, and private households, which total about 60 percent of EU emissions, are not included in EU ETS. So far experience of this scheme has revealed a number of shortcomings,[2] resulting in unimpressive carbon savings. It remains to be seen whether stricter caps and other reforms will improve the results during the next period after the conclusion of the Kyoto commitments in 2012 [17].[3]

For the future, experience shows that it is difficult to get sufficient reduction incentives via taxes on fossil fuels without politically unrealistic increases in CO_2 taxes. In addition, such taxes disadvantage low-income groups. Though EU ETS is expected to expand, hopes for what it can achieve are modest. To summarize, research shows that existing policies are unlikely to achieve sufficient savings. For example, a recent report from the Danish Economic-Environmental Council (DEEC) estimated that Denmark would not be able to reach the goal required by the EU Commission of 20 percent reduction in CO_2 emissions from the sectors outside the EU cap system without new and efficient schemes for these sectors [7]. The case for investigating new, supplementary policy ideas is clear.

In the new scheme for reducing GHG emissions based on PCA, every adult is allotted an equal, tradable ration of CO_2 emissions per year in connection with the consumption of a number of selected energy services for private households. Compared to the cap system used in the industrial sector with its emission caps on different industrial plants, PCA is applied to individual human beings to directly influence their behavior.

So far no country has introduced a PCA scheme, and only a few industrial countries are giving serious political consideration to the scheme. Thus it is still a proposal in its initial phase of investigation. This chapter aims at drawing attention to the potential of the PCA scheme and to outline the conditions for its promotion. The chapter's focus is on developed countries being the natural ground for the first experiments with PCA. The analysis is based on the potential for PCA within a single country; the cross-border effects among countries are not considered here.

The efficiency of a PCA system is dependent on the particular energy system in the country and on the institutional arrangements. The United

---

[2]For example, in connection with too-favorable caps and excessive use of flexible mechanisms such as joint implementation and clean developments mechanisms.

[3]These questions are also discussed in Chapter 2, by Chappin et al., in this volume.

Kingdom and Denmark have been chosen as case studies because both the energy situation and the institutional setup are quite different between the two nations. Results from these two countries are thus expected to illuminate a broad spectrum of possibilities for and barriers to introducing similar schemes in other countries.[4]

It is envisioned that in the initial phase of implementation, PCA may differ between countries, depending on the national commitments for reduction of CO_2 emissions and other factors. If the scheme is successful in industrial countries, it may be extended to the developing world. Ultimately, the personal caps should converge toward the same levels in accordance with the global equity principle.

Regardless of the details, any PCA scheme has to be evaluated against energy policy goals such as:

- *Eco-efficient* and *significant* with regard to being an efficient governance tool for the household reduction of greenhouse gas emission
- *Economically efficient* by reducing greenhouse gases at relatively low costs
- *Bureaucratically efficient* by having relatively low transaction costs
- *Innovation efficient* by promoting the development and implementation of an "optimal mix" of energy conservation and energy supply technologies
- *Democratically and educationally efficient* by inducing a better understanding of the greenhouse problems at the consumer level
- *Socially efficient* by reducing greenhouse gases in a socially balanced fashion

Section 4.2 describes the basic features of a national PCA scheme, and Section 4.3 provides two case studies for the United Kingdom and Denmark, followed by summary and conclusions.

4.2 Basic features of a national carbon allowance scheme

PCA is a policy scheme that would introduce a national cap on household and personal transport energy use by directly allocating a carbon cap to each citizen. All adults would receive an equal carbon allowance, which would be adjusted/reduced periodically to achieve stated goals. These allowances would

---

[4]Other work has also considered in some detail the applicability of PCA to the United States (Hillman, Fawcett, and Rajan, 2008).

be tradable, and enrollment in the PCA scheme would be mandatory. To pay an energy bill, put fuel into a car, or buy a plane ticket, each citizen would have to surrender carbon "credits" from their allowance as well as pay the financial cost. They would keep track of their carbon allowances by a carbon credit card and carbon account in the same way they keep track of their money by a money credit card and bank account.

If citizens used up all their initial allocation of credits, they could buy more at the market price. This market price could be high in situations where most people had little surplus left on their PCA cards or low if the majority significantly reduced their emissions. As the national carbon cap and personal allowance are reduced over time, people will have to adopt progressively lower carbon lifestyles via a combination of technology choices and behavioral changes.

For practical reasons, PCA would cover only "direct" energy uses by the householder for residential purposes and personal travel. It would not cover all other "indirect" uses of energy embodied in goods or services. The carbon impacts of direct energy use are well known, with the exception of the debate about the global warming impact of air travel, whereas the energy and carbon embodied in everyday goods and services depend on a large number of varying parameters in a complex way. It would simply be impossible at present to label food, clothing, and other goods with a carbon "price."[5] Indeed, in an age of globalized and fast-changing supply chains, it might never be possible to have reliable carbon labels on most products, let alone services. Therefore, while it might initially seem an attractive and effective idea, there is no suggestion that goods and services should be included within a PCA scheme. Instead the carbon used in their manufacture, transport, retail, and so on should be regulated via existing or new policies. These carbon emissions are outside the scope of the proposed PCA.

The ability to trade emissions is a key feature of a PCA scheme. Without trading, a considerable proportion of the population would immediately have to drastically change their lifestyles to live within their ration; this is unlikely to be either possible or democratically acceptable. Additionally, if trade were not allowed, experience with rationing systems has demonstrated that an illegal black market would quickly emerge. As discussed in later sections, trading is likely to lead to a transfer of money in general from richer to poorer

---

[5]There are contrary views, and carbon labels for products do exist. For example, the supermarket chain Tesco is currently testing carbon labels across four product ranges in the United Kingdom (Tesco, 2008). To put this in context, many Tesco stores stock over 40,000 products. This demonstrates the huge challenge comprehensive product labeling poses.

individuals and a transfer of carbon allowances in the opposite direction. Richer people travel more, have bigger cars and homes, and so on, resulting in bigger carbon footprints.

Introducing PCA would mark a profound shift in national priorities, signaling a permanent commitment to and a mechanism for reducing carbon emissions, with negligible uncertainty in the total emission reduction of CO_2 in the household sector in relation to "direct" energy consumption. To enable people to reduce their emissions over time, extensive supportive information and advice policies would be required. In addition, the current suite of policies designed to reduce electricity and energy-use/carbon emissions—labels, standards, subsidies, information, and so on—would need to be enhanced. If PCA were introduced, everyone would be explicitly included in reaching society's goal of reducing carbon emissions. Individuals would need to understand both their emissions and the options they have to reduce them, and society would need to prioritize enabling low-carbon lifestyles, for this policy—or any policy that tries to deliver significant sustained emissions reductions—to succeed.[6]

In principle, PCA could change people's relationships with their own carbon emissions, engender a greater sense of responsibility for and ability to reduce emissions, and drive a change in social norms to favor lower carbon lifestyles. Because there is already evidence that behavior can change in response to signals such as information and feedback [10], this is not an unfounded hypothesis. Indeed, unless individuals' relationships with their carbon emissions change, any significant carbon reduction policies seem unlikely to succeed.

The EU Emission Trading System (EU ETS) already covers the emissions of many larger commercial enterprises and electricity producers, and its extension to smaller enterprises and to the transport sector is currently being considered. If PCA were introduced in the present governance system, there would be an overlap or double-counting between the emissions included in EU ETS and PCA. Electricity production plants are part of the EU cap-and-trade system, and emissions from electricity used in households are also included in the PCA scheme. This would need to be taken into account by a coupling of the caps in the two systems. Indeed, for PCA to work well, carbon reductions would need to be correlated with reductions in the other sectors of the economy, whether or not these sectors were already covered by EU ETS.

---

[6]This is also discussed in Chapter 5, by Sovacool and Brown, in this volume.

Box 4.1 RELATED POLICY PROPOSALS

Other related policy proposals also aim to establish a national cap on carbon emissions in an equitable and effective way. Two of the best developed are *domestic tradable quotas*, also known as *tradable energy quotas*, or TEQs [20, 35], and *cap and share*, or C&S [3, 4].

For individuals, TEQ would be very similar to PCA except that air travel would not be included in the personal TEQ allowance. Organizations would have to buy emissions permits via a national auction. Many of the details of PCA have been developed by researchers on TEQ; the similarities of the policies are much more significant than their differences. Both TEQ and C&S are greater in scope than PCA; they cover all carbon emitted in the economy.

C&S differs considerably from PCA. In C&S each individual is given an annual certificate entitling that person to a share of national emissions. Individuals do not need these certificates to buy energy services or goods responsible for carbon emissions, but the organizations supplying them do. The idea is that (most) individuals will sell their certificates to energy suppliers, thereby getting additional income that will offset some or all of the additional cost of carbon-intensive goods and services. C&S therefore does not introduce personal responsibility for carbon emissions in the same way as PCA. Instead it relies largely on economic changes (in the price of direct and indirect carbon) to drive changes in personal behavior. It seems likely to be very different, and more limited, in its psychological impact than PCA.

As with PCA, both these policy ideas are still in their formative stage.

Would PCA be an effective, efficient, and equitable policy? This question is answered in the following with reference to the criteria established in the introduction:

■ *Significance.* As the case studies will demonstrate, a PCA policy would cover up to half the carbon and carbon-equivalent emissions in the economy—and it is therefore a significant policy.

■ *Costs and bureaucracy.* Though potential costs are an issue and are seen by the U.K. government as a significant barrier to PCA, there is evidence that the technology and institutions necessary for a successful scheme could be delivered [2, 26, 34, 36]. The costs of a PCA policy could be justified if it delivers significant savings that cannot be otherwise accessed, which is the proposition being explored in this chapter.

■ *Social equity.* PCA is an overarching policy instrument that aims to deliver a reduction in national emissions in an equitable way. Equity is delivered by allowing individuals to emit an equal quantity of carbon free of charge. This definition of equity matches that proposed within the "contraction and convergence" global approach [30] being championed by German Chancellor Angela Merkel, among others [29].

Although PCA delivers equity in principle, there are concerns about its potential impact on different individuals and sections of the population [33, 35]. Equal access to carbon allowances is not the same as equal access to energy, energy services, or the quality of life they enable. Technological, social, and infrastructure variations can ensure that very different levels of energy services and mobility are delivered for the same carbon emissions. For example, for household electricity use, efficient end-use equipment, well-insulated homes (e.g., where electricity is used for heating), lower carbon electricity sources, and renewable energy all create a disconnection between carbon emissions and energy services. It is this disconnection that offers people positive opportunities to reduce their carbon emissions without sacrificing energy services, which are important to them. However, it also means that giving people equal allowances will not result in "equality of sacrifice"—a notion of fairness that seems prevalent in popular debate but that is arguably impossible to deliver through any carbon reduction policy alone.

Initial U.K. research suggests that PCA could be a more socially progressive policy than carbon taxation [15, 25], but much would depend on how revenues from carbon taxation were recycled.

■ *Promoting innovation.* Under PCA there should be many opportunities for low-carbon innovation in products and services for householders. New business areas should emerge (e.g., personal carbon managers, low-carbon household renovation companies), whereas some high-carbon businesses (e.g., long-haul airlines) will decline. In addition to PCA, governments would be likely to continue to develop innovation policies and support—and these, in addition to the declining national carbon cap, might be more likely to directly influence future innovation than PCA itself.

■ *Educating and engaging the population.* The introduction of PCA should result in a much better-informed population who were motivated to reduce their personal carbon emissions. The public should be convinced that PCA does not necessarily imply a lower quality of life but will lead to significant cost savings. This is essential if PCA is to work and to retain political support.

In principle, then, PCA could be effective, efficient, and equitable. The following sections use experience from the United Kingdom and Denmark to look in more detail at how PCA would operate in different national contexts.

4.3 U.K. case study

This case study gives a thumbnail sketch of carbon emissions and personal energy use in the United Kingdom before moving on to present more detailed research evidence relevant to PCA. Using the evidence available, the likely effects of PCA and its prospects for adoption in the United Kingdom are considered.

4.3.1 Background on U.K. Energy Use and Carbon Emissions

The United Kingdom, with a population of 60 million, is in the top 10 carbon dioxide–emitting nations, with per capita carbon emissions from fossil fuels more than twice the global average [28]. In 2006, carbon dioxide emissions (as reported for Kyoto purposes) totaled 555 $MtCO_2$ in the United Kingdom [11], which equates to around 9.4 tCO_2 per capita.

Since 1990, carbon dioxide emissions from the United Kingdom have decreased, primarily due to a change in electricity generation fuels—a significant switch from coal to gas—and a reduction in energy use in the business sector. As shown in Figure 4.1, net emissions of carbon dioxide fell 6 percent between 1990 and 2006[7] [11].

Combined with a reduction in other greenhouse gases, this means that the United Kingdom is very likely to meet its Kyoto target of a 12.5 percent reduction by 2010. However, the United Kingdom will not meet its national target of a reduction in carbon emissions by 20 percent from 1990 to 2010. (See also the section on the United Kingdom in Chapter 15, by Nigel Cornwall.)

Personal energy use in the United Kingdom can be briefly described. For U.K. householders, the major fuel used for space and water heating is natural gas, delivered through a pipe network. Natural gas is used by around four fifths of households, with the remainder largely using either electricity or oil for heating. There is little use of either solid fuel or district heating (combined heat and power). It is widely acknowledged that heat losses from the United Kingdom's old and inefficient housing stock are high compared with many other EU countries. For personal travel, the United Kingdom is a heavily

---

[7]These figures are *actual* emissions and do not adjust national totals to take into account any trading within EU ETS.

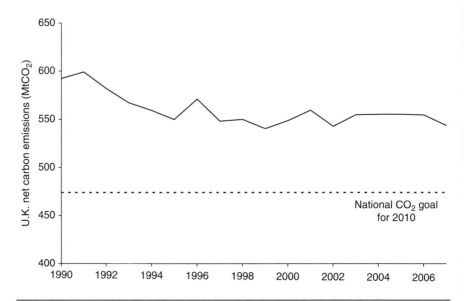

Figure 4.1 U.K. annual net carbon dioxide emissions, $MtCO_2$, 1990–2007. Source: Defra [11].

car-dependent society. Over four fifths of mileage traveled on land is undertaken by private car. U.K. residents also make strong use of air travel for leisure.

Despite some progress, there is little evidence that the United Kingdom is on the path to a sustainable low-carbon economy. Rather, a fortuitous switch to lower carbon fuels and a reduction in high-energy industries have enabled reductions in emissions to be made at a time when underlying economy-wide energy use, particularly for transport, is increasing.

4.3.2 PERSONAL DIRECT CARBON EMISSIONS IN CONTEXT

Using 2006 government figures [12] and excluding international aviation, PCA would cover 44 percent of the United Kingdom's carbon dioxide emissions. When carbon dioxide–equivalent emissions from international aviation are included,[8] PCA would cover 51 percent of the total—with the business transport, industrial, commercial, and public sectors being directly responsible for the remainder of national emissions.

---

[8]Carbon dioxide emissions from air transport have been multiplied by three, to include all the global warming potential of emissions from aircraft. The factor of three is based on IPCC research (RCEP, 2002) and has been supported by more recent analysis (Brand, 2006). However, it is at the higher end of multipliers used in the literature.

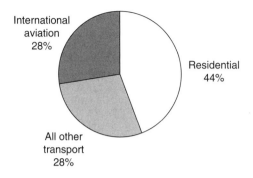

Figure 4.2   U.K. personal carbon emissions by category, 2006.
Source: Defra [12], with methodology from Hillman and Fawcett [22].

Emissions from residential energy use are the single most significant contribution, as shown in Figure 4.2.

4.3.3 SOCIAL DISTRIBUTION OF CARBON EMISSIONS

Initial U.K. research has shown that, on average, carbon emissions rise with income, but also that there is huge variation of emissions within each income decile [15]. Similar analysis undertaken by Fawcett [18] demonstrates how personal carbon emissions rise with income, with the average individual in the richest 10 percent of the population being responsible for almost 60 percent more carbon equivalent emissions in 2002–2003 than individuals from deciles one, two, and three (Figure 4.3).[9]

In general, poorer people could expect to be better off under PCA—because they are below-average carbon emitters and will have spare allowances to sell—but some will be considerably worse off. This is likely to be of concern to policymakers. It could be an issue that is more relevant in the United Kingdom than elsewhere, given the national prevalence of fuel poverty.[10] The situation in Denmark seems to be different, as discussed later.

The huge variations in personal carbon emissions have been further demonstrated by two studies. Carbon audits of 40 individuals undertaken in

---

[9]Note that the data for carbon-equivalent emissions from air travel is considerably less reliable than for domestic energy and motor fuel, for which data is based on expenditure surveys.
[10]A household in fuel poverty is defined as one that would need to spend more than 10 percent of its income to obtain adequate energy services. Government figures suggest that 2.5 million U.K. households were in fuel poverty in 2005 (Defra and BERR, 2007). However, there are fears that 2007–2008 energy price rises may have increased this number by up to 1 million.

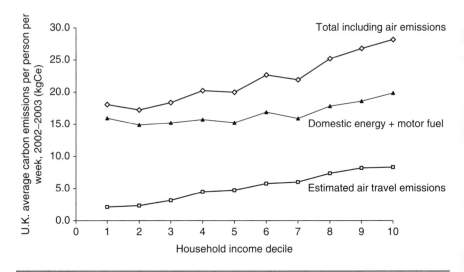

Figure 4.3 U.K. average weekly carbon emissions per person from private transport, gas and electricity use, and estimated air travel emissions by household income decile, 2002–2003.
Source: Fawcett [18].

2003–2004 showed that their total personal emissions varied by a factor of 12 [18]. Of those who traveled by air, the highest emitter was responsible for 46 times the emissions of the lowest. Surface travel emissions varied by a factor of 19 and household energy emissions per individual by a factor of 9. [6] calculated the carbon emissions from all personal travel of several hundred U.K. households. The data showed huge variations in personal travel emissions. The top tenth of emitters were responsible for 43 percent of total sample emissions, whereas the lowest tenth were responsible for just 1 percent.

To summarize, initially more people would benefit than lose from PCA in the United Kingdom—that is, more people would be below than above their allowance. Also, individual experience of the PCA would be highly variable, something that could be equally true of increased CO_2 taxation.

4.3.4 U.K. POLITICAL AND PUBLIC INTEREST IN PCA

PCA has attracted some political interest in recent years.[11] For example, it was promoted by the then Secretary of State for the Environment in 2006–2007 in

---

[11]An important expression of this interest was the introduction of a private member's bill (i.e., not government legislation) to Parliament in July 2004, "to introduce a domestic trading scheme for carbon emissions, to set a national ceiling for carbon emissions and for connected purposes" (Anon, 2004). The bill had a first reading and was debated but was not adopted as legislation.

the context of all parts of society needing to make a contribution to reducing carbon emissions:

Individuals can play an important part, too, and that is why I have led the debate about personal carbon allowances and so-called carbon credit cards, which could help individuals to see how they can make a contribution that will help the environment and themselves [31].

Recently both the key government department for this policy, the Department for the Environment, Food, and Rural Affairs (Defra), and an influential committee of members of Parliament have published reports regarding PCA. As a result of its research, Defra concluded that:

... while personal carbon trading remains a potentially important way to engage individuals ... it would nonetheless seem that it is an idea currently ahead of its time [13: 4].

Their key concerns were public acceptability and costs, with doubts whether these issues could be resolved satisfactorily. Defra concluded that the government should remain engaged in the debate around PCA, but that further work should be taken forward by academics and research organizations and not the government itself.

However the MPs' Environmental Audit Committee, which published its report a month later, was more supportive of PCA and indeed "regretted" Defra's decision to wind down further research work on PCA. Their inquiry concluded that:

Personal carbon trading could be essential in helping to reduce our national carbon footprint. Further work is needed before personal carbon trading can be a viable policy option and this must be started urgently, and in earnest [16: 3].

This committee is an important and influential one, and its findings on PCA made headline news in the United Kingdom. It remains to be seen how the committee's recommendations translate into action, but it seems fair to say that on balance PCA has gone up the political and research agenda.

There has been practical demonstration of some public support for the idea of PCA in the form of a growing network of CRAGs—Carbon Rationing Action Groups (www.carbonrationing.org.uk). The principles adopted by CRAGs are identical to those that inform PCA. The network was founded in early 2006 and now comprises over 30 groups throughout the United Kingdom and beyond. Experiences from the CRAG network are being investigated by a number of researchers, but this work is ongoing and has not yet been published.

A recent report has suggested that an important next step could be a research trial of PCA [19], whereas other work has outlined a 5-year research program into PCA and related policies [33].

The United Kingdom has a good track record in bold statements and commitments to reducing its carbon and greenhouse gas emissions but will fail to meet some important targets (e.g., reducing carbon dioxide emissions by 20 percent between 1990 and 2010). There is every reason to believe that without much more radical policy, the United Kingdom will not significantly reduce its carbon emissions. A growing body of work demonstrates how the United Kingdom could become a lower-carbon society. One recent example, focusing on the housing sector, demonstrates how carbon dioxide emissions could be cut up to 80 percent by 2050 using established technologies [5]. However, without a framework policy such as PCA, it seems highly unlikely that such a lower-carbon future will be achieved. PCA could be adopted in the United Kingdom, but the steps needed to introduce such a significant policy change have not yet been fully researched, and the political and social case for its adoption is still to be made.[12]

4.4 Danish case study

Denmark has a techno-institutional energy system that is different from the system in the United Kingdom—technically, due to better insulated houses, the large share of wind energy and cogeneration of heat and electricity; institutionally, due to high fixed tariffs and higher energy taxes. As a consequence, the starting point for the evaluation of possibilities and problems for a PCA system in Denmark is different from that of the United Kingdom.

Another difference is that the question of PCA has not been on the Danish policy agenda to the same extent as in the United Kingdom, neither in relation to scientific evaluations nor in relation to political discussions and proposals.

4.4.1 BACKGROUND ON DANISH ENERGY USE AND CO_2 EMISSIONS

Denmark has 5.1 million inhabitants and had a CO_2 emission per capita of 10.3 tons in 2006. An active energy policy promoting cogeneration and renewables, especially wind power, and energy conservation from the early

---

[12]See also Chapter 15, by Cornwall, on the British electricity industry.

1990s until 2002 resulted in a reduction in CO_2 emissions of 16 percent from 1990 to 2004.[13]

Since 2002 the present conservative-liberal Danish government has changed the national energy policy to rely strongly on market forces. This has resulted in a drastic reduction of the rate of growth of renewable sources in the Danish energy system. Since 2004, the net increase in land-based wind power capacity has been close to zero, and CO_2 emissions have increased 4 percent from 2004 to 2007. As a consequence, Denmark has severe problems in fulfilling both national and international commitments on reduction of CO_2 emissions.

In December 2009 Denmark will host the UN Climate Change Conference (COP15) with the goal of reaching a post-Kyoto agreement. Against this background, the majority in the Danish Parliament confirmed a new energy strategy in 2008 [9] for the period up to 2011. Moreover, a well-documented analysis has shown that it is technologically possible and economically favorable to reduce emissions of Danish greenhouse gases 60 percent by 2030 compared to 1990 [8]. This opens up policy options for alternative proposals potentially including PCA.

4.4.2 THE DANISH ENERGY SYSTEM IN RELATION TO PCA

To describe the potential impact of a Danish PCA scheme, one has to evaluate the possibilities of Danish households for active participation in the reduction of CO_2 emissions. In principle, people can decide to cut the number of tourist trips they make to distant countries, and most people can decide to reduce the average room temperature in their house and to make more short trips on bicycle instead of by car. Most people can also afford to buy low-energy lamps and low-energy refrigerators and washing machines when the old ones have to be replaced.

Improved insulation of dwellings (refitting), on the other hand, often requires better information and economic support from government or stricter government regulations, but there are no basic technological barriers for promoting such activities.

In principle, these considerations are valid for both the United Kingdom and Denmark. But there is an important difference between the two countries in the energy supply system. More than 60 percent of Danish households get their heat and electricity from cogeneration plants combined with district heating pipes, whereas this energy supply system is negligible in

---

[13]This number does not include CO_2 emissions from Danish-owned commercial transport.

the United Kingdom. Therefore, it is relevant to take a closer look at the special problems related to PCA in connection with cogeneration of heat and electricity.

The reason for the extensive penetration of cogeneration and district heating in Denmark is that the primary fuel is used about 30 percent more efficiently in this system compared to separate production of electricity in condensing plants and heat in boilers.

Cogeneration and district heating systems require relatively large initial investments for the heat pipes, but in the long perspective both fuel and money economies are better than separate production—especially with increasing fuel prices.

In relation to Danish households and PCA, however, it is a severe barrier that tariffs of the present system include a relatively large fixed part to compensate for the initial investment costs. This obviously reduces the incentive for households to conserve heat. A solution for this problem could be to reduce or eliminate the fixed part of the tariff by a balanced government policy with a transitional economic subsidy, especially to young district heating companies.

Another problem is that there is no generally accepted objective definition that links the fuel use in a cogeneration plant to either heat or electricity. If a Danish PCA scheme is introduced, it will be necessary to establish a clear definition of this division for consumers of heat and electricity from cogeneration plants.

It is clear from the preceding arguments that institutional and administrative changes are needed in the Danish energy system, with its high penetration of cogeneration and district heating, to make the PCA scheme acceptable and operational.

The fastest growth rate for CO_2 emissions is found in the transportation sector. At the same time, official data for CO_2 emissions from transport are rather confusing due to lack of clear account principles for emissions linked to international air traffic and international shipping.[14] In relation to a Danish PCA scheme, it would be relatively simple to account for the CO_2 emissions linked to the consumption of gasoline for cars. The emissions related to personal air traffic can be administrated by the flight companies. The question to be solved in this relation is which multiplying factor should be applied to the emissions at 10,000 meters above ground, where the greenhouse effect is amplified. A factor of two to three is often used, as mentioned earlier. The emissions from international shipping should not be included in the national PCA scheme.

---

[14]These emissions are left out of the Kyoto Protocol. However, the EU Commission is now considering how to include these sectors in a cap system.

4.4.3 CO_2 Emissions in the Danish Energy System

Table 4.1 illustrates the results of the Danish energy policy for the areas linked to household CO_2 emissions.

In 2006 the average CO_2 emission per person in Denmark was 10.3 tons, exclusive of emissions from shipping. Of these, 22 percent, or 2.15 tons per person, are linked to the household consumption of heat and electricity.

The relatively high reduction in Danish household carbon emissions in the period from 1990 to 2006 was accomplished by a combination of energy conservation and the massive introduction of wind power and cogeneration of electricity and heat. In 2007 wind power provided about 20 percent of the Danish electricity production.

These 2.15 tons of CO_2 emissions per person per year are one of the two targets for personal CO_2 quotas. That is the part we, as "directors" of our households, have the direct power to influence.

Table 4.1 CO_2 emission from various sectors in the Danish energy system, 1990 and 2006

	Mill. Tons of CO_2, 1990	Mill. Tons of CO_2, 2006	Relative Contribution (%) 2006	1990–2006 Change (%)
Energy sector, oil and gas sector	1.7	2.7	5	49.9
Transportation	**12.6** (16.3)	**16.0** (21.2)	**30**	**22.5**
Production	17.9	14.4	27	−21.2
Trade and service	10.7	7.8	15	−26.1
Households (heat and electricity)	**17.9**	**11.7**	**22**	**−34.7**
Total	60.8	52.6	100	−13.5

Source: Danish Energy Administration (Energistyrelsen), Annual Energy Statistics 2007.[15]

---

[15]The transportation numbers in parentheses in Table 4.1 illustrate the results based on a multiplication of the emissions from Danish-related air traffic by a factor of three. Emissions from Danish shipping are not included in Table 4.1.

In addition to the 2.15 tons of CO_2 linked to the household consumption of heat and electricity, the PCA scheme will also include personal transportation. A preliminary estimate is that between 6 and 9 million tons of the transportation emissions indicated in Table 4.1 are linked directly to household decisions. The exact value of this number is not known, and further investigations are needed to attain more exact results. However, at this early stage more exact numbers are not essential for the evaluation of a Danish PCA scheme.

Based on these estimated numbers for Denmark, the total annual CO_2 emissions per person directly linked to household decisions will amount to between 3 and 4 tons. These are the emissions that are susceptible to influence from a PCA system. They amount to between 30 percent and 40 percent of total Danish CO_2 emissions, which is lower than the estimation of a 50 percent share of total CO_2 emissions in the United Kingdom. This difference is partly due to a higher level of house insulation in Denmark and the introduction of a high proportion of cogeneration and wind energy in the Danish energy system and partly due to a lower estimate for emissions from household consumption of transportation.

4.5 Conclusion

This chapter has presented a scheme to motivate individuals to reduce their carbon emissions from electricity use and other sources. PCA is a policy option that could redefine the relationship between individuals and their carbon emissions as well as provide an effective and equitable cap on up to half of national emissions.[16] The necessary administrative technologies and information systems need to be developed to establish sufficient political support for PCA schemes.

The two case studies of the energy systems and institutional frameworks in the United Kingdom and Denmark have exposed important differences as well as similarities between the two countries. To succeed, the introduction of PCA must take into account the detailed technological and institutional setup of each nation.

To summarize the similarities first: Per capita carbon dioxide emissions differ little between Denmark and the United Kingdom. Both countries have reduced their carbon emissions since 1990, partly by decarbonizing their electricity supply—in Denmark by introducing cogeneration and wind power and in the United Kingdom by changing from coal to natural gas for

---

[16]In industrial countries, an estimated 30–50 percent of the national emissions would be covered by a PCA scheme.

power production. However, emissions reductions have not been sufficient to enable the two countries to meet important national and international targets. Therefore there is a need in each country for a new policy approach.

The case studies have also identified a number of differences. The percentage of national emissions that would be covered by PCA is higher in the United Kingdom than in Denmark—perhaps considerably higher. Different institutional reforms would be required in the two countries to enable consumers to react efficiently to a PCA scheme. In Denmark there is a need to establish institutions that allow PCA and a high degree of cogeneration and district heating to coexist successfully. Thus, fixed tariffs should be avoided or drastically reduced, and formal rules have to be introduced, accounting for the division of emissions related to heat and electricity. In addition, the Danish national legislation should give "open access" for renewable energy supply in areas with cogeneration and district heating. In the United Kingdom there is a need to learn from Danish successes in the introduction of wind power and cogeneration. These factors emphasize that a PCA system is embedded in institutions and existing policy frameworks that differ from country to country. A PCA system cannot stand alone.

A full program of research covering issues of equity, effectiveness, cost, and public and political acceptability is required to more fully explore PCA. In addition, investigation is needed into the coupling between general energy policy and PCA and into the question of whether PCA schemes should be purely national or are better established on an EU-wide basis. Consideration of a multinational scheme would raise many additional issues beyond those covered in this article.

It is clear that the United Kingdom and Denmark, along with almost all industrialized countries, need more radical approaches to carbon reduction if they are to contribute significantly to international efforts to avoid dangerous climate change. PCA is a promising approach that urgently requires a much greater research effort to see if it could be an important part of future carbon and energy policy.

ACKNOWLEDGMENTS

The contributions of Frede Hvelplund and Niels I. Meyer have been supported economically by the research project Coherent Energy and Environmental System Analysis (CEESA), partly financed by The Danish Council for Strategic Research.

REFERENCES

[1] Anon. Domestic tradable quotas (carbon emissions) bill. 53/3. London: The Stationery Office; 2004.

[2] Anon. Are consumers ready for green credit cards? Energy Informer 2007; September:3.

[3] Anon. How cap and share works. Published at: www.capandshare.org/howitworks.html; 2008 (accessed April 2008).

[4] Barnes P. Carbon capping: a citizen's guide. Minneapolis: Tomales Bay Institute; 2007.

[5] Boardman B. Home truths: A low-carbon strategy to reduce U.K. housing emissions by 80% by 2050. Oxford: Environmental Change Institute, University of Oxford; 2007.

[6] Brand C. Personal travel and climate change: exploring climate change emissions from personal travel activity of individuals and households. Unpublished Ph.D. thesis. Oxford: Oxford University Centre for the Environment; 2006.

[7] Danish Economic-Environmental Council. Energy and Environment 2008. Copenhagen, Denmark; 2008.

[8] Danish Engineering Society. Energy for the Future: Energy Plan 2030. Copenhagen, Denmark; 2006.

[9] Danish Parliament. Agreement on Danish Energy Policy for 2008–2011. Copenhagen, Denmark, February 21; 2008.

[10] Darby S. The effectiveness of feedback on energy consumption: A review for DEFRA of the literature on metering, billing and direct displays. Oxford: Environmental Change Institute, University of Oxford; 2006.

[11] Defra. U.K. climate change sustainable development indicator: 2007 greenhouse gas emissions, provisional figures. Statistical release 88/08. London: Department for Environment, Food and Rural Affairs and National Statistics; 2008a.

[12] Defra. U.K. climate change sustainable development indicator: 2006 greenhouse gas emissions, final figures. Statistical release 25/08. London: Department for Environment, Food and Rural Affairs and National Statistics; 2008b.

[13] Defra. Synthesis report on the findings from Defra's pre-feasibility study into personal carbon trading. London: Department for Environment, Food and Rural Affairs; 2008c.

[14] Defra, BERR. The U.K. fuel poverty strategy: Fifth annual progress report. London: Department for Environment, Food and Rural Affairs and Department for Business, Enterprise and Regulatory Reform; 2007.

[15] Ekins P, Dresner S. Green taxes and charges: reducing their impact on low-income households. York, UK: Joseph Rowntree Foundation; 2004.

[16] Environmental Audit Committee. Personal carbon trading. London: The Stationery Office; 2008.

[17] EU Commission. Questions and Answers on the Commission's proposal to revise the EU Emissions Trading System. MEMO/08/35. Bruxelles; January 2008.

[18] Fawcett T. Investigating carbon rationing as a policy for reducing carbon dioxide emissions from U.K. household energy use. London: The Bartlett Faculty of the Built Environment, University College; 2005.

[19] Fawcett T, Bottrill C, Boardman B, Lye G. Trialling personal carbon allowances. London: U.K. Energy Research Centre; 2007.

[20] Fleming D. Energy and the common purpose. Descending the energy staircase with tradable energy quotas (TEQs). London: The Lean Economy Connection; 2005.

[21] Hansen J, Sato M, Kharecha P, Russel G, Lea DW, Siddall M. Climate change and trace gases. Phil. Trans. R. Soc. A 2007;365:1925–54.

[22] Hillman M, Fawcett T. How we can save the planet. London: Penguin; 2004.

[23] Hillman M, Fawcett T, Rajan CS. How we can save the planet: preventing global climate catastrophe. New York: Thomas Dunne Books/St. Martin's Griffin; 2008.

[24] IPCC. Climate change 2007: The physical science basis. Contribution of Working Group I to the fourth assessment report of the Intergovernmental Panel on Climate Change. Solomon S, Qin S, Manning M, Chen Z, Marquis M, Averyt KB, et al. [eds]. Cambridge, U.K., and New York: Cambridge University Press; 2007.

[25] Keay-Bright S, Fawcett T, editors. Taxing and trading: Debating options for carbon reduction. Meeting report. London: UK Energy Research Centre; 2005.

[26] Lane C, Harris B, Roberts S. An analysis of the technical feasibility and potential cost of a personal carbon trading scheme: A report to the Department for Environment, Food and Rural Affairs. Accenture with the Centre for Sustainable Energy (CSE). London: Defra; 2008.

[27] Lenton T, Held H, Kriegler E, Hall JW, Lucht W, Rahmsstorf S, et al. Tipping elements in the Earth's climate system. PNAS 2008;105(6):17986–93.

[28] Marland G, Boden T, Andres RJ. Global, regional and national CO_2 emissions. In: Trends: a compendium of data on global change. Carbon Dioxide Information Analysis Centre, Oak Ridge National Laboratory, U.S. Department of Energy; 2007.

[29] Merkel A. Speech in the Leaders' Dialogue on "The Economics of Climate Change." New York; September 25, 2007.

[30] Meyer A. Contraction and convergence: the global solution to climate change. Totnes, UK: Green Books; 2000.

[31] Miliband D. Environment, food and rural affairs: House of Commons Debate. London: Hansard; Column 990. 14 December 2006.

[32] RCEP. The environmental effects of civil aircraft in flight. London: Royal Commission on Environmental Pollution; 2002.

[33] Roberts S, Thumim J. A rough guide to individual carbon trading: the ideas, the issues and the next steps. Bristol and London: Centre for Sustainable Energy and DEFRA; 2006.

[34] RSA. The carbon card pilot. Published on the Web: www.rsacarbonlimited.org/article.aspa?PageId=917&NodeId=1 (accessed September 08). London: RSA; 2008.

[35] Starkey R, Anderson K. Domestic tradable quotas: A policy instrument for the reduction of greenhouse gas emissions. Norwich: Tyndall Centre for Climate Change Research; 2005.

[36] Starkey R. Personal communication about meetings held with IT companies. 2007.

[37] Tesco. Carbon labelling explained. Published on the Web. www.tesco.com/greenerliving/cutting_carbon_footprints/carbon_labelling.page?; 2008 (accessed October 2008).

Addressing Climate Change: Global vs. Local Scales of Jurisdiction?

Benjamin K. Sovacool

National University of Singapore

Marilyn A. Brown

Georgia Institute of Technology

Abstract

This chapter assesses the advantages and disadvantages of tackling climate change through local, bottom-up strategies as well as global, top-down approaches, arguing that each has distinct costs and benefits. The chapter also explores how local and global scales might be integrated into a single and effective policy framework, incorporating the advantages of decentralization and local action (efficiency through diversity and innovation, flexibility, and accountability) along with the advantages

Doi: 10.1016/B978-1-85617-655-2.00005-5.

of centralized and national action (consistency, efficiency from economies of scale, and equity).

5.1 Introduction

Policymakers and regulators who want to address greenhouse gas emissions and tackle climate change have proposed a multitude of mechanisms. Reports about policy tools such as carbon cap-and-trade systems, carbon taxes, quotas, and credits multiply. Some analysts recommend as alternatives per capita carbon allowances (see Meyer et al., Chapter 4, in this volume), technology development and transfer policies [3], alteration of energy policy (see Fraser and Hrab, Chapter 13, in this volume), carbon-neutral bonding for industrial facilities, complete government phase outs of greenhouse gases, and establishing property rights for the atmosphere.

Yet one important question underlying all these proposals is as simple as it is frequently unasked: At what scale should these policies be implemented? Should local cities, citizens, mayors, grassroots activists, town councils, or smaller organizations take charge in pushing climate policy? Or should it be international actors, national politicians, and other stewards of big government who do so?

Policy analysts have spent decades developing criteria for selecting between alternative policy pathways (Dunn, 2008). Three of the more important selection criteria include *effectiveness*—whether the policy results in the achievement of a valued outcome or objective; *efficiency*—the amount of effort required producing a given level of effectiveness; and *equity*—which involves the concepts of distributive justice and fairness. These criteria have been used extensively to evaluate alternative energy and climate policies [7, 8, 12, 15]. The focus of these evaluations, almost without exception, is on the choice of policy mechanism. What is missing from this literature is an appreciation of the importance of geographic scale.

Since other chapters of this volume focus primarily on the *type* of policy intervention needed to fight climate change, this chapter is focused exclusively on the *scale* at which climate policies should be implemented. This is because *what* we regulate is almost as important as *how* we regulate it. The chapter begins by noting that top-down, global, far-reaching (and sometimes homogeneous) approaches to combating climate change must be blended with bottom-up, local grassroots schemes. The chapter discusses the benefits of local action often emphasized by activists, nongovernmental organizations (NGOs), and environmental lawyers, and then contrasts their views with those of economists, political scientists, and other academics who argue in favor of global action. The chapter concludes by calling for an integration of local *and* global scales into a single policy framework to maximize the efficacy of climate mitigation and adaptation efforts.

▌ 5.2 The pros and cons of local and global scales

Democratic activists, environmental lawyers, and nongovernmental organizations often argue that the local scale is the best way to approach the climate crisis. A long line of legal theorists have argued that bottom-up, decentralized, local approaches to environmental policy maximize the protection of the environment ([1, 2, 6, 11, 24, 25, 26, 35]; Adler, 2006) and offer an optimal way to fight climate change [13, 14, 30].

The most basic benefit emphasized by these advocates concerns the principle of *efficiency through innovation and diversity.* Local action can provide opportunities for experimentation in designing policy, since the existence of many individuals acting at once promotes competition and innovation [28]. Localizing environmental decision making provides for inter-jurisdictional competition that can optimize environmental policy and create "laboratories of democracy" as they experiment in crafting better policies. In addition to fostering innovation, local action can enable a more rapid response to changing needs and circumstances. Just as smaller ecological subsystems change according to a faster dynamic than do the larger ecosystems of which they are a part [23], one can argue by analogy that local policy adaptations can occur more rapidly than national or global policy change. In the United States, for instance, federal energy legislation has tended to occur at sluggish 10- to 15-year intervals, with major bills in 1978, 1992, 2005, and 2007.

Proponents of the local scale also argue that it promotes greater *flexibility.* Global, top-down international regulatory systems are often unable to incorporate all the specific, detailed, temporal, and geographic information necessary to design optimal policies. This "knowledge problem" needs local and regional responses from people familiar with their own conditions. A corollary of this argument is that action at smaller scales promotes administrative efficiency since state and local agencies are more agile and adaptive than federal or national ones, thus being more able to formulate solutions well tailored to local needs and preferences. Advocates of localism have called national and international policy making an "affront to nature," because ecological systems are intrinsically variegated and diverse. Failure to take into account local environmental conditions, tastes, preferences, and economic conditions leads to a one-size-fits-all prescription that is more often "one-size-fits-nobody" [1].

Recognizing the need to expose the unique climate change vulnerabilities faced by individual states, the National Conference of State Legislatures has published data describing each state's situation and has noted that every state faces a different set of climate perturbations and forecast costs [21]. For example, the forecast 1 to 3 foot increase in sea level would have multiple costly impacts for states such as New Jersey, including $790 million to protect the residents of Long Beach Island and even greater investment to protect the rails, bridges, and tunnels connecting New Jersey with New York City. Managing the Great Lakes water decline (from accelerated evaporation as the result of rising temperatures) will cause significant costs for states such as Michigan, Illinois, and Ohio. If Great Lakes water levels decrease as expected, system connectivity along the Great Lakes/St. Lawrence route could decline by 25 percent, resulting in an annual economic loss of $4 billion in foreign trade in and out of Michigan. Disruptions to agricultural systems in Midwestern states such as Kansas and shrinking water supplies in Nevada and Colorado represent another distinct set of statewide vulnerabilities.

Even urban environments have highly variable circumstances leading to wide-ranging per capita carbon footprints. Almost all the metropolitan areas east of the Mississippi and south of New York State have very large carbon footprints, due in part to the dominance of coal-based electricity, sprawling land use, and a lack of public transit. In contrast, most of the metropolitan areas with the smallest per capita footprints are located in the western and northeastern parts of the country, where low-carbon electricity is most dominant, urban areas tend to be more compact, and rapid transit systems are most developed [5]. Thus, there is great variation in metro-area vulnerabilities to climate policies and carbon markets based on the heterogeneity of their per capita carbon emissions. Figure 5.1 illustrates this diversity.

Confirming this point, in their research interviews in Kansas, North Carolina, Ohio, and Pennsylvania, Kates and Wilbanks found a widespread preference for state and local regulatory oversight rather than federal governance or international jurisdiction [19]. People shared a belief that state and local regulators were more trustworthy, capable of understanding local problems and resources, and approachable. Their findings highlight that bringing climate policy decisions closer to local citizens improves *accountability* and enhances participation regarding those decisions. The argument is based in part on civil republicanism, or the idea that participation in local government is desirable for instilling civic virtue in American citizenry. Localized decision making allows for a closer fit between policies and preferences, giving individuals the

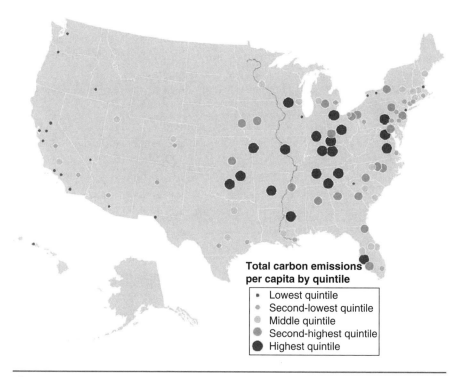

Total carbon emissions per capita by quintile

- Lowest quintile
- Second-lowest quintile
- Middle quintile
- Second-highest quintile
- Highest quintile

Figure 5.1 Per capita carbon footprints for the 100 largest metropolitan areas in the United States (reflecting transportation and residential energy use). In 2005, all but one of the 10 largest per capita emitters (Oklahoma City) were located west of the Mississippi River. The three metropolitan areas with the smallest footprints—Honolulu, Hawaii; Los Angeles, California; and Portland, Oregon—averaged just 1.4 metric tons of carbon per person, whereas the three metropolitan areas with the largest footprints—Toledo, Ohio; Indianapolis, Indiana; and Lexington, Kentucky—averaged 3.4 metric tons per person— 240 percent greater.
Source: Brown, Southworth, and Sarzynski, 2008 [5].

option to sort themselves among jurisdictions based on which offers the most appealing mix of policies [28].

Another argument for local/state policy based on accountability and administrative efficiency rests on the conditions most likely to promote private cooperation and resolution of externality problems. Ronald Coase argues that private responses can be sufficient in situations where property rights are clearly defined and costless to enforce, where utility is linear in wealth, and where there is costless bargaining among participants [9]. Private cooperation appears to be most effective at small scales of interactions. Examples are neighborhood associations that sometimes agree on mutually restrictive covenants and individual developers who occasionally reach contractual agreements on light easement to deal with shadows cast by adjacent buildings [31].

In sum, those in favor of relying on the local/state scale of climate action believe that it best promotes efficiency through innovation and diversity, flexibility, and accountability.

Advocates of international action—often economists and political scientists—argue almost the exact opposite: They respond that global standards offer *consistency*. Starting with the work of Richard Stewart [29], these theorists have argued that national or international regulation is needed to fight environmental problems. Having one top-down climate policy can engender a more efficient regulatory regime than a multiplicity of state and local standards, which tend to heighten barriers between individual states and lead to inefficiencies. A global, single policy would create consistent and predictable statutes that manufacturers and industry can anticipate and deal with. A simple, clear, and precise climate change policy would minimize many of the transaction costs that arise with localized and divergent state actions.

Analysis conducted for the 2007 Intergovernmental Panel on Climate Change (IPCC) showed that mitigation costs would rise significantly without broad participation in a post-Kyoto agreement. To be most efficient, the price of carbon must be universal and harmonized across all countries and regions. For example, Weyant et al. estimated that 2010 carbon taxes (or the marginal cost of carbon emissions) in the United States would be half as costly with global carbon trading compared to merely trading among industrialized countries [33]. Restrictions on global carbon trading, in other words, increase both the carbon permit price and the relative economic costs to national GDP [22]. This point is illustrated in Figure 5.2.

Consider the current case of U.S. climate policy. Policy variations and fragmentations exist across regions, states, and localities. International corporations today are operating in a patchwork of markets, some with strong carbon constraints and others without any carbon regulations. This mosaic of divergent policies is particularly challenging to entrepreneurs who are striving to develop national markets.

A second, related advantage to federal, national, and global action concerns *efficiency from economies of scale*. Absent international action, local actors can duplicate each other or engage in time-consuming and complex negotiations to divide labor and resources. The drive toward the local scale generally weakens technical capacities such as data collection and research and development requiring large-scale scientific instruments. Because RD&D creates spillover benefits that cannot be fully captured by firms, the amount of privately funded RD&D will remain lower than optimal.

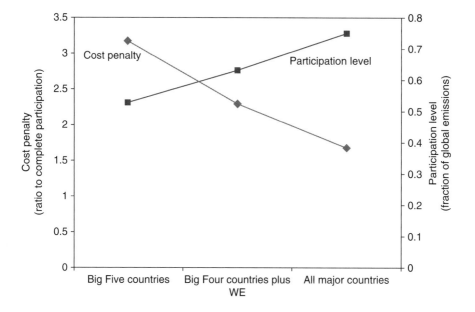

Figure 5.2   Cost penalty of limited participation in a global climate treaty.*
Big Five countries are the United States, China, Russia, India, and Germany.
Big Four countries plus WE are the Big Five countries and the Western European members of the EU.
All major countries includes the EU plus the Big Four countries and Brazil, Canada, Japan, Mexico, and South Africa.
Participation level is the fraction of 2005 global CO_2 emissions from the three different groups.
Cost penalty is the extra cost of CO_2 mitigation from partial participation relative to global participation. If only the Big Five countries are included, for instance, the cost penalty for achieving a given level of mitigation would be 3.2 times larger than with global participation.
Source: Based on data published in Nordhaus [22].

Yet there is generally high agreement among experts that innovation is needed to commercialize new technologies in the long term to stabilize greenhouse gas concentrations ([16], p. 622). Thus, government entities must subsidize the cost of creating climate technology solutions and must expedite bringing these innovations to market. There is a particularly strong role for federal or global RD&D funding because of the magnitude of the challenge and the need to avoid duplication.

Federal and global policies are able to address several *equity* issues that can prove difficult for localities and states. Local environmental decision making can create welfare losses and externalities—costs (or benefits) not fully internalized or priced by the existing market system. Empirical evidence, along with economic theory, suggests that states and localities as well as businesses and industries will underprovide public goods and positive externalities because they are not paid for them, but they will overproduce negative externalities because their cost is distributed throughout all of society. The problem with climate change is particularly acute: The rewards from restraints on greenhouse gases will come in the politically distant future while the costs will be incurred in the present. The wide distribution of expected but distant benefits in response to collective action provides an incentive for every country to encourage all to act but then shirk responsibility itself [10].

As a result, climate policy at the local scale risks creating spillovers and carbon leakage. Carbon is leaked when stricter emissions standards in one place encourage higher emissions elsewhere as the production of dirty goods moves to "safe havens" that have weaker environmental regulations [17]. Polluters can simply move to other places that do not have restrictive policies. National action, on the other hand, prevents such spillover effects. As one practical example, Weiner found that the Regional Greenhouse Gas Initiative (RGGI) has experienced "leakage" rates as high as 60 to 90 percent [32]. Power plants in adjacent states have actually increased their output to sell into the higher-priced RGGI electricity markets.

Thus, those in favor of relying on the national/global scale of climate action believe that it best promotes uniformity and consistency, along with economies of scale, and avoids spillover effects.

5.3 A policy framework for blending local and global scales

To promote efficiency through innovation and diversity, flexibility, and accountability alongside consistency, economies of scale, and equity requires that climate policy be designed to include the benefits of local/bottom-up and global/top-down scales of action without their costs and tradeoffs. To do so, we propose that policymakers implementing climate policy should:

■ Create a mandatory national climate target requiring greenhouse gas emissions reductions 80 percent below 2005 levels by 2050 and apply it to all upstream emitters of greenhouse gases (to ensure consistency)

■ Establish a national carbon credit trading scheme that allows emitters to use least-cost mitigation providers (to ensure flexibility)

■ Make greenhouse gas reduction targets meaningful (to ensure economies of scale)

■ Set gradual benchmarks (to minimize the risks of leakage)

■ Create strict penalties for noncompliance (to ensure accountability)

■ Enable and encourage individual states, counties, and cities to exceed the national standard (to ensure diversity).

This policy framework is elaborated in the following sections, in the context of the United States. However, to achieve broad global participation, all the world's rich countries should implement a similar framework quickly and simultaneously, while developing countries could join after one to three decades, as recommended by Nordhaus [22]. This would bring participation rates to 45 percent in 2020, 65 percent by mid-century, and 90 percent of countries by 2100.

5.3.1 CONSISTENCY

The developed world must adopt country-specific ceilings on GHG emissions to ensure that highly disruptive climate change does not occur [18]. To create predictability for investors and businesses, any target must apply to *all* upstream producers and emitters of greenhouse gases. We propose an upstream rather than downstream proposal merely because upstream mandates are much easier to monitor and offer policymakers wider coverage. An upstream mandate would apply to all fuel producers and would cover electric utilities, industries, and fuel suppliers—roughly 90 percent of greenhouse gas emissions—whereas a downstream mandate would require regulation of hundreds of millions of individual consumers, commercial enterprises, and businesses, a virtually impossible task [27]. The resulting uniformly stable target provides an equal playing field for all carbon-intensive industries and avoids creating perceived inconsistencies in regulation. An example would be requiring all refineries, natural gas processing facilities, power generators, iron and steel manufacturers, glass producers, cement makers, and other upstream industries—without exception—to meet a binding restriction on greenhouse gas emissions.

5.3.2 FLEXIBILITY

National and international climate regulators should also consider establishing a carbon credit-trading scheme that does not set geographic restrictions or limitations on the flow of credits, avoiding the creation of artificial winners and losers. Tradable credits are needed to function as a simple accounting system to prevent cheating and ensure compliance. They also provide considerable flexibility to utilities, retailers, and other upstream users. Upstream industries can rely on credits to reduce risks, since credits can be purchased to make up any shortfall or take care of any excess. Upstream industries located in areas that are carbon intensive are not "punished" because they have the ability to purchase credits from less carbon-intensive areas. Carbon credit trading schemes are becoming more popular and palatable among businesses. Four fifths of U.S. utility executives polled in 2007 expected mandatory emissions caps within a decade, and 10 companies—including Alcoa, Caterpillar, and DuPont—recently called on the U.S. Congress to set up a cap-and-trade system for greenhouse gases. Tradable credits would also ensure that a binding carbon cap-and-trade system would not need any formal expiration, since a "self-sunset" would take place once the market value credits have stabilized at or near zero, signifying that upstream facilities are all in compliance.

5.3.3 ECONOMIES OF SCALE

To ensure meaningful reductions in greenhouse gas emissions, any binding target must be large enough to achieve economies of scale in terms of technology production and supply chains as well as data collection, monitoring, and enforcement of the carbon policy. For instance, a national or international cap-and-trade system must ensure that the ultimate reductions of greenhouse gas emissions are sufficient enough to reduce the risks of climate change. Thus, we recommend that upstream facilities reduce greenhouse gas emissions at least 20 percent below 2005 levels by 2020, building to 80 percent below by 2050. This pace of mitigation exceeds the declaration made by leaders of the G8 in July 2008 to "move toward a carbon-free society" by seeking to halve worldwide emissions of heat-trapping gases by 2050. The G8's long-term climate goal is seen by many as inadequate to avoid potentially dangerous changes in the Earth's climate.

5.3.4 EQUITY

A well-designed carbon policy must also be gradual and set specific benchmarks so that compliance can be tracked and leakage minimized. Upstream sources, for instance, could be required to reduce emissions 20 percent below 2005 levels by 2020, 50 percent below by 2030, 65 percent below by 2040, and 80 percent below by 2050. By increasing the amount of greenhouse gas reductions slowly over time, the policy ensures that the carbon credit market will result in competition, efficiency, and innovation that will deliver credits and reductions at the lowest possible cost. A gradual phase-in provides time to set up standards for credit certification, monitoring, and compliance. It creates relative certainty and stability in the market for cleaner energy technologies by enabling long-term contracts and financing, in turn lowering costs. A gradual standard would also reduce credit prices and credit volatility, since emitters would realize they could implement reductions slowly over time. And it gives emitters an incentive to improve their competitive position in the market so that they have an interest in improving their efficiency of operation.

Equity concerns also underpin proposals to design an initial allocation of carbon allowances that recognizes (and roughly addresses) the disparate costs imposed by a carbon cap-and-trade program. For example, the National Commission on Energy Policy (NCEP) recommends that half of the allowance pool should be distributed in a manner that fairly addresses the cost concerns of low-income households as well as affected industries such as suppliers of primary fuels, the electric power sector, and energy-intensive manufacturers [20]. A recent study estimated, for instance, that the energy-efficiency gains required to offset the cost of climate policies in energy-intensive manufacturing would range from improvements of 14 to 34 percent by 2020 [34]. Iron and steel production, as well as paper and paperboard manufacturing, would require particularly large energy efficiency improvements. These disparities in energy use (and therefore associated greenhouse gas emissions) imply that equitable allocation mechanisms must be designed to mitigate excess costs in the short to medium term so that such industries will have an opportunity to transition to a low-carbon future while maintaining their competitiveness.

5.3.5 ACCOUNTABILITY

To minimize gaming, an effective carbon policy must have penalties for noncompliance equal to several times what it could cost to purchase carbon credits. A noncompliance penalty is needed to not only achieve meaningful reductions but to reduce the cost of achieving them. This is because, in part, emitters will base their investments in infrastructure on the certainty that the binding carbon mandate will be enforced. Automatic penalties imposed against emitters that fail to comply with the policy will give investors confidence that there will be potential buyers for carbon credits. Failure to create strict noncompliance penalties runs the risk of creating a "Catch-22" situation where emitters make an insincere effort to secure carbon credits from suppliers, and then—when they can find none or credits are too expensive—claim that none are available. Policymakers could then view the emitter's noncompliance as being in good faith, since there were no available credits for purchase, rather than seeing the situation as proof that the emitter never intended to comply. Thus, a high penalty level makes the policy self-enforcing by avoiding the need to resort to costly administrative and enforcement measures. The federal sulfur dioxide allowance trading program in the United States, under which an automatic \$2000/ton penalty (indexed to inflation) is imposed for each excess ton of SO_2 produced, for example, has had compliance rates above 99 percent since its inception.

5.3.6 Efficiency Through Innovation and Diversity

Any binding national or international carbon policy should set a minimum carbon cap-and-trade that prohibits states (or in this case, emitters that operate within and between states) from enacting weaker statutes but allows stronger ones. States, cities, municipalities, businesses, and even individuals should be free to exceed national or global standards as much as they wish. Setting a "floor" rather than a "ceiling" ensures that more aggressive state statutes are not precluded or restricted under a national standard. This type of compliance with local programs is often called *dual compliance* or *simultaneous compliance*. The national standard would only guarantee the promotion of a minimum level of greenhouse gas reductions. Such language should be clear and explicit in any national legislation so as to provide the maximum amount of clarity and predictability to emitters and investors and to avoid leaving the question open to political attacks during congressional or parliamentary deliberations. Examples of national environmental regulation in the United States that set a floor that the states could exceed include policies toward ambient air quality standards, low-emissions vehicles, hazardous waste, water quality, land reclamation, energy efficiency (e.g., appliance and building codes), acid rain, mercury emissions, and wetlands development [28].

5.4 Conclusion

This seeming tension between the costs and benefits of local and global, decentralized and centralized action suggests that whatever type of policy framework is proposed to address climate change, of equal importance is the scale at which we implement that framework.

Actions at local and global scales bring different sets of costs and benefits. Local action fosters diversity, which encourages innovation and experimentation. It ensures that policy mechanisms are flexible enough to adapt to local circumstances and needs, creating "ecologies of scale" that can maximize social welfare and minimize cost. Localism also tends to be more representative, creating variability in policy that better matches local interests and preferences. Such efforts are often the antidote to federal or globally imposed rigidity and "command-and-control"–style policy making.

Global action has its own unique set of advantages. It is the best way to provide consistency and minimize transaction costs among actors. Centralization creates better economies of scale in technology delivery, data collection, and research and development. Global action is the only way to ensure that all states bear the burdens of addressing climate change and to minimize free-riding and significant emissions leakage. See Table 5.1 for a summary of these costs and benefits.

Table 5.1 Costs and benefits from local/state and federal/global climate policy

Criteria	Local/Regional	Federal/Global
Favors Local/Regional Policy		
Efficiency through innovation and diversity	Encourages innovation and experimentation in designing policy and enables more rapid response to changing needs	Stifles innovation and experimentation; is prone to diseconomies of scale; changes slowly
Flexibility	More responsive and able to adapt to local conditions; promotes administrative efficiency	More uniform and rigid; tends to fail to account for local conditions
Accountability	Allows for closer fit between policies and preferences and affords option to sort between jurisdictions	Promotes "rent-seeking" behavior, which wastes resources trying to garner local advantages

(*Continued*)

Table 5.1 *(Continued)*

	Favors Federal/Global Policy	
Consistency	Building national markets for technology solutions is difficult when policies vary; local controls over major carbon emitters are often limited	Standardization minimizes transaction costs and policy uncertainties; captures long-distance influences on major emitters
Efficiency from economies of scale	Inefficient due to redundancies of R&D efforts and data collection systems	Better matched to promote economies of scale and avoid redundancies
Equity	Vulnerable to free ridership and carbon emissions leakage	Minimizes free ridership and carbon emissions leakage

Which brings us to a most important lesson: Any well-designed climate policy must incorporate the benefits of decentralization and local action (diversity, flexibility, and accountability) along with the benefits of centralized and national action (consistency, economies of scale, and the minimization of spillovers and leakage). Some scholars of governance have called this a *polycentric* approach to policy implementation, as it blends different actions at multiple scales together [4]. When applied to climate policy, this lesson suggests that the most effective approach would set a mandatory and binding cap on upstream greenhouse gas emitters, establish a national carbon credit trading scheme, commit to meaningful reductions in emissions, set gradual benchmarks, create strict penalties for noncompliance, and enable individual states and local actors to exceed national requirements. The global must be made local, and the local global.

References

[1] Adler JH. Jurisdictional Mismatch in Environmental Federalism. New York University Environmental Law Journal 2005;14:130–5.
[2] Adler JH. When Is Two a Crowd? The Impact of Federal Action on State Environmental Regulation. The Harvard Environmental Law Review 2007;31:67–114.

[3] Aldy JE, Stavins RN. The Role of Technology Policies in an International Climate Agreement. Cambridge, MA: The Harvard Project on International Climate Agreements; September 2008.

[4] Andersson KP, Ostrom E. Analyzing Decentralized Resource Regimes from a Polycentric Perspective. Policy Sci 2008;41:71–93.

[5] Brown MA, Southworth F, Sarzynski A. Shrinking the Carbon Footprint of Metropolitan America. Washington, DC: Brookings Institute Metropolitan Policy Program; 2008.

[6] Butler HH, Macey JR. Externalities and the Matching Principle: The Case for Reallocating Environmental Regulatory Authority. Yale Journal of Regulation 1996;14:23–66.

[7] Congressional Budget Office (CBO). Policy Options for Reducing CO_2 Emissions. Washington, DC: Congressional Budget Office; 2008.

[8] Chen C, Wiser R, Mills A, Bolinger M. Weighing the Costs and Benefits of State Renewable Portfolio Standards in the United States: A Comparative Analysis of State-Level Policy Impact Projections. Renewable and Sustainable Energy Reviews 2008.

[9] Coase R. The Problem of Social Cost. Journal of Law and Economics 1960;3(1):1–44.

[10] Cooper RN. Toward a Real Global Warming Treaty. Foreign Affairs 1998;77(2):66–79.

[11] Etsy D. Revitalizing Environmental Federalism. Mich Law Rev 1996;95:570–87.

[12] Geller H. Energy Revolution. Washington, DC: Island Press; 2003.

[13] Giuliano G. The Changing Landscape of Transportation Decision Making. Los Angeles: University of Southern California; 2007.

[14] Giuliano G. States, Regions, Locals, and the Feds: The Next Authorization. Los Angeles: University of Southern California; 2008.

[15] Greene DL, Patterson PD, Singh M, Li J. Feebates, Rebates and Gas-Guzzler Taxes: A Study of Incentives for Increased Fuel Economy. Energy Policy 2005;33:757–75.

[16] Intergovernmental Panel on Climate Change. Climate Change 2007: Mitigation of Climate Change. Cambridge, UK: Cambridge University Press; 2007.

[17] Kallbekken S. Why the CDM Will Reduce Carbon Leakage. Climate Policy 2007;7:187–211.

[18] Karp LS, Zhao J. A Proposal for the Design of the Successor to the Kyoto Protocol. Cambridge, MA: The Harvard Project on International Climate Agreements; September 2008.

[19] Kates RW, Wilbanks TJ. Making the Global Local: Responding to Climate Change Concerns from the Ground Up. Environment 2003;45(3):12–23.

[20] National Commission on Energy Policy. Energy Policy Recommendations to the President and 110th Congress. Washington, DC: National Commission on Energy Policy; 2007.

[21] National Conference of State Legislatures. Economic and Environmental Costs of Climate Change: An Overview. www.ncsl.org/print/environ/ClimatechangeOver.pdf; 2008.

[22] Nordhaus WD. Major Issues in the Economic Modeling of Global Warming. Presentation at the Workshop on Assessing Economic Impacts of Greenhouse Gas Mitigation. Washington, DC: The National Academies; October 2–3, 2008.

[23] Norton BG, Ulanowicz RE. Scale and Biodiversity Policy: A Hierarchical Approach. Ambio 1992;21:244–9.

[24] Revesz RL. Rehabilitating Interstate Competition: Rethinking the Race-to-the-Bottom Rationale for Federal Environmental Regulation. New York University Law Review 1992;67 (December): 1210–54.

[25] Revesz RL. Federalism and Interstate Environmental Externalities. University of Pennsylvania Law Review 1996;144:2341–416.

[26] Revesz RL. Federalism and Environmental Regulation: A Public Choice Analysis. Harv Law Rev 2001;115 (December): 553–80.

[27] Ross M, Smith A. Upstream Versus Downstream Implementations of Carbon Trading Systems. New York: Charles River Associates; 2002.

[28] Sovacool BK. The Best of Both Worlds: Environmental Federalism and the Need for Federal Action on Renewable Energy and Climate Change. Stanford Environmental Law Journal 2008;27:397–471.

[29] Stewart RB. Pyramids of Sacrifice? Problems of Federalism in Mandating State Implementation of National Environmental Policy. Yale Law Journal 1977;86:1196–272.

[30] Vandenbergh MP, Steinemann AC. The Carbon-Neutral Individual. New York University Law Review 2007;82:1673–745.

[31] Weimer D, Vining AR. Policy Analysis: Concepts and Practice. 4th ed. New York: Prentice Hall; Efficiency Losses of Negative and Positive Externalities 2005; 91–7.

[32] Weiner JB. 2007; Think Globally, Act Globally: The Limits of Local Climate Policies. University of Pennsylvania Law Review 1965.

[33] Weyant JP, de la Chesnaye FE, Blanford GJ. Overview of EMF-21: Multigas Mitigation and Climate Policy. The Energy Journal 2006;27:1–32.

[34] Yudken JS, Bassi AM. Climate Policy and Energy-Intensive Manufacturing: Impacts and Options. Washington, DC: National Commission on Energy Policy; 2009.

[35] Zerbe RO. Optimal Environmental Jurisdictions. Ecology Law Quarterly 1974;4:193–245.

Part 2

The Solutions

Eliminating CO$_2$ Emissions from Coal-Fired Power Plants

Chapter 6

Klaus S. Lackner and A.-H. Alissa Park

Earth and Environmental Engineering and Lenfest Center for Sustainable Energy, Columbia University

Bruce G. Miller

EMS Energy Institute, Penn State University

Abstract

Stabilizing the atmospheric concentration of carbon dioxide is possible only if carbon dioxide emissions from all sources are stopped. Fossil fuels can still provide energy, but the carbon dioxide resulting from the combustion must be captured and stored safely and permanently. This chapter reviews carbon capture and storage (CCS), lays out a number of novel concepts for coal-fired power plants that do not release carbon dioxide into the atmosphere, and discusses these concepts in the context of retrofits of existing power plants, in the development of advanced power plant designs, and in conjunction with adequate carbon dioxide storage options.

6.1 Introduction

The recent and very sudden rise in the cost of energy has once again demonstrated the importance of energy to the world economy. While more than 2 billion people have been working themselves out of poverty, world energy demand has been growing rapidly, with liquid fuel and electricity demands leading the way. Fossil fuel resources are not running out [97], but their use produces carbon dioxide (CO_2), which, as the most important greenhouse gas, drives climate change [17, 51, 52]. For example, the way electricity is made from coal today may cease to be an option.

A serious curtailment of fossil energy use due to climate change concerns would have dramatic consequences. Because 85 percent of all commercial energy is derived from fossil carbon resources, eliminating their use would precipitate an energy crisis of unprecedented proportions. The recent high prices of oil, coal, and gas were caused by much less severe supply constraints than those that could be precipitated by climate change restrictions.

Carbon capture and storage (also known as CCS) offers a way out of this dilemma. It is not the use of fossil fuels *per se* that needs to be curtailed, but rather the emission of the CO_2 that is produced in their use. Capturing CO_2 and storing it safely and permanently eliminate the most critical environmental impact associated with the use of fossil fuels.

CCS needs to be applied to a large fraction of all fossil fuel use, and it needs to be developed soon. Roughly half the CO_2 emitted at a given moment will leave the atmosphere in a matter of years, mostly by being absorbed into the ocean. Approximately one quarter of the CO_2 will stay behind for many thousands of years. On a century scale, one may consider that roughly 40 percent of CO_2 remains in the atmosphere. Thus, one can think of the atmospheric sink as a finite resource whose size can be measured in gigatons of CO_2 [14]. Roughly, it will take 5 Gt of carbon (18 Gt of CO_2) for every ppm in the atmosphere.[1] Hence, the world is roughly 300 Gt of carbon away from the frequently cited stabilization goal of 450 ppm of CO_2 in the atmosphere [102]. With current emissions of around 8 Gt of carbon per year, it will take about 40 years to finish off the remaining "carbon pie" [14]. Every Gigawatt of coal

---

[1]Without ocean uptake, it would be 2 Gt of carbon for every ppm in the atmosphere.

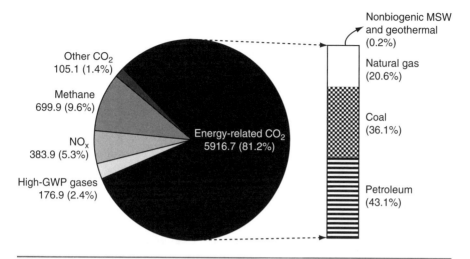

Figure 6.1 U.S. greenhouse gas emissions by gas (left) and U.S. CO_2 emissions from the energy sector (right), 2007 (million metric tons CO_2 equivalent). Source: Data obtained from EIA [26].

power locks in an emission of half a Gigaton of CO_2 over its 70-year life span. Figure 6.1 shows the estimated U.S. CO_2 emissions in 2007 as well as the emission of other greenhouse gases.

Over the past few years technologies have been developed to capture CO_2 and store it safely and permanently [51]. Initially, CO_2 capture was considered the difficult part of the problem.[2] Today public discourse has made it clear that developing the necessary storage capacity poses the more serious challenge. We will argue in this chapter that capture from all sources is possible and that together the various storage options have the capacity to store all the CO_2 that might result from a continued reliance on fossil fuels.

The electricity sector is very carbon intensive. Half of all electricity produced in the United States comes from coal-fired power plants [29, 30]. In China about 80 percent of electricity is generated from coal (Chapter 17 on China in this volume; and Deutsh and Moniz [25]). Coal is a relatively inexpensive fuel and therefore it is the fuel of choice in the electricity sector, which, unlike the transportation sector, is not tied to a particular energy source. However, coal is the most carbon-intensive fuel, and therefore it contributes disproportionately to CO_2 emissions. In a carbon-constrained world this provides a driver toward eliminating coal. At the same time, coal plants offer a technologically easy

---

[2]For a dissenting view, see Lackner et al., 1995.

target for the implementation of CCS, since they represent very large and very concentrated sources of emissions.

Thus, once CCS starts in earnest, it likely will start at coal-fired power plants, even though the first examples of CCS were in smaller and even easier targets.[3] In the United States, the Great Plains Synfuel Plant in North Dakota is collecting CO_2 from a coal gasification plant [83]. These small targets of opportunity provide more concentrated streams of CO_2 at higher pressures than would be obtained from a coal plant.

There are other opportunities for CO_2 capture—for example, in steel plants, cement plants, refineries, and ammonia plants. Nevertheless, these opportunities represent niche markets that are dwarfed by the emissions at power plants, which are the focus of this book. Electric power generation is the first large target for CCS. However, in the long run, one will have to address all emissions and thus consider options for either eliminating the emissions from cars and airplanes or developing means of recovering this CO_2 after it has been emitted to the atmosphere.

CCS can be viewed as an extension of previous efforts to develop cleaner, more efficient power plants, a process that in the United States has been referred to as *clean coal development*. However, the need to remove CO_2 from the plant's emissions changes the problem significantly. On the one side it is more challenging, but the need to capture CO_2 also offers new opportunities for managing pollutants such as SO_x, NO_x, mercury, and fine particulates. With CO_2 completely removed from the effluent, one can control emissions from the power plant by completely eliminating gaseous effluents to the atmosphere and instead capture the pollutants in solid and liquid streams that are kept completely out of the atmosphere.

We begin in Section 6.2 with a brief conceptual overview of CCS. Sections 6.3 and 6.4 outline options for capturing the emissions from power plants. Section 6.3 focuses on retrofitting existing power plants; Section 6.4 looks at various options for new greenfield installations. Section 6.5 on zero-emission coal plants summarizes a potential paradigm shift that can arise in the development of new power plants designed to collect their own CO_2. With complete CO_2 capture, it becomes possible to completely eliminate the smoke stack of a power plant and have no emissions to the atmosphere. Section 6.6 introduces the various options for long-term storage of CO_2.

---

[3]The Statoil facility in the North Sea that is putting 1 million tons of CO_2 underground every year operates on CO_2 that needed to be separated from natural gas extracted at the same site. This separation would be necessary even if there were no desire to store the resulting CO_2.

6.2 The basics of carbon capture and storage

The principle of CCS is simple. If production of CO_2 cannot be avoided, CO_2 must be kept out of the atmosphere. It must be captured and stored safely and permanently. CCS is a novel technology that could eliminate all CO_2 emissions. This is important because over the next 50 to 100 years, it will become necessary to collect all CO_2 from all fossil fuel use. Slowing down the rise in CO_2 concentration substantially will require a worldwide reduction in emissions by a factor of three. With a future world population of 10 billion people, a reduction by a factor of three from the 2000 level would result in a per capita emission allowance of 4 percent of the present per capita emission in the United States. This makes it necessary to collect virtually all emissions from all sources. If the reliance on fossil fuels persists throughout the century, CCS will have to cope with CO_2 generation much in excess of current emissions. Only if fossil fuels are replaced with other sources of energy, such as solar or nuclear energy, can the large-scale implementation of CCS be avoided.

Thus a likely outcome is CCS at a rate which is measured in tens to hundreds of gigatons of CO_2 per year. Century-scale storage could easily exceed several thousand gigatons, and eventually CCS could approach a scale where the accumulated CO_2 storage will be comparable to the size of the original fossil fuel resources.

Carbon capture and storage is not limited to coal plants. Since CO_2 emissions are not harmful locally, it is not necessary that the actual CO_2 of a particular process or plant is captured; the capture of any equivalent amount of CO_2 that is on its way to the atmosphere or already has reached the atmosphere would be equivalent. Consider, for example, the consumption of biomass as fuel. This will lead to the production of CO_2, but this CO_2 is not a greenhouse problem, since the carbon has just recently been collected by growing plants from the atmosphere. There is no need to capture and store CO_2 emission from biomass-fueled power plants. On the other hand, in an economy in which CCS has been implemented, it may prove far easier to collect the CO_2 from a biomass-burning power plant than from an airplane flying with conventional jet fuel. It therefore may well make economic sense to capture CO_2 from a biomass plant that does not really need to be captured and let the CO_2 released by the jet engine enter the atmosphere. The entire transaction can be made carbon neutral. Biomass does not offer the only avenue toward recovering CO_2 from the air. Rather than using biomass to capture CO_2 from the atmosphere, it is also possible to use engineered sorbents to collect CO_2 directly from the air [28, 59, 64, 110].

In most processes that consume fossil fuels, power plants included, the CO_2 tends to be produced in a dilute stream. Combustion processes typically utilize air, and hence the CO_2 content of the flue gas cannot exceed the oxygen content of the air. Because the combustion of hydrocarbons, including coal, oxidizes not only carbon but also hydrogen (H_2), and because the combustion is usually performed with excess oxygen, actual CO_2 concentration in flue gas is even lower. For coal-fired power plants, the levels range between 10 and 15 percent; for natural gas turbines, they are in the range of 3 to 5 percent. Higher concentrations can be found in the off-gases of cement plants and steel mills. The highest concentrations, nearly pure CO_2, can be found in ammonia production and certain refinery operations, including the removal of CO_2 admixtures from natural gas that often contains some CO_2.

It is conventional to categorize CO_2 capture as post-combustion, pre-combustion, or integrated into the design. Just as it is possible to scrub the exhaust of a power plant for SO_2, it is also possible to scrub the exhaust for CO_2. This approach is known as *post-combustion capture*. The second option, known as *pre-combustion capture,* is to remove the carbon upstream before the fuel enters the combustion process. For example, it is possible to produce H_2 from coal in an integrated gasifier combined cycle plant. The H_2 is combusted in a gas turbine, while the H_2 production process results in CO_2 prior to the combustion step. Finally, there is a large and diverse class of options that can generically be summarized as integration of power generation and CO_2 capture. A good example is the use of oxy-fuel combustion, but plants that produce electricity in solid oxide fuel cells oxidizing CO to CO_2 also belong in this class. Such classifications may become blurred over time because advanced power plants are likely to include aspects of all three approaches.

Once the CO_2 has been collected, it needs to be stored safely and permanently. One of the earliest suggestions has been to dispose of the CO_2 in dilute form in the ocean. Marchetti [75], for example, suggested dissolving the CO_2 into the dense saline outflow from the Mediterranean Sea, which flows from the Strait of Gibraltar into the abyssal plains of the deep North Atlantic. Though popular in the mid-1990s, ocean disposal has generally fallen into disfavor, mainly because it leaves many environmental concerns unanswered [51]. Today the most popular suggestion is to inject the CO_2 into deep underground formations, where it can remain virtually indefinitely. Concerns over the long-term stability of these formations have led us to consider mineral sequestration as yet another option [62]. In this case, the CO_2 is reacted with an alkaline mineral to form chemically stable carbonates. At present the cost of mineral sequestration is too high by about a factor of five for it to play a major role [51].

Apart from safety and economics, for any storage option one will need to consider the size of the storage capacity, the permanence of the storage, and the environmental consequences. If maintaining access to fossil fuels is the goal of CCS, the size of the fossil fuel resource sets the storage capacity. Unless fossil fuels are phased out immediately, the storage capacity will need to be at minimum 100 years of current consumption, or approximately 800 Gt of carbon.

Total storage capacity sets the minimum lifetime requirement for storage. Seepage from storage must be small enough so that it does not become a major source of emission. Annual leakage from storage in excess of 100 million tons of carbon appears problematic. Emissions in excess of 1 Gt per year are certainly not acceptable. Hence, even at a relatively small total storage target of 800 Gt of carbon, the minimum storage lifetime is between 800 and 8000 years. Only a scenario of rapidly phasing out fossil fuels would justify shorter storage lifetimes. However, one might question the rationale of creating a large new infrastructure centered on CCS if its immediate phaseout is already ordained. However, it is possible to start with a small storage capacity and short storage times, as long as it is understood that additional advances in capacity and lifetime are required.

The outlook for CCS also will shape the emphasis on retrofitting existing power plants vs. developing new power plants. CCS as a long-term option heightens the focus on new plant design. In a transition scenario, retrofitting existing power plants would be of greater importance. Particularly for coal-fired power plants, a retrofitting strategy is problematic because power plants were optimized under the assumption that fuel is inexpensive. However, capture and storage costs are effectively fuel costs, even though they arise separately. Capture and storage costs are unlikely to be much less than \$30/ton of CO_2. In effect, sequestration would add \$80 to the cost of a ton of coal.[4] The additional cost of CCS would drive toward a design optimum at a much higher efficiency, while capture and storage of CO_2 consumes energy and hence downgrades the already low efficiency of an old power plant. Thus, retrofitting old coal-fired power plants will unavoidably result in suboptimal power plant designs.

By contrast, novel power plant designs that integrate CCS from the beginning offer high efficiency. The difference can easily be a factor of two in the amount of CO_2 that will have to be stored per unit of electricity generated. As a result, the incremental cost of CCS per unit of electricity is only half that of an existing power plant. In addition, it is possible to integrate CCS with dramatic reductions in conventional pollutants, which further strengthens the case for developing advanced power plant designs. Cost estimates for CCS vary, depending on the specific technology, on the circumstances of a

---

[4]We assume 30 GJ per metric ton of coal, 500 kJ of heat of combustion per mole of CO_2 produced, and a cost of \$30/ton of CO_2.

particular plant, and on assumptions made about technology development. Typical numbers range from \$20 to \$100 per ton of CO_2, the CO_2 output varies from 3.5 to 8 kg/MJ, and conversion efficiencies are 30 to 70 percent. The high end requires innovative new power plant designs.[5]

6.3 Carbon capture technologies: Retrofit options for coal-fired power plants

Technologies for capturing CO_2 from gas streams have been used for many years to produce a pure stream of CO_2 from natural gas or industrial processing for use in the food processing and chemical industries. Methods currently used for CO_2 separation include:

■ Physical and chemical solvents, particularly monoethanolamine (MEA)
■ Various types of membranes
■ Adsorption onto solids
■ Cryogenic separation

These methods can be used on a range of industrial processes; however, their use for removing CO_2 from high-volume, low-CO_2 concentration flue gases, such as those produced by coal-fired power plants, is more problematic. The high capital costs for installing post-combustion separation systems to process the large volume of flue gas is a major impediment to post-combustion capture of CO_2. In addition, a large amount of energy is required to release the CO_2 from solvents or solid adsorbents after separation. Major technical and cost challenges need to be overcome before retrofit of existing power plants with post-combustion capture systems becomes an effective mitigation option. Capture technologies will likely improve in the coming years, thereby

---

[5]For a retrofit it is easy to calculate the cost of CCS and to determine the amount of CO_2 avoided. For advanced designs these seemingly simple accounting issues become more complex. For example, a power plant that has eliminated the flue stack and thus has no SO_x and NO_x emissions and also completely avoids the release of fine particulates into the atmosphere, while it captures all its CO_2 emissions, can allocate the cost of the advanced design in an arbitrary manner among the various benefits, making it impossible to uniquely determine the cost of CO_2 capture. There is a tendency in the literature to make a clear distinction between CO_2 captured and CO_2 avoided and to insist that assessments of various technologies should be clear on this distinction. There is a difference between CO_2 captured and CO_2 avoided, but more advanced power plant designs might not lend themselves to such simplistic analysis (IPCC, 2005; Rubin et al., 2007).

improving economics. This section discusses the status of technologies that are currently available for CO$_2$ capture and that could potentially be incorporated into currently operating power plants. We also point to those technologies that still are under research and development.

Post-combustion CO$_2$ capture and oxy-combustion technology provide retrofit options for existing coal-fired power plants. Table 6.1 lists a number of possibilities to perform post-combustion CO$_2$ capture [58]. Note, however, that some of the technologies in Table 6.1 are more suited for pre-combustion applications (e.g., physical absorption such as Rectisol and Selexol [13]) since they require concentrated CO$_2$ streams. This section focuses on options suitable for retrofits and oxy-combustion. Oxy-combustion modifies the combustion process so that the flue gas has a high concentration of CO$_2$. Also, an emerging technology for removing CO$_2$ from ambient air is presented. Biomass cofiring is discussed, since this is a retrofit option for CO$_2$ reduction in which coal and biomass are cofired, thereby reducing the carbon footprint of the power plant. The section concludes with a discussion of the economics of retrofitting.

Table 6.1 Technology options for post-combustion capture

Chemical absorption	Amines	■ Primary ■ Secondary ■ Tertiary ■ Sterically hindered
	Alternative solvents	■ Ammonia ■ Alkali-compound ■ Amino salt
Physical absorption		■ Rectisol ■ Selexol ■ Press. water
Adsorption		■ PSA ■ TSA ■ Solid sorbents
Alternative approaches	Membranes	■ Gas absorption membrane
	Cryogenic	■ Distillation ■ Frosting
	Others	■ Ionic liquid

6.3.1 POST-COMBUSTION CO_2 CAPTURE

Post-combustion CO_2 capture mainly applies to coal-fired power plants but can also be applied to gas-fired combustion turbines. In a typical coal-fired power plant, fuel is burned with air to produce steam, which drives a turbine to generate electricity. The exhaust gas, or flue gas, consists of mostly nitrogen, some CO_2, O_2, moisture, and trace impurities. Separating the CO_2 from this gas stream is challenging because the CO_2 is present in dilute concentrations (i.e., 13–15 percent of volume in coal-fired systems and 3–4 percent in natural gas-fired systems) and at low pressure (100–175 kPa). In addition, trace impurities (particulate matter, sulfur dioxide, and nitrogen oxides) in the flue gas can degrade sorbents and reduce the effectiveness of some CO_2 capture processes. Compressing the captured CO_2 from atmospheric pressure to pipeline pressure ($\approx$ 14 MPa) represents a large auxiliary power load on the overall power plant system. In spite of these difficulties, post-combustion capture has the greatest near-term potential for reducing CO_2 emissions, because it can be retrofitted to existing units that generate about 67 percent of the CO_2 emissions in the power sector [34]. Retrofitting will primarily be of interest in power plants with advanced efficiencies to compensate for the large efficiency loss from post-combustion capture. For this reason, capture-ready fossil fuel plants, which are intended to be retrofitted with a post-combustion capture process some time after the start of initial operation, could play an important role in the near future.

Post-combustion CO_2 capture options range from state-of-the-art amine-based systems to emerging technologies under development. Of the technologies presented in this section, chemical absorption technologies using amine or alternative solvent scrubbing are the nearest to commercial use in utility boilers. The others are emerging technologies at various stages of development. A rough ordering of the technologies from nearest to commercialization in utility boilers to longer lead times until commercialization are amine solvents, aqueous ammonia, chilled ammonia < solid sorbents, membranes < ionic liquids, and metal organic frameworks.

Since the partial pressure of CO_2 in the flue gas of fossil fuel-fired power plants is low, technologies driven by high CO_2 partial pressure differentials, such as physical solvents or membranes, are not applicable for post-combustion capture [58]. Only chemical solvents show an absorption capacity large enough to be applicable for CO_2 capture for CO_2 concentrations typical of coal-fired power plants.

6.3.1.1 Amine-based liquid solvent systems

A system for post-combustion capture of CO_2 by a chemical solvent is shown in Figure 6.2. The flue gas is usually cooled before entering the absorber column at the bottom. As the flue gas rises, the CO_2 is absorbed by a solvent in

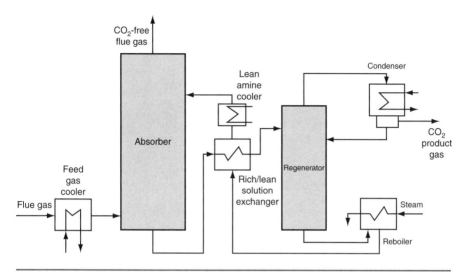

Figure 6.2   Schematic diagram of a post-combustion capture plant with chemical absorption.

counter-current flow. The CO$_2$-free gas is vented to the atmosphere. At the bottom of the absorber, the CO$_2$-rich solvent is collected and pumped through a heat exchanger, where the CO$_2$-rich solvent is preheated before entering the regenerator. In the regenerator, the CO$_2$-rich solvent flows downward. While it's flowing down, the temperature increases, thereby releasing CO$_2$, which rises to the top of the column and is removed.

Most commonly used for CO$_2$ capture is monoethanolamine (MEA). Amines are organic compounds with a functional group that contains nitrogen as the key atom. Structurally amines resemble ammonia, where one or more H$_2$ atoms are replaced by an organic substituent. Primary amines arise when one of the three H$_2$ atoms in ammonia is replaced by a carbon chain, secondary amines have two H$_2$ atoms replaced, and tertiary amines have all three H$_2$ atoms replaced [23].

CO$_2$ solvent extraction is based on the reaction of a weak alkanolamine base with CO$_2$, which is a weak acid, to produce a water-soluble salt. Chemical absorption with amines has been used for CO$_2$ separation in several industrial applications and in slipstream tests on coal-fired boilers; however, no amine-based CO$_2$ absorption system has been fully adapted to complete flue gas treatment in a large coal-fired utility boiler that would produce around 700 metric tons CO$_2$ per hour for a state-of-the-art 1000 MW$_e$ plant. Amine-specifically MEA, based systems are the leading technology for retrofit applications, since they are applicable for low-CO$_2$ partial pressure streams and provide recovery rates of up to 98 percent at a product purity of 99 vol%. Disadvantages of MEA systems include high process energy consumption, degradation of the solvent due to SO_x and NO_x forming nonregenerable, heat-stable salts, degradation of the solvent in the presence of oxygen, and high corrosion potential of the solvent [58]. However, improvements to amine-based systems for post-combustion CO_2 capture are being pursued by a number of developers, including Fluor, Mitsubishi Heavy Industries, and Cansolv Technologies [34].

6.3.1.2 Aqueous ammonia process

Ammonia-based wet scrubbing is similar to amine systems with a process scheme similar to that shown in Figure 6.2. The main difference to amine systems is the precipitation of crystalline solid salts at temperatures below 80°C and the relatively high vapor pressure of ammonia, which requires regeneration of the solvent at very high pressures to minimize ammonia slip. In the absorber column, CO_2 reacts with ammonium carbonate to form ammonium bicarbonate, which partly precipitates in the absorber as a crystalline product [58]. This promotes the absorption process and increases the CO_2 capacity. The CO_2-rich slurry, consisting mainly of ammonium bicarbonate, is pumped through a heat exchanger to the high-pressure regenerator.

Cycling between bicarbonate and carbonate, the regeneration energy required is approximately 30 percent of that in an MEA system. Ammonia-based absorption has a number of other advantages over amine-based systems, such as the potential for high CO_2 capacity, lack of solvent degradation during absorption/regeneration, tolerance to oxygen in the flue gas, and potential for regeneration at high pressure. It also offers multipollutant control by reacting with sulfur and nitrogen oxides to form fertilizer (ammonium sulfate and ammonium nitrate) as a salable by-product.

Disadvantages of the process include ammonia's high volatility, which often results in ammonia slip into the exit gas and the requirement that the flue gas be cooled to the 15–25°C range to enhance CO_2 absorptivity of the ammonia compounds. Also, ammonia is consumed through the irreversible formation of ammonium sulfates and nitrates as well as removal of HCl and HF.

A variation of the aqueous ammonia process is under development by ALSTOM and is called the *chilled ammonia process*. This process uses the same chemistry, but the absorber is kept at a significantly lower temperature, i.e., 0–10°C. It differs from the aqueous system in that no fertilizer is produced and a slurry of aqueous ammonium carbonate/bicarbonate and solid ammonium bicarbonate is circulated to capture CO_2 [34]. The main advantages of this process configuration are a smaller volume flow, less power duty for the blower, and less ammonia slip in the absorber [58]. Technical hurdles include cooling the flue gas and absorber to maintain operating temperatures below 10°C, mitigating the ammonia slip during absorption and regeneration, achieving 90 percent removal efficiencies in a single stage, and avoiding fouling of heat transfer and other equipment by ammonium bicarbonate deposition [34].

6.3.1.3 Alkali carbonate-based systems

Alkali carbonate systems take advantage of the same carbonate/bicarbonate chemistry as the ammonia system. A major advantage of carbonates over amine-based systems is a significantly lower energy requirement for regeneration. There are several processes (e.g., Benfield™ and Catacarb™) for the removal of CO_2 from high-pressure gases utilizing aqueous potassium carbonate (K_2CO_3). They are used in industries that produce natural gas and H_2 for ammonia synthesis [58]. These processes are attractive because they are not expensive and they use nonvolatile, nontoxic compounds. Currently, there are no plants using potassium carbonate for CO_2 removal from flue gas, but this option is being explored. Sodium carbonate solutions are also studied, since sodium carbonate solutions are nonhazardous and nonvolatile and have low corrosion rates.

6.3.1.4 Solid sorbents

Solid particles can be used to capture CO_2 from flue gas through chemical absorption, physical adsorption, or a combination of the two. Possible configurations for contacting the flue gas with solid particles included fixed, moving, and fluidized beds. A number of solids can be used to react with CO_2 to form stable compounds at one set of operating conditions and release the adsorbed CO_2 at another set of conditions.

In physical adsorption, the CO_2 is captured by a molecular sieve—for example, activated carbon or zeolite. In the regeneration phase, the CO_2 is released in a pressure swing or a temperature swing. *Pressure swing adsorption* (PSA) has been given more attention than *temperature swing adsorption* (TSA) due to the longer cycle times needed for heating during regeneration.

Research is also under way into investigating a dry, inexpensive, regenerable, supported sorbent, sodium carbonate (Na_2CO_3), which reacts with CO_2 and water to form sodium bicarbonate ($NaHCO_3$) [34]. A temperature swing is used to regenerate the sorbent and produce a pure CO_2/water vapor stream.

6.3.1.5 Membranes

Membrane-based CO_2 capture uses permeable or semipermeable materials that allow for the selective transport and separation of CO_2 from flue gas. Key technical challenges to the use of membrane systems are processing large flue gas volumes, relatively low CO_2 concentration, low flue gas pressure, flue gas contaminants, and the need for high membrane surface area [26].

Membranes for post-combustion capture are gas-liquid separation membranes. The purpose of the membrane is to keep the liquid and the gas separate. CO_2 diffuses from the gas stream through the membrane and is then absorbed by the solvent. This type of system prevents cross-contamination and loss of absorption liquid and eliminates design limitations from hydrodynamic constraints encountered in a free flow [58].

6.3.1.6 Metal organic frameworks

Metal organic frameworks (MOFs) are a new class of hybrid material built from metal ions with well-defined coordination geometry and organic bridging liquids [34]. They are extended structures with carefully sized cavities that can adsorb CO_2. High storage capacity is possible, and the heat required for recovery of the adsorbed CO_2 is low.

6.3.1.7 Cryogenics

The use of cryogenics for removing CO_2 has been considered for nearly 20 years [58]. Three methods being investigated include (1) compressing the flue gas to ≈ 7.5 MPa with water used for cooling/condensing and removing the CO_2; (2) compressing the flue gas to 1.7–2.4 MPa at $-15\sim +20°C$, dehydrating the feed stream with activated alumina or a silica gel dryer, and distilling the condensate in a stripping column; and (3) cooling the flue gas stream to condense CO_2 at atmospheric pressure. The latter method is also known as *frosting* or *antisublimation* (i.e., phase change of a gas to a solid) on a cold surface. For the removal of 90 percent of the CO_2 from a coal-fired flue gas, the sublimation temperature ranges from $-103°C$ to $-122°C$.

6.3.1.8 Ionic liquid systems

Ionic liquids are a broad category of salts, typically containing an organic cation and either an inorganic or organic anion. They are generally liquid at room temperature, nonvolatile, thermally stable, and nonflammable. Ionic liquids can dissolve gaseous CO_2 and are stable at temperatures up to several hundred degrees centigrade. Their good temperature stability offers the possibility of recovering CO_2 from the flue gas without having to cool it first. Also, since ionic liquids are physical solvents, little heat is required for regeneration. Due to their nonvolatility, solvent slip is basically eliminated. At this time, however, ionic liquids most suited for CO_2 separation have only been synthesized in small quantities in laboratories, and major developmental work is under way [4, 8, 11, 111].

6.3.2 OXY-FUEL COMBUSTION

Oxy-fuel combustion is an emerging approach to CO_2 capture. Oxy-fuel combustion is an integrated, rather than a post-combustion, process. In an oxy-fuel plant, coal combustion occurs in an oxygen-enriched (i.e., nitrogen depleted) environment, thereby producing a flue gas comprising mainly CO_2 (up to 89 vol%) and water. The water is easily separated and the CO_2 is ready for sequestration [32].

The most common oxy-fuel process involves the combustion of pulverized coal in a mix of nearly pure oxygen (greater than 95 percent) and recycled flue gas. A schematic is provided in Figure 6.3. Almost pure oxygen from a cryogenic air separation unit (ASU) is mixed with recycled flue gas ($\approx 2/3$ of the flue gas flow from the boiler) prior to combustion. Mixing recycled flue gas with the oxygen is necessary because currently available materials of construction cannot withstand the high temperature that would result from combustion in pure oxygen. To maintain the oxygen/recirculated flue gas flame so that the pulverized coal oxy-combustion has heat transfer characteristics similar to that of an air-fired system, an oxygen level of about 30–35 vol% is required [27].

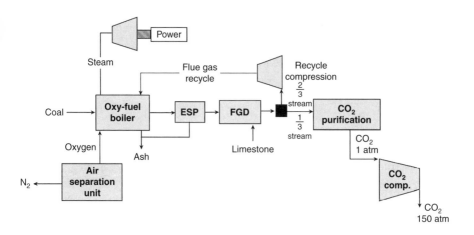

Figure 6.3 Schematic diagram of a pulverized coal oxy-fuel combustion system. Source: Modified from Figueroa et al. [34] and Kather et al. [58].

Power plants retrofitted for oxy-combustion would likely operate with oxygen levels that keep combustion conditions near those of air-fired systems. New oxy-combustion plants or repowered plants will not necessarily be constrained, and smaller boiler equipment will increase efficiency. Oxygen content may be adjusted to increase radiative heat transfer at the expense of convective heat transfer zones farther from the flame. By contrast, in air-fired systems approximately half the heat transfer occurs in the boiler and the balance of the heat transfer is convective in the backpasses. In both retrofit and new systems, the changes inherent in oxy-combustion affect many parameters, including flame behavior, heat and mass transfer, combustion gas chemistry and behavior, char burnout, and slag development, chemistry, and deposition.

Oxy-combustion is a promising technology with a significant amount of research and development focused on it; serious interest in designing and constructing new power plants with this technology has led to a number of demonstration projects, including the 30 MW oxy-fuel pilot plant in the "Schwarze Pumpe" (Black Pump) district in Spremberg, Germany [33]. However, there are still many technical (and economic) issues that need to be addressed. These can be categorized into combustion technology, steam generator design, CO_2 purification, and overall process issues [58].

In the area of combustion technology, the optimal oxygen concentration and the feed rate of gas need to be determined. High oxygen levels raise compressor demand, whereas low oxygen levels lead to corrosive atmospheres and may result in incomplete burnout. Reliable mixing of oxygen, flue gas, and coal is important for a good performance of the boiler and to avoid zones of incomplete combustion. Combustion behavior of coals in O_2, CO_2, and H_2O atmospheres has been studied in the laboratory, but more information is needed on full-scale systems. The formation of SO_x, NO_x, and other pollutants needs to be examined in more detail. Fouling and corrosion in oxy-fuel combustion environments is expected to be more severe due to the higher loadings of CO_2 and SO_x. Decades of research on air-blown combustion have no counterpart in oxy-fuel combustion.

In redesigning the steam generator, the effects of flue gas composition on heat transfer need further study. The atmosphere of the oxy-fuel–fired boiler chamber is different from that of an air-fired boiler, which leads to different flame characteristics and radiation heat transfer behavior. The heat capacity and density of CO_2 is higher than that of N_2, which results in a smaller furnace cross-section to maintain similar flue gas velocities. CO_2 and H_2O will increase the radiative heat transfer, but it is not known how significant this is compared to the radiative contribution from coal and fly ash particles.

One of the most critical parameters is the temperature of operation, which can be controlled in an oxy-fuel boiler. The temperature dictates the cross-section and materials of construction of the flue gas ducting, length of flue gas ducting, and power requirements for recirculating the flue gas. Defining the optimum temperature is dependent on the overall plant configuration and is crucial for the overall economics.

For producing a clean, liquefied stream of CO_2, a better knowledge of the phase equilibria of flue gas mixtures is important. Demands on CO_2 purity need to be better understood, and in situations where high-purity CO_2 is required, CO_2 cleanup options need to be investigated. A better understanding of gas-liquid solution kinetics is necessary for the development of improved liquefaction and flue gas cleaning equipment. Corrosion issues need to be understood. Similarly, the fate of contaminants during humidification must be better understood because many will be dissolved into the condensing water stream, and knowing their effect on effective CO_2 liquefaction is important.

Finally, many issues in system design have not been fully explored. The oxy-fuel process incurs large energy penalties from the ASU and liquefaction equipment. Efficiency improvements must be realized through process integration, to make the economics more favorable. The development of highly efficient ASUs specifically designed for power plants in excess of 400 MW_e is necessary.

Current coal mills use air for coal drying and transport, which is not acceptable for oxy-fuel systems. Recycled flue gas can be used, but this may require the flue gas to be dried and stripped of sulfur compounds to reduce corrosion in the mill and to achieve proper coal drying, which in turn results in increased process costs. A preferred option is the development of a mill that can use the recycled flue gas without prior removal of the sulfur compounds and water vapor.

Air in-leakage is typical for boilers operating at slightly subatmospheric pressure, adding nitrogen, which in turn is converted to NO_x to the gas mixture. In conventional power plants, air in-leakage is in the range of 4–10 percent of the flue gas volume, which is problematic when pure CO_2 streams are the target. It has been reported that a 75 percent reduction in air leakage should be achievable for an oxy-fuel retrofit and an air in-leakage rate of 2 percent of the flue gas should be feasible for new boilers being designed for oxy-fuel operation [58].

There is growing interest in oxy-fuel combustion in circulating fluidized bed (CFB) boilers, especially since CFB boilers have several inherent advantages over conventional combustion systems, including fuel flexibility, low NO_x emissions, *in situ* control of SO_2 emissions, excellent heat transfer, high combustion efficiency, and good system availability [69, 81]. Many of the challenges for pulverized coal oxy-fuel boilers apply to CFB boilers. CFB oxy-fuel combustion is considered at the same development stage as pulverized coal oxy-fuel combustion; however, since there are $\approx$ 1300 utility coal-fired boilers compared to fewer than 100 coal-fired CFB boilers [80], much of the larger-scale research and development has been on the pulverized coal oxy-fuel systems.

6.3.3 Air Capture: Collecting CO_2 from Ambient Air

The capture of CO_2 directly from the air—*air capture,* for short—is possible and has been demonstrated [28, 59, 64, 68, 110]. Air capture offers an interesting alternative to retrofitting a particular power plant by collecting an equivalent amount of CO_2 directly from the air. It is possible to reduce CO_2 emissions from a power plant simply by collecting an equivalent amount of CO_2 directly from the atmosphere in a different location and without interfering with the operation of the existing plant.

Air capture, like flue gas scrubbing, relies on the use of CO_2 sorbents. The process has two distinct stages: the contacting of the gas stream with the sorbent material and the sorbent recovery step, in which CO_2 and sorbent are separated again. Because the atmospheric concentration of CO_2 is much lower than in the flue stack of a conventional coal-fired power plant, the collector by necessity is much larger. Nevertheless it is still surprisingly small, and consequently it can be made affordable [64].

The optimal chemical binding strength of a sorbent for air capture is not much larger than for capturing CO_2 in a flue stack. The binding energy depends only logarithmically on the CO_2 concentration at the exit of the scrubber. For air capture, the exit concentration hardly differs from the inlet concentration, since it is sufficient to collect just some of the CO_2. In a power plant setting it is, however, necessary to collect all the CO_2. In the air collector the exit concentration may be 300 ppm; in the flue stack it should be 10,000 ppm or lower. The difference is small enough that most sorbents that can scrub flue stacks also remove CO_2 from the air. Indeed, the carbonate/bicarbonate chemistry mentioned earlier for scrubbing flue stacks has also been made to work for ion exchange resins that can extract CO_2 from air [68].

Even with CO_2 as dilute as in the air, the major effort in capture is focused on sorbent recovery, which is very similar to the sorbent recovery in a flue stack. The initial step of contacting the air is small compared to the subsequent sorbent recovery step but large compared to the contacting step in a flue stack. Even for air capture, the absorption cost does not amount to more than 20 percent of the total. As a result, CO_2 capture from the air could be performed at costs not much higher than those at a power plant.

Direct capture of CO_2 from the air can be performed anywhere because the air mixes fast and concentrations do not vary much from location to location.

As a result, it is possible to eliminate the transport of CO_2 to the disposal site and use remote disposal sites that generally will offer lower disposal costs. Therefore, air capture could become competitive with retrofitting existing power plants. The development of air capture would provide a more versatile option than developing retrofit options for old power plants. Air capture also offers the possibility of coping with the CO_2 emissions from cars and airplanes and thus is of interest far beyond its possibilities for retrofitting old power plants. Air capture would create a market for CO_2 credits that are available to all at a price that is not much different from the effective cost of CO_2 scrubbed from a flue stack. This view could prove wrong, either if air capture developments are not moving forward or if the demand for capture in the transportation sector grows so rapidly that it prevents the price of air capture from falling.

6.3.4 Biomass Cofiring

Cofiring is a near-term, low-cost option (compared to most capture technologies) for efficiently and cleanly converting biomass to electricity by feeding biomass as a partial substitute fuel in coal-fired boilers. Since biomass combustion is carbon neutral, cofiring reduces a plant's CO_2 emissions. If combined with capture and storage, it can reduce emissions to become net negative. Biomass cofiring has been demonstrated successfully in over 150 installations worldwide and has resulted in some commercially operating units using a variety of feedstocks in all boiler types commonly used by electric utilities, including pulverized coal, cyclone, stoker, and FBC boilers. Common feedstocks include woody and herbaceous materials, energy crops, and agricultural and construction residues [81, 106]. Extensive demonstrations and tests have confirmed that biomass energy can provide as much as 15–20 percent of the total energy input with only fuel feed system and burner modifications.

Cofiring biomass with coal offers several environmental benefits, including reductions in SO_2, NO_x, and greenhouse gas emissions. CO_2 emissions from coal-fired power plants are essentially reduced by the percentage of biomass cofired on a heat input basis. When biomass has been cofired at rates of 5 percent and 15 percent by heat input, greenhouse gas emissions have been reduced by 5.4 percent and 18.2 percent, on a CO_2-equivalent basis, respectively [74]. In studies comparing carbon emissions in producing biofuels and electricity from biomass, it has been shown that carbon displacement is significantly less when producing liquid fuels [86]. Thus, from the standpoint of reducing carbon emissions, it is better to use biomass to produce electricity. This would especially be the case if carbon were captured and sequestered from the biomass.

Biomass cofiring is a proven technology, but there are technical issues that must be considered when cofiring biomass with coal. Specifically, this includes fuel handling, storage, and preparation; NO_x formation; carbon conversion; ash effects such as deposition and corrosion; impacts on selective catalytic reduction and other downstream processes; and loss of sales of fly ash for beneficial uses. These issues are dependent on fuel characteristics, operating conditions, and boiler design. Proper combinations of coal and biomass and operating conditions can minimize or eliminate most impacts for most fuels.

6.3.5 Economics of Retrofitting

Technical issues for retrofitting coal-fired power plants for carbon capture were presented in the previous sections. In this section, the economics of retrofitting plants for carbon capture are addressed. Capturing CO_2 from power plants is expensive, and one can find a wide range of cost estimates for both retrofit situations and new plant construction. Increased costs are due to parasitic energy loss and an increase in capital costs, thereby resulting in an overall cost increase for electricity. Numbers vary from 5–30 percent parasitic energy losses, 35–110 percent increase in capital cost, and 30–80 percent increase in the cost of electricity [35]. The primary cost issues and their effect on the cost of electricity are summarized in this section.

6.3.5.1 Post-combustion CO_2 capture economics

The U.S. DOE performed a study to evaluate the technical and economic feasibility of various levels of CO_2 capture for retrofitting an existing pulverized coal-fired power plant using advanced amine-based capture technology [92]. Impacts on plant output, efficiency, and CO_2 emissions, resulting from the addition of CO_2 capture systems on an existing coal-fired power plant, were considered. Cost estimates were developed for the systems required to produce, extract, clean, and compress CO_2, which would then be available for sequestration or other uses.

Adding CO_2 capture technology impacts net plant output and thermal efficiency, as illustrated in Figure 6.4. For an amine scrubber, the efficiency decrease is essentially a linear function of CO_2 recovery level. Similarly, the specific investment cost is also a nearly linear function of CO_2 recovery level, with specific investment costs increasing from $540–1319/kW_e$ as CO_2 capture level increases from 30–90 percent (see Figure 6.5). The specific investment cost is a nearly linear function of CO_2 recovery level. In all cases studied, significant increases to the levelized cost of electricity (LCOE) are incurred as a

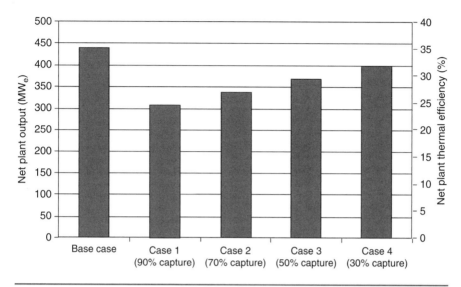

Figure 6.4 Plant performance impact of retrofitting a pulverized coal-fired plant at various levels of carbon capture.
Source: Ramezan [92].

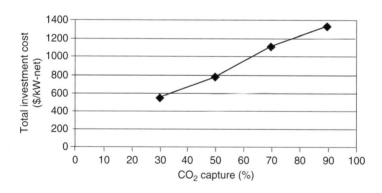

Figure 6.5 Effect of CO$_2$ capture rate on the total investment cost for retrofitting a pulverized coal-fired plant.
Source: Ramezan [92].

result of CO$_2$ capture, which is illustrated in Figure 6.6. For the ranges studied, the incremental LCOE is most impacted by the following parameters, in given order: CO$_2$ by-product selling price, CO$_2$ capture level, solvent regeneration energy, capacity factor, investment cost, and make-up power cost.

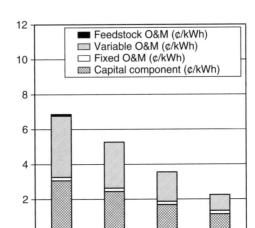

Figure 6.6 Incremental levelized cost of electricity for retrofitting a pulverized coal-fired plant (Conesville #5 unit in Ohio) using advanced amine-based capture technology.
Source: Ramezan [92].

ALSTOM, the developer of the chilled ammonia process described previously, performed cost-of-electricity comparisons for several CO_2 capture technologies in an 800 MWe pulverized coal-fired power plant [85]. Keeping in mind that economy of scale has a major influence on cost of electricity, Figure 6.7 shows that there are incremental reductions in cost of electricity compared to amine, future amine, and chilled ammonia.

These cost estimates are broadly in line with the IPCC [51]. *The MIT Study on the Future of Coal* estimates the cost of CO_2 avoided at a pulverized coal plant at $40/ton before transport and storage [25]. It also suggests an increase in the price of electricity of 4¢/kWh.

The U.S. DOE performed an assessment of the prospects of retrofitting existing coal-fired power plants for CO_2 capture and sequestration using a generic model of retrofit costs as a function of basic plant characteristics such as heat rate [39]. The cost for CO_2 retrofit included direct costs (capital and O&M), indirect costs (capacity and heat rate penalties), and a nominal cost for transportation, injection, measurement, monitoring, and verification. The penetration of retrofits across the fleet was not observed at any significant

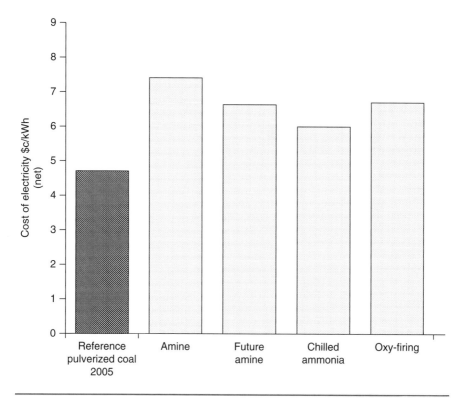

Figure 6.7 Cost-of-electricity comparison for various CO_2 capture technologies. Source: Modified from Otter [85].

extent until carbon emission allowance prices exceeded $30 per ton of CO_2 equivalent. Retrofits did not occur through 2030 at $30 per ton of CO_2 equivalent, but about one third of the starting coal fleet capacity was retrofitted at $45 per ton of CO_2 equivalent and about half at $60 per ton of CO_2 equivalent.

6.3.5.2 Oxy-fuel combustion economics

As a retrofit option for existing coal-fired boilers, economic evaluations have shown that oxy-coal combustion retrofits for CO_2 capture are less expensive than other technologies such as amine scrubbing. This is illustrated in Figure 6.7, showing the cost of electricity net in the ALSTOM study, and Figure 6.8, showing the percent increase in COE, from a DOE study [21] that compared current state amine scrubbing with current state oxy-fuel firing in a supercritical boiler and an ultrasupercritical boiler.

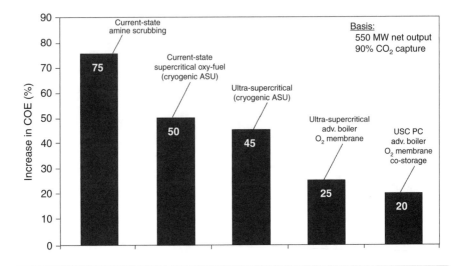

Figure 6.8   Oxy-fuel combustion costs.
Source: Ciferno [21].

There is significant interest in further reducing oxy-fuel combustion costs. This interest is illustrated in Figure 6.8, where the last two bars show a reduction in cost as oxygen membranes are developed to reduce the cost of oxygen production and costorage of CO_2 and SO_x. Similarly, costorage is being explored where other pollutants are condensed in the CO_2 stream when the CO_2 is used for enhanced oil recovery (EOR). Because CO_2 is in a liquid state for EOR, it provides a sink for other pollutants, and a small amount of SO_x in the CO_2 stream is acceptable because SO_x benefits the EOR process by improving oil miscibility [112]. The integrated emissions control approach has the potential to significantly reduce the costs related to pollution control and compensate for the increased investment for CO_2 capture. This technology is applicable to a new plant construction and might benefit a retrofit as well.

For comparison, the MIT study [25] appears more optimistic, suggesting a cost increase from 4.8¢/kWh to 7.0¢/kWh moving from a supercritical power plant without CO_2 capture to a supercritical oxy-fuel plant with 90 percent capture. Again, this study leaves out the cost of transport and storage.

6.3.5.3 Economics of biomass cofiring

Biomass cofiring is technically feasible; however, costs are still a major barrier to increased coal and biomass cofiring. Cofiring economics depends on location, power plant type, and the availability of low-cost biomass fuels. Biomass is not economically competitive for electricity production in the current energy market except in niche applications where fuel costs are low or negative, such as integrated pulp and paper facilities [99]. Biomass is generally more expensive than coal; the delivered cost of biomass typically ranges from $1–4 per GJ, whereas coal costs around $1 per GJ. Delivered costs for dedicated energy crops can be significantly higher [86].

A typical cofiring installation includes modifications to the fuel-handling and storage systems, possibly burner modifications, and sometimes boiler modifications or separate biomass feed injection into the boiler. Costs can increase significantly if the biomass needs to be dried, reduced in size, or pelletized or if the boiler requires a separate feeder. Retrofit costs can range from $100–300/kW of biomass generation in pulverized coal boilers. Cyclone boilers offer the lowest-cost opportunities, as low as $50/kW [50].

Fuel supply is the most important cost factor. Costs for biomass depend on many factors such as climate, closeness to population centers, and the presence of industries that handle and dispose of wood. Low price, low shipping cost, and dependable supply are critical. Usually the cost of biomass must be equal to or less than the cost of coal, on a heat basis, for cofiring to be economically feasible. This, however, could change if CCS legislation becomes enacted in the United States, for example.

Although coal and biomass cofiring is generally not cost competitive in the current energy market, cofiring offers benefits that could result in more widespread applications under certain policy scenarios. For instance, the waste-to-energy technology has recently recaptured significant attention by representing carbon-neutral energy sources, and a large portion of solid municipal wastes (SMW) is bio-based. Thus, the use of biomaterials in SMW could be a good starting point for the biomass cofiring scenario in terms of the economic point of view. This is especially true as a near-term CO$_2$ mitigation strategy.

6.3.5.4 Economics of other systems

For many of the more advanced systems, it is not possible to give accurate cost estimates. Nevertheless, they will have to compete against systems discussed above and against nuclear power. CCS in any form is not likely to be introduced on a large scale unless price increases can be held to about 2–3¢/kWh. At higher prices, coal-based electricity is unlikely to compete with nuclear energy. To set the scale, 1¢/kWh at a coal-fired power plant would pay for $8–22 per ton of CO$_2$, depending on the efficiency of the power plant (25–70 percent, respectively). For efficient power plant designs, it appears very feasible to collect the CO$_2$ within a budget of $60/ton of CO$_2$.

An interesting opportunity for cost reductions could arise if systems are not designed to collect all the CO$_2$ that is produced at a power plant. Though this cannot solve the long-term problem of creating deep reductions, it could provide an easier start. Of interest here could be very weak sorbents such as sodium or potassium carbonate that could lead to easy capture and sorbent recycling, resulting in an exit CO_2 concentration that could be as high as 10 percent of the gas flow.

Air capture can deal with CO_2 emissions at sites that are particularly difficult to retrofit. Here we point to the underlying logic that suggests that CO_2 capture is not significantly more expensive than simple scrubbing systems at flue stacks. Estimates of the cost range from $30 per ton of CO_2, which would be very competitive, to about $200 per ton, which would greatly limit the applicability of the technology.

Future power plant designs are likely more costly than current plants, but they will combine low fuel consumption with a nearly complete elimination of air pollution and a CO_2 stream ready for disposal [109]. It has been estimated that their cost could be completely justified just in terms of pollution abatement. However, the cost of these plants is still highly speculative and much research needs to be done before they can be implemented.

6.4 Integrated CO_2 capture designs

In Section 6.3 we focused on technologies for retrofitting existing power plants. Here we discuss technologies that show promise for future generations of power plants. A number of these technologies are built around H_2 production from fossil fuels. H_2 production facilities typically use fossil fuels as energy input and can be designed to capture CO_2 for disposal. This is already done in gasification units operating in refineries. These plants remove CO_2 from the gas stream simply to obtain pure H_2. Gasification typically results in a mixture of CO and H_2, syngas. To obtain pure H_2, the carbon monoxide (CO) contained in the syngas is reacted with steam. Under the right conditions and in the presence of catalysts, the oxygen from the water is transferred to the CO, producing CO_2. This reaction is known as the water-gas-shift (WGS) reaction.

Conventionally, CO_2 is separated from H_2 by chemical or physical sorbents or by traditional membrane separation, as described in the previous section. These processes are often costly and highly energy intensive and generally require low operating temperatures. The efficiency of the H_2 production in the preceding gasification-WGS system with a conventional CO_2 separation unit is reported to be around 64 percent [101].

A number of advanced H_2 production schemes that integrate decarbonization and H_2 production more tightly are in various stages of development and demonstration. One such approach is to use limestone sorbents to capture CO_2 from the gas mixture. Limestone, or calcium carbonate, is calcined to form calcium oxide, which is then used to react with and continuously remove the CO_2 product formed during the WGS reaction. Because the CO_2 absorption proceeds at very high temperatures, the substantial heat of carbonation is useful and is actually incorporated into the overall process [7, 49, 72]. This basic idea has led to development of a number of integrated carbon-capture processes that enhance H_2 production via the carbonation-calcination loop of the calcium-based sorbents, such as the Zero Emission Coal Alliance (ZECA) process [109], Hypr-RING process [71], ALSTOM process [3], GE process [70], and calcium looping process [89].

Another approach to producing carbon-free H_2 employs a different looping technology based on an oxidation-reduction cycle. These technologies, including coal-direct chemical looping and syngas redox processes, use looping particles made from metal or a low oxidation state metal oxide as oxygen carriers. The metal reacts with steam to form the metal oxide, and it is converted back into the metal in the presence of a carbonaceous fuel that needs the oxygen for its combustion. By separating the process steps into two reactors, one for the oxidation of carbonaceous fuels (e.g., CO, coal, and biomass) and the other for the reduction of steam, inherent gas separation can be achieved while producing high-purity H_2. The exit stream of the fuel reactor contains only sequestration-ready CO_2 along with water, which can easily be condensed out. The advantages of this approach to H_2 production are that it eliminates the separation of H_2 from the fuel gas and does not require the cleanup of fuel gas prior to H_2 production [77]. Moreover, chemical looping processes produce a sequestration-ready CO_2 stream at high pressure, which eliminates a costly pressurization step.

In what follows, a number of integrated technologies developed by incorporating CO_2 capture with H_2 production are discussed. Once produced, the H_2 can be used in various chemical processes or converted to electricity using an energy conversion system such as a solid oxide fuel cell. In the integrated gasifier combined cycle (IGCC) plant, the H_2 is used in a conventional gas turbine to produce electricity [67].

6.4.1 INTEGRATED GASIFIER COMBINED CYCLE PLANT WITH PRE-COMBUSTION CO_2 CAPTURE

Gasification technologies have been used in the refining industries for many decades. The output of a commercial gasifier can be used in combined cycle gas-fired power plant that produces electricity in a combustion turbine and in a secondary steam cycle that utilizes the waste heat from the high-temperature combustion turbine. Such IGCC plants have high efficiency and thus have been studied for decades, but they have not yet captured the power generation market. Similarly, WGS reactors to produce H_2 from syngas have also been applied in the petrochemical industry. As a result, the first advanced power plant designs with CCS start from a conventional IGCC plant and add a shift reactor to produce H_2, which becomes the fuel for the combined cycle plant. After the CO_2 is separated from the H_2 stream, it is cleaned up and made pipeline ready. Among currently available technologies, oxy-fuel combustion and IGCC designs vie for the lowest-cost implementation of carbon capture and storage. In the MIT study (2007), the IGGC system with CCS resulted in the lowest-cost CO_2 capture, raising the cost of electricity in comparison to a supercritical pulverized coal plant by less than 1.7¢/kWh.

6.4.2 THE HYPR-RING PROCESS

The HyPr-Ring process developed in Japan provides an alternative implementation for the gasifier in an IGCC plant. CO_2 separation and gasification are tightly integrated. The reactors are fluidized bed reactors capable of excellent heat and mass transfer. As illustrated in Figure 6.9, coal is fed into the gasifier along with steam and lime (CaO) as the solid chemical sorbent removing product CO_2 from the gasifier. Removal of CO_2, together with the large amount of heat released in the absorption, drives the reaction toward pure H_2, which is the output of the HyPr-Ring. The H_2 can be used to fuel a combined cycle plant using H_2 as fuel.

The gasification reaction is driven forward by the removal of CO_2 and by the large amount of heat generated in the CO_2 absorption of lime. The regeneration of the lime occurs in a separate reactor vessel at a higher temperature, where fuel, some of the solid ash particles carried over from the gasifier and pure oxygen, are added to create heat and to drive the CO_2 off the sorbent. The HyPr-Ring process produces a stream of pure H_2, where some of the chemical energy of the H_2 is derived from the heat of carbonation. The estimated H_2 production efficiency of the HyPr-Ring process is 77 percent [71].

The main issues related to the further development of the HyPr-ring process are the high operating pressure ($\sim$7 MPa), the use of excess steam, and the purity of the final product H_2 [71]. Furthermore, at high temperatures and long holding times, there is considerable solid-solid interaction between the calcium-based sorbent particles and the minerals in the coal ash. This reduces the CO_2 sorption capacity of the sorbent. Steam also increases the interaction between the sorbent and the coal minerals at high temperatures and pressures [70].

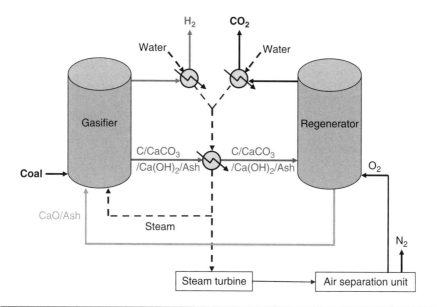

Figure 6.9   Schematic diagram of HyPr-RING process.
Source: Adapted from Lin et al. [70].

6.4.3 THE ZECA PROCESS

Figure 6.10 illustrates the ZECA process developed by Los Alamos National Laboratory. This process, which was proposed about the same time as the HyPr-Ring process, also produces H_2. But in the ZECA process the H_2 is an intermediate product; the end product is electricity. Gasification and CO_2 absorption have been broken into two distinct processes. Coal is gasified under H_2-rich conditions to produce mainly methane. In the carbonation vessel, methane is reformed with steam to produce H_2, and the lime (CaO) in the vessel is carbonated and thus absorbs the CO_2 produced during reforming [113]. The substantially exothermic carbonation provides the energy necessary for reforming [113]. The product H_2 is converted to electricity in a solid oxide fuel cell. The spent lime sorbent is regenerated in the calcination vessel. The heat for the calcination is produced in the fuel cell and transferred through a heat exchange loop. Solid oxide fuel cells are particularly attractive because they integrate electricity production with air separation, avoiding another costly process step. The ZECA design aims at 70 percent conversion efficiency while delivering CO_2 in a concentrated stream [65]. It has no flue stack and thus truly qualifies as a zero-emission power plant.

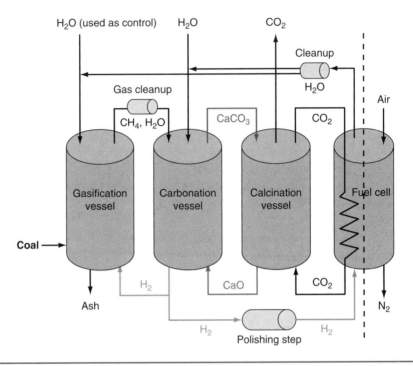

Figure 6.10 Schematic diagram of the ZECA process.
Source: Adapted from Ziock et al. [113].

6.4.4 THE ALSTOM PROCESS

ALSTOM's Hybrid Combustion-Gasification Chemical Looping Coal Power Technology also uses solids to transfer heat and oxygen. The ALSTOM process can operate in three different configurations: (1) indirect coal combustion for heat generation, (2) coal gasification for producing syngas, and (3) coal gasification for producing H_2 [3]. For the first and second configurations, a single chemical loop of CaS-$CaSO_4$ is used; for the third configuration a CaO-$CaCO_3$ loop is added.

In the first configuration, two reactors are used with calcium sulfate ($CaSO_4$) as a looping medium. Coal reduces the calcium sulfate to calcium sulfide in the first reactor while forming a high-purity CO_2 stream. The calcium sulfide product is combusted in air to calcium sulfate in the second reactor. A portion of the combustion heat drives the coal gasification in the first reactor, while the rest is used to produce high-temperature, high-pressure steam for electricity generation. Though similar to the first configuration, the second configuration feeds much more coal and steam into the first reactor.

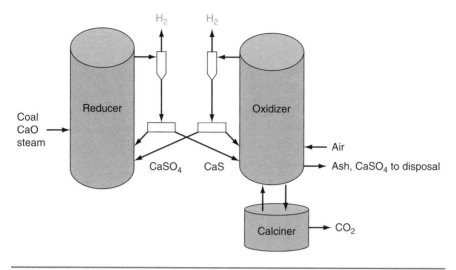

Figure 6.11 ALSTOM's hybrid combustion-gasification process for H$_2$ production.

In the configuration shown in Figure 6.11, pure H$_2$ is produced by deploying lime as a CO$_2$ sorbent. The reducer reactor removes CO$_2$ from the WGS reaction with the help of lime, and the third reactor is added to calcine limestone back to lime. The heat integration of this configuration includes the utilization of the heat generated from the combustion of calcium sulfide in the first chemical loop in the regeneration of the calcium-based sorbent in the second chemical loop [2].

6.4.5 THE CALCIUM LOOPING PROCESS

Figure 6.12 illustrates a gasification process integrated with the calcium looping process developed by OSU. The concept of this integrated technology is similar to those of the previously discussed processes. Lime (CaO) is again used as a CO$_2$ sorbent. It also will capture sulfur by forming calcium sulfide. The key improvement is the particle design. The engineered sorbent uses surface modifiers to provide stable mesoporous morphological properties that allow for high CO$_2$ capture and sulfur removal capacity while producing H$_2$ with greater than 95 percent purity, even without WGS catalysts. With catalysts the product stream exceeds 99 percent H$_2$. At laboratory scale, the engineered mesoporous Casorbent has shown recyclability over 100 or more cycles while maintaining its reactivity [31, 54]. At this time, this technology is being demonstrated at the pilot scale with DOE support.

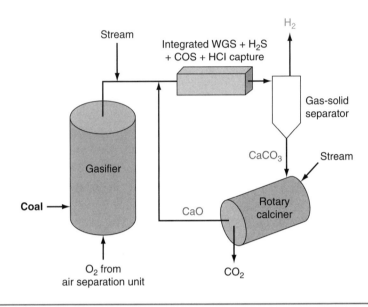

Figure 6.12 Schematic diagram of calcium looping processes for H_2 generation.

6.4.6 THE GE FUEL-FLEXIBLE PROCESS

General Electric Co. (GE) has also developed an integrated process that is known as the fuel-flexible process. This process takes a variety of carbonaceous feedstocks such as coal and biomass, and produces H_2 and electricity in adjustable ratios [93]. This technology involves two chemical loops, the carbonation-calcination, and the oxidation-reduction loops, using three fluidized-bed reactors (see Figure 6.13). In the gasifier, coal is partially gasified with steam to form syngas containing H_2 and CO. CO_2 is concurrently captured by the calcium-based sorbents. Similar to the ZECA and ALSTOM processes, here the depletion of CO_2 drives the WGS reaction toward producing H_2. Moreover, sulfur in the coal is captured by the same calcium-based sorbent-forming $CaSO_3$. As a result, a high-purity H_2 stream is obtained from the gasifier [93].

This scheme fully integrates the carbon capture and the air separation into the gasification system to produce pure H_2 while generating a sequestration-ready CO_2 stream. However, the challenge lies in handling the solids. By connecting three fluidized beds in a series, there is a potential problem of the intermixing of particles within three reactors. As the particles from three reactors are mixed with each other, the purity of the product stream of each

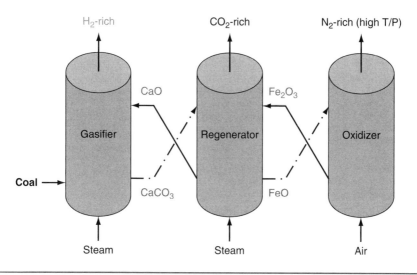

Figure 6.13   Schematic diagram of a fuel-flexible gasification-combustion process.

reactor will be compromised. In addition, a portion of the solids in the chemical loops needs to be purged to avoid ash accumulation and to maintain solid reactivity [93]. The overall energy conversion efficiency for the GE fuel-flexible process is estimated to be 60 percent [94].

6.4.7 THE COAL-DIRECT CHEMICAL LOOPING REFORMING PROCESS AND THE SYNGAS REDOX PROCESS

Compared to the previously discussed multiloop processes, the coal-direct chemical looping reforming (CLR) process and the syngas redox (SGR) process developed by OSU simplify the reaction scheme by utilizing only one chemical loop, the oxidation-reaction cycle, as illustrated in Figure 6.14. In both cases, a specially tailored, highly reactive metal oxide particle that can undergo multiple (>100) reduction-oxidation cycles is used to convert carbonaceous fuels (i.e., coal for the CLR process and syngas for the SGR process) into H$_2$.

These chemical looping processes are based on reactions similar to the steam-iron process developed in the 1970s for production of H$_2$ from producer gas [9, 18, 38, 57]. In this case the metal oxide was iron oxide. The single-pass conversion efficiency of the steam-iron process was low. The iron oxide particles had poor recyclability and oxygen-carrying capacity. As a result, the economic feasibility of the steam-iron process was inferior to steam reforming of natural

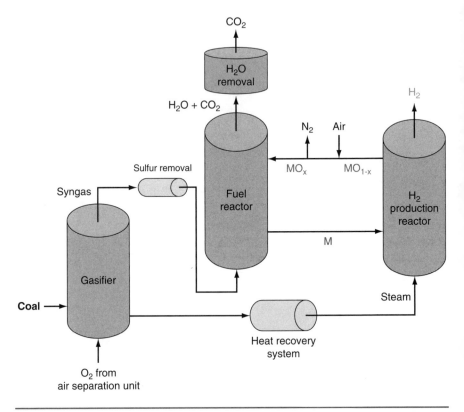

Figure 6.14 Schematic of the syngas chemical looping process for H_2 production from coal.

gas, which took over the market for H_2 production. In recent years, the oxidation of various metals (e.g., NiO, Fe_2O_3, CuO, MnO_x, Co_2O_3) has been extensively investigated to improve the oxygen-carrying capacity of the chemical looping sorbent [19, 22, 53, 56, 73, 78, 104]. With improved reactor and particle designs, these chemical looping processes have potential for commercialization, and pilot-scale demonstrations are currently under way in the United States, Europe, and Asia. The H_2 conversion efficiencies for the CLR and SGR processes are estimated to be >80 percent and 74 percent, respectively [42, 105, 107].

6.4.8 THE MEMBRANE PROCESS

Most of the processes discussed so far drive the WGS reaction to produce H_2 by removing CO_2 from the product stream with a chemical sorbent. Another way of shifting the equilibrium toward H_2 is to employ membranes that let either H_2 or CO_2 pass. However, currently existing membranes are not capable of completely separating CO_2 or H_2 from the mixture, and a considerable pressure drop across the membrane leads to the costly operation and poor process economics [95].

The recent development of advanced membranes has resulted in development and testing of experimental high-temperature CO_2 membranes that operate in the same temperature range as that of the WGS reaction [20, 108]. Better H_2 membranes are also being developed. In 2006, the National Energy Technology Laboratory and the Oak Ridge National Laboratory demonstrated the cost reduction of 25 percent for the H_2 production via coal gasification equipped with a porous H_2 membrane technology [103].

Membrane technologies will require better materials, higher permeabilities, and better selectivities, particularly at high-temperature conditions. Cost and reliability represent major challenges. Membranes would be particularly useful if they could be deployed directly at reactor conditions. For H_2 production, a process simulation indicates that utilizing hot gas cleanup units and advanced H_2 separation membrane with the ability to operate at 600°C could increase the coal gasification energy conversion efficiency for electricity and H_2 coproduction by ~11 percent compared to steam methane reforming processes [103].

6.5 The zero-emission concept

Advanced power plant designs, beginning with oxy-fuel combustion, have in common that they do not need a flue stack to release gaseous combustion products to the atmosphere. As long as CO_2 is emitted to the atmosphere, air emissions are impossible to avoid. No matter what other changes were built into such a power plant design, it would be necessary to clean the CO_2 from other pollutants before one could release it to the atmosphere. If one collects the CO_2 and avoids bringing in nitrogen from the air, the remaining material output of an electric power plant is fully condensable. Hence it is possible to eliminate the flue stack and stop all emissions to the atmosphere. There are still waste products—for example, sulfur and nitrogen compounds produced from the sulfur and nitrogen present in the coal. Also, large amounts of liquid water will be discharged as well as ash in coarse and fine particulate form. However, there is no gas stream that will carry these materials to the atmosphere.

We refer to such a plant as a *zero-emission plant*. The ZECA plant is an example of such a design; a well-designed oxy-fuel combustor could also operate as a zero-emission plant. Most plants that claim to be zero emission or near zero emission, including DOE's Future Gen plant, still funnel H_2-rich combustion gases through gas turbines. In that case combustion gases are released directly to the atmosphere and thus by our definition would not qualify as zero-emission plants.[6] Here we limit the term *zero emission* to designs that, according to their blueprints, release no emissions into the air.

The advantage of a zero-emission plant is that it anticipates future tightening of air emission standards. Rather than scrubbing harder and harder to accommodate tightening limits on SO_x, NO_x, and other air pollutants, this plant from the beginning has no air emissions. The need to eliminate CO_2 emissions allows this change in paradigm. In a world without carbon constraints one may end up collecting the CO_2 simply to avoid the cleanup costs, but the economics will favor scrubbing unless emission standards become extremely difficult to meet.[7]

Conceptually, an oxy-fuel power plant provides a simple path toward zero emissions. Combustion with pure oxygen or a mixture of combustion products and oxygen will result in an exhaust stream that is dominated by CO_2 and H_2O, and all constituents, with the exception of the CO_2, will be readily condensable. The remaining gaseous waste products, such as argon that is carried into the plant with the oxygen, are so small that scrubbing them clean is a relatively easy task. Although current oxy-fuel plants are not yet aiming for zero emissions, it is clear that this goal could be achieved in future designs.

Most of the advanced designs discussed in the previous section can be developed into zero-emission plants. The ZECA plant explicitly started with this goal [109]. Solid oxide fuel cells, which combine the generation of electricity and the separation of oxygen from air, provide a particularly straightforward approach to zero-emission designs. However, the cost of fuel cells is still too high to make these power plant designs practical.

6.6 Carbon storage options

Disposal or long-term storage of CO_2 is the final step in carbon management. Although the cost of storage accounts only for a small portion of the total cost [51], it may well be the most difficult step in the chain of carbon management. The challenge will grow over time as the volumes that will need to be stored grow rapidly. At present the world is releasing about 8.5 Gt/year of carbon,

---

[6]The term *clean coal* is used on occasion to describe advanced coal-fired power plants. However, this term was coined in reference to plants that aggressively remove sulfur and nitrogen oxides from the flue gas stream. Hence, *clean coal* in general does not imply CO_2 capture, nor does it suggest the elimination of the flue stack.

[7]In developing the ZECA plant (see Section 6.4.3) some years ago, we noticed that EPA estimates for dealing with PM 2.5 constraints could raise costs by similar amounts to capturing the CO_2.

or about 30 Gt of CO$_2$, into the environment. As mentioned before, storage volumes over the century could easily amount to several thousand gigatons of CO$_2$. Many different technologies have been suggested for long-term CO$_2$ storage. These include storing CO$_2$ in the ocean, in geological formations, and through mineral sequestration, which involves the formation of solid carbonates from CO$_2$ and minerals.

Power plant designs and CO$_2$ storage are usually very distinct operations. The goal of the power plant operator is to deliver a stream of concentrated CO$_2$, which typically will have to be pipelined to the site of disposal. Any capture scheme can be combined with any disposal scheme. There are a few exceptions in which the sorbent used in the power plant is directly disposed of.[8]

A special category is biological sequestration that does not use captured CO$_2$ but instead CO$_2$ directly extracted from the atmosphere. Biological sequestration might be considered as an alternative to retrofitting, but it does not provide means of CO$_2$ storage for CO$_2$ that is directly derived from power plants, nor does it itself provide long-term storage of the carbon.

Although the CCS technology is still too new to settle on a single approach, any storage option that would be useful beyond small niche markets must possess the following properties: have a large capacity, be safe, be environmentally benign, and guarantee the long-term stability of storage. Storage lifetime constraints will become more stringent over time.

6.6.1 Ocean Sequestration

Since 1977, when the first concept of direct ocean sequestration of CO$_2$ was introduced [75], significant research efforts have taken place. The ocean is a large natural sink for CO$_2$ and spontaneously absorbs a significant fraction of the fossil CO$_2$ emissions. In equilibrium, the partitioning of CO$_2$ between ocean and atmosphere is roughly 4 to 1 [16]. It would, however, take thousands of years for this equilibrium to be reached naturally. Even though the ocean contains 39,000 Gt of carbon as carbonate and bicarbonate, its uptake capacity is far smaller. An addition of 1000 Gt of carbon would already severely change ocean chemistry [66].

In spite of these limitations, estimates for ocean storage capacity in excess of 100,000 billion GtC have been cited [45]. Such numbers would be feasible only if alkalinity (e.g., NaOH) were added to the ocean to neutralize the

---

[8]Usually disposal of the sorbent is either too costly or defeats the purpose of CCS. For example, lime sorbents are made from limestone and in their production produce more CO$_2$ than they bind. Thus, they would not store any net CO$_2$ if they were used only once.

carbonic acid. Over thousands of years, the dissolution of calcareous oozes at the bottom of the ocean could provide such alkalinity [6, 15, 46].

A number of CO_2 injection methods have been investigated. One is to dilute the dissolved CO_2 at a depth below the mixed layer where carbon can be stored for decades to centuries, although capacities are limited. Another approach is to form lakes of CO_2 at the bottom of the ocean. Below 2700 meters, the density of the compressed CO_2 is higher than that of seawater, and therefore, CO_2 sinks to the bottom. In addition, CO_2 will react with seawater to form a solid clathrate, which is an ice-like cage structure with approximately six water molecules per CO_2 [24].

Estimates of the cost of the direct disposal of CO_2 in the oceans range from $1–6 per ton of CO_2 [36] to $5–15 per ton of CO_2 [44, 51]. The main hurdle to the acceptability of ocean storage is the environmental concerns related to long-term chronic issues of altering the ocean chemistry as well as on the local effects of low pH (as low as 4) and its effect on marine organisms (e.g., stunting coral growth). It does not help to replace one environmental problem with another.

6.6.2 GEOLOGICAL SEQUESTRATION

Underground storage in geological formations is the most developed disposal option for CO_2. Its main advantages are existing knowhow and low cost. For many decades, injection of CO_2 has been used for enhanced oil recovery (EOR). EOR capacity has been estimated at 20 to 60 GtC [47, 51, 82]. In West Texas, approximately 20 million tons of CO_2 are consumed in tertiary oil recovery [100].[9] These projects and similar projects in Canada demonstrate that the disposal of CO_2 is practically feasible.

Similar to EOR, high-pressure CO_2 can also be used for enhance methane recovery from unmineable coal beds [82, 90]. This method is attractive in the sense that most of the injected CO_2 will be immobilized by either physical or chemical absorption on the coal surface.

Depleted oil and gas wells provide another reservoir with an estimated sequestration capacity of 200–500 GtC. Large underground reservoirs such as deep aquifers represent the best long-term underground geologic storage option. Such aquifers are generally saline and are separated from shallower aquifers and surface water supplies. The estimated potential

---

[9]These injections do not qualify as sequestration, since the CO_2 has been extracted from underground wells about 500 miles away.

storage capacities of these aquifers in the United States and Canada have been estimated between 900 and 3000 billion Gt of CO_2 [10, 82].

Saline aquifer injection has been demonstrated by Statoil at the Sleipner West gas reservoir in the North Sea. In this project, approximately a million tons of CO_2 are injected annually into an aquifer under the sea floor [1]. This project shows that injection into deep aquifers is economically feasible. Unlike enhanced coal bed methane recovery, where CO_2 is physically or chemically absorbed, in aquifer sequestration, the injected CO_2 will likely persist as super-critical CO_2 without being fully dissolved. Hence, there may be greater potential for leakage [37, 51].

The uncertainty in geological storage lies in the long-term stability of reservoirs [12, 51]. The central question is long-term safety and long-term leakage rate, which ultimately will define the capacity of these technologies. It is likely that there will be a large number of sites at which CO_2 can be stored safely and for all practical purposes permanently. Therefore geological sequestration will make great contributions to solving global climate change issues [5]. There are some concerns that the actual cost may rise because public acceptance will demand more monitoring than is currently envisioned. Both near-surface and in-depth monitoring technologies will play key roles in the success of the geological sequestration method.

6.6.3 MINERAL SEQUESTRATION

As the amount of stored CO_2 increases, it becomes progressively more difficult to guarantee that physical barriers will prevent CO_2 from returning to the atmosphere. A chemical conversion to a thermodynamically lower and therefore stable state thus would be desirable and is indeed possible. The formation of carbonates from silicates is well known as *geological weathering*. In nature this is a very slow process, but thermodynamically, CO_2 can be bound as a carbonate. In many instances these carbonates dissolve in water, but some, such as magnesium or calcium carbonates, are remarkably stable as solids [62, 96].

Even during the geological sequestration described in the previous section, CO_2 can react with surrounding minerals and ions in ground water to form various carbonates. Since these carbonates are thermodynamically stable, immobilization of carbonation truly provides a permanent disposal of CO_2. Unfortunately, the reaction between CO_2 and minerals is slow, and thus the portion of carbon storage via mineralization is usually very limited in geological sequestration, although some researchers have suggested potential natural carbonation pathways that are rapid enough to sequester a large quantity of CO_2 [60]. Similarly, the formation of carbonates in basalts has been researched [76, 79] in the Columbia River Basin [82], in Iceland [84], and in India [55].

In the absence of geological storage sites that allow for fast carbonate formation, it is possible to use above ground industrial processing to speed up the carbonation process in a chemical reactor [62]. This is the third storage option described in the *IPCC Special Report on Carbon Dioxide Capture and Storage* [51]. Mineral rock for above ground sequestration is readily available. Peridotite and serpentinite rocks rich in magnesium exist in amounts far exceeding fossil carbon resources [40, 41].[10] During carbon mineral sequestration processes, Mg in olivine or serpentine minerals is leached out into the aqueous phase and reacts with dissolved CO_2 to form $MgCO_3$. The main challenges of this storage method have been the slow dissolution kinetics and large energy requirements associated with the mineral processing. However, researchers' latest advancements have shown a great potential to reduce the cost of the *exsitu* mineral carbonation [43, 87, 88, 91].

Although the cost of mining, mineral preparation, and tailing disposal is well known and relatively inexpensive, the chemical reaction process is still too expensive for a practical implementation[11] [51, 63]. Total costs need to be reduced from current estimates of $80 to $100 per ton of CO_2 [51] to more acceptable numbers below $30 per ton [62].

6.7 Conclusion

CCS could provide a solution to the climate change problem, but the introduction of this novel technology is a major undertaking. From a technology development point of view, one must distinguish between retrofitting existing power plants and designing new power plants with CO_2 capture fully integrated. Although retrofitting requires compromises that will lead to power plants that are far from optimal, fully integrated designs can combine CO_2 capture with

---

[10]There are large deposits in the United States and Puerto Rico (Goff and Lackner, 1998; Krevor et al., 2008). There are also large deposits of these minerals in Canada and parts of Europe (Krevor et al., 2008).

[11]Since carbonic acid has proven too weak to dissolve serpentine or olivine with sufficient speed, some of the more promising approaches use mixtures of weak acids that enhance the dissolution of serpentine and olivine while still allowing for their recovery at minimal energy cost (Park et al., 2003; Park and Fan, 2004). Particularly, a pH swing process that dissolves Mg at a low pH and precipitates $MgCO_3$ under a high pH could offer a viable approach for mineral sequestration while producing value-added solid products (Park and Fan, 2004).

the elimination of all other emissions to the atmosphere. Radically new designs could be ultra-efficient and extremely clean and could deliver CO_2 for permanent storage.

The use of fossil fuels unavoidably produces excess CO_2 that under climate change constraints must not be released to the atmosphere. Storing thousands of billions of tons of CO_2 for many centuries presents a formidable challenge to engineering, science, and public policy. If the *MIT Study on the Future of Coal* is correct, CCS will compete with oil shipping and coal mining in the size of its operation. Though daunting, this is unavoidable, in that we are dealing with the back end of the same carbon cycle. Maintenance and operation of the world's energy infrastructure represent a few percent of the world's GDP; closing the carbon cycle will be a good size fraction of that number and has been estimated around 1 percent of the world's GDP [102].

Underground injection of CO_2 is feasible today at an affordable price. Thus, there is no obstacle to starting immediately. It appears likely that additional storage options need to be tapped to provide storage to match the scale of the fossil carbon resource. Mineral sequestration and the storage of CO_2 under deep ocean floors, where CO_2 is denser than the surrounding pore waters [48], offer large additional reservoirs, but these technologies are in their infancies and require further development.

Existing power plants could collect at least some of their CO_2 immediately. Even if retrofitting proves uneconomical in many instances, new power plants could be designed to capture all their CO_2. Air capture technology, because it can be introduced without affecting the existing infrastructure, could offer an alternative, if it is developed to its full potential. Because units can be small, development could be quite fast; commercial applications could be ready in a matter of years rather than decades.

Capture at new integrated power plants could essentially decarbonize the entire coal-fueled power plant sector. Though we did not discuss gas- or oil-fired power plants, it is clear that these could also be decarbonized in a similar manner.

From these discussions emerges a cautiously optimistic view on managing the world's CO_2 emissions. Continued reliance on fossil fuels is a possible option, but the current technology needs to be augmented by CCS. The introduction of such sweeping changes can be managed. Though the required changes are technically demanding, guiding this transformation is mostly a challenge to public policy. In the case of geological storage, the missing components are regulatory frameworks and legal frameworks that assign ownership, liability, and benefits from carbon management. Developing advanced power generation options will ultimately result in far cleaner electric power that can be generated from even the lowest grades of coal.

REFERENCES

[1] AGI (American Geological Institute). Demonstrating carbon sequestration, Geotimes 2003; www.agiweb.org/geotimes/mar03/feature_demonstrating.html#netl

[2] Andrus Jr HAE, Chiu JH, Stromberg PT, Thibeault PR. Alstom's hybrid combustion-gasification chemical looping technology development. In: Proc. 22nd Annu. Int. Pittsburgh Coal Conference, Pittsburgh, PA; September 2005.

[3] Andrus Jr HAE, Chiu JH, Stromberg PT, Thibeault PR. Hybrid Combustion-Gasification Chemical Looping Coal Power Technology Development. ALSTOM technical report for project DE-FC26-03NT41866; 2006.

[4] Anthony JL, Aki SNVK, Maginn EJ, Brennecke JF. Feasibility of Using Ionic Liquids for Carbon Dioxide Capture. Int J Env Tech & Management 2004;4.

[5] Apps JA. A Review of Hazardous Chemical Species Associated with CO_2 Capture from Coal-Fired Power Plants and Their Potential Fate in CO_2 Geologic Storage. Berkeley, CA: LBNL; 65. LBNL-59731. 2006.

[6] Archer D, Kheshgi H, Maier-Reimer E. Multiple Timescales for Neutralization of Fossil Fuel CO_2. Geophys Res Lett 1997;24(4):405.

[7] Balasubramanian B, Ortiz AL, Kaytakoglu S, Harrison DP. Hydrogen from Methane in a Single-Step Process. Chem Eng Sci 1999;54:3543.

[8] Bates ED, Mayton RD, Ntai I, Davis Jr JH. CO_2 Capture by a Task-Specific Ionic Liquid. J Am Chem Soc 2002;124:926–7.

[9] Benson EH. Method for the Production of a Mixture of Hydrogen and Steam. U.S. patent 3,421,869, 1969.

[10] Bergman PD, Winter EM. Disposal of CO_2 in Aquifers in the US. Energy Convers Mgmt 1995;36:523.

[11] Blanchard LAHD, Beckman EJ, Brennecke JF. Green Processing Using Ionic Liquids and CO_2. Nature 1999;399:2.

[12] Blunt M, Fayers FJ, Orr FM. Carbon Dioxide in Enhanced Oil Recovery. Energy Convers Mgmt 1993;34:1197.

[13] Breckenridge W, Allan H, James OYO. Use of SELEXOL Process in Coke Gasification to Ammonia Project. In: Proc. of Laurance Reid Gas Conditioning Conference, Norman, Oklahoma; March 2000.

[14] Broecker WS. Response to "CO_2 Emissions: A Piece of the Pie," Science 2007;316(5826):830

[15] Broecker WS, Takahashi T. Neutralization of Fossil Fuel CO_2 by Marine Calcium Carbonate. In: Andersen NR, Malahoff A, editors. The Fate of Fossil Fuel CO_2 in the Oceans. New York: Plenum Publishing Corp; 1978.

[16] Butler JN. Carbon Dioxide Equilibria and Their Applications. Michigan: Lewis Publisher; 1991.

[17] CCSP (U.S. Climate Change Science Program and the Subcommittee on Global Change Research). Abrupt Climate Change: Final Report, Synthesis and Assessment Product 3.4. Reston, VA: U.S. Geological Survey; 2008.

[18] Charles HW. Hydrogen manufacture. U.S. patent 3,027,238, 1962.

[19] Cho P, Mattisson T, Lyngfelt A. Comparison of Iron-, Nickel-, Copper- and Manganese-Based Oxygen Carriers for Chemical-Looping Combustion. Fuel 2004;83(9):1215.

[20] Chung SJ, Li D, Park JH, Ida J-I, Kumakiri I, Lin JYS. Dual-phase inorganic metal-carbonate membrane for high temperature carbon dioxide separation. Ind Eng Chem Res 2005;44:7999.

[21] Ciferno J, Klara J, Wimer J. Outlook for Carbon Capture from Pulverized Coal and Integrated Gasification Combined Cycle Power Plants. In: Proc. of the Sixth Annu. Conference on Carbon Capture and Sequestration; May 7–10, 2007.

[22] Corbella BM, De Diego LF, García-Labiano F, Adánez J, Palacios JM. Performance in a Fixed-Bed Reactor of Titania-Supported Nickel Oxide as Oxygen Carriers for the Chemical-Looping Combustion of Methane in Multicycle Tests. I & EC Research 2006;45(1):157.

[23] Davidson RM. Post-combustion carbon capture from coal fired plants: solvent scrubbing. International Energy Association Clean Coal Centre; 2007.

[24] Dendy E, Sloan J. Fundamental Principles and Applications of Natural Gas Hydrates. Nature 2003;426:353.

[25] Deutch J, Moniz E. The Future of Coal: Options for a Carbon-Constrained World. Cambridge: MIT Press; 2007.

[26] DOE (U.S. Department of Energy). DOE to Provide $36 Million to Advance Carbon Dioxide Capture. www.netl.doe.gov/publications/press, July 31, 2008a.

[27] DOE (U.S. Department of Energy). Oxy-fuel Combustion Fact Sheet. www.netl.doe.gov, August 2008b.

[28] Dubey MK, Ziock H, Rueff G, Elliott S, Smith WS, Lackner KS, et al. Extraction of Carbon Dioxide from the Atmosphere through Engineered Chemical Sinkage. Preprints of Symposia - American Chemical Society, Division of Fuel Chemistry 2002;47(1):81–4.

[29] EIA (Energy Information Administration). International Energy Outlook-2006. Washington, DC: Office of Integrated Analysis and Forecasting, U.S. Department of Energy; Report # DOE/EIA-0484, June 2006.

[30] EIA (Energy Information Administration). Annual Energy Review 2007. Washington, DC: Office of Energy Markets and End Use, U.S. Department of Energy; Report # DOE/EIA-0384, June 2008.

[31] Fan L-S, Iyer M. Coal Cleans up Its Act. TCE 36, October 2006.

[32] Fan Z, Seitxer A, Hack H. Minimizing CO2 Removal Penalty in Oxyfuel Combustion. In: Proc. of the 32nd Int. Technical Conference on Coal Utilization & Fuel Systems. Coal Technology Association; 2007.

[33] Farzan H, Vecci S, McDonald D, McCauley K, Pranda P, Varagani R, et al. State of the Art Oxy-Coal Combustion Technology for CO$_2$ Control from Coal-Fired Boilers: Are We Ready for Commercial Installation? In: Proc. of the 32nd Int. Technical Conference on Coal Utilization & Fuel Systems. Coal Technology Association; 2007.

[34] Figueroa JD, Fout T, Plasynski S, McIlvried H, Srivistava RD. Advances in CO$_2$ capture technology—The U.S. Department of Energy's Carbon Sequestration Program. Int J of Greenhouse Gas Control 2008;9–20.

[35] Fout TE. Carbon capture r&d: doe/netl r&d program. In: Proc. of the 7th Annu. Conference on Carbon Capture and Sequestration; 2008.

[36] Freund P, Ormerod WG. Progress toward storage of carbon dioxide. Energy Convers Mgmt 1997;38(Suppl.):S199.

[37] Gale J. Geological Storage of CO$_2$, What do we know, where are the gaps and what more needs to be done? Energy 2004;29:1329–38.

[38] Gasior SJ, Forney AJ, Field JH, Bienstock D, Benson HE. Production of synthesis gas and hydrogen by the steam iron process—Pilot-plant study of fluidized and free-falling beds. Bureau of Mines Report of Investigations 1961;5911:49.

[39] Geisbrecht RA. Retrofitting Coal-Fired Power Plants for Carbon Dioxide Capture and Sequestration—Exploratory Testing of NEMS for Integrated Assessments. DOE/NETL-2008/1309, 2008.

[40] Goff F, Lackner KS. Carbon dioxide sequestering using ultramafic rocks. Environ Geosci 1998;5(3):89.

[41] Goldberg P. Mineral sequestration team activities, introduction, issues & plans. Powerpoint presentation for Mineral Carbonation Workshop, August 8, 2001, sponsored by NETL; www.netl.doe.gov/publications/proceedings/01/minecarb/goldberg.pdf; 2001.

[42] Gupta P, Velazquez-Vargas L, Li F, Fan L-S. Chemical Looping Reforming Process for the Production of Hydrogen from Coal. In: Proc 23rd Annu Int Pittsburgh Coal Conference, Pittsburgh, PA; September 2006.

[43] Hänchen M, Prigiobbe V, Storti G, Seward TM, Mazzotti M. Dissolution Kinetics of Fosteritic Olivine at 90–150 °C including Effects of the Presence of CO_2. Geochim Cosmochim Acta 2006;70:4403.

[44] Herzog H, Caldeira K, Reilly J. An Issue of Permanence: Assessing the Effectiveness of Ocean Carbon Sequestration. Climatic Change 2003;59(3):293–310.

[45] Herzog H, Drake E, Adams E. CO_2 Capture, Reuse, and Storge Technologies for Mitigating Global Climate Change—A White Paper. Final Report, DE-AF22-96PC01257. Cambridge, MA: MIT Energy Laboratory; January 1997.

[46] Hoffert MI, Wey Y-C, Callegari AJ, Broecker WS. Atmospheric Response to Deep Sea Injections of Fossil Fuel CO_2. Clim Change 1979;2:53.

[47] Holloway S. Storage of Fossil Fuel–Derived Carbon Dioxide beneath the Surface of the Earth. Ann Rev of Energy and the Environ 2001;26:145.

[48] House KZ, Schrag DP, Harvey CF, Lackner KS. Permanent Carbon Dioxide Storage in Deep-Sea Sediments. Proc Natl Acad Sci U S A 2006;103(33):12291–5.

[49] Hufton JR, Mayorga S, Sircar S. Sorption-Enhanced Reaction Process for Hydrogen Production. AIChE J 1999;45:248.

[50] Hughes E. Biomass Cofiring in the U.S.—Status and Prospects. In: Proc of the Third Chicago/Midwest Renewable Energy Workshop; June 19–20, 2003.

[51] IPCC (Intergovernmental Panel on Climate Change). Carbon Dioxide Capture and Storage. Cambridge: Cambridge University Press; 2005.

[52] IPCC (Intergovernmental Panel on Climate Change). Climate Change 2007: Synthesis Report. Cambridge: Cambridge University Press; 2008.

[53] Ishida M, Takeshita K, Suzuki K, Ohba T. Application of Fe_2O_3-Al_2O_3 Composite Particles as Solid Looping Material of the Chemical-Loop Combustor. Energy & Fuels 2005;19(6): 2514.

[54] Iyer MV, Gupta H, Sakadjian B, Fan L-S. Multicyclic Study on the Simultaneous Carbonation and Sulfation of High-Reactivity CaO. I & EC Research 2004;43:3939.

[55] Jayaraman KS. India's carbon dioxide trap. Nature 2007;445:350.

[56] Jin H, Ishida M. A New Type of Coal Gas Fueled Chemical-Looping Combustion. Fuel 2004;83(17–18):2411.

[57] Katell S, Faber JH, Wellman P. An economic evaluation of hydrogen production by the continuous steam-iron process at seven atmospheres. Bureau of Mines Report of Investigations (No. 6089), 1962:13.

[58] Kather A, Rafailidis S, Hersforf C, Klostermann M, Maschmann A, Mieske K, et al. Research and Development Needs for Clean Coal Deployment. International Energy Association Clean Coal Centre; 2008.

[59] Keith DW, Minh H-D, Stolaroff JK. Climate Strategy with CO_2 Capture from the Air. Climatic Change 2006;74(1–3):17–45.

[60] Kelemen PB, Matter J. In Situ Carbonation of Peridotite for CO_2 Storage. Proc Natl Acad Sci 2008;105(45):17295–300.

[61] Krevor SC, Graves CR, Van Gosen BS, McCafferty AE. Mapping the Mineral Resource Base for Mineral Carbon Sequestration in the United States. In: Proc. of ACEME08—2nd Int. Conference on Accelerated Carbonation for Environmental and Materials Engineering, Rome, Italy; October 1–3, 2008. p. 123–8.

[62] Lackner KS, Wendt CH, Butt DP, Joyce EL, Sharp DH. Carbon Dioxide Disposal in Carbonate Minerals. Energy 1995;20(11):1153–70.

[63] Lackner KS, Butt DP, Wendt CH, Goff F, Guthrie G. Carbon Dioxide Disposal in Mineral Form, Keeping Coal Competitive. LA-UR-97-20941997. Los Alamos, New Mexico: Los Alamos National Laboratory report; 1997.

[64] Lackner KS, Ziock H-J, Grimes P. Carbon Dioxide Extraction from Air: Is It an Option? In: Proc. of the 24th Annu. Technical Conference on Coal Utilization and Fuel Systems; 1999.

[65] Lackner KS, Ziock H-J. The us zero emission coal alliance. VGB Powertech 2001;12:57.

[66] Lackner KS. Carbonate Chemistry for Sequestering Fossil Carbon. Ann Rev Energy Environ 2002;27(1):193.

[67] Lackner KS, Park A-HA, Fan L-S. Carbon Dioxide Capture and Disposal, Carbon Sequestration. In: Lee S, editor. Encyclopedia of Chemical Processing. Dekker Encyclopedias; 2005. p. 305–15.

[68] Lackner KS. Thermodynamics of the Humidity Swing Driven Air Capture of Carbon Dioxide. GRT Report (manuscript in preparation).

[69] Liljedahl GN, Turek DG, Nsakala N, Mohn NC, Fout TE. Alstom's Oxygen-Fired CFB Technology Development Status for CO$_2$ Mitigation. In: Proc. of the 31st Int. Technical Conference on Coal Utilization & Fuel Systems, Coal Technology Association; May 21–25, 2006.

[70] Lin S, Suzuki Y, Hatano H, Harada M. Developing an Innovative Method, hypr-RING, to Produce Hydrogen from Hydrocarbons. Energy Convers Mgmt 2002;43:1283.

[71] Lin S, Harada M, Suzuki Y, Hatano H. Process Analysis for Hydrogen Production by Reaction Integrated Novel Gasification (hypr-RING). Energy Convers Mgmt 2005;46:869.

[72] Lopez-Ortiz A, Harrison DP. Hydrogen production using sorption enhanced reaction. Ind Eng Chem Res 2001;40:5102.

[73] Lyngfelt A, Leckner B, Mattisson T. A Fluidized-Bed Combustion Process with Inherent CO$_2$ Separation, Application of Chemical-Looping Combustion. Chemical Engineering Science 2001;56(10):3101.

[74] Mann MK, Spath PL. A Life Cycle Assessment of Biomass Cofiring in a Coal-Fired Power Plant. Clean Prod Processes 2001;3:81–91.

[75] Marchetti C. On Geoengineering and the CO$_2$ Problem. Climatic Change 1977;1:59.

[76] Matter JM, Takahashi T, Goldberg D. Experimental Evaluation of In Situ CO$_2$-Water-Rock Reactions during CO$_2$ Injection in Basaltic Rocks: Implications for Geological CO$_2$ Sequestration. Geochem Geophys Geosyst 2007;8:Q02001.

[77] Mattisson T, Johansson M, Lyngfelt A. Multicycle Reduction and Oxidation of Different Types of Iron Oxide Particles—Application to Chemical-Looping Combustion. Energy and Fuels 2004;18(3):628.

[78] Mattisson T, Johansson M, Lyngfelt A. The Use of NiO as an Oxygen Carrier in Chemical-Looping Combustion. Fuel 2006;85:736.

[79] McGrail BP, Schaef HT, Ho AM, Chien Y-J, Dooley JJ, Davidson CL. Potential for Carbon Dioxide Sequestration in Flood Basalts. J Geophys Res 2006;111:B12201.

[80] Miller BG. Coal Energy Systems. Amsterdam: Elsevier Academic Press; 2005.

[81] Miller BG, Tillman DA, editors. Combustion Engineering Systems for Solid Fuel Systems. Amsterdam: Elsevier Academic Press; 2008.

[82] NETL (National Energy Technology Laboratory). Carbon Sequestration Atlas of the United States and Canada. 2nd ed. Department of Energy, Office of Fossil Energy; November 2008.

[83] NRC (National Research Council—Committee on Coal Research, Technology, and Resource Assessments to Inform Energy Policy). Coal Research and Development to Support National Energy Policy. Washington, DC: The National Academies Press; 2007.

[84] Oelkers EH, Gislason SR, Matter J. Mineral Carbonation of CO_2. Elements 2008;4:333–7.

[85] Otter N. The Deployment of Clean Power Systems for Coal. In: Proc of the Conference on Clean Coal Technology; May 15–18, 2007.

[86] Overend RP, Milbrandt A. Tackling Climate Change in the U.S.: Potential Carbon Emissions Reductions from Biomass by 2030. In: Kutscher CF, editor. American Solar Energy Society; 2007.

[87] Park A-HA, Jadhav R, Fan L-S. CO_2 Mineral Sequestration, Chemical Enhanced Aqueous Carbonation of Serpentine. Canadian J Chem Eng 2003;81(3–4):885.

[88] Park A-HA, Fan L-S. CO_2 Mineral Sequestration, Physically Activated Dissolution of Serpentine and ph Swing Process. Chem Eng Sci 2004;59(22–23):5241.

[89] Park A-HA, Lackner KS, Fan L-S. Carbon sequestration. In: Gupta RB, editor. Hydrogen Fuel: Production, Transport, and Storage. Boca Raton: CRC Press; 2008. p. 569–601.

[90] Parson EA, Keith DW. Fossil Fuels without CO_2 Emissions. Science 1998;282(6):1053.

[91] Penner L, O'Connor WK, Gerdemann S, Dahlin DC. Mineralization Strategies for Carbon Dioxide Sequestration. In: Proc. of 20th Pittsburgh Coal Conference, Pittsburgh, PA; September 15–19, 2003.

[92] Ramezan M, Skone TJ, Nsakala N, Liljedahl GN, Gearhart LE, Hertermann R, et al. Carbon Dioxide Capture from Existing Coal-Fired Power Plants. DOE/NETL-401/110907, 2007.

[93] Rizeq G, Kulkarni P, Wei W, Frydman A, McNulty T, Shisler R. Fuel-Flexible Gasification-Combustion Technology for Production of H_2 and Sequestration-Ready CO_2. Annu Technical Progress Report to DOE 2002.

[94] Rizeq G, West J, Frydman A, Subia R, Zamansky V, Das K. Advanced Gasification-Combustion Technology for Production of Hydrogen, Power and Sequestration-Ready CO_2. Presented at Gasification Technologies 2003, San Francisco, CA; October 12–15, 2003.

[95] Roark SE, Mackay R, Sammells AF. Hydrogen Separation Membranes for Vision 21 Energy Plants. Proc of the Int Tec Conf on Coal Utilization & Fuel Systems 2002;27(1):101.

[96] Robie RA, Hemingway BS, Fischer JR. Thermodynamic Properties of Minerals and Related Substances at 298.15K and 1 bar (10^5 Pascal) Pressure and at Higher Temperatures. US Geological Bulletin 1452, Washington, DC; 1978.

[97] Rogner H-H. An assessment of world hydrocarbon resources. Annu Rev Energy Environ 1997;22:217–62.

[98] Rubin ES, Chen C, Rao AB. Cost and Performance of Fossil Fuel Power Plants with CO_2 Capture and Storage. Energy Policy 2007;35:4444–54.

[99] Robinson AL, Rhodes JS, Keith DW. Assessment of Potential Carbon Dioxide Reductions Due to Biomass-Coal Cofiring in the United States. Environ Sci Technol 2003;37(22):5081–8.

[100] Ruether J, Dahowski R, Ramezan M, Schmidt C. Prospects for Early Deployment of Power Plants Employing Carbon Capture. Pittsburgh, PA: National Energy Technology Laboratory; 2002.

[101] Simbeck D, Chang E. Hydrogen Supply, Cost Estimate for Hydrogen Pathways: Scoping Analysis. USDOE NREL/SR-540-32525, July 2002.

[102] Stern N. The Economics of Climate Change: The Stern Review. Cambridge University Press; 2007.

[103] Stiegel GJ, Ramezan M. Hydrogen from Coal Gasification, An Economical Pathway to a Sustainable Energy Future. Int J of Coal Geology 2006;65:173.

[104] Stultz SC, Kitto JB. Steam—Its Generation and Use. 40th ed. Barberton: Babcock and Wilcox Company; 1992.

[105] Thomas T, Fan L-S, Gupta P. Combustion Looping Using Composite Oxygen Carriers. U.S. Patent Application Serial No. 11/010,648, 2004.

[106] Tillman DA, Harding NS. Fuels of Opportunity: Characteristics and Uses in Combustion Systems. Amsterdam: Elsevier Academic Press; 2005.

[107] Velazquez-Vargas LG, Li F, Gupta P, Fan L-S. Hydrogen Production from Coal derived Syn Gas Using Novel Metal Oxide Particles. In: Proc. 23th Annu. Int. Pittsburgh Coal Conference, Pittsburgh, PA; September 2006.

[108] Wade JL, Lackner KS, West AC. Transport Model for a High Temperature, Mixed Conducting CO$_2$ Separation Membrane. Solid State Ionics 2007;178(27–28).

[109] Yegulalp TM, Lackner KS, Ziock H. A Review of Emerging Technologies for Sustainable Use of Coal for Power Generation. Int J of Surface Mining, Reclamation and Environment 2001;15(1):52–68.

[110] Zeman FS, Keith DW. Carbon neutral hydrocarbons. Philos Transact Royal Soc, A Math Phys Eng Sci 2008;366(1882):3901–18.

[111] Zhang JM, Zhang SJ, Dong K, Zhang YQ, Shen YQ, Lv XM. Supported Absorption of CO$_2$ by Tetrabutylphosphonium Amino Acid Ionic Liquids. Chem Eur J 2006;12:7.

[112] Zheng L, Tan Y, Pomalis R, Clements B. Integrated Emissions Control and its Economics for Advanced Power Generation Systems. In: Proc. of the 31st Int. Technical Conference on Coal Utilization & Fuel Systems, Coal Technology Association; May 21–25, 2006.

[113] Ziock H-J, Lackner KS, Harrison DP. Zero Emission Coal Power, a New Concept. In: Proc. of First National Conference on Carbon Sequestration, www.netl.doe.gov/publications/proceedings/01/carbon_seq/2b2.pdf, May 14–17, 2001.

The Role of Nuclear Power in Climate Change Mitigation

Geoffrey Rothwell, Ph.D.
Department of Economics, Stanford University

Rob Graber, Vice President
EnergyPath Corporation

Chapter 7

Abstract

Some believe that a nuclear power renaissance will lead to a low-carbon future. Nuclear generation emits no greenhouse gases (GHGs), and, unlike most renewable technologies, it operates at near full capacity. However, after Chernobyl, the United States and Europe rejected investment in new nuclear generation. Further, emerging economies have chosen to build coal-fired power plants. By 2008 coal plants represented over 40 percent of global electricity-generating capacity. As a result, GHG emissions are predicted to double by 2030, even with aggressive new nuclear plant construction. Until the cost of GHG is reflected in fossil-fired electricity generation, nuclear power might not make an appreciable difference in global GHG emissions.

7.1 Greenhouse gas emissions and nuclear power

Some believe that nuclear power provides a good path to a low carbon-electricity future. Two influential environmentalists, Patrick Moore, cofounder of Greenpeace International, and James Lovelock, creator of the Gaia ecosystems theory, have recently endorsed nuclear power:

I believe the majority of environmental activists ... fail to consider the enormous and obvious benefits of harnessing nuclear power to meet and secure America's growing energy needs. These benefits far outweigh any risks. There is now a great deal of scientific data showing nuclear power to be an environmentally sound and safe choice.
—Patrick Moore, before the U.S. Senate Committee on Energy and Natural Resources, April 28, 2005

By all means, let us use the small input from renewables sensibly, but only one immediately available source does not cause global warming and that is nuclear energy.
—James Lovelock, The Independent, May 24, 2004

Why have these two leading environmentalists, and others like them, taken this view? Nuclear plants emit few greenhouse gases (GHGs) over their 40- to 60-year lifetimes and no direct carbon dioxide (CO_2) emissions during their operation.[1] Their availability is now over 90 percent, at least twice the availability of solar and wind. Further, they are a direct replacement for coal plants, the technology that produces 75 percent of global GHG emissions in the electricity sector [6]. The global electricity sector is the largest contributor to GHG emissions, accounting for about 33 percent of the world's total in 2007, as shown in Table 7.1. Transportation is second at 19 percent.

---

[1]Production of nuclear fuel does produce a small amount of GHG if enrichment is done with diffusion technology Meier, (2002).

Table 7.1 Global CO_2 emissions (excludes land use, forestry, and waste), 2007

Sector	Millions of Metric Tonnes of CO_2	(%)
Electricity	10,763	32.6
Transportation	6377	19.2
Manufacturing and construction	5582	16.8
Heat	5066	15.3
Other fuel combustion	3896	11.7
Industrial processes	1455	4.4
Total	33,141	100

Source: WRI [32].

The following equation is an identity showing the drivers of GHG emissions:

$$GHG_i^E = \frac{GHG_i^E}{Electricity_i} \times \frac{Electricity_i}{Energy_i} \times \frac{Energy_i}{GDP_i} \times \frac{GDP_i}{Capita_i} \times Population_i$$

where:

GHG_i^E = GHG emissions from electricity production in country i

$\dfrac{GHG_i^E}{Electricity_i}$ = GHG intensity of electricity production in country i

$\dfrac{Electricity_i}{Energy_i}$ = Electricity intensity of country i

$\dfrac{Energy_i}{GDP_i}$ = Energy intensity of country i

$\dfrac{GDP_i}{Capita_i}$ = GDP per capita of country i

Table 7.2 shows these variables for a sample of four countries. We make three observations:

■ In terms of GHG intensity of electricity production (Column 2), Russia's very low figure is because Russia generates 40 percent of its electricity from natural gas rather than coal. Russia has nearly a quarter of the world's natural gas reserves.

Table 7.2 GHG emissions components by country

Country	GHG_i^E	$\dfrac{GHG_i^E}{Electricity_i}$	$\dfrac{Electricity_i}{Energy_i}$	$\dfrac{Energy_i}{GDP_i}$	$\dfrac{GDP_i}{Capita_i}$	Population$_i$
	(1)	(2)	(3)	(4)	(5)	(6)
	Million Tonnes CO$_2$	Tonnes CO$_2$/MWh	MWh/ MMBtu	Thousand BTU/2005$	2005$/ Capita	Millions
China	1936.2	0.93	34.72	8.25	5591.1	1298.85
India	532.1	0.84	40.69	4.68	3086.1	1075.47
Russia	418.8	0.47	29.6	21.26	9811.2	143.51
United States	2541.4	0.64	39.65	8.54	40,098.2	293.03
ROW	7864.6	0.87	37.51	7.62	8927.0	3557.87

Source: EIA [5], CIA [2], and WRI [32].

- India's relatively low energy intensity of GDP value (Column 4) indicates an energy-efficient economy, but this is related to geography. Warm-weather countries typically have lower GDP intensities, and this is amplified in the case of India by cities that are close together. These are opposite conditions facing Russia, where a cold climate and far-flung cities cause the country to have high energy intensity.

- GDP per capita in Column 5 is the most compelling indicator in the table. These data are a measure of social welfare, and it is the intention of all three non-U.S. countries to address the imbalance. For all their economic growth (~8–10 percent per year), China and India are still poor countries relative to the developed world. For instance, China's GDP/capita is seven times smaller than the United States. Were China to grow its economy to resolve this disparity, it could generate seven times the current GHG emissions from electricity production, thus accelerating climate change.

Can nuclear power save the world from climate change? There were 440 nuclear plants operating in 31 countries at the end of 2007. Although this is a large number, they represent only about 15 percent of the world's electricity-generating capacity [6]. Coal plants comprise 41 percent of electricity production globally and are the dominant generation technology in the world [6]. Investment in nuclear power must be quick if nuclear power is to catch up to coal.

Table 7.3 provides a snapshot of the current status of nuclear power.[2] Column 1 shows the total number of operating nuclear plants; Column 2 shows the total *capacity* of operating nuclear units. Column 3 accumulates country capacities starting with the highest. It shows that the countries with the five highest nuclear capacities represent nearly 70 percent of all the nuclear capacity in the world, and the top 10 comprise 85 percent. Nuclear's percentage of "country total electricity generation," shown in Column 4, demonstrates wide variation. Although the United States has the highest installed nuclear capacity, nuclear generates slightly less than 20 percent of U.S. electricity. In France, however, with the second highest capacity, nuclear generates nearly 80 percent of electricity. At the other end of the spectrum, India and China depend on nuclear for only about 2 percent of their electricity; both countries are highly dependent on coal for electricity generation. Currently, there are 31 nuclear plants under construction in the world.

Offsetting the positive aspects of nuclear power are some challenges. These include high (at-risk) capital costs, long licensing and construction cycles, and high-level waste (HLW) and spent nuclear fuel (SNF) storage and disposal. There are also concerns about security and proliferation of nuclear material.

---

[2]The data in Table 3 are from the EnergyPath Nuclear Database. The database is a compilation of the status of all nuclear plants in the world, whether operating, retired, under construction, planned, or proposed. The database is maintained by EnergyPath Corporation. See EnergyPath (2008).

Table 7.3 Status of nuclear plants by country, December 31, 2007

Country	# Operating Nuclear Units	Capacity of Operating Units (GWe)	Cum % of World Nuclear Capacity	% of Country Electricity	# Nuclear Units Under Construction
	(1)	(2)	(3)	(4)	(5)
United States	104	100.5	27.0	19.8	0
France	59	63.2	44.0	78.6	1
Japan	56	47.9	56.8	26.4	2
Russia	31	21.7	62.7	15.5	5
Germany	17	20.3	68.1	27.5	
South Korea	20	16.8	72.6	35.9	3
Ukraine	15	13.1	76.1	47.9	
Canada	18	12.6	79.5	14.7	
United Kingdom	19	10.5	82.3	20.0	
Sweden	10	10.0	85.0	50.3	
China	11	8.6	87.3	2.3	6
Belgium	7	5.8	88.9	56.3	
Taiwan	6	4.9	90.2	18.8	2
India	17	3.7	91.2	2.4	5
ROW*	50	32.8		14.2	7
Total	440	372.4	100.0	15.7	31

ROW operating units: Argentina (2), Armenia (1), Brazil (2), Bulgaria (2), Czech Republic (6), Finland (4), Hungary (4), Lithuania (1), Mexico (2), Netherlands (1), Pakistan (2), Romania (2), Slovakia (5), Slovenia (1), South Africa (2), Spain (8), Switzerland (5). ROW construction units: Argentina (1), Bulgaria (2), Finland (1), Iran (1), Slovakia (2).
Source: EnergyPath [7].

Taking into account nuclear's current unresolved issues would likely reduce the potential for nuclear power to positively impact GHG emissions. So, for the purposes of assessing the *maximum* potential impact of nuclear power on GHG emissions, it will be assumed that these issues can be successfully resolved.

The alternative to building a nuclear plant is generally to build a coal plant, currently using pulverized coal technology, but possibly in the future using "clean" coal technology with sequestration of CO_2. Both are baseload technologies, that is, they run at near full capacity. Although natural gas produces lower levels of GHG than coal, natural gas is not as easy to deliver and store as coal. Some natural gas-fired plants are sometimes baseloaded, but they are usually used to follow demand for electricity by ramping up and down. Furthermore, natural gas supplies are interruptible, as Chile and the Ukraine discovered when Argentina and Russia, respectively, shut off their natural gas imports. This chapter assumes that nuclear power plants can substitute directly for coal plants. On this basis, nuclear power could offset about 20 percent of total global GHG emissions that would otherwise be emitted.

Using the U.S. Energy Information Administration's (EIA) 2008 *International Energy Outlook* (IEO) reference scenario, "business as usual," current trends in global generating capacity additions will increase GHG emissions from electricity to about twice the 2007 levels by 2030, as shown in Figure 7.1.

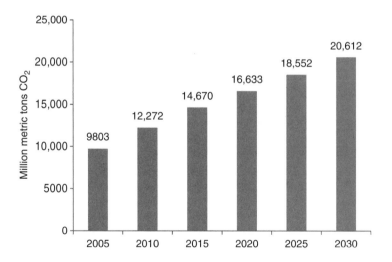

Figure 7.1 Global CO_2 projections from electricity ($MtCO_2$).[3]
Source: EIA [6] and EPA [9].

---

[3]The data in Figure 7.1 was calculated using the projections of electricity by fuel from EIA (2008) and CO_2 emission rates for coal (2249 lbs. CO_2/MWh) and natural gas (1135 lbs. CO_2/MWh) in EPA (2007).

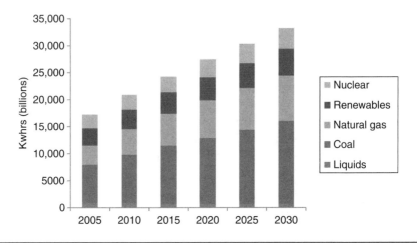

Figure 7.2 EIA global electricity growth forecast (billion kWh), 2005–2050. Source: EIA [6].

As shown in Figure 7.2, this increase in GHG emissions is predicated on rapid growth in coal and natural gas-generating resources to fuel global economic growth, particularly in China and India.[4] This growth is expected to outpace the potential nuclear construction rate so that the percentage of electricity generated by nuclear declines from 15 percent to 11 percent in spite of the fact that electricity production from nuclear power plants increases over 42 percent. Electricity from renewables, including hydro, declines from 18 percent to 15 percent. Coal increases its overall share of global electricity production to 46 percent by 2030.

The 2008 EIA Reference Case [6] reflects a more optimistic view of nuclear power capacity increases relative to earlier forecasts. The 2008 Reference Case projects nuclear capacity in 2025 to be 31 percent, higher than the 2005 forecast [5]. However, as Figure 7.1 illustrates, even with this more optimistic view of nuclear power's prospects, GHG emissions still double from 2005 to 2030. With this background, this chapter discusses whether there is a policy by which nuclear power can play a more pronounced role in GHG emissions control than envisioned by forecasting agencies such as the EIA.

Section 2 focuses on the cost of nuclear power, particularly its capital costs. Section 3 discusses nuclear power's critical growth regions, where nuclear power can achieve more meaningful GHG reductions. Section 4 looks at the contribution that nuclear power can make to GHG emissions control under

---

[4]For a discussion on China, refer to Lewis et al., Chapter 17 in this volume.

various scenarios using more aggressive but credible construction forecasts. The final section presents the chapter's conclusions.

7.2 The cost of nuclear power

Obviously, cost plays a large role in the choice of generating technologies (e.g., see Rothwell [26]). And, at least in the United States, nuclear power had, until very recently, priced itself out of the power markets because nuclear's capital costs ballooned for plants whose construction was completed between 1982 and 1996. This is also true to a lesser extent in Europe, except that political considerations have played an equally negative role following the Chernobyl nuclear accident in 1986. However, countries in East Asia, such as Japan, Taiwan, and South Korea, with few indigenous energy resources, have continued to build nuclear plants, although at a reduced rate. The significant capital cost increases experienced in the United States were avoided in those countries, largely because there was more standardization and fewer changes in regulation.

When nuclear power plant capital costs and associated uncertainty increased, electric utilities turned to coal [24]. Coal technology was mature and less expensive, and there was not the concern about GHG emissions as there is today. But coal-fired power plant capital costs escalated as SO_x and NO_x emissions came under scrutiny by environmental agencies. However, the cost of GHG control will greatly increase the cost of coal-fired generation [16].[5] Moreover, this cost is uncertain because "clean" coal technologies are neither technically nor commercially proven. This fact makes coal plant construction currently risky. Coal-fired power plants are being cancelled or deferred. Many U.S. electric utilities and independent developers are choosing to build nuclear baseload capacity, even though the capital costs are likely to be high, especially for the first few plants. Part of the reason for this construction is that there are federal subsidies for the first few new nuclear plants as part of U.S. energy policy, discussed in Section 3. Regardless of the costs, state regulatory commissions are approving the selection of the nuclear alternative (e.g., FPSC [11]).

The initial indications were that this emerging second round of nuclear construction would be considerably cheaper than the first round (1957–1996), but the cost of global commodities, such as steel, copper, and cement, have risen due to high global demand and higher transportation costs [1]. There is also

---

[5]For a discussion of clean coal technology, refer to Chapter 6, by Lackner et al., in this volume.

a global labor shortage owing to an enormous backlog of global engineering, procurement, and construction (EPC) projects.

However, it should be emphasized that it is the total cost of generating electricity that dictates the choice of generating alternatives, not just capital costs. Although nuclear's capital costs might be high relative to other baseload alternatives, the cost of generating electricity can be lower, especially as prices for coal and natural gas increase.

Nevertheless, nuclear's high capital costs are relevant in that they limit the number of nuclear plant developers that have the assets to qualify for financing, which could limit nuclear power plant construction. The 2008 recession and credit crisis is likely to have an influence on nuclear's prospects because of the technology's large capital requirements. Nuclear plants nearing construction are likely to be delayed as the financial crisis works itself out. However, most nuclear plants, as in the United States, are in the early phases of licensing, and licensing could take 4 to 5 years. Licensing costs represent less than 2 percent of total plant funding requirements and are thus financially manageable by most developers, even without external funding.

Ultimately, nuclear power could be competing with "clean" coal technologies with sequestration of CO_2. These technologies are expected to capture about 90 percent of the CO_2 that would otherwise be emitted. Their ability to compete with nuclear will depend critically on their capital, O&M, and fuel costs, and these costs are too technically and commercially uncertain to adequately compare at this stage of their development. However, these technologies will have significantly higher generating costs than traditional pulverized coal plants.

As discussed in Section 3, overnight capital cost estimates of new nuclear plants in the United States could be as high or higher than $5000/kWe, resulting in a new two-unit 2200 MWe plant costing $14 billion to $17 billion when fully constructed, not including interest during construction or escalation [4]. Estimates of capital costs for Asian plants do not reflect the U.S. and European experience, although the rising cost of commodities will likely affect these plants as well. In China, announced overnight capital costs of a Russian-built plant were only $1600/kWe in 1998 [21]. The two European plants now under construction, in Finland and France, have been announced to be $2400/kWe and $2600/kWe, respectively[6] [15]. Recent South Korean nuclear plants have announced overnight capital costs of $1800/kWe. Japanese plants have ranged between $1800/kWe and $2800/kWe [15]. Note that all the nuclear plants being built in the United States and Europe are newer designs representing the next generation of nuclear technology without licensing or

---

[6]The Finnish plant is a turnkey plant experiencing substantial cost and schedule overruns. The latest projections are that the final costs could be closer to $3375/kWe.

construction experience. The nuclear plants in Japan and South Korea are mature, standardized designs with substantial experience.

7.3 Nuclear's critical growth regions

If nuclear power is to play a significant role in global GHG emission abatement, there are four countries that are crucial: China, India, the United States, and Russia. These four countries will be referred to as *critical growth regions*, or CGRs: These countries are the most likely to build nuclear power plants in substantial numbers. Countries that have traditionally favored nuclear power but that are not included, such as France, Japan, and South Korea, will build nuclear plants but with slower growth rates. The United States is included because nuclear power is likely to increase its share of the electricity markets disproportionately as concern over coal-fired plant emissions intensifies, a trend already in progress. The same is true for Russia, which is likely to emphasize nuclear power domestically to allow the export of more natural gas.

It might be surprising that Europe is not considered a CGR. With 186 nuclear reactors in 16 countries, Europe is currently dependent on nuclear power to a large extent. France has the second largest nuclear fleet in the world. However, following the Chernobyl accident in 1986, a near moratorium on nuclear power has descended on Western European countries. Germany, Spain, and Sweden have made political commitments to shut down existing nuclear plants before the end of their lives. Italy has already shut down its nuclear power industry. There are programs to expand nuclear power in some Eastern European countries, but these programs could result in only a small number of nuclear reactors. There are currently two nuclear reactors under construction in Finland and France. But this is far fewer than were experienced in the 1970s and 1980s (Figure 7.3). Since then and until recently, a very large and well-organized antinuclear coalition has evolved and has been effective in discouraging nuclear plant construction.

There are, however, signs that the current antinuclear sentiment in Europe could be changing. The U.K. government, long discouraging nuclear power, has recently changed position and is now encouraging nuclear power because of its concerns over GHG emissions. This is a pattern that is starting to emerge in other European countries—for example, Germany, Italy, Spain, and Sweden could reconsider nuclear power—but there are no solid indications of a "nuclear renaissance."[7]

---

[7]In September 2008, British Energy agreed to be acquired by French nuclear giant EDF. If the deal is completed, this sets the stage for a significant new nuclear build in the United Kingdom.

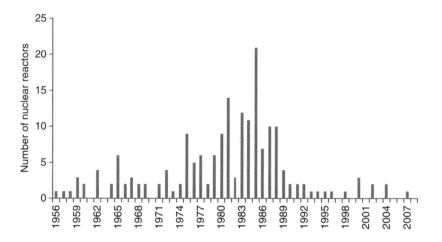

Figure 7.3 Western Europe nuclear construction, 1956–2008.
Source: EnergyPath [7].

7.3.1 THE UNITED STATES

With 104 operating reactors, the United States has the largest nuclear fleet in the world and could be experiencing a "nuclear renaissance" [29]. As Figure 7.4 shows, the first beginning ended badly. By the late 1980s nuclear plant construction costs had risen to levels exceeding $3000/kWe, driven by the post–Three Mile Island escalation of capital costs, lengthening construction cycles,

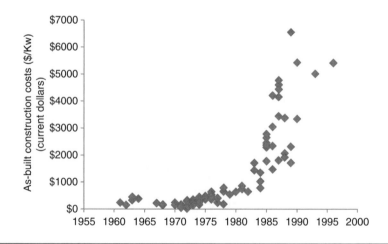

Figure 7.4 As-built construction costs ($/kWe; nominal dollars), 1955–2000.
Source: EnergyPath [7].

and high interest rates. As mentioned in Section 7.2, the costs shown in Figure 7.4 are considerably higher than what was projected when construction began. Because of these higher-than-expected capital costs leading to (1) the largest bond default in U.S. history, by the Washington Public Power Supply System, and (2) some state rate commissions' disallowance of cost overruns into rate bases, financial markets were closed to nuclear power starting in the mid-1980s and continuing until today. Table 7.4 shows current estimates of nuclear capital costs, including those of two U.S. electric utilities planning to build nuclear plants.

It is instructive to compare current nuclear plant capital cost estimates from Table 7.4 with those of the last nuclear construction period. Figure 7.5 shows

Table 7.4 Recent capital cost estimates, 2003 and 2007

Source	Year	Overnight Capital Cost Estimates	As-Spent Capital Cost Estimates
Moody's	2007		$5000–6000/kW
Keystone	2007	$3600–4000/kW	
FPL	2007	$3456–4540/kW	$6256–8005/kW
Progress Energy	2003		$6300/kW

Source: Cummins [3], FPL [12], Keystone [17], and Moody [20].

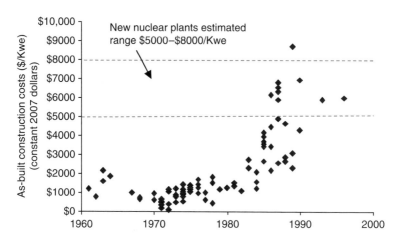

Figure 7.5 U.S. nuclear plant as-built construction costs ($/kWe; in constant 2007 dollars), 1960–2000.
Source: EnergyPath [7].

the as-built construction costs in constant 2007 dollars. The horizontal lines represent the range of expected as-built capital costs for the first new U.S. plants from Table 7.4 and show that the current estimates are at the high end of the as-built capital costs of the previous nuclear construction period. Actual capital costs will not be known until the first plants of the second construction period are ordered and constructed. Follow-on plants (after the first four to six for each new nuclear technology) are expected to have lower capital costs due to learning and series effects.

As a component of U.S. energy policy, there have been legislative attempts to revive the nuclear industry, but financial markets have remained skeptical and unenthusiastic. Thus, to revive the nuclear power alternative, federal energy policy has increasingly focused on opening financial markets to nuclear power. Although not specifically intended to address nuclear's financial issues, the first of these, the Energy Policy Act of 1992 (EPAct92), had the opposite effect on nuclear power. This was because it initiated wholesale power market deregulation and left nuclear in the position of having significant above-market costs, which were believed to be unrecoverable in competitive electricity markets. A potential financial catastrophe was avoided when nuclear electric utilities agreed to deregulation with recovery of "stranded" costs within competitive power markets [28]. Recovery of these costs through charges to retail customers resulted in nuclear plants being able to compete effectively in wholesale power markets. With stranded costs recovered and much of the original capital investment amortized, modern nuclear plants are cash generators for their owners due to their low marginal operating cost.

Without addressing the problem of closed financial markets, EPAct92 did not result in new nuclear plants. However, there were two nonlegislative developments that did attempt to address nuclear's financing issues. First, the U.S. Nuclear Regulatory Commission (NRC) introduced a new licensing process. This was because the earlier process was thought to be responsible for the lengthy delays in commercial operation after construction completion, because it required the licensee to initiate an operating license application only after the construction process was completed. This led to construction issues being readdressed in the operating license application, increasing the time required to complete the licensing process.

The new process combines the construction and operating licenses (COLs) into a single license issued at the beginning of the project. Thus, once the NRC approves the constructed plant as having met all acceptance criteria, the licensee can bring the plant into commercial operation without having to apply for an operating license. In addition, the new process introduces standardization of nuclear plant designs so that the NRC does not have to review each license as though it were a new design. Also, with an Early Site Permit (ESP) a licensee can license a specific location for a nuclear power plant without having to specify the reactor design and without making a commitment to build immediately. The ESP is good for at least 20 years.

The second significant development was the U.S. Department of Energy's (DOE's) Nuclear Power 2010 (NP2010) program, which envisioned an industry/government collaboration to develop advanced nuclear plants, with the government sharing the first-of-a-kind (FOAK) development costs. The vision was (1) to demonstrate the viability of the new licensing process and (2) to construct new plants by 2010 for $1500/kWe (overnight). As part of the new licensing process, a licensee would reference a design already certified by the NRC, thus saving time and cost over the old licensing process. Although successful in the development of a new generation of more advanced reactors, so-called Generation 3 nuclear power plants, NP2010 did little to restore access to financial markets. By 2005, reservations about the role that nuclear power would play in U.S. energy and climate change policy were dispelled by the Energy Policy Act of 2005 (EPAct05), which was the most pro-nuclear legislation in a half-century [8]. EPAct05 attempted to open financial markets to nuclear plants. Table 7.5 summarizes EPAct05's provisions.

With the additional knowledge that climate change legislation was certain (although the timing of it is still uncertain), these incentives resulted in the first nuclear orders in the United States in 35 years. In November 2007 a consortium led by NRG Energy, an independent power producer with $10 billion in assets, filed a COL application with the NRC. The loan guarantee provisions of EPAct05 had a positive impact on NRG and its financial partners. Also, filing as early as possible would ensure that the two-unit plant, South Texas Project 3 & 4, would receive the benefits of the production tax credits and standby support; to qualify for the production tax credits, a nuclear power plant developer must have filed a COL application by December 31, 2008.

After NRG Energy filed its COL application, the NRC was notified of the intent of other nuclear plant developers to file for COLs. Figure 7.6 shows the number of nuclear plants expected to go into commercial operation over the period 2008–2020 based on estimates of COL applications by the NRC [22].[8] Figure 7.6 is intended to demonstrate the lag time required to license and construct the first nuclear plants. Up to 23 plants could enter commercial operation from 2017 to 2019, if all plants are built. Because six or fewer will have access to the subsidies in Table 7.5, it is unlikely that all will finish the race by 2020.

---

[8]Estimates of license applications are from the U.S. NRC. The year of commercial operation is based on EnergyPath (2008) estimates of licensing and construction durations.

Table 7.5 Nuclear-related provisions, Energy Policy Act of 2005

Provision	Description	Limitations
Loan guarantees	Debt and interest (up to 80% of the project costs) are guaranteed by the U.S. Treasury	Total available for all new plants, $18,500 million (this amount was determined in 2007)
Standby support	Insurance against construction delays caused by the new one-step licensing process, up to $500 million for the first two plants and $250 million for the next four plants	No more than three advanced nuclear designs can participate
Production tax credits	Up to $18/MWh in tax credits for the first 8 years of operation	$7.5 billion for qualifying nuclear plants; PTC phases out as electricity price exceeds $80/MWh
Decommissioning funds taxation	Decommissioning funds taxed at reduced rates (20%)	
Liability limitation extension	Extends Price Anderson limits of accident liability through 2025	Nuclear industry not responsible for damage exceeding $10.6 billion as the result of a nuclear accident

Source: Energy Policy Act of 2005 [8].

7.3.2 CHINA

China currently operates 11 nuclear reactors comprising just over 2 percent of its total electricity generation. Over 77 percent of China's electricity is derived from coal, one of the highest of any country. This is the cause of China's high GHG intensity of electricity production, shown in Table 7.2. China has the most aggressive nuclear power development program in the world, as shown in Figure 7.7 ([27]; Lewis et al., [18] Chapter 17 in this volume). This figure shows

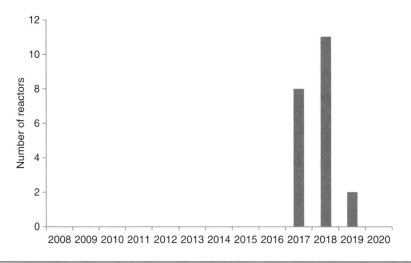

Figure 7.6 Expected number of U.S. nuclear plant commercializations based on projected COL applications, 2008–2020.
Source: EnergyPath [7] and NRC [22].

the expected operational nuclear capacity for the CGRs, according to the EIA Reference Case, in which China's growth rate for nuclear additions is over 8 percent per year; the U.S. growth rate is projected at 0.5 percent per year.

China has depended on foreign countries (e.g., France, Russia, and Canada) for its nuclear reactors, but it is developing an indigenous capability

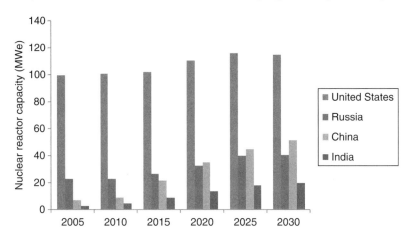

Figure 7.7 EIA Reference Case of nuclear reactor capacity for critical growth regions, 2005–2030.
Source: EIA [6].

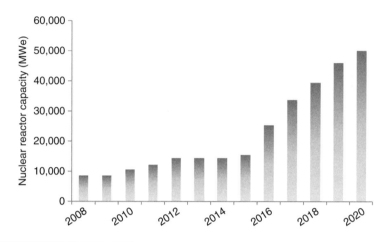

Figure 7.8 China annual nuclear construction (in MWe), 2008–2020. Source: EnergyPath [7].

as the result of technology transfers from nuclear suppliers providing nuclear reactors to China. The current planned nuclear capacity on a year-to-year basis for China is shown in Figure 7.8.

7.3.3 INDIA

India is one of the world's fastest-growing economies, has the world's second largest population, and is the second largest consumer of coal for electricity production in the world after China. India also has an aggressive nuclear program, although only 6 GWe of nuclear capacity are currently operating. These are mostly small reactors (~200 MWe) that employ natural uranium and heavy water moderation. The most important aspect of the Indian program is that it has been effectively quarantined from the world nuclear community because (1) it has refused to sign the Nuclear Nonproliferation Treaty (NPT) and (2) it has acquired nuclear weapons capability as a direct result of its commercial nuclear program.[9] The initial two reactors acquired by India were based on U.S. BWR technology requiring enriched uranium, but all subsequent nuclear designs were based on the use of natural uranium. However, this should change with the conclusion of the U.S.-India Nuclear Supply Treaty that could open the way for India to partially join the world nuclear

---

[9]India is one of only three countries that have not signed the NPT. The others are Israel and Pakistan.

community under partial IAEA safeguards.[10] Also, there are two reactors of Russian design, pressurized light water reactors known as water-water electricity reactors (WWERs), with fuel supplied by Russia, starting in 2008 and 2009; there are plans for two additional WWERs to be completed by 2017.

Further, India is developing a unique indigenous nuclear capability tailored to its objective of being energy independent. This strategy relies on India using its large thorium resource base, the second highest in the world after Australia, with about 25 percent of the world's total thorium reserves. India has only limited resources of uranium, which would not support a long-term nuclear program using its current nuclear technologies. India's strategy is to develop a three-stage nuclear program that strives for energy independence by maximizing the energy production from its indigenous uranium and thorium resources. However, this strategy puts emphasis on technologies that are not yet commercially proven and in some cases are technically unproven.

The three-stage program is shown in Figure 7.9 and comprises:

- Pressurized heavy water reactors (PHWRs)
- Fast breeder reactors (FBRs)
- Advanced heavy water reactors (AHWRs)

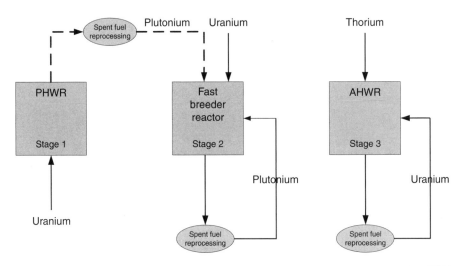

Figure 7.9 India three-stage nuclear program.
Source: Based on Raj [23].

[10]The International Atomic Energy Agency (IAEA), an international organization established in 1957, seeks to promote the peaceful uses of atomic energy. It has since evolved into the world's nuclear "watchdog" and reports to both the U.N. General Assembly and Security Council.

The Stage 1 inefficiently uses only about 1 percent of the potential energy from uranium. However, it does produce small amounts of plutonium that are needed to start up Stage 2, shown by the dotted line. The FBRs (Stage 2) use uranium much more efficiently and produce large amounts of plutonium for reprocessing and recycling. Finally, Stage 3 will use India's large thorium resources with reprocessing and recycling.

The centerpiece of this three-stage program is the fast breeder reactor (FBR). This strategy is risky because FBRs have not achieved commercial and technical success.[11] It also depends on economically reprocessing spent fuel, already being done commercially (although not necessarily economically) in France and the United Kingdom and is under development in Japan.[12] Moreover, the first prototype FBR is not expected to begin operation until after 2010. The first AHWR will probably not be operational until after 2020. Figure 7.10 shows India's expected nuclear capacity through 2020.

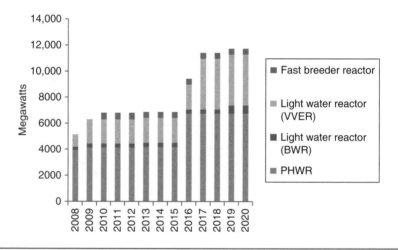

Figure 7.10 Projected India nuclear capacity (MWe), 2008–2020. Source: EnergyPath [7].

---

[11]There have been 11 breeder reactors in the world; eight have been shut down and one never started. They have a history of technical disappointment and have the highest capital costs of any reactor type.

[12]Reprocessing has been shown not to be economically justified until uranium prices exceed $300/lb. of U_3O_8 (Fetter et al., 2003).

7.3.4 RUSSIA

Russia is included as a CGR because it is embarking on an ambitious program to greatly increase nuclear generation, partly to maximize its exports of natural gas. The aim of Russian strategy is to increase the percentage of electricity generated from nuclear plants to 25 percent from the current 15 percent by 2030. This amounts to about 26 new nuclear plants during the next 20 years [30]. Russia's nuclear program, initially with a rapid construction rate, nearly came to a halt after the Chernobyl nuclear accident in 1986 and the subsequent dissolution of the Soviet Union in 1991. Figure 7.11 shows nuclear capacity growth from 1956 to 2008. Since 1991, Russia has added only three nuclear reactors. The current fleet is about 23 GWe.

Russia's decision to rapidly construct nuclear power plants after allowing its current nuclear fleet to deteriorate reflects conditions in the global natural gas and oil markets as well as the fact that Russia's economic growth rate has averaged over 7 percent per year since the financial crisis of 1998. As one of the largest natural gas exporters in the world, GazProm can get up to five times more for its natural gas in export markets than by selling it in domestic electric markets. Currently natural gas comprises about 40 percent of Russia's generation (with coal providing another 24 percent and hydro 17 percent).

Russia also intends to expand its current status as a leading exporter of nuclear power plants, with plans to export up to 60 nuclear plants in the next

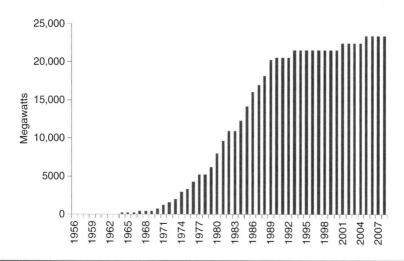

Figure 7.11 Russia nuclear capacity (MWe), 1956–2008.
Source: EnergyPath [7].

20 years. It is currently building nuclear plants in India, China, Iran, and Bulgaria after having upgraded its standard WWER design to a level of safety commensurate with nuclear designs of Western countries. Russia is also planning to build FBRs and currently has one of the two operating breeder reactors in the world (the BN-600). An advanced version of its FBR technology is expected to enter commercial operation in 2012. This also entails an advanced program to fabricate mixed-oxide fuel (MOX) for FBRs.[13] Figure 7.12 shows the expected nuclear growth in Russia through 2020. As is evident, the WWER technology is the primary growth technology. The RBMK technology (the technology employed in the Chernobyl reactors) is not being developed further.

7.3.5 OTHER MARKETS

Although not included in the category of critical growth regions, several other countries—notably Japan, South Korea, and Canada—have large nuclear programs that will contribute to the established global nuclear base [25].

Although plagued by incidents that have marred its nuclear safety record, Japan is continuing to rely on nuclear power almost by necessity, given the

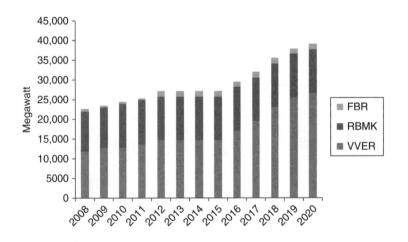

Figure 7.12 Russia expected nuclear growth (MWe), 2008–2020. Source: EnergyPath [7].

---

[13]MOX comprises two fissile elements (uranium and plutonium) in an oxide matrix. The fuel currently used in the vast majority of nuclear plants worldwide employs uranium only in an oxide matrix.

country's virtually nonexistent energy resource base. Also, Japan is home to three of the six major nuclear reactor suppliers, including the parent of U.S.-based Westinghouse Nuclear Systems, one of the largest nuclear suppliers in the world.[14] All three Japanese nuclear suppliers are very active in Western and Asian nuclear export markets, particularly in the United States. The Japanese electric power sector is privately owned, and Tokyo Electric Power (TEPCO) has one of the largest privately owned nuclear fleets in the world.

South Korea also has aggressive nuclear plans, although with a smaller capacity than Japan, but for the same reason: lack of any substantial indigenous resource base to support the country's heavily industrialized economy. The country depends extensively on oil and natural gas imports and is the fifth largest importer of petroleum in the world. Coal is still the leading fuel for the generation of electricity, followed by natural gas, with nuclear power coming in third. Although the country has modest coal reserves (relative to consumption), it is still a net importer of coal. The electric power sector is currently being privatized, but privatization is moving slowly.

Canada has one of the world's richest energy resource bases and a modest nuclear base. Nevertheless, the country is planning to expand its nuclear base of PHWRs, a technology that it pioneered starting in 1944 and has exported to Argentina (one unit), China (two units), India (two units), Pakistan (one unit), Romania (two units), and South Korea (four units). Its PHWR technology is seen as an alternative to the much more globally prevalent LWR.[15] Nuclear capacities for the three countries are shown in Figure 7.13.

7.4 Nuclear power's impact on GHG emissions

The technologies used to generate electricity in the CGRs are shown in Figure 7.14, along with the rest of the world (ROW) and Europe. Nuclear power comprises about 15 percent of world total electricity generation, with about 20 percent of the world total generated in the United States and Europe. Russia, China, and India's nuclear generation account for about 1 percent of the world's total generation. Major nuclear producers in ROW are Japan and South Korea (2.4 percent) and Canada (<1 percent). Figure 7.14 also

---

[14]The six nuclear power exporters are Atomic Energy of Canada, Ltd. (AECL), Areva (France), Atomstroyexport (Russia), GE-Hitachi (United States/Japan), Mitsubishi Heavy Industries (Japan), and Toshiba (Japan). South Korea's Doosan Heavy Industries may soon export nuclear power plants.

[15]Canada's PHWRs, known as Canadian Deuterium Uranium (CANDUs), are noteworthy because they employ natural uranium fuel and can be refueled during operation. The use of natural uranium avoids the expensive process of enrichment, always required in LWRs.

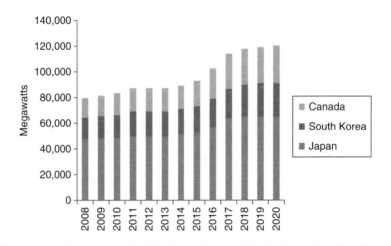

Figure 7.13 Japan, South Korea, and Canada nuclear capacity (MWe), 2008–2020.
Source: EnergyPath [7].

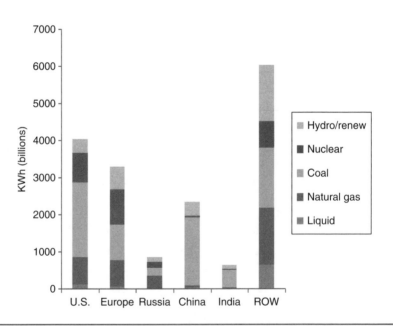

Figure 7.14 Sources of electricity by country and fuel type (billions of kWh), 2005.
Source: EIA [6].

shows the predominance of coal in the CGRs, particularly in China and the United States, which together account for 40 percent of the world's coal reserves.

As shown in Figure 7.15, the 2030 relative shares of fuels in the CGRs do not change significantly. The EIA's Reference Case assumes that current laws and policies remain unchanged through 2030 and so do not account for any potential impacts from a GHG reduction agreement after 2009.

Although there is additional nuclear power in each of the CGRs, coal remains the chief resource for power generation. This is reflected in global GHG emissions from electricity in EIA's Reference Case shown in Figure 7.16. Also shown in Figure 7.16 are the GHG emissions that could result if nuclear power was not present in the EIA forecast. GHG emissions will continue to climb through 2030, despite the projected growth in nuclear power, and are nearly twice as high in 2030 as they were in 2005. Nuclear power, though playing a role, does not eliminate a large amount of GHG emissions in the EIA Reference Case.

Figure 7.17 compares the EIA Reference Case with three other nuclear capacity forecasts from 2005 to 2030:

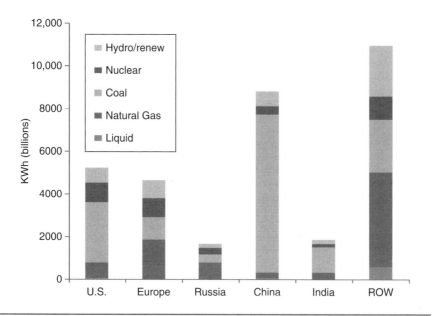

Figure 7.15   EIA Reference Electricity Case by country and fuel type (billions of kWh), 2030.
Source: EIA [6].

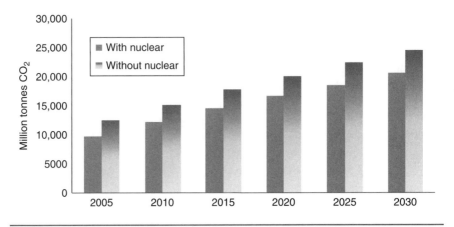

Figure 7.16 GHG emissions EIA Reference Case (MtCO$_2$), 2005–2030.
Source: EIA [6] and EPA [9].

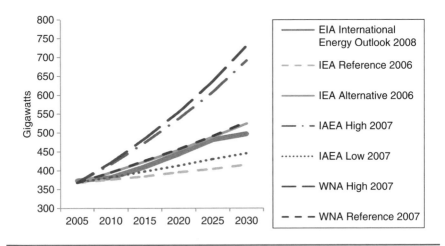

Figure 7.17 Comparison of global nuclear capacity forecasts (GWe), 2005–2030.
Source: EIA [6], IAEA [13], IEA [14], and WNA [31].

- The International Atomic Energy Agency (IAEA) 2007 High and Low forecasts
- The International Energy Agency (IEA) 2006 Reference and Alternative scenarios
- The World Nuclear Association (WNA) 2007 High and Reference forecasts

The solid thick line shows the EIA Reference Case, which is a little lower than midway between the highest and lowest forecasts. With disparate forecasts, as shown in Figure 7.17, it is evident that a wide range of views exist regarding the future role of nuclear power.

To analyze the impact of a higher nuclear capacity forecast on EIA projections, we chose the highest forecast, that is, the WNA's 2007 High Forecast, and substituted this forecast into the EIA world electricity forecasts, reducing the coal capacity by an amount equal to the increase in nuclear capacity. The GHG emissions were recalculated using EPA estimates of carbon intensity for coal and natural gas [9]. Figure 7.18 shows the results.

Even with the much higher WNA forecast, GHG emissions continue to increase over the period, although some progress is achieved in reducing absolute levels of GHG emissions. However, what is apparent is that even at the highest nuclear capacity forecast level, there is only a moderate reduction in GHG emissions in spite of the fact that the WNA nuclear forecast is 50 percent higher than the EIA Reference Case. So, starting today and building at the highest rates envisioned by reputable forecasting agencies, including agencies that promote nuclear power, does not result in an appreciable reduction of GHG emissions. The reason that nuclear power by itself is not capable of significantly reducing GHG emissions is primarily that the nuclear power capacity in 2005 was small relative to the huge coal-fired capacity and because China and India will continue to rely extensively on traditional coal to power their fast-growing economies, at least until "clean" coal technologies can be

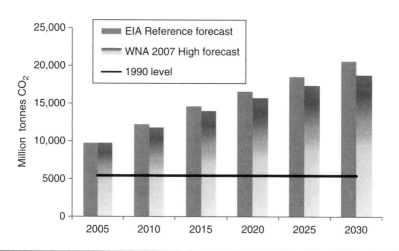

Figure 7.18   GHG emission comparison (MtCO$_2$), EIA Reference versus WNA High 2005–2030.
Source: EIA [6], EPA [9], and WNA [31].

implemented in the future. In 2005 nuclear comprised only 15 percent of global electricity, whereas coal and natural gas comprised over 60 percent. This might not be the case had new nuclear operation continued at its 1985 level, as shown in Figure 7.19. In 1985 44 GWe of new nuclear capacity was placed into commercial operation globally. From 1983 through 1988 an average of more than 25 GWe of new nuclear capacity was added annually. For the past 23 years, nuclear plants have represented a decreasing share of the world's new generation. This led to nuclear increasing its share of world generation through 1996 (17.6 percent) and decreasing thereafter, as shown by the solid line in Figure 7.20.

Figure 7.20 shows nuclear power's share of world electricity, assuming:

- The maximum nuclear capacity construction (in 1985)
- The average nuclear capacity construction from 1983 to 1988

These assumptions were used to forecast future GHG emissions from 1985 forward. The share of nuclear power stabilizes at 46 percent of world generation in 2005 in the case of the maximum construction and at 30 percent in the

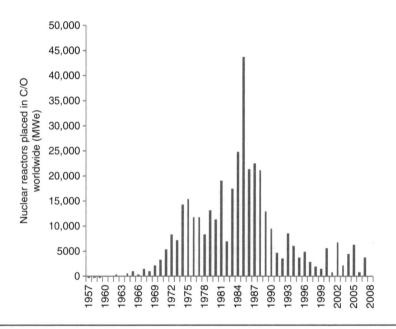

Figure 7.19 Nuclear reactors placed in commercial operation, worldwide (MWe), 1957–2008.
Source: EnergyPath [7].

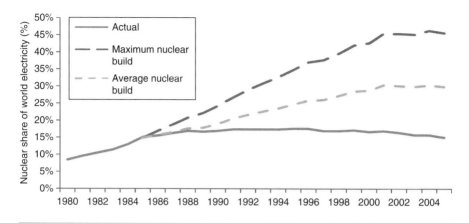

Figure 7.20 Nuclear share of world electricity generation (%), 1980–2005. Source: EnergyPath [7].

case of the average nuclear capacity construction. Using these construction assumptions through 2030 results in the GHG emissions shown in Figure 7.21.

Using either the "average" new construction or the "maximum" new construction assumption results in considerable reductions. In the case of the maximum new construction, the 1990 levels of GHG emissions from electricity might have been achieved by 2030. Therefore, to significantly influence GHG emissions, future construction rates must be at least as high as the average construction rate from 1983 to 1988 through 2030 and beyond—that is, there must be 25 to 40 GWe of new nuclear capacity coming into operation as soon as possible, and it must continue through 2030.

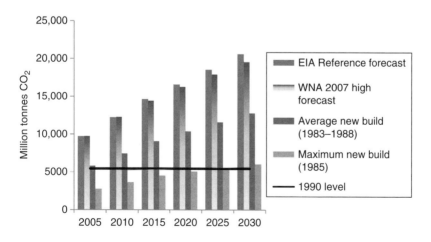

Figure 7.21 GHG emissions under all scenarios (MtCO$_2$), 2005–2030. Source: EIA [6], EnergyPath [7], and WNA [31].

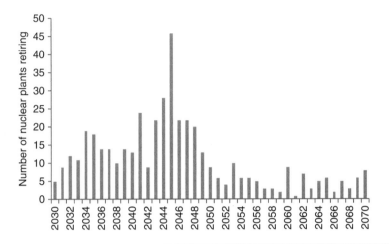

Figure 7.22 Retiring nuclear plants, 2030–2070.
Source: EnergyPath [7].

However, another potential problem emerges following 2030. Between 2030 and 2070, 454 operating nuclear plants could be retired, or about 10 per year. This is based on a 60-year lifetime for all operating plants, where most require relicensing to operate after the first 40 years of life—for example, Belgium mandates that all nuclear plants must be retired after 40 years. Thus, to sustain any GHG emissions advantages of nuclear power after 2030, these plants would need to be replaced in addition to the nuclear plants required to meet increasing electricity demand. Figure 7.22 shows the number of nuclear reactors that would need to be replaced after 2030.

7.5 Conclusion

The role of nuclear power in abating GHG emissions in the global electricity generation sector might not have the immediate impact that its supporters claim unless investment in nuclear power capacity starts immediately and is sustained through 2030. Like renewables, nuclear power produces no GHG emissions during operation, but there are too many global carbon dioxide–emitting generation sources. It will take decades for these plants to be replaced by cleaner technologies, such as "clean" coal, nuclear, or renewables.

In response, all critical growth regions are proposing aggressive new nuclear construction. However, new nuclear power plants are large investments and can require up to 10 years to license and construct, although this is likely to decrease to 5 years with learning. Capital costs will remain an issue, especially in the United States, until several plants of each new technology are constructed and Nth-of-a-kind costs can be observed.

Thus, though nuclear power can make a contribution toward stabilizing GHG emissions, new nuclear construction plans could fall short of what is needed to make a substantial impact on global GHG emissions. Until GHGs are accounted for in electricity costs and prices, it is unlikely that national programs will aggressively invest in nuclear power continuously in the coming decades.

On the other hand, nuclear power is the only central-station, GHG-free alternative that could replace global ever-growing, ever-polluting coal-fired capacity. If utilities and nations are prepared to significantly increase their investment in nuclear power plant construction, nuclear power is capable of making an important contribution to GHG reduction and climate stabilization.

REFERENCES

[1] Chupka MW, Basheda G. Rising Utility Construction Costs: Sources and Impacts. The Brattle Group, Prepared for the Edison Foundation (September); 2007.

[2] CIA. CIA World Factbook. U.S. Central Intelligence Agency. https://www.cia.gov/library/publications/the-world-factbook/; 2005.

[3] Cummins WE, Corletti MM, Schulz TL. Westinghouse AP1000 Advanced Passive Plant. Westinghouse Electric Company, LLC Proceedings of ICAPP '03 Cordoba, Spain, May 7, 2003, Paper 3235, Table 1.

[4] Dewhurst M. Financial Perspectives on Nuclear Power in a Consolidating Electric Power Industry. FPL Group, Platts 3rd Annual Nuclear Energy Conference, Washington, DC, 2007 (February).

[5] EIA. U.S. Energy Information Administration, International Energy Outlook 2005. U.S. Department of Energy, DOE/EIA-0383(2005); 2005.

[6] EIA. U.S. Energy Information Administration, International Energy Outlook 2008, U.S. Department of Energy, DOE/EIA-0484; 2008.

[7] EnergyPath. EnergyPath Nuclear Database. 2008.

[8] Energy Policy Act of 2005, U.S. 109th Congress, Public Law 109-58 http://frwebgate.access.gpo.gov/cgi-bin/getdoc.cgi?dbname=109_cong_public_laws&docid=f:publ058.109

[9] EPA. Clean Energy: Air Emissions. U.S. Environmental Protection Agency; 2007.

[10] Fetter S, Bunn M, Holdren J, van der Zwann B. The Economics of Reprocessing vs. Direct Disposal of Spent Nuclear Fuel, Project on Managing the Atom. Belfer Center for Science and International Affairs, John F. Kennedy School of Government, Harvard University, Final Report (December); 2003.

[11] FPSC. FPSC Approves Need for Two New Nuclear Units at FPL Turkey Point Facility. State of Florida Public Service Commission, News Release (3/18/2008), www.psc.state.fl.us/home/news/index.aspx?id=376; 2008.

[12] FPL. Direct testimony of Steven D. Scroggs, Florida Power and Light Company, before the Florida Public Service Commission, Docket No. 070650-EI, October 16, 2007; 2007.

[13] IAEA. Energy, Electricity and Nuclear Power Estimates for the Period 2005–2030. Vienna: International Atomic Energy Agency, www.iaea.org; 2007.

[14] IEA. World Energy Outlook. Paris: International Energy Agency, www.worldenergyoutlook .org; 2006.

[15] Joskow P. Prospects for Nuclear Power: A U.S. Perspective. University of Paris-Dauphin, http://econ-www.mit.edu/files/1187; 2006.

[16] Johnstone B (Saskatchewan News Network). Plant's Clean Coal Retrofit Would Cut Power Output. Star Phoenix, www.canada.com/saskatoonstarphoenix/news/business/story.html?id = 3482511b-2636-4619-a1fe-82afe31ec3d1; March 6, 2008.

[17] Keystone. Nuclear Power Joint Fact Finding. Colorado: The Keystone Center, www .keystone.org; June, 2007.

[18] Lewis J, Xiliang Z, Chai Q. China. Chapter 17 in this book.

[19] Meier PJ. Life Cycle Assessment of Electricity Generation Systems and Applications for Climate Change Policy Analysis. Fusion Technology Institute, University of Wisconsin (Madison); UWFDM-1181 August, 2002.

[20] Moody's 2007. New Nuclear Generation in the United States: Keeping Options Open vs. Addressing an Inevitable Necessity. Moody's Corporate Finance; October 2007.

[21] NEI. Chinese Finally Sign Up for the WWER-91. Nuclear Engineering International; March, 1998.

[22] NRC (U.S. Nuclear Regulatory Commission). Expected New Nuclear Power Plant Applications. www.nrc.gov/reactors/new-reactors/new-licensing-files/expected-new-rx-applications.pdf; 2008.

[23] Raj B. Knowledge Management in a Nuclear Research Center. Presented at International Conference on Knowledge Management in Nuclear Facilities, International Atomic Energy Agency, Vienna; 18–21 June, 2007.

[24] Rothwell GS. Electric Utility Power Plant Choice under Investment Regulation. Ph.D. Dissertation, Berkeley: Department of Economics, University of California; 1985.

[25] Rothwell GS. Nuclear Energy in Asia, editor. Special Issue of the Pacific and Asian Journal of Energy 1998;8:1.

[26] Rothwell GS. Nuclear Power Economics. In: The Encyclopedia of Energy. Academic Press/ Elsevier Science; 2003a.

[27] Rothwell GS. Standardization, Diversity, and Learning in China's Nuclear Power Program, History Matters: Essays on Economic Growth, Technology, and Demographic Change. Stanford University Press; 2003b.

[28] Rothwell GS, Gomez T. Electricity Economics: Regulation and Deregulation. IEEE Press/ John Wiley; 2003.

[29] Rothwell GS, Graber R. Will There Be a Nuclear Renaissance in the U.S. by 2020? Stanford Institute for Economic Policy Research, Stanford University; 2008.

[30] Uranium-stocks.net. Putin Signs to Double Russian Nuclear Power. www.uranium-stocks .net; July 17, 2007.

[31] WNA. The Global Nuclear Fuel Market: Supply and Demand 2005–2030. London: World Nuclear Association, www.world-nuclear.org; 2007.

[32] World Resources Institute (WRI). Climate Analysis Indicator Tool. http://cait.wri.org/cait .php; 2008.

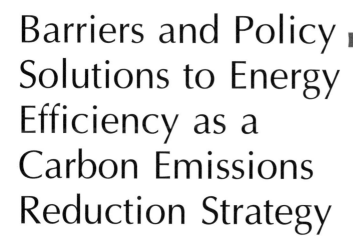

Barriers and Policy Solutions to Energy Efficiency as a Carbon Emissions Reduction Strategy

Bill Prindle
ICF International

Jay Zarnikau
Frontier Associates

Erica Allis
Consultant at the UN Environment Program

Abstract

Energy efficiency is a low-cost, rapidly deployable, and large-scale energy resource. Reducing growth in energy demand is essential to any clean energy strategy: without efficiency advances, clean energy supplies might not keep up with demand, and carbon emissions could continue to grow. Despite efficiency's benefits, many of the policy mechanisms intended to reduce greenhouse gas emissions will not fully value energy efficiency in proportion to its economic potential unless policy and market barriers are reduced. The authors describe efficiency's potential contribution to reducing carbon emissions, identify policy and market barriers to efficiency investments in a climate policy context, and outline policy and program solutions.

8.1 Introduction

The potential contribution of energy efficiency toward reducing electricity demand growth and greenhouse gas emissions is vast. A number of nations, states, provinces, and local governments have sought to include energy efficiency as a key component of their climate change mitigation strategies. Yet achieving reductions in energy use through efficiency measures requires overcoming a number of formidable economic and institutional barriers. The introduction of energy efficiency resources into cap-and-trade mechanisms presents additional and unique challenges related to policy design.

For the purposes of this chapter, *energy efficiency* is defined as the application of technologies and facility operating practices to deliver energy services at a reduced rate of energy commodity consumption. Energy services encompass lighting, space conditioning, shaft power, electronic functions, industrial processes, and other end uses of energy in buildings and industrial facilities. Energy commodities include electricity or other energy types used to drive the device or system that delivers the energy service. It may be noted that economists tend to define energy efficiency much differently—as the level of energy usage associated with performing an overall task at a minimum cost.[1] In many cases, a higher level of energy use may be "efficient," in an economic sense, if it reduces overall costs by reducing the need for capital, labor, or other inputs or fosters increases in productivity. This metric is known as energy *intensity*— the amount of energy used for a unit of economic activity. Efficiency and intensity are generally the inverse of each other—increased efficiency reduces intensity—although the relationship is more complex in that intensity can also include other factors such as structural changes in the economy and the changes in product mix and the movement of manufacturing activities induced by growing international trade.

---

[1] Berndt E. Aggregate Energy, Efficiency, and Productivity Measurement. Annual Review of Energy 1985;3:225–73.

This chapter is organized into the following sections:

■ Section 8.2 describes the size and characteristics of the energy efficiency resource.

■ Section 8.3 defines barriers to efficiency investment, in terms of both policy design and market barriers.

■ Section 8.4 provides an overview of government policies that have sought to include energy efficiency as a component of their climate change mitigation strategies, including a detailed analysis of the challenges inherent in the integration of energy efficiency into the proposed cap-and-trade programs.

■ Section 8.5 presents several cases based on actual policy and program experience.

■ Section 8.6 is a summary of findings and conclusions.

8.2 The magnitude of the energy efficiency resource

Energy efficiency can be treated in analytical and policy terms as an electricity resource, since the need for supply-side energy resources can be deferred or displaced by the deployment of high-efficiency technologies in new and existing buildings and industrial facilities and through changes in energy consumption patterns or practices. Efficiency has occurred and continues to occur, to an extent, through market forces, in response to high energy prices, technology advances, and other factors. Studies have shown, however, that potential exists for substantial additional investment.

The International Energy Agency has recently projected the contribution that energy efficiency can make to greenhouse gas emission reduction.[2] Figure 8.1 illustrates the potential for energy efficiency to reduce demand growth in IEA member countries.

Efficiency also provides a major share of the GHG emission reductions in typical economy-wide policy/technology scenarios. The IEA's Alternative Policy Scenario, summarized graphically in Figure 8.2, shows efficiency providing the majority of emission reductions.[3]

In the United States, energy efficiency efforts since the 1970s have played a major role in reducing overall energy intensity. A recent American Council

[2]International Energy Agency. Energy Use in the New Millennium: Trends in IEA Countries; 2007.
[3]International Energy Agency. World Energy Outlook; 2006.

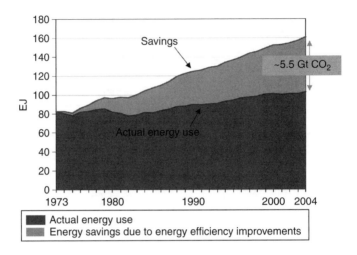

Figure 8.1 Effects of energy efficiency on GHG emissions: United States, Japan, Australia, Germany, France, the United Kingdom, Spain, Poland, the Netherlands, and Belgium, 1973–2004. Energy values in exajoules.
Source: IEA (2007).

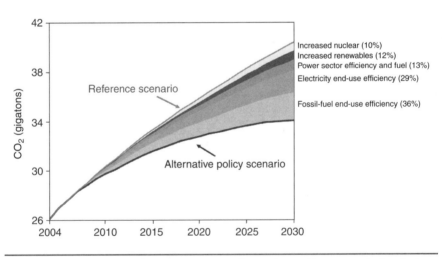

Figure 8.2 International Energy Agency alternative policy scenario (emissions on Y-axis expressed as gigatons of CO_2).
Source: IEA (2007).

for an Energy-Efficient Economy (ACEEE) study notes that U.S. energy intensity, defined as Btu per dollar of GDP, as of 2008 has been slashed to half the 1970 level, from 18,000 Btus to about 8900 Btus.[4] However, a fuller disaggregation of the change in energy consumption into various components (e.g., changes in industrial structure, changes in product mix, and changes in intensity), though beyond the scope of this analysis, would produce a more complex picture.

Another recent study concludes that states with aggressive energy efficiency efforts have reduced their rate of growth in electricity demand by about 60 percent, relative to the growth that would have occurred absent programs.[5] A leading energy economist found evidence that states with strong commitments to energy efficiency successfully reduced commercial and industrial electricity intensity through the promotion of energy-efficient technologies and practices, helping to retain economic activity with the state, although gains in the residential sector were not as apparent.[6] A new index from the U.S. DOE suggests that energy intensity dropped by 10 percent from 1985 to 2004, with the greatest gains occurring in the industrial sector.[7] However, there is some evidence that these figures may overstate energy efficiency achievements, due to the manner in which various types of energy resources are aggregated by using energy-content values rather than economic metrics.[8] Although there is some evidence that the cost-effectiveness and impacts of energy efficiency programs have been overstated,[9] other studies refute those claims.[10] The analytic record supports a general observation that although other factors matter, efficiency has accounted for the majority of energy intensity improvements.

---

[4]Ehrhardt-Martinez K, Laitner J. The Size of the U.S. Energy Efficiency Market: Generating a More Complete Picture. ACEEE; 2008.

[5]Berry D. The Impact of Energy Efficiency Programs on the Growth of Electricity Sales. Unpublished manuscript, March 2008.

[6]Horowitz M. Changes in Electricity Demand in the United States from the 1970s to 2003. The Energy Journal 28:93–119.

[7]U.S. Department of Energy. U.S. Energy Intensity Indicators: Highlights of Energy Intensity Trends — Total Energy. Available at: http://intensityindicators.pnl.gov/total_highlights.stm (last visited September 7, 2007).

[8]Zarnikau J. Will Tomorrow's Energy Efficiency Indices Prove Useful in Economic Studies? The Energy Journal 1999; and When Different Types of Energy Resources Are Aggregated for Use in Econometric Studies, Does the Aggregation Approach Matter? Energy Economics 1999.

[9]Joskow P, Marron DB. What Does a Megawatt Really Cost? Evidence from Utility Conservation Programs. The Energy Journal 1992;13(4):41–74; and Loughran D, Kulick J. Demand-Side Management and Energy Efficiency in the United States. The Energy Journal 2004;25(1):19–41.

[10]Auffhammer M, Blumstein C, Fowlie M. Demand-Side Management and Energy Efficiency Revisited. The Energy Journal 2008;29(3):91–104.

The quantification of energy efficiency potential is typically performed by comparing a "business-as-usual" forecast of energy consumption to the level that would occur if all end users adopted higher-efficiency technologies.[11] Potential studies are typically categorized into three types: technical, economic, and achievable potential.

Technical potential represents the savings that are technically feasible regardless of the cost of energy efficiency measures. These savings are typically calculated on an instantaneous basis (assuming that all end-use equipment is immediately replaced with best available efficient equipment) or on a phase-in basis (assuming that equipment is replaced with the most efficient equipment readily available in the marketplace at the end of the useful life of the existing equipment). *Economic potential* refers to the subset of technical potential that can be realized cost-effectively, typically based on a life-cycle cost basis, where the net present value of energy savings is compared to the net present value of measure costs. Finally, *achievable* or *market potential* is an estimate of the energy efficiency savings that can be reasonably expected from programs and policies, taking into account constraining market realities.

One proxy for achievable efficiency potential can be based on the measured impacts of best-practice programs and policies in use today. The National Action Plan for Energy Efficiency reports that energy efficiency programs are realizing significant energy savings in other areas of the country. Savings "on the order of 1 percent of electricity and natural gas sales" are "helping to offset 20 to 50 percent of expected growth in energy demand in some areas."[12]

Eleven studies examined in an ACEEE study suggest that very substantial technical, economic, and achievable energy efficiency potential are available in the United States.[13] Across all sectors, these studies show a median technical potential of 33 percent for electricity and 40 percent for gas, and median economic potentials for electricity and gas of 20 percent and 22 percent, respectively. The median achievable potential is 24 percent for electricity (an average of 1.2 percent per year) and 9 percent for gas (an average of 0.5 percent per year).

McKinsey Global Institute suggests that the global growth in energy demand could be cut in half over the next 15 years from energy efficiency projects with an internal rate of return of 10 percent or more.[14] Figure 8.3

[11]Meier A, Wright J, Rosenfeld A. Supplying energy through greater efficiency: the potential for conservation in California's residential sector, 1983.

[12]NAPEE, p. ES-4.

[13]Nadel S, Shipley A, Neal Elliott R. The Technical, Economic, and Achievable Potential for Energy Efficiency in the US: A Meta Analysis of Recent Studies. In: 2004 ACEEE Summer Study on Energy Efficiency in Buildings, August 2004.

[14]McKinsey Global Institute. Curbing Global Energy Demand Growth: The Energy Productivity Opportunity, May 2007.

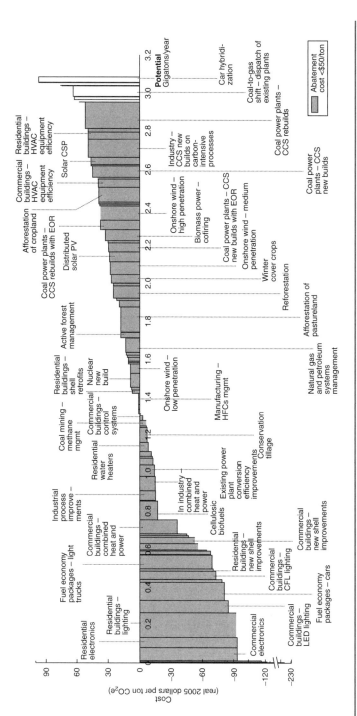

Figure 8.3 U.S. midrange carbon dioxide abatement cost curve, 2030.
Note: The McKinsey report only examines a scenario through 2030. NRDC recommends a goal of 80 percent emission reductions by 2050.
Source: McKinsey analysis, NRDC (2007).

illustrates this analysis, based on a subsequent McKinsey analysis for the Natural Resources Defense Council.[15]

8.3 Integration of energy efficiency into climate change mitigation policies

8.3.1 BARRIERS AND CHALLENGES IN REALIZING ENERGY EFFICIENCY POTENTIAL FOR CARBON EMISSION REDUCTIONS

Energy efficiency's potential for reducing electricity demand growth, reducing CO_2 emissions, and reducing the cost of climate policies has been well established.[16] Few analysts challenge the economic value efficiency can bring to solving the climate problem. However, there is less agreement on how efficiency can best be tapped for this purpose. One school of thought holds that a carbon cap-and-trade policy will, in and of itself, generate an optimal pattern of investment as covered entities seek the lowest-cost technologies for meeting their compliance requirements. The authors of this chapter take a different view: Because of the structural particulars of cap-and-trade systems, energy efficiency is not likely to be widely used as an emission reduction measure that is traded in carbon markets, as discussed here. And even if the structural issues in cap-and-trade design could be resolved, several factors would continue to limit efficiency investment. The weakness of net elasticity effects, the continued presence of large end-use market barriers, and utility regulatory barriers based on long-held institutional practices would continue to hamper optimal efficiency investment. This chapter describes these barriers and their implications, and offers policy options that can reduce or eliminate these barriers.

8.3.2 POLICY BARRIERS

8.3.2.1 Cap-and-trade policy design issues (see also Adib et al.)

The emissions cap-and-trade policy designs most often proposed to reduce carbon emissions do not, in and of themselves, provide sufficient impetus for the level of efficiency investment needed to realize their full benefits. These

---

[15]Duke R, Lashof D. The New Energy Economy: Putting America on the Path to Solving Global Warming. Natural Resources Defense Council; 2007.

[16]See U.S. Environmental Protection Agency. National Action Plan for Energy Efficiency. Available at: www.epa.gov/cleanenergy/energy-programs/napee/index.html; International Energy Agency. Promoting Energy Efficiency Investments. Available at: www.iea.org/Textbase/publications/free_new_Desc.asp?PUBS_ID = 2009; 2008.

limitations stem principally from the disparity between the economic location of CO_2 emissions caps and the location of energy efficiency resources. This asymmetry between the location of the caps and the location of efficiency investment opportunities creates serious structural problems that can keep efficiency-based emissions reductions from participating in carbon trading markets. This would severely inhibit the level of efficiency investment under a cap-and-trade scheme and would thus increase the overall economic impact of the policy. In simple terms, benefits are disconnected from the costs or accrue to different parties.

Most cap-and-trade policies in place or under discussion today place emissions caps "upstream," or at the energy production or conversion level. In the power sector, caps are likely to be placed at the power plant level. End-use efficiency potential, by contrast, is realized "downstream" in individual buildings served by electric distribution utilities. The arguments typically made for upstream caps are that they are administratively simpler and economically more efficient than downstream approaches. Administrative simplicity stems in part from the fact that it is easier to measure emissions from a well-known and well-monitored set of smokestacks than to estimate the greenhouse gas emissions associated with the production and transportation of a consumer product or service. Further, if the administering agency can deal with fewer entities, operation of the cap-and-trade system is arguably simpler. Fawcett et al. take this principle further by discussing personal carbon allowances (PCAs) that would push caps all the way downstream to individual consumers.

However, placement of the cap upstream makes it difficult for covered entities to invest in downstream energy use reductions, because such reductions are treated as "indirect" emission reductions and so are not generally accepted as tradable allowance credits. The indirect-reduction issue refers to the fact that even if energy users reduce consumption, upstream emitters possess no fewer allowances and thus can operate high-emitting sources longer or can sell the unused allowances over the term of the compliance period. From an emissions trader's viewpoint, end-use energy consumption can be affected by many factors, including weather and economic conditions, beyond any effects energy efficiency investment might have. Emissions traders therefore do not view end-use efficiency as a basis for carbon allowance credit that is tradable in a market operating under an upstream cap.

Some jurisdictions have recognized that downstream caps, at least in the power sector, can offer advantages, especially in terms of encouraging energy efficiency investment. In Oregon, for example, the Governor's Carbon Allocation Task Force recommended a "load-based" cap design in which the emissions cap would be placed on distribution utilities rather than on power generators.[17] This load-based cap approach recognizes that if the entity that directly delivers energy to an end user is required to hold carbon emission allowances, the number of allowances it must hold are *directly* related to the amount of electricity it sells to customers. Therefore, if the distribution utility assists customers in reducing electricity usage by a certain amount, it can claim direct emission reductions as a result.

This load-side cap approach has garnered some attention in part because with CO_2, the lowest-cost emission reductions are typically not at the point of emission, or the "smokestack," as has been the case with other U.S. Clean Air Act pollutants. Sulfur dioxide, particulates, and nitrogen oxides can be controlled by fuel switching and flue gas scrubbing technologies relatively cost-effectively. But carbon capture and sequestration from power generators is more complex and expensive; it is therefore more appropriate to consider cap-and-trade designs that more effectively capture the lowest-cost emission reduction technology opportunities. A load-based cap is one way to accomplish such an objective. The combination of prevailing economic theory and the "weight of tradition" appears to keep upstream caps the most likely policy design approach.

Even if a cap-and-trade design were modified such that a generator could directly claim emission reductions from customer end-use efficiency gains, other barriers would serve to discourage the use of this option. First, reducing customer usage reduces generators' revenue; in the case of unregulated power generators, there is no way to compensate generation owners for such financial losses. Unregulated generators would see a double risk in this respect: (1) they would lose marginal revenues from the kWh sales involved in individual transactions of this kind, and those marginal prices typically would be many times the allowance value of the same kWh, and (2) if overall energy sales fell far enough in the overall market, generators would see marginal prices fall overall, affecting revenue and profit for their entire sales base.

Distribution utilities, however, are typically regulated by state or local governments, and rate-making practices can be modified to align regulated utility financial interests with the public interests of energy and climate policy.[18] Examples are provided here of states that have taken policy steps to address these issues.

---

[17]Nelson, H. Cost and Design Considerations for Reducing Carbon Dioxide from Oregon's Power Sector: A Report to the Oregon Carbon Allocation Task Force; 2006.

[18]National Action Plan for Energy Efficiency. Aligning Utility Incentives with Investment in Energy Efficiency. Prepared by Val R. Jensen, ICF International. Available at: www.epa.gov/eeactionplan; 2007.

It can also be difficult for upstream entities to reach across markets to effectively identify, aggregate, and market such reductions. Because of the transaction costs associated with bundling together large numbers of individual efficiency investments, this challenge would be economically inefficient for upstream entities without direct links to the affected markets. Distribution utilities, on the other hand, have direct contact with virtually every building and industrial facility in their service area through customer account records and other information sources. They have personnel and other infrastructure for managing large-scale efforts, which can reduce transaction costs, making them a much more suitable candidate for delivery of efficiency resources. This tends to support the case for distribution utilities as the most effective aggregators of efficiency resources.

8.3.2.2 Utility regulatory barriers

The regulation of investor-owned electric utility rates in the United States and many other nations has a long-standing practice of *volumetric pricing,* in which utility costs are recovered in proportion to electricity sales. Each kilowatt hour sold thus carries a small share of the utility's fixed and variable costs. Volumetric rate-making has several advantages, including conveying price signals to customers that encourage the wise use of electricity. If fixed costs were recovered in lump-sum fashion in monthly customer charges, for example, the price assigned per kilowatt hour would be lower, and customers would be encouraged to use more energy.

Volumetric pricing, however, has the disadvantage of creating disincentives to significant utility investment in energy efficiency. When efficiency programs reduce electricity sales below forecast levels, utilities fail to recover the portion of their fixed costs that would have been recovered through those "lost revenues." The impacts on revenues and profits can get to be significant if efficiency program impacts grow large.[19]

This lost-revenue problem can be addressed through rate-making mechanisms, often referred to as *revenue-decoupling* policies. In simple terms, decoupling typically involves creating a true-up mechanism so that any revenues lost (or gained) in relation to original forecasts can be recovered (or refunded) through an annual rate adjustment mechanism. The details of decoupling mechanisms are important to get right so that the effects of weather, energy efficiency, economic conditions, and other factors are treated accurately and fairly.

Although decoupling can remove a major disincentive to utility investment in efficiency, it is only one of three elements of regulatory policy that affect

---

[19]Ibid.

utilities' financial interest in energy efficiency investment. Cost recovery is a basic issue for utilities; costs must be recoverable in a timely manner for the utility to be willing to make such investments. Most states have established mechanisms that adequately address cost recovery. The third element of regulatory policy that must be addressed is utility incentives: Beyond recovering costs and keeping revenues whole, utilities must be able to earn a reasonable return on efficiency investments if they are to become willing partners in large-scale efforts to stimulate customer energy efficiency investment. Some U.S. states have begun to address this issue, though only a few have fully implemented incentive policies. Incentives instituted to date tend to take the form of paying utilities a percentage of the net benefits from efficiency programs or allowing a specified rate of return on approved program costs.

8.3.3 MARKET BARRIERS

Even if carbon cap-and-trade policies did not contain structural barriers to end-use efficiency investment, large and persistent existing market barriers would stifle investment in many major end-use markets. Although there is ample literature on numerous barriers,[20] there are three principal types of barriers that classical economists commonly recognize as valid considerations in policy discussions:

- ◼ *Principal-agent problems.* Also called *split-incentive barriers*, principal-agents problems occur where one entity (the agent) makes investment decisions on behalf of another (the principal) whose outcomes affect the principal's total cost of owning, occupying, and operating a building or an energy-using product or system. In end-use energy markets, homebuilders typically act as agents for homeowners, making decisions on home energy efficiency that affect energy costs for the life of the building or its energy systems. Landlords for residential and commercial rental property play similar roles. In large organizations, purchasing departments act as

---

[20]For more discussion of barriers, see Sorrell S, O'Malley E, Scott S, Schleich J. The Economics of Energy Efficiency: Barriers to Cost-Effective Investment. Cheltenham: Edward Elgar. Available at www.e-elgar.co.uk/bookentry_main.lasso?id = 2607; 2004. And Bjornstad D, Brown M. A Market Failures Framework for Defining the Government's Role in Energy Efficiency. JIEE 2004–02. Joint Institute for Energy & Environment. Available at: www.jiee.org/pdf/ 2004_02marketfail.pdf; 2004. And Sathaye J, Bouille D. Barriers, Opportunities, and Market Potential of Technologies and Practices. Chapter 5 of Climate Change 2001: Mitigation. A Report of Working Group III of the Intergovernmental Panel on Climate Change. Available at: www .grida.no/climate/ipcc_tar/wg3/pdf/5.pdf; 2000.

agents for facility operators. In each case, the agent tends to seek minimum capital costs for the building or energy system, because it does not pay the ultimate energy bill.

■ *Information-cost or transaction-cost problems.* In most energy end-use markets, individual transactions on the purchase of energy-using technologies are small, and consumers and business owners lack the specialized knowledge needed to identify the "optimum" energy performance. Energy is also competing with other attributes in such products, such as size, performance, ease of use, color, and convenience. The result is that underinformed decision makers in these small transactions are not willing or able to incur the cost (often expressed as time) of researching energy performance characteristics of competing models to make economically optimum decisions. By contrast, energy supply projects are typically much larger in scope, and their energy performance is vital to their economic value. Such large projects can and do incur the transaction costs (engineering studies, financial analysis, etc.) needed to reach an optimum decision. This leads to a pattern of chronic underinvestment in energy efficiency in the millions of end-use equipment decisions made every year.

■ *Failure to internalize externality costs.* Economists have long recognized that some costs do not appear in market prices for commodities or goods and that these costs can ultimately reduce overall welfare in the economy. In the context of this book, environmental damage is the classical externality cost. Over the past several decades, the electricity industry has begun to internalize the costs of air pollution and other environmental externalities by investing in pollution control equipment and by switching to lower-emission fuels and technologies. Greenhouse gases are the newest externality issue, with CO_2 the main pollutant of concern in the electricity industry, although sulfur hexafluoride (SF_6) and other GHG molecules also occur in the electricity sector. Both cap-and-trade policies and carbon taxes would serve to internalize these externalities. However, in the following section the authors discuss the limits of pricing carbon as an emission reduction strategy and as a stimulus for efficiency investment.

The amount of energy use affected by market barriers can be substantial. ACEEE recently managed a research effort funded by the International Energy Agency in which five country teams used a common methodology to calculate the amount of energy that is isolated from the effects of price signals by measurable market barriers in common building end-use markets. The analysis showed that for residential end uses such as heating, cooling, and hot water, up to 50 percent of annual energy use in those markets is affected by the principal-agent barrier alone.[21] Table 8.1 illustrates some of these key findings.

Because home builders make decisions on insulation levels, window efficiency, heating and cooling system efficiency, and hot water equipment, their decisions have long-lasting effects on home energy usage. In making their

Table 8.1   Summary of IEA market barrier case study results

Case Study Type: End-Use Market	Barrier Type	Country	Energy Use Affected by Barrier (%)
Residential refrigerators	Principal-agent	United States	25.2
Residential water heating	Principal-agent	United States	77.0
Residential water heating	Principal-agent	Norway	38.3
Residential water heating	Information	Norway	85.0
Residential space heating	Principal-agent	United States	47.5
Residential space heating	Principal-agent	Netherlands	46.1
Commercial leased space	Principal-agent	Netherlands	40.0
Commercial leased space	Principal-agent	Norway	80–90
Commercial leased space	Principal-agent	Japan	60
Vending machines	Principal-agent	Japan	44
Vending machines	Principal-agent	Australia	80

Source: Prindle et al., 2007. Quantifying the Effects of Market Failures in the End-Use of Energy. American Council for an Energy-Efficient Economy, final draft report to the International Energy Agency.

---

[21]Prindle et al. Quantifying the Effects of Market Failures in the End-Use of Energy. American Council for an Energy-Efficient Economy, final draft report to the International Energy Agency; 2007.

choices of building design, equipment, and other features, these "agents" are motivated primarily to minimize the capital costs of the buildings they construct. They are *not* motivated to minimize life-cycle costs and are especially not motivated to reduce energy and other operating costs, because the benefits of those cost savings go to the "principals"—building owners and occupants. In the United States, most buildings are built on a speculative basis—that is, the builder typically does not have a contract with a specific buyer before making design decisions. This keeps most buyers and occupants from having a role in these key design decisions. Because of this structural "principal-agent" problem, builders chronically underinvest in efficiency.

New buildings are a critical market for climate policy because they typically represent the majority of new electricity loads on utility systems and thus account for most of the growth in carbon emissions in the power sector. It is also true that the time of design and construction offers the most cost-effective opportunity to achieve major efficiency gains and carbon reductions. Simple choices such as the orientation and geometry of the building, its envelope design, lighting systems, and mechanical systems can be combined in whole-building analyses to produce buildings that use less than half the energy of standard-practice buildings, often at little or no added capital costs. Unless climate policy can overcome this kind of market barrier, these critical opportunities will continue to be lost, and the effectiveness and cost of climate policy will be needlessly compromised.

8.3.4 The Limits of Price Elasticity Effects

Many climate policy analysts, reflecting a classical economics view, assert that internalizing externality costs by pricing carbon is a sufficient climate policy in and of itself. This assertion subsumes the idea that pricing carbon is sufficient to drive economically optimal investments across all markets. Although the authors do not disagree with the importance of pricing carbon, experience suggests that carbon pricing alone will not be sufficient to tap many of the most cost-effective energy efficiency resources. In our view, energy prices would have to rise to very high levels to drive the levels of efficiency investment needed to meet the climate challenge.

The key assumption underlying the assertion that "pricing carbon is enough" is that the price elasticity of demand in energy markets is a force sufficient to produce an optimal pattern of investment in energy-using technology. However, this argument is flawed in that it relies too much on just one element of economic price theory: the price elasticity of demand. Though price elasticity remains a valid principle—that is, markets do ultimately respond to higher energy prices by reducing demand to some degree—demand elasticity is only one element of price theory. At least two other elements should be considered in seeking to gauge the realistic response of energy markets to price signals: income elasticity of demand and cross-elasticity of demand. These two elements tend to blunt or displace the expected effects of price elasticity:

- *Income elasticity of demand* refers to the effect in which rising personal incomes drive rising demand for energy services. In thriving economies, rising incomes have driven demand for larger homes, larger vehicles, larger appliances, and a proliferation of new end uses for electricity. This income elasticity effect works opposite to price elasticity; for many households, especially those in middle- and upper-income brackets, prices would have to rise very substantially to offset the income elasticity effect.

- *Cross-elasticity of demand* refers to the effect in which rising energy prices result in little change in energy consumption but larger reductions in consumption of other goods. This effect has been documented regularly in the economic press in the past several years: Each time gasoline prices have spiked, retail sales have dropped noticeably. Consumers drive to shop but take home fewer discretionary purchases. Similar effects occur in electricity markets, especially for smaller customers. Cross-elasticity thus tends to displace the effect of rising prices onto other goods. This causes economic damage but blunts the effect of price elasticity on energy usage.

Even if price elasticity effects worked as strongly as would be expected or predicted, the absolute level of the price impacts that can be expected from carbon prices in the United States will likely be very limited for many years. In U.S. policy circles today, carbon prices above \$10–20/ton are rarely discussed, and "safety valve" provisions are strongly advocated to keep prices below this level. The price elasticity effects of carbon prices at this level would likely be very small. Prices of \$10/ton for CO_2 would add less than one cent, or less than 10 percent, to the average price of electricity in the United States. One cent is less than half the increase in average electricity prices in U.S. markets over the past 7 years. U.S. gasoline price increases since 2002 are equivalent to about \$200/ton in carbon prices, and demand is decreasing only marginally. A carbon price of \$10–20/ton might increase gasoline prices 10–20 cents per gallon, which cannot be expected to have a strong price elasticity effect, since this is within the annual range of price volatility in U.S. energy markets. In summary, price elasticity effects do occur, but they are not likely to show strong enough near-term effects to drive the levels of efficiency investment that analysts suggest are warranted.

8.4 Policies and programs to overcome barriers and market imperfections

Several policy options can be used to address the problems described in the previous section and thus to bring more of the emission reduction and cost-containment benefits of energy efficiency to climate policies. This section divides these policy options in two parts: (1) policy options within the design of climate policies themselves, and (2) complementary energy policies, which are not directly linked to climate policies but which serve to reduce the cost and speed the pace of compliance with climate policies.

8.4.1 CLIMATE POLICY OPTIONS

In a cap-and-trade policy framework, we identify three principal categories in which policy elements can be focused to produce the maximum value from energy efficiency resources: allowance auction policies, allowance allocation policies, and allowance set-asides.

8.4.1.1 Allowance auction policies

The climate policies enacted and under serious discussion today all include provisions to auction significant fractions of carbon allowances rather than giving allowances to emitters for free. The free-allocation approach, whose roots come from the Title IV sulfur dioxide provisions of the Clean Air Act Amendments of 1990, was based in part on the assumption that the point of regulation—i.e., the smokestack—would also be the focus of the lowest cost emission reductions. In the case of sulfur dioxide as well as nitrogen oxides, this assumption has proven to be largely true. However, the SO_2 precedent for free allowance allocation does not fully apply in the CO_2 context, for two main reasons:

- Much of the lowest cost CO_2 emission reductions are not expected to be found at the smokestack level, that is, at the point of regulation. Energy efficiency is a perfect example of this misalignment between the point of regulation and the location of the lowest cost emissions abatement technologies. As discussed earlier, generation owners today largely operate in unregulated power markets, as opposed to the situation 20 years ago, when the SO_2 program was legislated and when most power plants were owned by vertically integrated companies and regulated by state utility commissions. Generation owners today are thus less equipped than in the past to go after end-use efficiency gains, even if they could overcome the indirect-emissions problem that would keep efficiency credits from trading in carbon markets. And as discussed earlier, generators must not only pay for the allowance value of efficiency-based carbon reductions; they must also lose the marginal price value of the lost sales associated with the energy savings, creating an extra disincentive.

■ Free allocations do not reduce the ultimate consumer costs of electricity and in fact increase net revenues to generators when they operate in unregulated wholesale markets. In such markets, carbon prices become bundled into bid prices, and all generators receive revenues based on these prices, whether or not they have compliance costs. This can result in a net gain in total revenue for generators as a whole, although generators with higher CO_2 emission rates will incur higher compliance costs than those with low or no compliance costs. Since a large fraction of U.S. power sales go through unregulated wholesale markets, policymakers have ruled that it does not always make economic sense to give all allowances to generators for free. In the Regional Greenhouse Gas Initiative (RGGI), a 10-state power-sector cap-and-trade system in the Northeast, expert analysis found that generators would gain about $1 billion in revenue annually as carbon allowance prices would become embedded in wholesale prices, regardless of any compliance costs. This analysis was borne out by press reports from the United Kingdom and Germany in 2006, showing that power generators were experiencing additional revenues from carbon prices.

The issue of free allocation vs. auction is complicated in the United States by the fact that about half of the states have restructured electricity markets and half do not. In the states that do not have access to fully functioning unregulated wholesale markets, generators have a stronger argument for free allocation. RGGI states, however, are all served by unregulated wholesale markets. This issue would not be the same in states where utilities maintain vertically integrated ownership of generation and in which state utility commissions regulate those plants' rates. This issue continues to generate controversy at the federal level, where reconciling these two different generation market perspectives is not easy. Harris and Stern and others address these issues in more detail.

CO_2 cap-and-trade policies that have ruled on this issue have taken the following approaches:

■ In RGGI, the model rule requires that states auction at least 25 percent of all allowances and use the funds for energy efficiency and other low-carbon technologies.[22] This policy was set in recognition of the two issues discussed earlier. To date, all the states that have issued draft or final rules under RGGI have elected to auction all or nearly all of their allowances. This high-percentage trend in part reflects the fact that all the RGGI states are served by unregulated wholesale power markets, wherein free allowance allocation can result in additional revenues to generators.

■ In U.S. federal legislation, S. 2191, America's Climate Security Act,[23] the last climate bill to be voted on as of this writing, contained an auction policy that begins at 26.5 percent in 2010 and rises to 69.5 percent in 2049. Of this auction fraction, more than half would go to energy technology deployment, and energy efficiency would be allowed to participate in the zero- or low-carbon technology element, which could receive about one third of the technology deployment funds. According to analysis from the Center for Clean Air Policy, the technology deployment funds from allowance auctions would be about $16 billion in 2012 (assuming $16/ton carbon prices), rising to about $64 billion in 2049 (assuming $96/ton).[24] If efficiency were to receive one third of deployment funds, that would exceed $5 billion in 2012 and $20 billion in 2049. Current federal spending on efficiency is under $1 billion, and state-utility spending on efficiency is currently about $2 billion, so these auction funds could nearly double efficiency investment in 2012.

8.4.1.2 Allowance allocation policies

As discussed earlier, some climate policies have followed the path of allocating at least some portion of allowance to entities other than direct emitters. The logic for this is essentially twofold: Some entities receive allocations to mitigate their costs under the policy, and some entities are seen as better able to secure the lowest cost carbon emissions.

In S. 2191, for example, significant allocations go to gas and electric distribution utilities (load-serving entities; 11 percent throughout the compliance period) and to states (10.5 percent throughout the compliance period). In both cases, funds from the sale of these allowances could be used for energy efficiency. The total value of these allowances would be about $20 billion in 2012 and nearly $40 billion in 2049. As in the auction proceeds provisions, there is language in some sections that explicitly encourages funds to be used

---

[22]www.rggi.org/modelrule.htm
[23]http://thomas.loc.gov/cgi-bin/query/D?c110:1:./temp/~c110CkI2OO::
[24]Center for Clean Air Policy. Summary of Allowance Allocation and Value—S. 2191, the Lieberman-Warner Climate Security Act of 2008; 2008.

for efficiency, but other sections are less clear. These details would need to be refined in implementation rules if energy efficiency is to compete effectively among the many possible uses of such funds.

A key related issue is the allowed uses of funds from allocated allowances. S. 2191, for example, allows distribution utilities to use allowance proceeds to rebate the electricity rate impacts of allowance prices directly back to customers. Though this sounds like a fair and logical policy, analysis in the RGGI states has shown that investing the same allowance proceeds in energy efficiency results in much greater electric bill reductions than simply rebating the money as customer bill credits. In RGGI, the analysis showed that investing in efficiency would reduce average customer bills by 3 to 12 times as much as simply rebating allowance auction proceeds.[25] It is thus important to consider whether allowance proceeds will be used in ways that will provide maximum cost containment benefits, which could mean channeling the money toward technologies such as energy efficiency.

An unresolved issue with direct-allocation policies is the degree of specificity with which the policy should stipulate allowable uses or performance criteria. Without some such stipulations, carbon allowance allocations could become just another public funding mechanism, subject to local priorities. This could divert allowance funds from their intended purpose of reducing or mitigating the cost of compliance. One way to address this issue would be to create energy performance targets for states and utilities that receive allocations. More than 15 U.S. states have Energy Efficiency Resource Standard (EERS) policies (see the following discussion) that set specific performance targets for utility efficiency initiatives. Allowance allocations could be made contingent on state or utility adoption of such targets, coupled with periodic reporting of attainment.

8.4.1.3 Allowance set-asides

In U.S. clean air policies, policymakers have often created *set-asides,* in which small fractions of allowances are removed from emitters' allocations and reserved for designated uses. The Title IV Clean Air Act Amendments of 1990, which created the current sulfur dioxide cap-and-trade system, set aside the Conservation and Renewable Energy Reserve. In State Implementation Plans (SIPs) for federal nitrogen oxides regulations, several states have set aside small percentages of allowances, typically 5 percent or less, for energy

---

[25]Prindle et al. Energy Efficiency's Role in a Carbon Cap-and-Trade System: Modeling Results from the Regional Greenhouse Gas Initiative. American Council for an Energy-Efficient Economy, Report no. E064; 2006.

efficiency and renewable energy projects that can document their emission-reduction performance.

Reviews of these programs have shown them to be of limited effectiveness.[26] The set-aside mechanism requires individual entities to develop projects, go through substantial administrative review, and wait substantial periods of time before receiving emission allowances. Moreover, the net effect of allowance values on energy efficiency projects has been modest. Set-asides thus appear to provide little effective incentive for private entities to develop low-carbon projects, solely on the promise of receiving emission allowances in the future.

By contract, auction or direct-allocation policies provide funds that can be used to support efficiency and other low-carbon investments up front. By administering these funds through effective program designs, the funds can be targeted to create maximum private investment leverage.

Complementary energy policies

Policymakers in the United States and the European Union and other carbon policymakers are choosing to pursue complementary efficiency (among other clean energy policies) in parallel with cap-and-trade programs, to achieve the maximum cost-containment benefits that energy efficiency can offer to climate policy. These policymakers appear to recognize that broad cap-and-trade systems, though necessary for economy-wide coverage and flexibility, do not necessarily engage the most cost-effective energy technologies simply by setting emissions caps and creating carbon prices. Complementary policies add value to cap-and-trade systems because they can get at the markets that are most affected by market barriers and can thus compensate for the limitations of a high-level cap-and-trade system. The U.S. policy debate appears to be heading toward a hybrid mix of broad cap-and-trade climate policy and targeted energy policies that harvest the lowest-cost emissions reductions, with the goal of minimizing the cost of meeting national climate goals. This hybrid approach is reflected in the California Clean Air Resources Board's draft scoping plan for implementing the state's AB 32 legislation; this plan obtains the great majority of carbon emission reductions from complementary policies, with the remainder coming through the cap-and-trade system.[27] Harris and Stern also cover this topic.

---

[26]Prindle el al., 2005. Cleaner Air Through Energy Efficiency: Analysis and Recommendations for Multi-Pollutant Cap-and-Trade Policies. American Council for an Energy-Efficient Economy, Report no. U043.

[27]CARB. 2008. Climate Change Draft Scoping Plan: A Framework for Change. California Clean Air Resources Board.

Energy efficiency resource standards (EERS) and renewable portfolio standards (RPS)

States have developed renewable portfolio standards (RPSs) and energy efficiency resource standards (EERSs) since the late 1990s, in part to compensate for the loss of utility investment in energy efficiency and renewable energy that accompanied electricity restructuring that deregulated retail electricity markets in some 25 states, as well as federal legislation that opened up generation and transmission markets to competition in many regions. RPSs were developed first, with more than half the states now having specific targets for utility generation or purchase of designated renewable energy resources. Targets are typically set as a percentage of total electricity sales; targets run well over 20 percent in the most aggressive states, with target dates in the 2015–2025 range for ultimate attainment.

EERSs are designed to set simple, long-term energy savings targets for utilities, either as a percentage of sales or load growth or as a fixed set of units of energy or demand. The timeframe in EERS requirements can extend 15 years or longer—some state EERSs set in recent years extend through 2025. This approach sets clear, long-term goals for efficiency programs, giving utilities clear planning guidelines as well as a long-term planning goal. Some 15 states and three European Union nations have instituted these policies, which set numerical energy savings targets for utilities to meet through customer efficiency investments, combined heat and power, and other efficiency measures. ACEEE has documented these policies in multiple reports and white papers.[28] Figure 8.4 illustrates the states that have adopted EERS. These policies are being implemented aggressively in many of these states, with utility commission orders, utility program plans with large budgets, and large procurement actions for program delivery services.

EERSs have also been acted on in the U.S. Congress. In August 2007, the House of Representatives passed a bill with a Renewable Electricity Standard that included both renewable energy and energy efficiency. Efficiency was allowed to account for up to 27 percent of the total resource requirement. In 2020, the House RES would have required 15 percent of electricity sales from covered utilities to come from eligible resources; efficiency would have been allowed to provide up to 4 of the 15 required percentage points, or 27 percent of the total.

Although the RES was not included in the final version of the Energy Independence and Security Act that was enacted in December 2007, ACEEE and others conducted in-depth analysis of the benefits and costs of this combined

---

[28]Nadel S. Energy Efficiency Resource Standards: Experience and Recommendations. American Council for an Energy-Efficient Economy, Report no. E063; 2006.

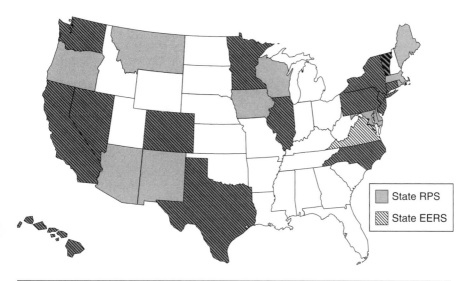

Figure 8.4 States with energy efficiency resource standards or renewable portfolio standards.
Note: New Mexico has adopted an RPS, and Maryland and Ohio enacted EERSs since this map was created.
Source: ACEEE (2007; www.aceee.org/energy/state/2pgEERS.pdf).

RES-EERS policy. The ACEEE study examined the RES as passed by the House, an EERS policy with a 10 percent savings target for electric utilities and 5 percent for gas utilities in 2020, and a "15–15" policy, with 15 percent targets for renewables and for electricity in 2020. ACEEE's study used the IPM model operated by ICF International, a widely used model in federal and state air-quality policy analysis for the power sector.[29] IPM calculates the effects of policies on capacity additions, power sales, power prices, and emissions of various pollutants.

ACEEE's analysis of the House RES included an assessment of the RES in the context of a climate policy similar to S. 2191. Though the analysis showed positive effects on electricity prices in a business-as-usual context, the RES showed even greater benefits when modeled against a climate policy backdrop. Figure 8.5 illustrates these price effects in a climate policy context. As the figure indicates, EERS-RES policies can reduce electricity prices significantly. The 15–15 EERS-RES policy reduced wholesale prices by about 18 percent in 2025 in a climate framework. These price reductions, plus reductions in consumer electricity bills from efficiency investments and other net economic

---

[29]www.icfi.com/Markets/Energy/doc_files/IPM-global-a4.pdf

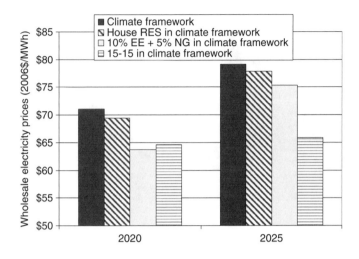

Figure 8.5 EERS-RES effects on reducing electricity prices in a climate policy framework, 2020 and 2025.
Source: Prindle et al. (2007), *Assessment of the House Renewable Electricity Standard and Expanded Clean Energy Scenarios.* American Council for an Energy-Efficient Economy, Report no. E079.

effects, would provide up to $600 billion in net benefits to consumers in 2030.[30]

This analysis shows that energy efficiency in the form of an EERS or in combination with RES can provide very effective cost-containment benefits for federal climate policy.

Compliance with EERS targets is in its early stages. Some of the targets we've mentioned are higher than utility programs have attained in the past. In recognition of this, some states are including nonutility policies in their overall strategies for meeting these goals; building codes and appliance standards are getting attention in this regard. Some states, such as Connecticut, are using "white certificates" market mechanisms, in which customers or third parties can bring forward projects with verifiable savings and sell compliance credits to utilities. Electricity prices are also rising in many states, and the emergence of smart meters and other smart-grid technologies are creating additional opportunities for utility customers to better manage their usage. It may take a wide range of these options to achieve the significant shifts in electricity usage that these targets represent.

---

[30]Prindle et al. Assessment of the House Renewable Electricity Standard and Expanded Clean Energy Scenarios. American Council for an Energy-Efficient Economy, Report no. E079; 2007.

Public benefits funds (PBFs) and demand-side management

Demand-side management, or DSM, developed in the 1980s as the electricity industry's approach to helping customers manage their energy demand (both peak demand and total usage). DSM was practiced in various forms in most states by the 1990s, and DSM spending exceeded $2 billion annually by the early 1990s. DSM was regulated by state utility commissions and delivered primarily by regulated utilities. As electricity restructuring spread to about half the U.S. states during the later 1990s, DSM spending fell precipitously.

Public benefits funds (PBFs) originated in the mid-1990s in states seeking new funding mechanisms for energy efficiency and renewable energy as their utility industries were restructured. Unlike DSM, PBFs were collected through utility bills and often administered by utilities, but the PBF itself was a public fund. As such, PBFs focused more on overall market transformation than on acquiring specific levels of efficiency resources.

PBFs differ from EERSs in that EERSs set goals in terms of energy savings, whereas PBFs set funding levels, creating pools of funds that then support efficiency programs or renewable investments. PBFs and EERSs can work in concert, as in California, where PBFs provide part of the funding utilities rely upon to reach their EERS goals. Some 20 states operate PBFs.

PBF administration can be carried out by utilities, state agencies, and third parties. Although the majority of state PBF programs are utility administered, some states (New York and Wisconsin) use state agencies to administer the programs, and some (Oregon, New Jersey, Delaware, and Vermont) use third-party contractors or nonprofit organizations. All these administrative models have been shown to work effectively if the programs are well designed and well managed and if there is strong policy support and sustained funding.

Utility regulatory treatment of demand-side resources

Whether policymakers pursue EERS, PBF, DSM, or other utility-sector approaches to encouraging energy efficiency, other state regulation of electricity markets and franchised utilities is a critical lever for encouraging clean energy development, including demand-side resources. Several regulatory issues must be addressed if utilities are to deliver demand-side resources effectively to their customers. In this section we discuss four such issues:

■ *Resource acquisition and procurement rules.* In many states, integrated resource plans (IRPs) are filed in a self-contained process that is not directly linked to the processes by which resource acquisition decisions are actually made. In others, utilities operate under rules that define the kinds of resources that can be acquired and in some cases set priorities for procuring clean energy resource options prior to conventional supply options. It is important that states rationalize not only their resource planning but also the resource acquisition process so that the state enjoys the best balance of resource deployment. This set of issues can be simpler in states where electricity markets have not been restructured and where electric utilities have remained vertically integrated, regulated monopolies. In restructured states, these issues are more complicated in that states typically do not regulate generation resources. Some of the resource acquisition options in use by states today include these:

- *Portfolio management.* This is a generic term for state initiatives that engage distribution utilities in the role of "portfolio manager" for the state's electricity customers, even though the utility does not provide integrated generation, transmission, and distribution service as in the past. The term evolved as restructured states experienced limitations in competitive markets' ability to deliver the range of benefits expected from restructuring. Portfolio management can take many forms, but its essence is to re-establish a comprehensive resource management role for regulated utilities, and can lead to expanded roles in the acquisition of demand-side and supply-side resources.[31]

- *Loading order.* In California and Connecticut, state regulations define specific sequences in which utilities acquire resources. Demand-side resources are to be acquired first, up to cost-effectiveness limits. Defined renewable resources are to be pursued next, followed by conventional generation if needed.[32] These are long-term resource development policies—they generally do not affect system operational practices, though they do affect the portfolio of resources that system dispatchers have at their disposal for daily and seasonal use.

- *Procurement of default service.* Distribution utilities in restructured states typically provide "default" electricity service to customers who do not choose unregulated power suppliers for their generation service. In most cases, the vast majority of smaller customers have ended up taking default service. This places a major implicit burden on the distribution utility, which is still state regulated, to make good resource decisions for customers. Some states have considered requiring a mix of resources to be acquired in the default service process, since most utilities solicit bids for electricity supply. Such a mix could include demand-side bidding for a portion of default service needs, procurement of renewable supplies,

---

[31] RAP site: www.raponline.org/Pubs/PortfolioManagement/SynapsePMpaper.pdf
[32] California Energy Commission. Implementing California's Loading Order for Electricity Resources. Report no. 400–2005–043; 2005.

and a mix of conventional supplies including mixed-length power purchase contracts.[33]

■ *Cost-recovery methods for DSM programs.* Utility costs for demand-side
programs must be recovered from customers. Whereas supply-side investments are typically capitalized and recovered over long periods, demand-
side program costs have typically been smaller in absolute size and thus
could be treated as annual expenses or could be capitalized. If they're capitalized, recovery periods are typically quite short compared to supply
investment amortization periods. States must also decide whether to
recover costs across all customer classes or allocate and recover program
costs by customer class. U.S. EPA has provided detailed resources on this
and the two related issues that follow.[34]

■ *Revenue stability mechanisms.* Historical rate-making practices have linked
utility revenues to the amount of electricity sold. Although utility revenues
and profits are calculated to provide an approved rate of return on assets,
the revenue requirement is typically divided by forecast kWh sales. This
means that if kWh sales deviate from forecast, the utility may experience
surpluses or deficits in revenues and profits in a given period. In this
approach, if DSM programs reduce kWh sales, utilities typically see revenue shortfalls because they don't fully recover the fixed costs allocated to
each kWh of sales. This creates a major disincentive for utility DSM investment. States have sought to correct this problem largely by separating utility revenues from kWh sales. This can be done through "decoupling"
mechanisms, which true-up revenues each year, adjusting for changes in
kWh sales so that any revenue surplus or shortfall is made up in the next
year through a minor rate adjustment. Some four states use electricity
decoupling today, and many more use decoupling for natural gas utilities;
Figure 8.6 illustrates this trend. Some utilities have proposed to address
this problem by placing a higher portion of fixed costs per customer in
the fixed customer charge portion of electric bills so that changes in electricity sales affect only the utility's variable costs. However, this approach
reduces customer incentives to save energy by reducing the electric bill
impact of increasing energy usage, and states have generally not approved
this method.[35]

---

[33][ME PUC] Maine Public Utilities Commission. Docket No. 2006–411. July 26, 2006.
[34]National Action Plan for Energy Efficiency. Aligning Utility Incentives with Investment in
Energy Efficiency. Prepared by Val R. Jensen, ICF International, www.epa.gov/eeactionplan;
2007.
[35]Ibid.

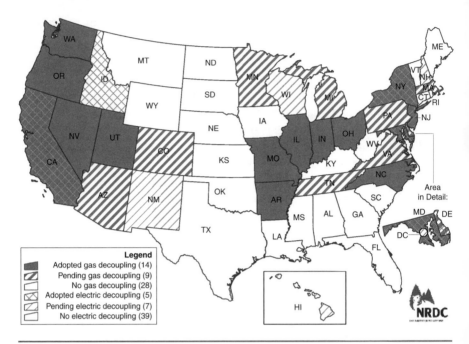

Figure 8.6   Gas and electric decoupling in the United States, February 2008. Source: NRDC.[36]

■ *Shareholder incentive mechanisms.* Making DSM financially attractive to investor-owned utilities is a three-part problem. In addition to cost recovery and revenue stability, utilities need to be able to earn profits on DSM investments comparable to the profits they are allowed to earn on supply investments. States have used three main bases to provide shareholder incentives in this context:

• *Rate of return on program costs.* This approach treats program costs as a regulatory asset and calculates a rate of return comparable to that allowed on other utility assets. In some cases, states allow a "bonus" rate of return on DSM program assets to send a clear signal to utilities that DSM resources have preferred value. It is also possible to link the rate of return to the level of savings achieved, such that bonus rates of return are paid based on utilities' exceeding energy savings benchmarks.

• *Share of net benefits.* Some states calculate the net benefits of DSM programs by subtracting costs from total benefits and then pay utilities a

---

[36]Natural Resources Defense Council. Compiled by Amelia Nuding; February 2008.

share of this net amount. This approach encourages utilities to maximize the net energy savings from their programs rather than just their costs.

- *Share of avoided costs.* Another approach is to use the avoided cost associated with saved kWh from demand-side programs and pay utilities a share of this amount. This method provides utilities a clear and consistent price signal for their programs and can net them substantial profit margins if program costs are kept well below the payment level. For example, if avoided cost is calculated at 5 cents per kWh and the utility can deliver savings for 3 cents per kWh, this would nominally provide a 60 percent return on program costs. However, if lost revenues were factored in, the net return would be substantially lower. And if the utility's costs were higher or savings were lower than expected, their revenue could turn negative. Because of these uncertainties, states and other parties are studying this approach in detail.

Appliance efficiency standards

Appliance standards, enabled by the National Appliance Energy Conservation Act of 1987 and the Corporate Average Fuel Economy Standards, enabled in 1975 and updated in the Energy Independence and Security Act of 2007, have been very effective in the United States. Appliance standards address the two most prominent structural market barriers in U.S. residential and commercial end-use markets: the principal-agent problem and the information-cost or transaction-cost barrier. As discussed earlier, the principal-agent problem affects some 50 percent of the heating and cooling energy use in U.S. homes, because builders pick heating and cooling equipment and are not motivated to invest in models with the greatest energy savings. Standards are typically set on the basis of the lowest life-cycle cost and thus serve to overcome this key barrier.

The information-cost or transaction-cost barrier affects millions of purchase decisions by consumers and small businesses, where the cost of finding and evaluating the information needed to make economically optimal decisions is typically too high relative to the size of the transaction. A good example is external power supplies, the "bricks" or "wall-warts" that come with most electronic devices. The standby power they consume is small per unit, such that it's not worth a consumer's time to try to find a more efficient unit. But the growing number of such devices per home and the huge collective numbers of devices make them a significant source of energy waste. Appliance standards are thus an appropriate policy solution for such technologies.

Appliance standards, which now cover dozens of products, have already saved U.S. consumers and businesses more than $50 billion in net energy savings. In 2030, standards enacted through 2007 are expected to save energy users a net $242 billion. By 2030, these standards will offset more than 10 percent of forecast electricity usage and more than 15 percent of expected peak demand.[37] Table 8.2 summarizes these and other data.

Given the well-established benefits of appliance standards, it is not surprising that S. 2191 as debated in committee in the fall of 2007 contained the appliance standards provisions from the House version of the energy bill then in progress. Since the Energy Independence and Security Act ultimately included the standards provisions when it was enacted in December 2007, the standards provisions in S. 2191 were rendered moot. However, it is quite feasible to include standards as complementary policies in future versions of climate legislation, either for newly covered products or for upgraded standards on products already covered by federal law.

Building energy codes

As discussed earlier, new construction markets are among the most severely affected by market barriers, notably the principal-agent barrier, because builders are not motivated to invest the extra design time and capital to optimize energy efficiency for the building's life cycle. New buildings are also the largest source of new energy consumption and associated carbon emissions in most electricity systems, making it all the more imperative that this market be addressed by public policy.

Moreover, new buildings offer the most cost-effective design solutions, allowing dramatic reductions in energy use and carbon emissions for modest

Table 8.2 Savings from U.S. appliance standards enacted to date, 2000–2030

Year	Electricity (TWH)	Peak Demand (MW)	Total Energy (Quad)	Carbon (MMTC)	Net Savings ($B)
2000	88	21,000	1	25	—
2010	272	73,128	4	66	—
2020	514	166,945	6	111	—
2030	572	181,040	7	226	242

Source: Nadel et al. Leading the Way: Continued Opportunities for New State Appliance and Equipment Efficiency Standards. American Council for an Energy-Efficient Economy, Report no. A062; 2006.

---

[37]Nadel et al. Leading the Way: Continued Opportunities for New State Appliance and Equipment Efficiency Standards. American Council for an Energy-Efficient Economy, Report no. A062; 2006.

and even neutral capital costs. Through careful, whole-building design approaches, energy loads can be reduced so substantially that heating/cooling systems can be markedly downsized; this downsizing reduces capital costs, offsetting added costs for better insulation, windows, lighting, and so on.

Building energy codes have been adopted by most states, using the International Energy Conservation Code (IECC) and other model codes. In the 2007 energy bill debate, the House passed a provision that would have required the federal government to seek improvements of up to 50 percent in the stringency of the IECC and, if the IECC did not produce such gains, to develop a federal model code for states to adopt. This language, though it did not make it into the final energy bill, was included in S. 2191. Codes thus can and should be considered as part of the national climate policy toolkit. Additional information on building codes can be found via the Building Codes Assistance Project[38] and the New Buildings Institute.[39]

Labeling and rating policies

Governments around the world have used energy efficiency labeling and ratings for appliances and buildings since the 1970s. Virtually all the Organization for Economic Cooperation and Development (OECD) countries have some sort of voluntary labeling or rating programs covering some set of end-use energy efficiency technologies. The Collaborative Labeling and Standards Project (CLASP)[40] and the International Energy Agency[41] maintain information on these efforts.

In the United States, the Federal Trade Commission (FTC) administers a mandatory labeling system (using the yellow Energy Guide labels) that was authorized in the 1970s. During the 1990s, voluntary labeling arose under the ENERGY STAR® brand, administered by U.S. EPA and U.S. DOE. Scores of products, from light bulbs to entire buildings, now use the ENERGY STAR® label. EPA and DOE actively support the ENERGY STAR® label through a network of programs administered by states, utilities, retailers, manufacturers, and other partners. Sales of ENERGY STAR® labeled products exceeded $100 billion in 2004.[42]

ENERGY STAR® and other labeling efforts support a range of energy rating systems used to develop the technical information behind the label. For new homes, the Residential Energy Services Network (RESNET)

---

[38][BCAP] Building Codes Assistance Project. Available at: www.bcap-energy.org
[39][NBI] New Buildings Institute. Available at: www.newbuildings.org
[40][CLASP] Available at: www.claspacpline.org/clasp.online.worldwide.php
[41][IEA] Available at: www.iea.org
[42]Ehrhardt-Martinez K, "Skip" Laitner JA. The Size of the U.S. Energy Efficiency Market: Generating a More Complete Picture. American Council for an Energy-Efficient Economy, Report. No. E083; 2008.

maintains a consensus-based set of technical standards that help ensure uniformity and consistency among energy performance ratings of whole buildings. RESNET standards are used in voluntary programs that go beyond building energy codes, including the EPA ENERGY STAR® Homes program, the U.S. Department of Energy's Building America program, and evolving "green building" programs such as the U.S. Green Building Council's Leadership in Energy and Environmental Design (LEED) program.

Research, development, and demonstration (RD&D)

RD&D is the "pipeline" that keeps new technologies coming into the market to drive future waves of investment in clean energy. RD&D is sometimes thought of as a federal government role, but federal clean energy RD&D has fallen by some 50 percent in real dollars since the 1970s. State spending on efficiency and renewables has more than made up the difference, including state RD&D efforts. Several states operate their own RD&D institutions, often in cooperation with state universities, federal agencies, and the private sector. State-based efforts can ensure an appropriate focus on technologies most suitable to the state's resource and economic mix and can also help incubate new enterprises that create in-state "green-collar" jobs. A good source of information on current RD&D efforts in the states is the Association of State Energy Research and Technology Transfer Institutions (ASERTTI).[43]

▌ 8.5 Conclusion

This chapter provides a discussion of the potential benefits energy efficiency can offer climate policy, especially in terms of reducing the cost of CO_2 emission reductions. It also assesses the barriers that efficiency faces in the design of climate policies themselves, in the marketplace, and in other regulatory terms. It offers policy solutions that can overcome many if not all of these barriers.

Based on this analysis and a review of the literature as well as policy experience to date, the authors conclude that energy efficiency is an essential part of the climate solution. It is difficult to imagine attaining deep reductions in greenhouse gas emissions without first realizing deep cuts in the growth of energy use. Energy efficiency gains of the magnitude needed to solve the climate problem will not occur through market forces alone, although markets are the channels through which efficiency will occur. Policies are needed, both

---

[43][ASERTTI] Association of State Energy Research and Technology Transfer Institutions. Available at: www.asertti.org

within and outside climate regulations, to gain the maximum benefit from the efficiency resource.

Key policy recommendations include:

■ Use greenhouse gas allowance auctions to generate revenue that subsequently supports accelerated investment in efficiency and other low-carbon technologies.

■ Use allowance allocation policies to direct allowances to entities that can best use them for efficiency investment; these entities include distribution utilities and states.

■ Institute energy policies that drive additional energy efficiency in parallel with climate policies. These include building energy codes, appliance efficiency standards, and utility efficiency programs.

■ Reform utility regulation to align utility shareholders' incentives with those of customers so that utilities have a direct business case for substantial investment in demand-side resources, including energy efficiency.

Chapter 9

Wind Power: How Much, How Soon, and at What Cost?

Ryan Wiser

Lawrence Berkeley National Laboratory

Maureen Hand

National Renewable Energy Laboratory

Abstract

The global wind power market has been growing at a phenomenal pace, driven by favorable policies toward renewable energy and the improving economics of wind projects. Going forward, utility-scale wind power offers the potential for significant reductions in the carbon footprint of the electricity sector. Specifically, the global wind resource is vast and, though accessing this potential is not costless or lacking in barriers, wind power can be developed at scale in the near to medium term at what promises to be an acceptable cost.

Doi: 10.1016/B978-1-85617-655-2.00009-2.

9.1 Introduction

The challenges of combating global climate change are enormous, and there is no single panacea. Instead, as suggested throughout this book, an assortment of technologies will need to be deployed, infrastructures reconfigured, and behaviors altered to slow and ultimately reverse rising CO_2 emissions.

Renewable energy has helped meet the energy needs of humans for millennia, and the world's renewable energy resources are enormous. Though truly comparable data on renewable resource potential do not exist, Table 9.1 shows that the world's resource potential far exceeds global primary energy supply, currently around 470 exajoules (EJ), and that this potential has barely begun to be tapped [37]. In principal, at least, dramatically increased use of renewable energy could go a long way toward reducing energy-sector carbon emissions.

Unfortunately, however, renewable resources are often diffuse, location dependent, and variable. The diffuse and location-dependent nature of the resources sometimes makes them prohibitively expensive to employ, whereas resource variability creates concerns about grid integration and system reliability. In part as a result of these factors, recent growth in the use of renewable electricity, though significant, is not on the scale needed to make a large dent in the climate problem. At the end of 2006, for example, aggregate worldwide renewable electricity capacity stood at 980 GW, predominantly from

Table 9.1 Global renewable technical potential and use[1]

Energy Source	Estimated Available Energy Resource (EJ/yr)	Annual Use of Energy Resource in 2005 (EJ)
Hydro	62	25.8
Wind	600	0.95
Biomass	250	46
Geothermal	5,000	2
Solar electricity	1650	0.23
Ocean	7	Insignificant

Source: IPCC [37]; simplified by authors.

[1]Other efforts to compile resource potential for a variety of renewable sources, with sometimes stark differences from those shown here, include de Vries et al. (2007), Hoogwijk (2004), Johansson et al. (2004), REN21 (2007), UNDP (2000), WBGU (2004), and World Energy Council (2007). Continued work is needed to develop resource potential estimates that use a common methodology and that are therefore truly comparable.

Table 9.2 Global installed renewable energy capacity

Generation Technology	Installed Capacity, End of 2006
Large hydropower	770 GW
Small hydropower	73 GW
Wind	74 GW
Biomass	45 GW
Geothermal	9.5 GW
Solar electricity	8.2 GW
Ocean	0.3 GW

Source: Ren21 [52].

hydropower plants (Table 9.2). This is up from 943 MW in 2005 but represents a year-on-year growth of just 4 percent. In total, renewable electricity met 18.4 percent of global electric power demand in 2006. If one excludes large hydropower, however, this figure drops to just 3.4 percent [52], hardly a dent given the enormity of the challenge that climate change presents [36].[2] Clearly, for renewable energy to be a significant contributor to reducing global carbon emissions, a step-change in growth is required.

The aim of this chapter is to discuss the important role that wind power *could* play in achieving carbon emissions reductions, as well as some of the barriers to that outcome. Wind power is a mature, zero-emission technology that offers an immediate option for reducing the carbon footprint of the electricity sector. In good wind resource regimes, its costs are comparable to fossil-fuel generation, adequate wind resources are available throughout the globe, and there are no insurmountable technical barriers to dramatically increased deployment of this technology. Along with other important near-term strategies, increased deployment of wind can help buy time as newer technologies are developed (e.g., low-cost solar, carbon sequestration, biofuels) or as other technologies seek greater public acceptance (e.g., nuclear power). And, because wind power represents a relatively low-cost carbon abatement option, its potential for carbon emissions reductions extends beyond the near term. Indeed, wind offers at least one important "wedge" for reducing carbon

---

[2]To achieve stabilization at 535–590 ppm CO_2-equivalent, for example, Hansen et al. (2008) estimates that CO_2 emissions must peak from 2010–2030 and then drop to -30 percent to $+5$ percent of 2000 levels by 2050. A number of governments have expressed the desire to keep concentrations of CO_2 well below this level, requiring even more dramatic reductions in CO_2 emissions, and some scientists have called for stabilization levels as low at 350 ppm (Hansen et al., 2008).

emissions [56], and the analysis presented in this chapter suggests that a significant expansion of wind deployment can be achieved at what many would consider to be an acceptable cost.

This chapter begins with an overview of the global wind power market, emphasizing historical growth trends, cost comparisons, and forecasts for future growth. To assess the feasibility of achieving even higher levels of wind power penetration, the chapter then highlights an analysis of the technical and economic viability of wind energy meeting 20 percent of U.S. electricity needs by 2030. Similar analyses conducted on a global basis are summarized, and the potential role of wind in meeting worldwide electricity needs is discussed. The chapter ends with a summary of what is needed to achieve these higher levels of wind power deployment. Though wind power is the exclusive focus of this chapter, many of the other renewable energy sources may also play important (and in some cases leading) roles in combating climate change; solar electricity, geothermal, and hydropower are discussed further in Chapters 10, 11, and 12, respectively.

9.2 The global wind power market

Wind energy has been used for millennia, but the use of wind to generate electricity on a commercial scale only began in earnest in the 1980s. Since California's initial foray into large-scale wind deployment early in that decade, wind power has come a long way.

Today, a standard, land-based wind turbine stands on a tower of 60–100 meters in height, with a rotor diameter of 70–100 meters and a power capacity of 1.5 to 3 MW. Turbines installed offshore can be even larger. Leading global wind turbine manufacturers include major industrial firms such as Vestas, General Electric, Gamesa, Enercon, Suzlon, and Siemens [5]. Increasingly, the developers and owners of wind projects are major electric utilities and investment firms.

Global wind power capacity is growing at a rapid pace and, as a result, wind power has quickly become part of the mainstream electricity industry. In 2007, roughly 20 GW of new wind capacity was added globally, yielding a cumulative total of 94 GW (Figure 9.1). Since 2000, cumulative wind capacity has grown at an average annual pace of 27 percent. The vast majority of this capacity has been located on land; offshore wind capacity surpassed 1 GW at the end of 2007, with accelerated growth expected in the future, especially in Europe [5, 21].

Although European countries such as Germany, Denmark, and Spain have led the charge over most of the past decade, the United States became the fastest-growing wind power market in 2005, followed by European stalwarts

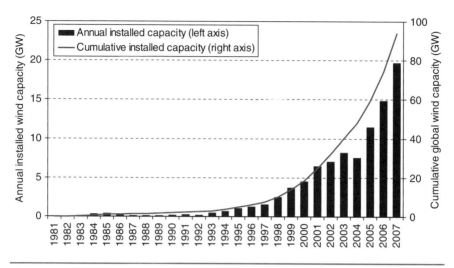

Figure 9.1 Growth in global wind capacity, 1981–2007.
Source: Earth Policy Institute, BTM Consult, American Wind Energy Association.

Table 9.3 International rankings of wind power capacity, 2007

Incremental Capacity (MW)		Cumulative Capacity (MW)	
United States	5329	Germany	22,277
China	3287	United States	16,904
Spain	3100	Spain	14,714
Germany	1667	India	7845
India	1617	China	5875
France	888	Denmark	3088
Italy	603	Italy	2721
Portugal	434	France	2471
United Kingdom	427	United Kingdom	2394
Canada	386	Portugal	2150
Rest of world	2138	Rest of world	13,591
Total	19,876	Total	94,030

Source: Wiser and Bolinger [63].

Germany and Spain as well as the up-and-coming Asian markets in China and India (Table 9.3). With major development now occurring on several continents, wind power is becoming a truly global electric generation resource.

In both Europe and the United States, wind now represents a major new source of electric capacity additions. From 2000 through 2007, wind was the

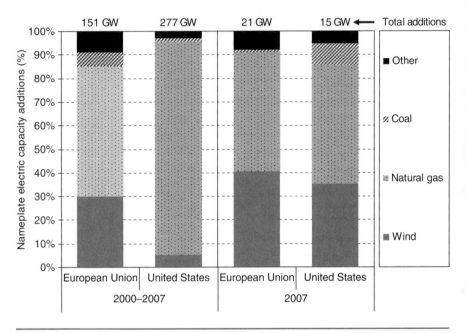

Figure 9.2 Relative contribution of generation types to capacity additions in the United States and the European Union, 2000–2007.
Source: Wiser and Bolinger [63], EWEA [18].

second largest new resource added in the United States (5 percent of all capacity additions) and the EU (30 percent of all capacity additions), in terms of nameplate capacity, behind natural gas but ahead of coal (Figure 9.2). In 2007, 35 percent of all capacity additions in the United States and 40 percent of all additions in the EU came from wind power [18, 63].

As a result of this expansion, some countries are already successfully utilizing wind as a significant contributor to overall electricity supply. Denmark, for example, generates roughly 20 percent of its electricity from wind, whereas Spain is at 12 percent, Portugal at 9 percent, Ireland at 8 percent, and Germany at 7 percent (Figure 9.3). On a global basis, however, wind is still an emerging player, serving 1.2 percent of total worldwide electricity needs [63].

Though the wind capacity installed by the end of 2007 is able to contribute just 1.2 percent of the U.S. electricity supply, the United States is one of the most dynamic markets for wind. The United States has led the world in wind capacity additions for 3 years running (2005–2007). The 5.3 GW of wind installed in 2007 represented 27 percent of the worldwide wind market in that year and was more than double the previous U.S. installation record set in

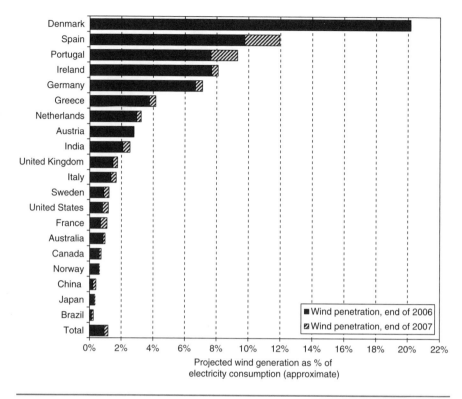

Figure 9.3 Approximate wind power penetration in the 20 countries with the greatest installed wind capacity, 2006 and 2007.
Source: Wiser and Bolinger [63].

2006. Cumulatively, at the end of 2007, nearly 17 GW of wind capacity was installed in the United States. And, with at least 225 GW of in-development wind projects in transmission interconnection queues at the end of 2007, the U.S. wind market is poised for continued strong growth in the years ahead [63].

Driving the growth both in the United States and globally are a number of factors, including promotional policies [28, 44, 48]. In the major European markets, including Germany and Spain, aggressive feed-in tariff programs have been predominant, offering wind power owners standardized and known payment streams [47]. In other markets in Europe, the United States, and elsewhere, renewables portfolio standards (RPSs) have come into vogue, requiring electricity suppliers to meet a specified and growing percentage of their load with renewable electricity [60]. And, in many countries, a mixed set of policies has been used to good effect, including RPS programs and tax incentives in the United States; RPS, feed-in tariffs, and tax incentives in India; and RPS, feed-in tariffs, and auctions in China. In many countries, the current reality and/or future prospect of carbon regulations has also helped motivate wind development.

Though promotional policies differ and healthy debate exists over the relative merits of different approaches, a key finding is that policy continuity and market stability are of utmost importance. The U.S. market, for example, has been hampered by the boom-and-bust cycle of wind project development caused by short-term extensions of that country's production tax credit (PTC) for wind [62].[3] More generally, experience also shows that promotional policies must generally be backed by other enabling policies, such as proactive transmission planning for wind, and siting and permitting procedures that allow development to proceed without undue impediments.

Of equal importance to the aforementioned policy drivers has been the improved underlying economics of wind power relative to fossil fuels, which is, in part, dependent on wind project performance and installed costs. In the United States, wind project performance has improved with time. In 2007, for example, wind projects in the United States produced power with an average capacity factor of just over 31 percent in aggregate. Those projects installed prior to 1998, however, maintained an average 2007 capacity factor of just 22 percent, whereas those projects installed from 2004 through 2006 averaged 34 percent, a significant improvement in project performance over time (upper graphic in Figure 9.4). These performance improvements may be attributable to several factors, including improved turbine design, larger turbines on taller towers, and improved siting and operations.

Trends in the installed cost of wind projects are more mixed. Specifically, the average installed cost for wind projects in the United States has increased significantly since the early 2000s, from $1300/kW in 2001 to roughly $1700/kW in 2007 (lower graphic in Figure 9.4) and to more than $2000/kW in 2008 (real 2007 dollars).[4] These cost pressures are not, however, unique to wind; the installed cost of other generating technologies has increased by a similar magnitude [7]. Moreover, even at more than $2000/kW in 2008, installed costs remain well below the $4000/kW average seen in the early 1980s.

---

[3]Since 1994, the PTC has offered new wind projects in the United States a 10-year, inflation-adjusted tax credit that stood at $20/MWh in 2007. The PTC has often been extended for relatively short periods of time, however, imposing significant uncertainty to the market.

[4]Increases in the installed cost of wind projects since the early 2000s are due to numerous factors, including the weakening of the U.S. dollar, increased materials and energy input prices, and an overall demand for wind turbines that exceeds available supply.

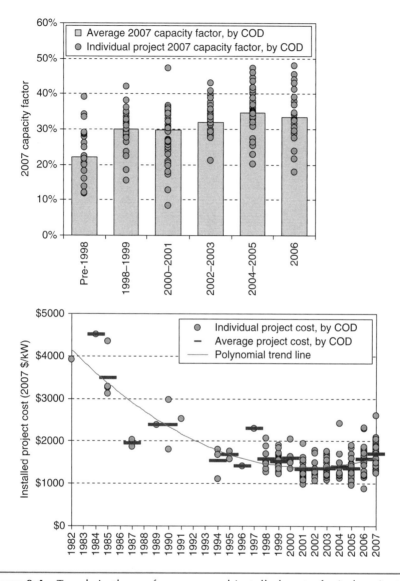

Figure 9.4 Trends in the performance and installed cost of wind projects in the United States, by date of commercial operation (COD), 1982–2007. Source: Wiser and Bolinger [63].

Figure 9.5 presents data on the resulting historical average price of wind power in the United States, from 1990 to the present, both with and without the federal PTC. The figure also provides data on the annual fuel and variable operating cost of natural gas plants over this period, conservatively assuming that wind operates to conserve fuel and variable O&M but that it does little to offset the need for new dispatchable power plants to maintain electricity reliability.[5] Finally, the figure provides data on the price of a flat block of power across the numerous wholesale market trading hubs in the country from 2003 through 2007; these data are lower than the average cost of operating natural gas plants because natural gas is not always the marginal resource.

Though clearly a simplified approach to comparing the economics of different energy sources, Figure 9.5 nonetheless shows that, since 2000, wind power with the PTC has often been economically competitive with other sources of electricity in the United States. Even without the PTC, the cost of wind has not been far out of line with fossil-fueled generation. Moreover, though a confluence of factors has put upward pressure on wind power prices since the early 2000s, those cost pressures have affected other generation technologies as well. As a result, the relative economic position of wind has not changed dramatically over this period.

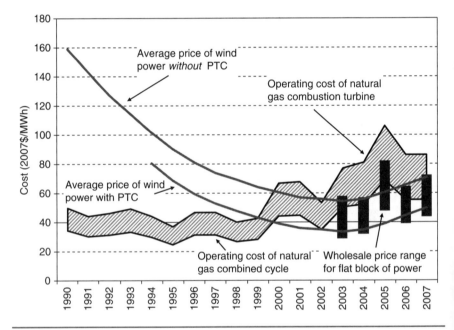

Figure 9.5 Comparative economics of wind, natural gas, and wholesale electricity prices in the United States, 1990–2007.
Source: Prepared by Berkeley Lab, based, in part, on data from EIA, NREL, and Wiser and Bolinger [63].

---

[5]In fact, a large number of studies have shown that wind does have some "capacity value," though the level of that value depends on many factors.

Given all these trends, wind capacity additions are likely to continue at a rapid clip, both in the United States and globally. Several studies have tried to predict future global wind capacity under what might be considered business-as-usual (BAU) conditions, where existing policies are maintained but not dramatically expanded. Figure 9.6 presents the results of several of these studies. Though a considerable range exists, reflecting the inherent uncertainties in such predictions, it appears that 300–700 GW of wind might be expected on a global basis in the 2015 to 2030 timeframe in a BAU scenario. To put this in context, 300 GW of wind in 2015 represents more than 3 percent of total expected worldwide electricity supply for that year, whereas 700 GW by 2030 represents roughly 6 percent of expected supply.

Simply assuming that wind offsets the use of new natural gas and coal power plants in equal proportions, global wind power capacity installed at the end of 2007 was already saving roughly 34 million metric tons of carbon equivalent (MMTCE) annually. Installed capacity of 300–700 GW would raise this figure to 130–300 MMTCE/yr. Though meaningful, even this growth is modest relative to what would be needed in a deep carbon-reduction scenario [36] and relative to wind's resource potential. The chapter therefore now turns to an analysis of the feasibility of achieving much higher levels of wind penetration. To provide a detailed example of such analysis, the next section

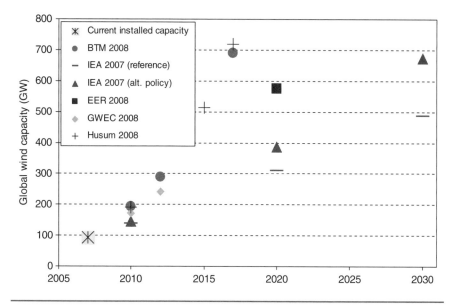

Figure 9.6 Projections for global wind capacity growth, 2005–2030. Source: Compiled by Berkeley Lab from the sources cited in the figure [5, 13, 21, 33, 34].

focuses on the feasibility of achieving 20 percent wind energy penetration in the United States. A later section summarizes similar studies conducted on a global basis.

9.3 Twenty percent wind electricity by 2030

In 2008, the U.S. Department of Energy, in collaboration with its national laboratories, the wind industry, and others, completed a major analysis of the technical and economic feasibility of wind power meeting 20 percent of the nation's electricity supply needs by 2030 [59]. As discussed here, that study finds that there are no insurmountable technical barriers to achieving 20 percent wind penetration by 2030 and that the economic costs of doing so are likely to be modest.

The key questions addressed by this analysis include:

- Is it technically feasible to achieve 20 percent wind energy by 2030?
- Does the nation have sufficient wind resources and land area?
- What are the technology and manufacturing requirements, and are they achievable?
- Can the electric network accommodate 20 percent wind energy, and how?
- What are the likely costs and benefits of achieving 20 percent wind penetration?

Previous studies that have estimated the potential and costs for rapid wind energy deployment in the United States have shown divergent results.[6] These results are the product of differing input assumptions and differing treatment of wind power in capacity expansion models. Modeling wind presents a unique challenge [49]. As a variable and nondispatchable generation resource, wind's distinctive characteristics require changes to modeling algorithms. The nation's most robust wind resources are often located at a distance from load centers, so the way new transmission is modeled is of critical importance. And because wind resources are not uniform across the nation, an appropriate modeling tool must incorporate geographic detail that can capture the nature of the underlying wind resource. Many previous efforts to evaluate wind's potential have largely ignored these details or have instead relied on crude

---

[6]For some examples, see Cryts et al. (2007), EIA (2008b), Hoogwijk et al. (2007), Key (2007), Kutscher (2007), Kyle et al. (2007), and Short et al. (2004).

approximations or supply constraints that intend to loosely account for wind's unique characteristics. Though these approaches are often taken in the name of simplicity, their application can impose poorly documented and potentially unrealistic constraints and costs on wind generation that other generation sources do not face.

Though no modeling tool is perfect, the analytic backbone of the U.S. DOE report was the Wind Energy Deployment Systems (WinDS) model.[7] WinDS is a capacity expansion model of the U.S. electricity system. It uses a detailed geographic information system (GIS) representation of the nation's wind resources and their proximity to existing transmission infrastructure and load centers. By dividing the country into 358 distinct regions, the model uniquely represents the geographic variation of wind resources, infrastructures, and loads. WinDS also models seasonal and diurnal wind resource variations that, combined with statistical algorithms, address the variable nature of wind energy and allow the model to incorporate the costs of integrating wind power into electric grids. In sum, WinDS provided a means of estimating the location of wind installations, the transmission infrastructure expansion requirements of those installations, and the composition of generation technologies needed to maintain reserve capacity requirements while meeting projected electricity demand.

To isolate the impacts, costs, and benefits of producing 20 percent of the nation's electricity from wind by 2030, two scenarios were contrasted. In one scenario, annual wind energy capacity and generation growth are prescribed such that wind supplies 20 percent of total U.S. electricity generation by 2030 ("20 percent wind" scenario). The other scenario assumes no additional wind capacity after 2006 ("no new wind" scenario). In both scenarios, conventional generation technologies compete for the nation's residual supply needs. To uniquely identify the costs and impacts of increased wind deployment, the modeling assumed no new policy incentives (e.g., carbon mitigation policies) and that the PTC was not available after 2008. Underlying the model were a large number of assumptions about the future cost and performance of electric generation technologies. Also, to efficiently accommodate the unique characteristics of wind, the analysis assumed that the electricity grid is operated within large, regional markets. New transmission lines are "built" by WinDS as needed, at costs that vary regionally. These latter assumptions for grid operations and transmission expansion, though not infeasible, would require the removal of significant institutional barriers to market integration and new transmission investment. For a complete account of the many assumptions behind the model, the reader is referred to the final report of the project [59].

---

[7]See www.nrel.gov/analysis/winds/

Not surprisingly, the study finds that reaching 20 percent wind energy would require a dramatic increase in wind capacity. In particular, based on assumed improvements in wind project performance over time and an assumption that transmission to access the nation's lowest cost wind resources can be built at will (if cost-effective), 305 GW of wind capacity would be needed by 2030 (Figure 9.7).[8] WinDS analysis projects that more than 50 GW of this capacity would be installed offshore. From 2018 to 2030, roughly 16 GW of wind would need to be installed annually, compared to the 5.3 GW added in 2007.

Though surely a daunting challenge, the analysis finds that the United States has vast wind resources, far exceeding what is needed to achieve 20 percent wind energy. Considering bus-bar economics alone, for example, and ignoring transmission and integration costs, Figure 9.8 provides a supply curve for wind using cost and performance assumptions from 2007, segmented by wind resource class and by onshore, shallow offshore, and deep offshore

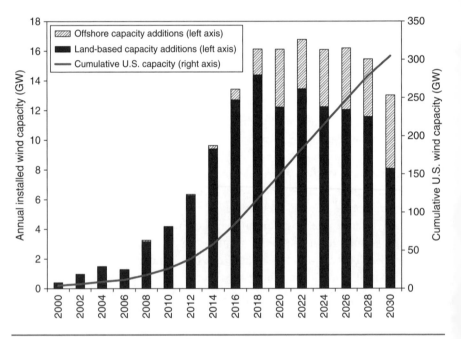

Figure 9.7 Annual and cumulative U.S. wind capacity installations in the "20 percent wind" scenario, 2000–2030.
Source: Derived from DOE [59].

---

[8]If transmission could not be so readily constructed or if state policies shifted development toward less cost-effective locations, the aggregate amount of wind capacity needed to achieve 20 percent wind would increase, as would the incremental cost of achieving that target.

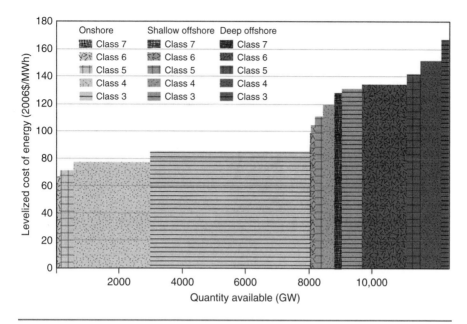

Figure 9.8 The supply curve for wind in the United States with today's technology.

Note: This figure presents the bus-bar economics of wind as a function of wind resource class, excluding the PTC as well as transmission and integration costs.

Source: Derived from DOE [59].

wind potential. Even assuming that the PTC is not available, the study finds that roughly 8000 GW of wind is potentially available at bus-bar generation costs of less than $85/MWh. This far exceeds the 305 GW of wind required in the "20 percent wind" scenario. Moreover, though not shown in the figure, even if the cost to connect to the existing transmission system or to connect to *nearby* load centers is considered, over 600 GW of wind generation potential remains available at a cost of less than $100/MWh.[9]

The analysis also shows that there are no fundamental physical limitations to raw material supplies (e.g., sand, cement, iron ore), though the potential for shortages of processed materials like fiberglass or steel does exist and may have to be overcome. This level of growth would also require a substantial labor force. A key challenge would be to maintain downward cost pressure on installed wind project costs while the industry rapidly expands.

---

[9]This 600 GW figure is extremely rough because it assumes that 10 percent of all existing transmission capacity is available for wind delivery and it completely ignores the very real possibility of new, longer-distance transmission.

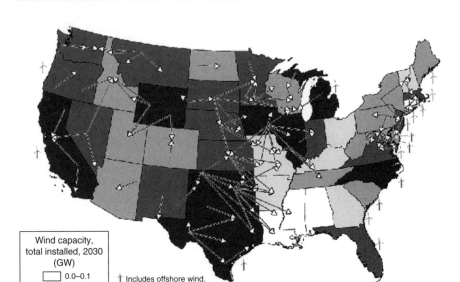

Wind capacity,
total installed, 2030
(GW)
☐ 0.0–0.1
▨ 0.1–1
▨ 1–5
▨ 5–10
■ >10

† Includes offshore wind.

—▷ Wind on new transmission lines

Figure 9.9   Installed wind capacity by state in 2030 and new transmission needs.
Source: Derived from DOE [59].

The study further concludes that, although sufficient land area exists, siting challenges will be significant. Figure 9.9 illustrates the potential location of the 305 GW of wind needed to achieve 20 percent wind energy, based on the economic optimization routine used by WinDS.[10] As shown, WinDS adds wind capacity in almost every state; development in the Southeast is limited due to limited on-shore wind resource potential.[11] Siting this quantity of wind capacity will pose a significant challenge, and the land area requirements are not insubstantial. Assuming that the projected land-based wind turbines are all located in one wind plant, for example, that plant would cover a square roughly 225 km (140 miles) on a side, or roughly 0.5 percent of total U.S. land area. Only 2–5 percent of this area would be occupied by turbine towers, roads, and the like, with the balance available for other uses such as farming

---

[10]The wind capacity levels projected by WinDS in each state depend on a variety of assumptions. In reality, the distribution of the 305 GW across states would differ from WinDS projections and would depend, in part, on state policies used to promote or restrict wind energy.
[11]Alaska and Hawaii were not included in the study, but each state possesses ample wind resources and is expected to install wind capacity in the future.

or ranching. Nonetheless, concerns about wildlife impacts, aesthetics, and other factors will surely need to be addressed to achieve this level of capacity additions.

Because the best wind resources tend to be located long distances from population centers, expansion of the nation's transmission infrastructure is an essential precondition for achieving high levels of wind penetration. Figure 9.9 presents one broadly suggestive scenario for the wind energy transfer between regions, based on WinDS output. The "20 percent wind" scenario projects the addition of approximately 12,000 miles of new transmission lines at a discounted cost of about $20 billion. Although the $20 billion cost estimate is not prohibitive from the perspective of the overall electric sector, the institutional barriers to planning, siting, and recovering the cost of this quantity of new transmission lines are not insubstantial, and transmission expansion may pose the most significant challenge to achieving high levels of wind generation.

The study also concludes that the integration of "20 percent wind" into U.S. electricity markets is both technically and economically feasible. In contrast to many other sources of electricity, wind power is inherently variable and non-controllable. In addition, though forecasting techniques are improving, there are limits to the level of those improvements. Despite these characteristics, a growing number of studies indicates that high levels of wind penetration can be accommodated as part of interconnected power systems.[12] Though accommodating wind requires modifications to power system planning and operations, the costs imposed by these changes are generally found to be below $10/MWh for wind penetrations of up to 20 percent. Spatial diversity in wind plants, wind output forecasting, control area coordination, liquid real-time markets, and quick-ramping fossil assets can all help alleviate the inherent variability in wind generation.

The WinDS model considers the cost of integrating wind into electricity systems as well as the cost of maintaining overall system capacity and reliability. One key output of the WinDS modeling is that the wind generation in the "20 percent wind" scenario offsets both coal and combined-cycle natural gas (Gas-CC) usage (see rightmost graphic in Figure 9.10) but also leads to a significant increase in quick-ramping gas combustion-turbine capacity (Gas-CT) to maintain system reliability (see leftmost graphic in Figure 9.10).[13] The U.S. DOE study concludes that additional storage is not strictly essential to achieving 20 percent wind penetration. Subsequent analysis confirms this result but also finds that the economical use of storage could reduce the costs

---

[12]For summary articles on this topic, see Gross et al. (2007) and Smith et al. (2007).
[13]These gas combustion-turbine plants operate very infrequently, as shown in the rightmost graphic in Figure 9.10.

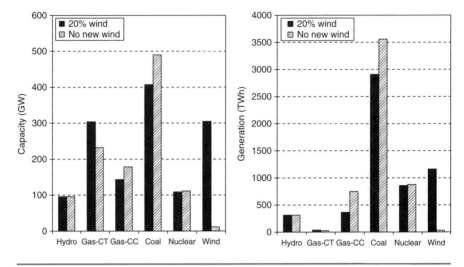

Figure 9.10 Electric sector capacity and generation in 2030. Source: DOE [59].

of maintaining grid reliability in high-wind penetration scenarios by, in part, reducing the need for gas combustion-turbine capacity [57].

At the heart of the analysis was an assessment of the economic feasibility of achieving 20 percent wind energy, and here the news is positive: If the significant institutional barriers can be overcome, the overall incremental costs of achieving 20 percent wind energy are projected to be relatively modest. Figure 9.11 shows the total estimated electric-sector costs for both the "20 percent wind" and the "no new wind" scenarios, considering capital costs, fuel costs, operation and maintenance costs, and transmission costs (embedded in these categories are also resource adequacy and integration costs). The "20 percent wind" scenario requires a larger capital investment, but that investment is offset by reduced fuel costs. In aggregate, the "20 percent wind" scenario imposes an estimated net discounted incremental cost of over $40 billion.[14] This implies an average incremental cost of wind (compared to the fossil generation deployed in the "no new wind" scenario) of $9/MWh, an average increase in retail electricity rates of $0.6/MWh, and a bill impact for an average household of roughly $0.50 per month. Subsequent technical analysis has explored the sensitivity of this cost to differing input parameters and has found the cost estimate to be relatively robust [4]. In particular, if conservative assumptions are

---

[14]WinDS is used to estimate generation and transmission capital expenditures as well as fuel and O&M costs through 2030, while extrapolations of fuel usage and O&M requirements are applied from 2030 to 2050. A 7 percent real discount rate is used to estimate present value figures.

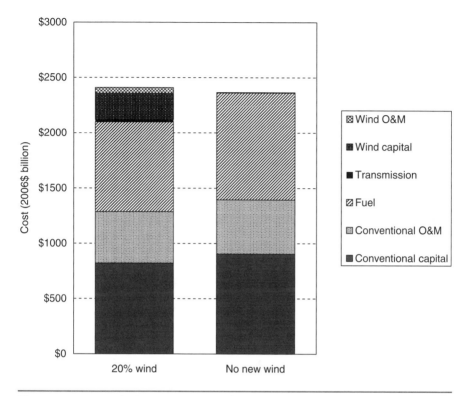

Figure 9.11   Total electricity-sector direct costs (discounted at 7 percent per year).
Source: DOE [59].

used, the estimated incremental cost of achieving 20 percent wind energy can, in the more extreme cases, more than double while, if more optimistic assumptions are used, aggregate electric-sector *savings* are sometimes predicted.

Of course, these direct costs are offset by a number of *possible* benefits, including lower fossil-fuel prices, environmental improvement, reduced water consumption, rural economic development, and employment opportunities in the renewable energy sector. Perhaps most important, achieving 20 percent wind energy would reduce the carbon footprint of the U.S. electricity sector. WinDS modeling shows a displacement of coal and natural-gas generation, yielding an annual reduction of 225 MMTCE by 2030, equivalent to roughly 20 percent of expected electricity-sector carbon emissions in 2030 in the "no new wind" reference case (Figure 9.12).

To put that figure in context, several legislative proposals were considered in the 110th U.S. Congress to limit economy-wide greenhouse gas emissions. The Energy Information Administration evaluated three of these

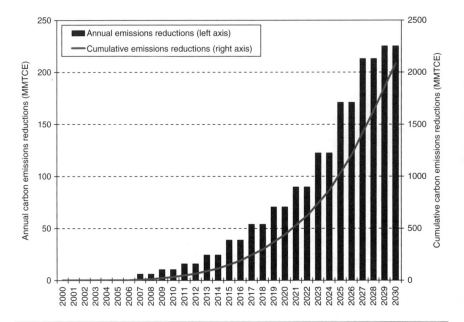

Figure 9.12 Carbon emissions reductions associated with 20 percent wind penetration, 2000–2030.
Source: DOE [59].

proposals [14–16], finding that emissions reductions would come *primarily* from the electric power sector and that electric-sector emissions in 2030 would fall to just *15–60 percent* of 2005 levels. As shown in Figure 9.13, the annual emissions reductions predicted to come from the "20 percent wind" scenario in 2030 represent a substantial down payment toward even these aggressive emissions reduction targets.

Moreover, it appears that the "20 percent wind" scenario represents a relatively low-cost option for reducing carbon emissions. If one allocates the full predicted incremental cost of achieving 20 percent wind ($40 billion plus) to the carbon savings predicted earlier, for example, one finds that the "20 percent wind" scenario yields carbon reductions at under $13/tCO$_2$-eq.[15] This figure is below the $20–80/tCO$_2$-eq carbon price range that IPCC [36] reports may be necessary by 2030 to achieve 550 ppm stabilization by 2100[16]

---

[15]Even in the more conservative cases presented by Blair et al. (2008), carbon reduction costs rarely exceed $30/tCO$_2$-eq.

[16]With induced technology changes, the IPCC reports prices from $5–65/tCO$_2$-eq. Again, note that some governments and scientists have called for carbon stabilization at well below 550 ppm, presumably requiring even higher carbon prices.

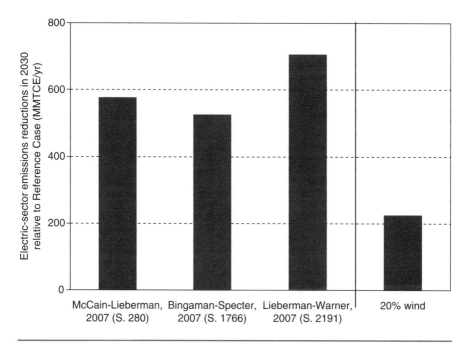

Figure 9.13 Comparing the carbon savings from 20 percent wind to proposed U.S. climate legislation.
Source: Compiled by Berkeley Lab, based on EIA [14–16].

and is also lower than the carbon prices of roughly $25–60/tCO$_2$-eq predicted by the EIA to be required in 2030 under climate legislation proposed during the 110th U.S. Congress [14–16].

9.4 The potential global role of wind power

The United States is not alone in its potential for substantial increases in wind power supply. In fact, a growing number of studies have sought to document the world's wind resources. Reporting on other studies, the IPCC [37] identifies an aggregate worldwide onshore wind power potential of roughly 600 EJ, which is more than three times greater than the current worldwide electricity supply.[17]

---

[17]Studies that have been conducted on the world's wind resource potential include Archer and Jacobsen (2005), Fellows (2000), Grubb and Meyer (1993), Hoogwijk et al. (2004), World Energy Council (1994), and WBGU (2004). Recent studies have often found an onshore wind resource potential that exceeds that reported by the IPCC (2007b); the IPCC figure also excludes offshore wind energy, which would add to the technical potential estimates.

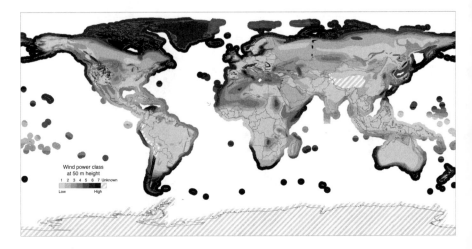

Figure 9.14 The global wind resource.
Source: Worldwide wind energy resource distribution estimates (PNL, 1981) and *Wind Energy Resource Atlas of the United States* (PNL, 1987).

The wind resource is not evenly distributed across the globe (Figure 9.14), however, and North America is particularly well endowed; two studies have estimated that North America hosts roughly 25 percent of the global onshore wind potential [27, 64], while another study finds that the United States alone holds 22 percent of the world's potential [31]. As such, the relatively low cost of wind as a carbon-reduction strategy for the United States should not simply be assumed to hold in every region of the globe [31]. Nonetheless, ample technical potential exists in most regions to enable high levels of wind penetration, even excluding the potential for offshore wind. As such, the size of the resource potential itself is unlikely to pose a significant *global* barrier to wind power expansion.

The United States is also not alone in conducting analyses of the feasibility of achieving higher levels of wind power deployment. In part relying on the resource data presented earlier, analyses of the technical and economic feasibility of achieving high levels of wind power penetration have been conducted in a number of countries and regions[18] as well as globally, using a diverse set of

---

[18]In this book, for example, Chapter 17 provides an assessment of the possible role of wind power (among other technologies) in meeting China's energy needs in both a business-as-usual and carbon-constrained future; other chapters include some discussion of the role of wind in the United Kingdom (Chapter 15) and New Zealand (Chapter 14).

Table 9.4   Recent global wind energy deployment scenarios

Study Sponsors	Wind Power as Proportion of Global Electricity Supply
IPCC [37]	7% by 2030
IEA [34]*	8.4% by 2030
IEA [35]	9% (ACT Map scenario) or 12% (BLUE Map scenario)
European Commission [17]**	13% by 2050
Edmonds et al. [12]***	14% by 2035
Shell [53]****	23% by 2050
WBGU [61]*	135 EJ by 2040 (13% of *total* energy supply)
Greenpeace and EREC [24]	23% by 2050
Greenpeace and GWEC [23]*****	11–18% (moderate) or 21–34% (advanced) by 2050
Greenpeace and EWEA [25]	12% by 2020

450 ppm carbon stabilization case.
**Carbon constraint scenario.*
***550 ppm carbon stabilization scenario; penetration levels increase to 2095.*
****Approximate percentage for "Blueprints" scenario.*
*****Penetration range within single scenario depends on assumed degree of energy efficiency.*

analytic tools and methods. A sample of recent global scenarios is summarized in Table 9.4.[19]

The range of approaches used to generate these scenarios is great, making suspect any comparison of the scenarios. Additionally, as noted earlier, existing modeling tools often do a relatively poor job at estimating the technical and economic viability of high levels of wind penetration due to the geographic and temporal characteristics of wind potential and production. As such, the global scenarios presented here should be considered, at best, indicative of wind's potential and should not be used as an estimate of the "optimal" or "maximum" level of wind production that might be feasible on a global basis.

Nonetheless, the summary of these scenarios provided in Table 9.4 suggests that wind penetration levels that approach 10 percent of global electricity supply by 2030 are feasible and that aggressive policies and/or technology

---

[19]For reviews of the renewable energy scenario literature more broadly, see Martinot et al. (2007) and Hamrin et al. (2007).

improvements may allow wind production to ultimately reach 20 percent or more of global supply. As with the U.S. DOE analysis presented earlier, most of these scenarios are not predicated on the widespread use of electricity storage to manage wind's inherent variability. Other strategies (e.g., flexible, fast-ramping combustion turbines) can also be used to manage wind's variability, and storage will have to compete with these options. If wind penetration levels exceed 10–20 percent, however, complementary storage technologies may be required to minimize cost.[20] Even at lower penetration levels, the availability of storage can ease the burden of integrating wind into electric power grids, though storage should only be deployed for this purpose if it is economical to do so relative to other integration strategies.

Pacala and Socolow [50, 56] describe a simplified scenario that may enable CO_2 stabilization at 500 ppm, initially requiring the displacement of 175 billion tons of carbon (GtC) over 50 years, or seven "wedges" of 25 GtC each. Though decomposing the carbon problem into such "wedges" has limitations, it nonetheless provides a useful heuristic for highlighting the importance of different technology solutions—including wind—in combating global climate change. In particular, the high-penetration wind power studies presented in this chapter suggest that wind power may be able to deliver one of these seven wedges in the medium term. Specifically, assuming that the "20 percent wind" scenario presented earlier for the United States achieves and then maintains 305 GW of wind from 2030–2050, that scenario is estimated to deliver more than 6 GtC emissions reductions through 2050 and to do so at a cost below that of many other carbon reduction strategies. By 2030, the U.S. electricity sector is expected to represent roughly 18 percent of global electricity generation. By simple extrapolation, if the world was also able to achieve a similar 20 percent wind penetration level, global cumulative carbon emission reductions from wind would exceed the size of a single 25 GtC wedge.[21]

---

[20]At higher levels of wind penetration, wind output may need to be curtailed during some portion of the year. At this point, the value of storage increases significantly. A variety of studies have looked at wind power penetrations in excess of 20 percent in certain regions, typically involving the use of storage, plug-in hybrid vehicles, demand response, and/or other technologies to manage the variability of wind power. For a sample of recent (and in some cases somewhat U.S.-centric) examples, see Benitez et al. (2008), Black and Strbac (2006), Cavallo (2007), DeCarolis and Keith (2006), Denholm (2006), Hoogwijk et al. (2007), Greenblatt et al. (2007), Kempton and Tomic (2005), Lund and Kempton (2008), and Leighty (2008).

[21]Clearly, this is a crude approach, since it assumes that the fossil generation offset by worldwide wind additions is similar to that estimated for the United States, and it ignores possible interactions between carbon reduction strategies. Nonetheless, the estimates provided here are reasonably consistent with the wind electricity wedge posited by Pacala and Socolow (2004) and Socolow and Pacala (2006).

9.5 Conclusion

The challenge of supplying energy reliably, securely, and economically while also tackling the threat of climate change is daunting. In concert with other technologies, however, wind power is ready today to begin confronting this challenge. Global wind resource potential is vast, and though accessing this potential is not costless or lacking in barriers, wind power can be developed at what many would consider to be an acceptable cost.

To grow wind at scale, however, a "business as usual" path will not do. In addition to the common needs of other low-carbon technologies, such as stable, long-term promotional policies and placing an economic price on carbon emissions, the following actions are likely to be needed:

1. Dramatically expanded transmission investments specifically designed to access remote wind resources.

2. Larger power control regions, better wind forecasting, and increased investment in fast-responding generating plants to more effectively integrate wind power into electricity grids.

3. Streamlined siting and permitting procedures that allow developers to identify appropriate project locations and quickly move from wind resource prospecting to project construction.

4. Enhanced research and development to lower the cost of offshore wind power and incrementally improve conventional onshore wind technology.

5. Investments in technologies and practices that may enable even greater penetration of wind, including storage, plug-in hybrids, demand response, and hydrogen production.

Long ago, Henry David Thoreau wrote: "First, there is the power of the Wind, constantly exerted over the globe. . . . Here is an almost incalculable power at our disposal, yet how trifling the use we make of it. It only serves to turn a few mills, blow a few vessels across the ocean, and a few trivial ends besides. What a poor compliment do we pay to our indefatigable and energetic servant!"[22] More than a century and a half later, we are now able to both calculate and efficiently utilize the power in the wind. Perhaps it is finally time to lean more heavily on this "indefatigable and energetic servant."

---

[22]Thoreau HD. Paradise (To Be) Regained (1843). In: The Writings of Henry David Thoreau, vol. 4. pp. 286–287, Houghton Mifflin (1906).

Acknowledgments

The work reported in this chapter was funded by the Wind & Hydropower Technologies Program, Office of Energy Efficiency and Renewable Energy of the U.S. Department of Energy under contract numbers DE-AC02-05CH11231 (Lawrence Berkeley National Laboratory) and DE-AC36-99-GO10337 (National Renewable Energy Lab). The authors thank the many individuals who commented on earlier versions of this paper or who contributed to the analysis and data on which this paper is based.

References

[1] Archer C, Jacobsen M. Evaluation of Global Wind Power. J Geophys Res 2005;110:D12110.

[2] Benitez L, Benitez P, van Kooten G. The Economics of Wind Power with Energy Storage. Energy Economics 2008;30:1973–89.

[3] Black M, Strbac G. Value of Storage in Providing Balancing Services for Electricity Generation Systems with High Wind Penetration. Journal of Power Sources 2006;62:949–53.

[4] Blair N, Hand M, Short W, Sullivan P. Modeling Sensitivities to the 20% Wind Scenario Report with the WinDS Model. NREL/CP-670-43511. Conference paper: WINDPOWER. Houston, Texas; 2008.

[5] BTM Consult. International Wind Energy Development: World Market Update 2007. Ringkobing. Denmark: BTM Consult; 2008.

[6] Cavallo A. Controllable and Affordable Utility-Scale Electricity from Intermittent Wind Resources and Compressed Air Energy Storage (CAES). Energy 2007;32:120–7.

[7] Chupka M, Basheda G. Rising Utility Construction Costs: Sources and Impacts. Prepared by The Brattle Group for The Edison Foundation; 2007.

[8] Cryts J, Derkach A, Nyquist S, Ostrowski K, Stephenson J. Reducing U.S. Greenhouse Gas Emissions: How Much at What Cost? McKinsey & Company; 2007.

[9] DeCarolis J, Keith D. The Economics of Large-Scale Wind Power in a Carbon Constrained World. Energy Policy 2006;34:395–410.

[10] Denholm P. Improving the Technical, Environmental, and Social Performance of Wind Energy Systems Using Biomass-Based Energy Storage. Renewable Energy 2006;31:1355–70.

[11] de Vries B, van Vuuren D, Hoogwijk M. Renewable Energy Sources: Their Global Potential for the First-Half of the 21st Century at a Global Level: An Integrated Approach. Energy Policy 2007;35:2590–610.

[12] Edmonds J, Wise M, Dooley J, Kim S, Smith S, Runei P, et al. Global Energy Technology Strategy: Addressing Climate Change. Phase 2 Findings from an International Public-Private Sponsored Research Program. College Park, Maryland: Joint Global Change Research Institute; 2007.

[13] Emerging Energy Research (EER). Wind Turbine Industry Steps Up to Global Demand. Emerging Energy Research press release on Global Wind Turbine Markets and Strategies. p. 2008–20; 2008.

[14] Energy Information Administration (EIA). Energy Market and Economic Impacts of S. 280, the Climate Stewardship and Innovation Act of 2007. SR/OIAF/2007-04. Washington, DC: Energy Information Administration; 2007.

[15] Energy Information Administration (EIA). Energy Market and Economic Impacts of S. 1766, the Low Carbon Economy Act of 2007. SR/OIAF/2007-06. Washington, DC: Energy Information Administration; 2008a.

[16] Energy Information Administration (EIA). Energy Market and Economic Impacts of S.2191, the Lieberman-Warner Climate Security Act of 2007. SR/OIAF/2008-01. Washington, DC: Energy Information Administration; 2008b.

[17] European Commission. World Energy Technology Outlook—2050—WETO H2. Brussels, Belgium: European Commission; 2006.

[18] European Wind Energy Association (EWEA). Pure Power: Wind Energy Scenarios up to 2030. Brussels, Belgium: European Wind Energy Association; 2008.

[19] Fellows A. The Potential of Wind Energy to Reduce Carbon Dioxide Emissions. Glasgow, UK: Garrad Hassan; 2000.

[20] Hansen J, Sato M, Kharecha P, Beerling D, Berner R, Masson-Delmotte V, et al. Target Atmospheric CO_2: Where Should Humanity Aim? Submission to Open Atmospheric Science Journal last revised June 2008: arXiv:0804.1126v2; 2008.

[21] Global Wind Energy Council (GWEC). Global Wind 2007 Report. Brussels, Belgium: Global Wind Energy Council; 2008.

[22] Greenblatt J, Succar S, Denkenberger D, Williams R, Socolow R. Baseload Wind Energy: Modeling the Competition Between Gas Turbines and Compressed Air Energy Storage for Supplemental Generation. Energy Policy 2007;35:1474–92.

[23] Greenpeace and Global Wind Energy Council (GWEC). Global Wind Energy Outlook 2006. The Netherlands and Brussels, Belgium: Amsterdam; 2006.

[24] Greenpeace and European Renewable Energy Council (EREC). Energy [R]evolution: A Sustainable World Energy Outlook. The Netherlands and Brussels, Belgium: Amsterdam; 2007.

[25] Greenpeace and European Wind Energy Association (EWEA). Wind Force 12: A Blueprint to Achieve 12% of the World's Electricity from Wind Power by 2020. Prepared for the Global Wind Energy Council; 2005.

[26] Gross R, Heptonstall P, Leach M, Anderson D, Green T, Skea J. Renewables and the Grid: Understanding Intermittency. Energy 2007;160:31–41.

[27] Grubb M, Meyer N. Wind Energy: Resources, Systems, and Regional Strategies. In: Johansson H, Kelly H, Reddy A, Williams R, editors. Renewable Energy: Sources for Fuels and Electricity. [Chapter 4]. Washington, DC: Island Press; 1993.

[28] Haas R, Meyer N, Held A, Finon D, Lorenzoni A, Wiser R, et al. Promoting electricity from renewable energy sources – lessons learned from the EU, U.S. and Japan. In: Sioshansi F, editor. Competitive Electricity Markets: Design, Implementation, Performance. [Chapter 12]. Oxford, UK: Elsevier; 2008.

[29] Hamrin J, Hummel H, Canapa R. Review of the Role of Renewable Energy in Global Energy Scenarios. Prepared for the International Energy Agency Implementing Agreement on Renewable Energy Technology Deployment. San Francisco, California: Center for Resource Solutions; 2007.

[30] Hoogwijk M. On the Global and Regional Potential of Renewable Energy Sources. Utrecht, The Netherlands: Utrecht University; 2004.

[31] Hoogwijk M, de Vries B, Turkenburg W. Assessment of the Global and Regional Geographic, Technical and Economic Potential of Onshore Wind Energy. Energy Economics 2004;26:889–919.

[32] Hoogwijk M, van Vuuren D, de Vries B, Turkenburg W. Exploring the Impact on Cost and Electricity Production of High Penetration Levels of Intermittent Electricity in OECD Europe and the USA, Results for Wind Energy. Energy 2007;32:1381–402.

[33] Husum WindEnergy (Husum). WindEnergy Study 2008: Assessment of the Wind Energy Market Until 2017. Prepared by Messe Husum in cooperation with Hamburg Messe on behalf of Husum WindEnergy; 2008.

[34] International Energy Agency (IEA). World Energy Outlook 2007. Paris, France: International Energy Agency; 2007.

[35] International Energy Agency (IEA). Energy Technology Perspectives 2008: Scenarios & Strategies to 2050. Paris, France: International Energy Agency; 2008.

[36] Intergovernmental Panel on Climate Change (IPCC). Climate Change 2007: Synthesis Report. Contribution of Working Groups I, II and III to the Fourth Assessment Report of the Intergovernmental Panel on Climate Change. Geneva, Switzerland: Intergovernmental Panel on Climate Change; 2007a.

[37] Intergovernmental Panel on Climate Change (IPCC). Climate Change 2007: Mitigation. Contribution of Working Group III to the Fourth Assessment Report of the Intergovernmental Panel on Climate Change. Cambridge, United Kingdom, and New York: Cambridge University Press; 2007b.

[38] Johansson T, McCormick K, Neij L, Turkenburg W. The Potentials of Renewable Energy. Thematic Background Paper for International Conference for Renewable Energies, Bonn, 2004.

[39] Kempton W, Tomic J. Vehicle-to-Grid Power Implementation: From Stabilizing the Grid to Supporting Large-Scale Renewable Energy. Journal of Power Sources 2005;144:280–94.

[40] Key T. Role of Renewable Energy in a Sustainable Electric Generation Portfolio. Presentation to EIA Energy Outlook, Modeling, and Data Conference; 2007 Washington, DC; 28 March; 2007.

[41] Kutscher C, editor. Tackling Climate Change in the U.S.: Potential Carbon Emissions Reductions from Energy Efficiency and Renewable Energy by 2030. American Solar Energy Society; 2007.

[42] Kyle P, Smith SJ, Wise MA, Lurz JP, Barrie D. Long-Term Modeling of Wind Energy in the United States. PNNL-16316. College Park, Maryland: Joint Global Change Research Institute; 2007.

[43] Leighty W. Running the World on Renewables: Hydrogen Transmission Pipelines and Firming Geologic Storage. International Journal of Energy Research 2008;32:408–26.

[44] Lewis J, Wiser R. Fostering a Renewable Energy Technology Industry: An International Comparison of Wind Industry Policy Support Mechanisms. Energy Policy 2006;35(3): 1844–57.

[45] Lund H, Kempton W. Integration of Renewable Energy into the Transport and Electricity Sectors Through V2G. Energy Policy 2008;36:3578–87.

[46] Martinot E, Dienst C, Weiliang L, Qimin C. Renewable Energy Futures: Targets, Scenarios, and Pathways. Annual Review of Environment and Resources 2007;32:205–39.

[47] Mendonca M. Feed-in Tariffs: Accelerating the Deployment of Renewable Energy. London, UK: Earthscan; 2007.

[48] Meyer N. Learning from Wind Energy Policy in the EU: Lessons from Denmark, Germany and Spain. European Environment 2007;17(5):347–62.

[49] Neuhoff K, Ehrenmann A, Butler L, Cust J, Hoexter H, Keats K, et al. Space and Time: Wind in an Investment Planning Model. Energy Economics 2008;30:1900–2008.

[50] Pacala S, Socolow R. Stabilization Wedges: Solving the Climate Problem for the Next 50 Years with Current Technologies. Science 2004;305(5686):968–72.

[51] REN21. Renewable Energy Potentials: Opportunities for the Rapid Deployment of Renewable Energy in Large Energy Economies. Intermediate Report to the 3rd Ministerial Meeting

of the Gleneagles Dialogue on Climate Change, Clear Energy and Sustainable Development. Paris, France: REN21 Secretariat; 2007.

[52] REN21. Renewables 2007 Global Status Report. Paris, France: REN21 Secretariat; 2008.

[53] Shell. Shell Energy Scenarios to 2050. The Hague, The Netherlands: Shell International; 2008.

[54] Short W, Blair N, Heimiller D. Projected Impact of Federal Policies on U.S. Wind Market Potential. Proceedings of the 2004 Global Wind Power Conference, Chicago, Illinois; March 29–31; 2004.

[55] Smith JC, Milligan M, DeMeo E, Parsons B. Utility Wind Integration and Operating Impact State of the Art. IEEE Transactions on Power Systems 2007;22(3):900–8.

[56] Socolow R, Pacala S. A Plan to Keep Carbon in Check. Sci Am 2006;295(3):50–7.

[57] Sullivan P, Short W, Blair N. Modeling the Benefits of Storage Technologies to Wind Power. NREL/CP-670-43510. Conference paper: WINDPOWER. Houston, Texas; 2008.

[58] United Nations Development Programme (UNDP). World Energy Assessment: Energy and the Challenge of Sustainability. New York: United Nations Development Programme; 2000.

[59] U.S. Department of Energy (DOE). 20% Wind Energy by 2030: Increasing Wind Energy's Contribution to U.S. Electricity Supply. DOE/GO-102008-2567. Washington, DC: U.S. Department of Energy; 2008.

[60] van der Linden NH, Uyterlinde MA, Vrolijk C, Nilsson LJ, Khan J, Åstrand K, et al. Review of International Experience with Renewable Energy Obligation Support Mechanisms. Energy Centre of the Netherlands. ECN-C-05-025; 2005.

[61] German Advisory Council on Global Change (WBGU). World in Transition: Towards Sustainable Energy Systems. London, England, and Sterling, Virginia: Earthscan; 2004.

[62] Wiser R, Bolinger M, Barbose G. Using the Federal Production Tax Credit to Build a Durable Market for Wind Power in the United States. The Electricity Journal 2007;20(9):77–88.

[63] Wiser R, Bolinger M. Annual Report on U.S. Wind Power Installation, Cost, and Performance Trends: 2007. DOE/GO-102008-2590. Washington, DC: U.S. Department of Energy; 2008.

[64] World Energy Council. New Renewable Energy Resources: A Guide to the Future. London, UK: Kogan Page Limited; 1994.

[65] World Energy Council. 2007 Survey of Energy Resources. London, UK: World Energy Council; 2007.

Solar Energy: The Largest Energy Resource

Chapter 10

Paul Denholm, Easan Drury, Robert Margolis, Mark Mehos

National Renewable Energy Laboratory

Abstract

The fraction of electricity generated by solar technologies is small but growing rapidly, with enormous potential to generate a large fraction of the world's electricity needs while significantly reducing global carbon emissions. Realizing this potential, however, will require overcoming both technical and economic barriers. In the short term, it will be important to decrease costs, improve solar conversion efficiency, and implement electricity rate structures that capture the time-varying value of solar-generated electricity. In the long term, challenges will include using material resources more efficiently, integrating intermittent PV electricity into the grid, and building transmission capacity for utility-scale solar generation systems linking areas with good solar resources to population centers.

10.1 Introduction

The earth receives more energy from the sun in one hour than the global population uses each year. Not only does solar energy have a larger technical potential than any other renewable energy resource, it is readily available in every inhabited environment. However, capturing this large but diffuse energy resource will require overcoming both technical and economic challenges.

The current fraction of electricity produced by solar technologies is very small relative to the size of the resources available, which is much greater than any foreseeable worldwide demand for electricity. Global electricity generation capacity in 2005 was 3872 GW, whereas the total generation capacity of photovoltaics (PV) and concentrated solar power (CSP) combined was roughly 9 GW by the end of 2007,[1] accounting for less than 0.1 percent of global electricity generation. The contribution from other direct solar technologies such as solar water heating, space heating, and building lighting is difficult to quantify but significantly larger, with current estimates of solar water heating at roughly 105 $GW_{thermal}$ [22].

Though starting from a small base, grid-connected solar generation is growing more quickly than any other renewable technology [22]. PV installations have grown more than 30 percent per annum for the past 15 years, with a 50 percent growth rate in 2007. Installed grid-connected PV accounted for more than 8 GW in 2007, with an estimated 3–4 GW installed in 2008. Thin-films production capacity more than doubled during 2007, growing from a niche market to a proven technology that accounts for more than 10 percent of the PV market. CSP has seen a large resurgence in investment and potentially rapid growth in the United States and Spain. Over 4 GW of CSP projects are under contract in the United States, driven primarily by favorable tax policies currently in place.

Solar technologies have the potential to significantly mitigate carbon emissions. Under the "wedge" scenario defined by Socolow and Pacala (2007), solar energy could reduce carbon emissions by 1 Gt C/year by 2050, representing about 20 percent of the electricity generation market. Achieving this level of penetration will, however, require continued significant growth in the global PV and CSP markets, which will need to be driven by price reductions and efficiency gains. Going beyond this level will bring additional challenges such as integrating intermittent

[1]This included roughly 8.5 GW of PV systems and 0.5 GW of CSP systems.

PV generation into the electricity grid and building additional transmission capacity linking CSP in areas with a good solar resource to load centers.

This chapter begins with an overview of the global solar resource and provides an overview of solar generation technologies, including cost and market growth trends. The chapter discusses barriers and opportunities associated with high penetration of solar technologies and the potential for carbon mitigation. The chapter includes technologies that directly harness solar energy; "indirect" solar generation, such as wind power and hydroelectric power, are covered elsewhere in this volume.

▌ 10.2 Solar resource and conversion technologies

10.2.1 THE SOLAR RESOURCE

The mean annual solar energy resource is frequently expressed as the amount of radiant energy received on a given surface area per unit time ($kWh/m^2/day$). This incident solar energy is shown globally in Figure 10.1.

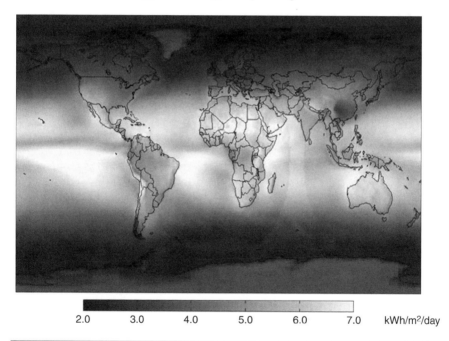

2.0   3.0   4.0   5.0   6.0   7.0 $kWh/m^2/day$

Figure 10.1 Global mean solar resource ($kWh/m^2/day$). Source: Adapted from the NASA/SSE satellite dataset.[2]

[2]Satellite-derived solar energy resource constrained by data from 1195 surface stations. The data are available at eosweb.larc.nasa.gov/sse/.

Solar energy contains a *direct component*, which is light from the solar beam, and a *diffuse component*, which is light that has been scattered by the atmosphere. This distinction is important because only the direct solar component can be effectively focused by mirrors or lenses. The direct component typically accounts for 60–80 percent of the total solar insolation in clear sky conditions and decreases with increasing humidity, cloud cover, and atmospheric aerosols such as dust or pollution plumes. Technologies that rely on the direct solar component such as CSP plants work best in areas with high direct normal irradiance, which generally limits their application to arid regions, as seen in Figure 10.2. Nonconcentrated solar technologies such as PV panels can use both the direct and diffuse solar components and are not as geographically limited in their application.

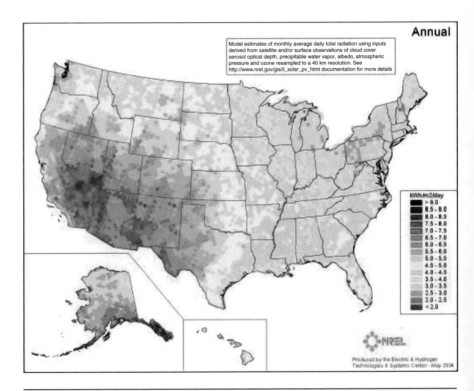

Figure 10.2   Direct normal solar insolation for a two-axis tracking concentrator in the United States.
Source: NREL.

10.2.2 SOLAR CONVERSION TECHNOLOGIES

Solar energy can be utilized to meet a variety of end uses. Figure 10.3 illustrates the range of solar technologies and methods used to capture solar heat and light and to generate electricity. Within each application, there are technologies that use either the natural solar intensity or use "concentrating" technologies to focus the direct component of solar radiation. Both the efficiency and economic viability of solar energy technologies depends, in large part, on the quality of the local solar resource.

Figure 10.3 indicates that there are a number of pathways to turn sunlight into useful energy. Electricity pathways include PV and CSP technologies; other solar technologies, including heating and lighting, do not produce electricity directly but may reduce the demand for electricity in buildings, effectively offsetting carbon-intensive electricity.

10.2.2.1 Photovoltaics (PV)

PV solar cells convert sunlight directly into electricity. Although PV cells are fabricated from a number of materials, they all generate electricity by enabling photons to "knock" electrons out of a molecular lattice, leaving a freed

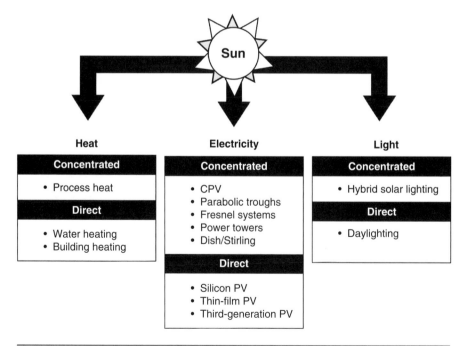

Figure 10.3 End uses for solar energy.

electron and "hole" pair that diffuse in an electric field where the charge carriers are separated, delivering a direct current (DC) to an outside circuit. The engineering challenge in all PV devices is to increase the absorptivity of the cell material and to decrease the rate of electron/hole recombination while keeping the cost low.

A PV module contains multiple PV cells enclosed in glass or other transparent protective material. In typical applications, modules are wired together to form an array. In a grid-connected system, an inverter is used to convert the DC output from the array to alternating current (AC) as required by the electric power grid. The power output of a PV module is defined as its peak DC output under standard test conditions (STC) of 1000 W/m^2 of solar intensity and a module temperature of 25°C. The module efficiency refers to the fraction of incident solar energy that is converted to DC electrical energy. In real-world applications, the module's DC efficiency is further reduced by 10–25 percent based on losses in the inverter, wiring, and module soiling such as dust or debris on the panel.

PV arrays can be either fixed or sun-tracking, with one or two axes of rotation (see Figure 10.4). Fixed PV systems are typically deployed on flat or tilted rooftops or on ground-mounted arrays tilted to the south to increase the amount of incident radiation. One-axis tracking systems follow the sun's east/west "motion" with fixed inclination, and two-axis tracking systems additionally track the sun's inclination. Tracking systems allow a given unit of PV capacity to generate more electricity at the expense of the mechanical tracking equipment.

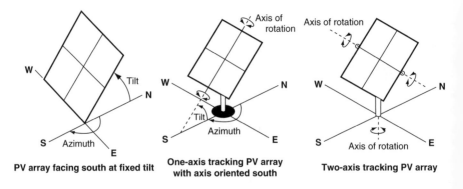

Figure 10.4 Fixed and tracking PV arrays.
Source: NREL.

There are several types of PV technologies currently in the market or under development, including:

- *Wafer-based crystalline silicon (c-Si)*. The most common material used to make PV is crystalline silicon, which is a semiconductor material similar to that used to make computer chips. Discrete solar cells are made from silicon wafers that have been cut from single-crystal silicon ingots or blocks of case-grown poly-silicon or grown in thin sheets. Positive and negative connections are made to the semiconductor to make a photovoltaic cell; these connections are linked to other cells to form a module. Typical crystalline silicon PV modules have a DC efficiency of approximately 15 percent at STC, with operating lifetimes of at least 25 years. Some crystalline silicon modules have efficiencies as high as 20 percent.

The most significant technical challenge faced by crystalline silicon is to reduce the cost of materials used in the relatively thick cells.[3] Periodic bottlenecks in the silicon supply chain have prompted new manufacturing techniques that use less silicon to produce wafers. These techniques include reducing material losses while pulling single-crystal ingots and investigating novel wafer liftoff techniques to eliminate kerf losses.[4]

- *Thin films*. Thin-film solar cells have a much higher rate of light absorption that allows the cells to be made roughly 100 times thinner than crystalline cells. Thin-film cells are made from layers of semiconductor material only a few micrometers thick, which are frequently manufactured using vacuum deposition. The most common thin-film technologies are amorphous silicon (a-Si), cadmium telluride (CdTe), and copper indium (gallium) diselenide (CIS or CIGS). Thin films are frequently deposited on a glass substrate, but some manufacturers use flexible substrates that can be incorporated into building materials, such as shingles. Thin-film panels have lower peak DC efficiencies ($\sim$7 percent for a-Si and $\sim$11 percent for CdTe) than crystalline silicon.

The main challenges faced by thin-film manufacturers are increasing device efficiency and reliability while maintaining low cost. This process involves refining techniques developed in laboratories to manufacture cells on a commercial scale while maintaining cell uniformity over large areas and generating high throughput rates. One advantage of thin-film PV is the

---

[3]The cells are thick so they can absorb a large fraction of solar photons. If a PV material can be made more absorptive, as with thin films, the cells can be cut much thinner while absorbing the same fraction of photons.

[4]Kerf losses represent the amount of material lost due to sawing or cutting the ingots into wafers. Diamond wire saws are currently used to minimize loss.

possibility of creating flexible modules that can reduce installation costs and increase the number of applications (e.g., building incorporated materials). The downside of flexible form factors is the difficulty of creating a robust encapsulation material that is transparent, durable, moisture proof, and affordable.

■ *Concentrating PV (CPV)*. CPV technologies use lenses or mirrors to concentrate sunlight 2 to 1000 times onto a high-efficiency silicon or multijunction[5] solar cell. The use of inexpensive materials such as glass and steel to focus sunlight reduces the amount of semiconductor material up to 1000 times. Recent efficiency gains of multijunction cells (up to 40 percent) offer the possibility of very high power density if manufacturability and reliability can be proven. This has renewed interest in CPV technologies in utility-scale or commercial systems.

CPV faces a number of engineering challenges, such as fabricating a low-cost and reliable tracking system to keep the concentrator pointed at the sun and a cooling system to dissipate heat away from the cells. Like other concentrated solar technologies, they perform best in locations with good direct normal solar insolation, generally limiting their application to arid regions. Large CPV systems have historically had to compete with the electricity generation rates in utility markets rather than the end-use electricity rates in residential or commercial markets, which has prevented CPV manufacturers from reaching the volume production required for this technology to gain a foothold.

■ *Other PV materials*. There are a number of advanced PV materials under development that are frequently referred to as *third generation*. These include dye-sensitized solar cells that use dye molecules in an electrolyte solution to absorb solar radiation and have demonstrated efficiencies of up to 11 percent. Organic solar cells are based on plastics that have semiconductor properties with demonstrated laboratory efficiencies of up to about 5 percent. Organic modules have the potential for low-cost manufacturing using existing printing and lamination technologies [23]. Quantum dots are nanospheres that have physical properties similar to both semiconductors and discrete molecules. They have the potential to be made inexpensively and absorb at multiple solar frequencies, but the technology has not yet produced efficient solar cells. Other

---

[5]Multijunction cells consist of different semiconductor layers stacked on top of each other that have unique energy band gaps that absorb different parts of the solar spectrum. This allows multijunction cells to convert more of the sun's energy to electricity, enabling them to reach higher DC efficiencies than conventional cells.

nanotechnologies such as carbon nanotubes are active fields of research and have the potential to supply efficient and inexpensive solar technologies in the future.

10.2.2.2 Concentrated solar power (CSP)

CSP plants produce electric power by first using mirrors to focus sunlight to heat a working fluid. Ultimately, this high-temperature fluid is used to spin a turbine or power an engine that drives a generator. CSP's main advantages are low capital cost and potential for integrating thermal storage, enabling dispatchable generation. CSP's disadvantages include the fact that it can only utilize the direct normal component of solar radiation, it has higher O&M expenses than PV, and it generally has to be deployed as large power plants (one to several hundred MW) in order to be cost competitive in utility-scale markets.

Several types of CSP are now being deployed, including:

■ *Parabolic troughs.* The predominant CSP systems currently in operation in the United States are linear concentrators using parabolic trough collectors. In such a system, the receiver tube runs along the focal line of each parabola-shaped reflector. The tube is fixed to the mirror structure and the heated fluid flows through and out of the field of solar mirrors to the generator, where it is used to create steam for a conventional turbine generator.

The key technical challenges for parabolic troughs include improving efficiency, further reducing component costs, and integrating efficient, low-cost, thermal storage that will enable dispatchable generation. The design point[6] solar to AC electric efficiency for current parabolic troughs is roughly 24 percent, which leads to mean annual operating efficiencies of 12–17 percent [29][7].

Figure 10.5 illustrates the basic principle of trough collectors. The remainder of the system, including the heat exchanger and generator, is illustrated conceptually in Figure 10.7.

■ *Solar power towers.* Solar power tower plants consist of a field of tracking mirrors, called *heliostats*, that focus solar intensity to greater than

---

[6]A design point efficiency represents the fraction of solar energy converted to electrical energy under design conditions that are frequently defined as a solar insolation of 950 W/m^2 with the sun directly overhead and the thermal generator operating at peak efficiency.

[7]The mean annual efficiency represents measured operating efficiency over the course of 1 or more years.

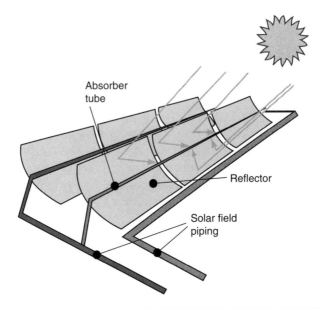

Figure 10.5 The parabolic trough principle.
Source: SolarPACES.

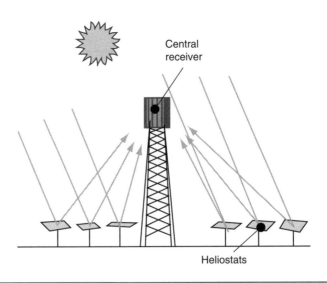

Figure 10.6 Power Tower.
Source: SolarPACES.

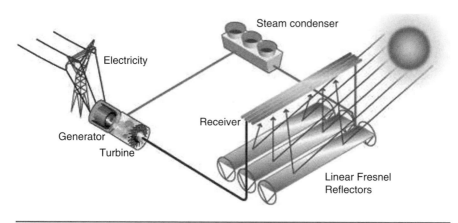

Figure 10.7   Linear Fresnel system with thermal storage and generator components.
Source: U.S. Department of Energy.

1000 times onto a centrally located receiver at the top of a tall tower. A heat-transfer fluid is pumped through the receiver, where it is heated and later used to generate steam, which powers a conventional turbine generator. Figure 10.6 illustrates the basic principle of solar power towers. Some advanced designs are experimenting with molten nitrate salt because of its superior heat-transfer and energy-storage capabilities. Individual commercial plants can be sized to produce up to 200 MW of electricity. Mean annual efficiencies could reach 15–20 percent [26] with the adoption of efficient thermal generators, new corrosion-resistant heliostats, and increased storage to limit parasitic energy use.[8]

■ *Linear Fresnel reflectors.* A linear Fresnel reflector plant uses a series of flat or shallow-curvature mirrors to focus light onto a linear receiver located at the focal point of the mirror array. Linear Fresnel systems have the potential for a lower up-front capital cost than parabolic dish systems because they use flat rather than deeply curved mirrors, and the mirrors are located close to the ground, giving them a lower wind profile. However, linear Fresnel systems have lower operating efficiencies than parabolic trough systems. Since there are no commercial Fresnel-based systems in operation,

[8]Electricity drawn from the grid during times when the plant is not producing electricity. The parasitic energy use will decrease as plant capacity factors increase and the plant is kept online longer.

it is unclear whether the lower up-front capital cost will outweigh the lower operational efficiency.

Figure 10.7 illustrates a conceptualized layout of a linear Fresnel plant, along with the steam turbine component that is common to trough and power tower technologies.

■ *Thermal energy storage.* Thermal energy storage (TES) has the potential to add significant value to trough, linear Fresnel, and power tower CSP plants by allowing electricity generation to be dispatchable and for utilities to use them as load-following generators. A near-term, high-temperature TES option has been developed for parabolic troughs with an oil-to-salt heat exchanger to transfer thermal energy from the solar field to a two-tank nitrate salt storage medium. The integration of TES has the potential to cost-effectively boost plant capacity factors from about 30 percent for solar-only operation to greater than 40 percent with dispatchable thermal storage.

Solar plant capacity factors could be further increased to over 60 percent by increasing the size of TES. Plant capacity factors can also be increased by hybridizing power plants with biofuels or natural gas to augment thermal generation. Several new parabolic trough systems under development are being integrated with combined-cycle natural gas plants. The use of TES could increase the use of CSP plants beyond "peaking" duty to intermediate load or even base-load applications, providing a large fraction of a region's electricity.

■ *Dish/Stirling systems.* Individual two-axis tracking dish/Stirling systems (see Figure 10.8) track the sun and focus the solar intensity to greater than 1000 times onto a receiver. A heat engine/generator, located at the focal point of the dish, converts the heat energy into electricity. The power output of current dish/Stirling units ranges from 3–25 kW. Dish/Stirling power plants may benefit from modular deployment, where power plants could be brought online or expanded quickly. Stirling engines offer high mean annual operating efficiency of roughly 22 percent [29] and tolerance of non-uniform flux distributions and do not require water for cooling the heat engine. One drawback of dish/Stirling systems is that they cannot generate dispatchable electricity since they lack thermal storage.

10.2.2.3 Solar water heating (SWH)

Solar hot water systems (see Figure 10.9) typically use unfocused sunlight to heat water directly or indirectly via a heat transfer fluid flowing through a collector. Common solar water heaters include evacuated tube collectors (47 percent of the existing market), flat plate collectors (33 percent), and

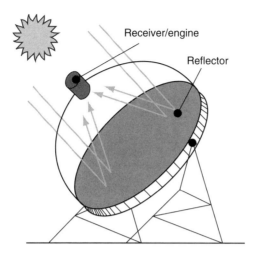

Figure 10.8   Dish/Stirling system.
Source: SolarPACES.

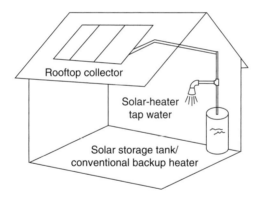

Figure 10.9   A simple solar water-heating system.
Source: NREL.

uninsulated plastic or rubber tube collectors (19 percent) [33][9]. Typical active solar water-heating systems pump water through the collector to a storage tank and then to the end use. More sophisticated solar water-heating systems use a closed loop with a thermal medium, typically water mixed with antifreeze, that pumps the thermal fluid from the solar collector through a heat exchanger in a water storage tank. The efficiency of solar water heaters, defined as the

[9]Simple tube collectors are used primarily to heat swimming pools.

fraction of solar energy captured as useful heat energy, ranges from 30–45 percent based on the complexity of the SWH system [2].

Although solar water-heating systems do not generate electricity, they reduce demand in regions where electricity is used to heat water. In the United States, 44 percent of water is heated using electricity, which accounts for 9 percent of residential electricity demand [7, 32]. In low geographical latitudes (between 40° N and S), 60–70 percent of domestic hot water can be supplied by solar heating systems, with water temperatures of 60°C [22]. The fraction of U.S. residential water heating that can be produced by current SWH technology is plotted in Figure 10.10.

10.2.2.4 Other direct solar technologies

In the United States, heating, ventilation, and air conditioning (HVAC) systems account for nearly 30 percent of the electricity used in residential buildings [32]. Active and passive solar heating and cooling systems can be used to reduce the energy used to heat and cool most buildings. Passive techniques include heating a thermal mass and storing daytime solar energy or designing selective shading. Active techniques include using fans to circulate warm or cool air, using pumped solar hot water systems to heat homes, designing movable shading, and using solar absorption chillers, which typically use hot water to drive a heat pump based on a close-cycle evaporative cooling process.

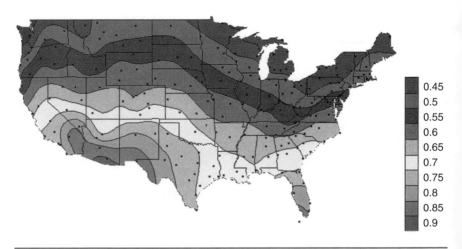

Figure 10.10 Simulated fraction of residential hot water heating that could be supplied by current solar water-heating technology.
Source: Adapted from Burch et al [1].

Solar lighting in buildings offsets electricity use directly by replacing artificial lighting and indirectly by reducing the need for air conditioning. When solar lighting features are properly implemented, they can reduce lighting-related energy requirements by 25 percent, reducing household electricity use by 2–3 percent [32]. Hybrid solar lighting (HSL) uses tracking mirrors to focus sunlight onto optical fibers that transmit light into a building's interior to supplement conventional lighting. In single-story applications these systems are able to transmit 50 percent of the direct sunlight received.

10.3 Solar technology cost and market trends

Solar energy technologies serve both the retail and wholesale electricity markets, with distributed PV, SWH, and other technologies competing against retail electric rates offered by utilities, and central PV and CSP competing against wholesale market rates. The cost competitiveness of each solar technology is, therefore, based on the relative cost of each technology in its associated market. Though the potential size of each solar market is driven by the cost competitiveness of each technology, the associated cost of each technology is linked to the size (cumulative production capacity) of each market. Many solar technologies, particularly PV, have exhibited substantial cost reductions over the past several decades, and all technologies are thought to have potentially significant opportunities for further cost reductions.

10.3.1 SOLAR PV

In 2006–2007, the cost of a residential grid-connected silicon PV system, including the inverter and installation, was about US\$8.00/W in the United States. A similar commercial PV system cost about US\$7.20/W,[10] where larger installations benefit from lower inverter and installation costs on a per-Watt basis [30, 34]. The total cost of a PV installation, illustrated in Figure 10.11, includes the PV module, DC/AC inverter, and installation. The "other" category shown includes regulatory fees (permits, interconnection, etc.) and business overhead (marketing, gross margin, etc.).

---

[10]Here and elsewhere, the prices cited for solar technologies are for the years noted. All cost comparisons need to consider the variations in underlying commodity prices, which are a significant driver for energy- and materials-intensive solar technologies.

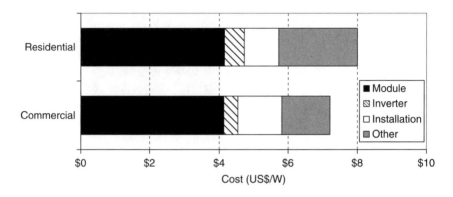

Figure 10.11 Price of PV components for residential and commercial systems, 2007.
Source: Wiser et al. (2008) [34] and U.S. DOE (2008) [30].

As indicated by Figure 10.11, module costs dominate the total cost of PV systems. Crystalline silicon modules cost about US$4.15/W [30]; thin-film modules cost approximately $3.50/W in nominal 2008 dollars [19]. The cost of crystalline silicon modules has remained fixed or increased since 2003 due to shortages in the polycrystalline supply [34].

The global PV market has grown rapidly over the past decade at a 30 percent annual mean rate. In 2007, PV production grew 50 percent, reaching 3.5 GW of manufactured capacity [19]. The top PV producers in 2006 were Sharp (Japan), Q-cells (Germany), Kyocera (Japan), Suntech (China), and Sanyo (Japan). First Solar was the top-producing U.S. company, ranking 13th globally [22]. Chinese production (370 MW) surpassed U.S. production (200 MW) in 2006. Taiwan doubled its production capacity to 180 MW in 2006. Many PV companies have announced plans to scale up manufacturing to 1 GW/year production targets. In anticipation of this new manufacturing capacity, more than 70 new silicon manufacturing facilities are under construction or in planning stages, and the poly-silicon market grew 30 percent in 2007 to capitalize on the supply bottleneck of silicon feed stocks [19]. Thin-film production grew 121 percent in 2007, increasing global production capacity to 400 MW. Thin films accounted for 12 percent of the PV market in 2007, and their market share is expected to grow as their technology and durability are further demonstrated.

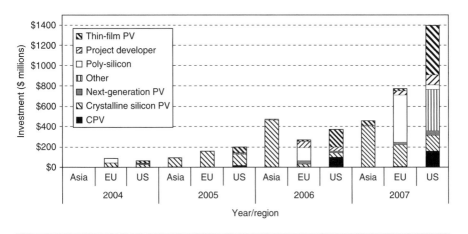

Figure 10.12 Global private investment in solar technologies by region, 2004–2007.
Source: NREL.

Investments in PV technologies have increased significantly in recent years (Figure 10.12). This new investment has the potential to drive down PV costs with improvements to technology and increased manufacturing capacity.

The top end-use markets for PV are shown in Figure 10.13. Germany and Japan have large, stable markets. The United States, Spain, Italy, France, and Greece are likely to become high-growth markets since they have strong solar resources and they have recently implemented significant capacity-based (U.S.) or production-based (Spain, Italy, France, Greece, and some U.S. states) incentives, discussed in Section 10.4. In the United States, the California utility Pacific Gas & Electric Company (PG&E) has signed a power purchase agreement backing 800 MW of utility-scale PV.

The price of PV systems has decreased significantly with cumulative production, and this trend is expected to continue. Figure 10.14 shows the decrease in PV module price with manufacturing experience, where the price of PV decreases about 20 percent for each doubling of installed PV capacity. The price reduction trend does not hold in 2005 since PV demand has outpaced supply. This has led to a supply-constrained PV market, where the retail price for crystalline silicon modules increased from 2003–2008 [19]. However, recent increases in PV manufacturing capacity will likely shift the market to demand-constrained when new production capacity comes online over the next few years. Thin-film PV modules are already less expensive than crystalline

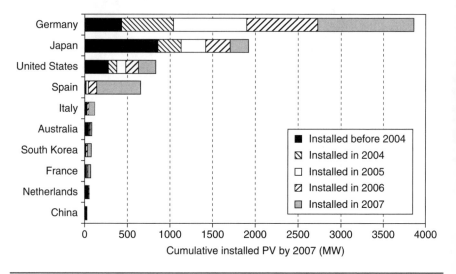

Figure 10.13 Cumulative PV installations for the top 10 countries through the end of 2007.
Source: Adapted from the IEA (2008) [16].

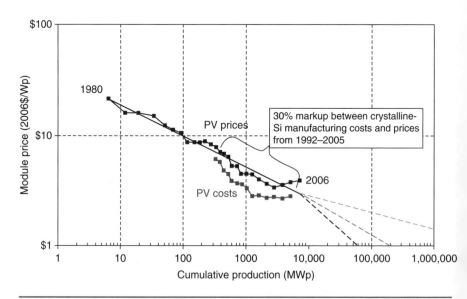

Figure 10.14 PV module learning curve, 1992–2005.
Source: NREL.

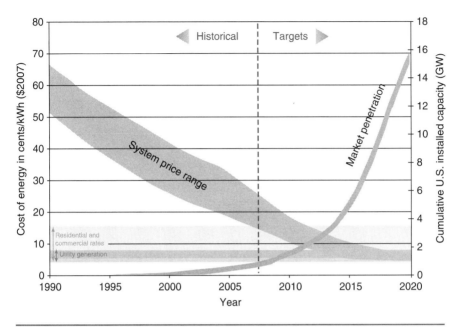

Figure 10.15 Projected levelized cost of energy (LCOE) and market penetration of PVs, 1990–2020.
Source: Adapted from U.S. DOE (2008) [30].

silicon modules, and there is even greater potential for "experience"-based cost reductions.

Continued decreases in module and overall system costs will increase market competitiveness for solar PVs, further increasing growth. This concept is illustrated in Figure 10.15, where the market penetration for PV is projected to grow significantly when solar-generated electricity becomes less expensive than utility rates.

The point at which solar electricity becomes cost-competitive with grid electricity is often referred as *grid parity* and may be possible for much of the world in the next decade if aggressive cost reductions are achieved and if the combination of decreased fossil fuel availability and carbon constraints increases the price of conventional electricity generation.[11] The increasing cost competitiveness of solar PV has resulted in anticipation of continued dramatic growth.

---

[11]In the United States, PV is close to or at grid parity in a few locations, primarily California, due to the combination of quality of solar resource, incentives, and high electricity prices. Very low electricity retail prices in some states such as West Virginia and Washington, combined with lower-than-average solar resources, will make it difficult for PV to reach grid parity in those states (U.S. DOE 2008).

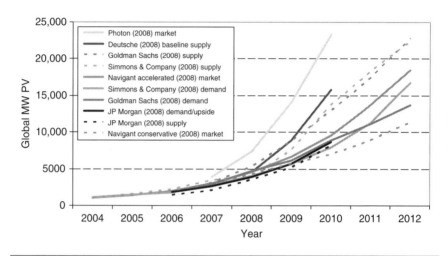

Figure 10.16 Global PV market projections, 2004–2012.
Source: NREL/company analyst reports/FACC.

Figure 10.16 plots the range of estimates from six PV market studies. Although there is significant spread in the market projections, even the most conservative estimate shows the PV market doubling from 2008 to 2012, with an annualized growth rate of 22 percent. The higher estimates show annualized growth rates up to 45 percent.

10.3.2 CSP

Reported costs for installed CSP plants in 2007 were roughly US$4.20/W for a 100 MW system built in California without thermal storage and US$4.90/W for the same system with six hours of thermal storage. Annual O&M fees for the system are estimated at US$67/kW/yr [27, 30]. Increases in manufacturing capacity and learning from experience are expected to reduce capital costs and O&M fees. One projection of these cost reductions estimates US$2.60/W for a 200 MW system without thermal storage and US$3.20/W for the same system with six hours of thermal storage by 2015 [27]. The cost reductions are driven by less expensive solar collectors, which benefit from manufacturing volume and economies of scale.

The costs for dish/Stirling and tower systems are much less certain due to the lack of data on commercial-scale systems. The current costs of dish/Stirling systems are high (US$8.6/W, [29]) but represent hand-built prototypes and are likely to drop significantly as manufacturing capacity is scaled up. One projection suggests US$3.30/W by 2011, with an associated LCOE of US $0.15/kWh [12]. Costs for towers are projected to be US$4.40/W by 2009, corresponding to an LCOE of US$0.20/kWh [12]. The accuracy of these projections will remain uncertain until utility-scale projects begin to come online.

There are three areas that have the potential to significantly drive down the cost of electricity generated by CSP systems:

- ■ *Technology development.* This includes increasing the efficiency and durability of specific technology components (receivers, concentrators, reflectors, and the balance of solar field), optimizing manufacturing techniques, and reducing O&M costs by simplifying systems and learning from experience.

- ■ *Volume production.* Increased CSP production brings significant cost reductions by decreasing the manufacturing cost of components and systems, driving down the cost of materials through bulk purchasing or vertically integrating the supply chain and standardizing construction.

- ■ *Scale-up in plant or project size.* Scaling up plant sizes allows CSP to benefit from economies of scale in equipment, systems operation, and management.

Possible cost reductions for parabolic trough systems based on these three areas are plotted in Figure 10.17.

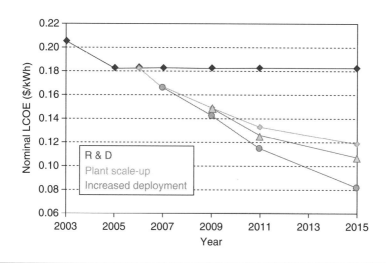

Figure 10.17 Possible cost reductions for parabolic trough systems resulting from continued R&D, plant scale-up, and increased deployment, 2003–2015. Source: Adapted from Mehos and Kearney [20].

Development and deployment of CSP technology in the 1980s were largely halted due to low fossil fuel prices. More recent increases in fuel prices have led to significant new investment and construction of multiple utility-scale plants. Commercial plants are currently operating in the United States (419 MW), Spain (11 MW), and Morocco (6 MW) [12]. CSP technology is now seen as having enormous growth potential, with a greater number of plants of increasing sizes (several hundred MW) planned or under construction. About 11 GW of CSP plants are planned or under construction in the United States (4800 MW), Spain (1850 MW), China (900 MW), Israel (250 MW), Egypt (150 MW), South Africa (150 MW), and Abu Dhabi (100 MW) and smaller plants in Australia, Algeria, Morocco, Iran, and Mexico. A number of proposed plants include thermal storage, such as the proposed 280 MW Solano plant in Arizona.

Beyond these proposed plants, significant growth in CSP markets is expected; industry professionals estimate that current manufacturing capacity could allow for up to 13,000 MW in the United States by 2015 [3]. The close proximity of regions with a high direct normal solar resource, such as countries in north Africa or southern Europe, to high load-use regions such as countries in northern Europe has led to proposals to build long-distance transmission lines.

CSP has the lowest cost of solar-generated electricity, but it must compete with electricity generation rates in utility markets rather than end-use electricity rates in residential or commercial markets.[12] The cost of CSP electricity has decreased with increasing installed capacity, and a current estimate is US$0.12/kWh in nominal 2007 dollars [29]. In parts of California, peaking electricity costs US$0.10–0.18/kWh and CSP is already cost effective without additional government subsidies [4]. If the forecast cost reductions (projected in Figure 10.17) are met, the CSP market will grow rapidly as it becomes a more cost-effective source of dispatchable electricity, especially in high-cost markets such as California.

10.3.3 Solar Water-Heating Costs and Market Trends

The cost of solar water-heating systems is highly variable based on the complexity, size, and location of the system. Typical solar water-heating systems in the United States had an electric-equivalent LCOE of 8–10 US¢/kWh in mild Sunbelt climates and 11–12 US¢/kWh in cool climates in 2002 [29].

---

[12]Because CSP and PV serve different markets, their potential market competitiveness cannot be easily compared.

A 2006 estimate of global installed solar water-heating systems was about 105 GW_{th}, producing an annual yield of 76,959 GWh_{th} in 2006[13] [33]. If only 20 percent of this water would have been heated using electricity, the electricity use avoided by solar water-heating systems is equivalent to all the electricity generated by PV and CSP installed in 2007. The majority of installed capacity was located in China (68 GW_{th}), Turkey (6.6 GW_{th}), Germany (5.6 GW_{th}), Japan (4.7 GW_{th}), and Israel (3.4 GW_{th}) [22], shown in Figure 10.18. Installed capacity grew 35 percent in China in 2006, and there has been significant growth in several EU countries, Turkey, Japan, Israel, Brazil, and the United States.

Solar water-heating capacity grew in all but four EU countries in 2007, bringing the total installed capacity to 15.4 GW_{th} [25]. In the United States, low electricity and natural gas prices, along with a limited customer awareness of SWH systems and a small installer base, have limited SWH growth. The SWH market could grow significantly with lower SWH costs and an increased availability of trained and licensed SWH installers and by increasing consumer awareness of the performance, cost, and benefits of SWH systems.

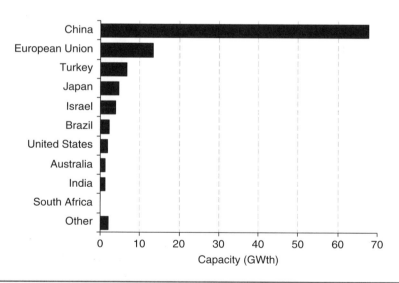

Figure 10.18 Cumulative solar water-heating installations for the top 10 countries as of 2006 (not including swimming pool heating). Source: Adapted from REN21 [22].

---

[13]Excludes solar water heaters used to heat swimming pools.

10.3.4 MARKET INCENTIVES

Since solar electricity has yet to reach grid parity in most regions of the world, incentives have been and will continue to be a major market driver. Incentives are used to decrease the up-front cost of solar technologies (tax incentives and rebates) or to increase revenues from PV generation (feed-in tariffs). Germany incentivizes PV by providing both low-interest-rate loans and a feed-in tariff [13]. Japan instituted a solar rebate program from 1980–2005. In the United States, the Emergency Economic Stabilization Act of 2008 extended a 30 percent tax credit for solar-generated electricity and solar water heat through the end of 2016. The United States also has a patchwork of state- and utility-based incentives, most notably in California and New Jersey.

Besides Germany, a number of European countries have introduced feed-in tariffs[14] to stimulate PV adoption, including Spain (41 €cents/kWh for 25 years, 33 €cents/kWh thereafter), Italy (44–49 €cents/kWh for 20 years), Greece (40–50 €cents/kWh), and France (30 €cents/kWh + 25 €cents/kWh if building integrated). Similar feed-in tariffs have been introduced for CSP in Spain (27 €cents/kWh for plants up to 50 MW for 25 years), Portugal (16–27 €cents/kWh for plants up to 10 MW), France (30–40 €cents/kWh), Greece (25–27 €cents/kWh), and Israel (US$0.16–0.20/kWh).

10.3.5 OVERCOMING MARKET BARRIERS TO WIDE-SCALE SOLAR DEPLOYMENT

It is critical to address the market barriers limiting the adoption of solar technologies at the same time that the technical barriers are addressed. For photovoltaic systems, the principal market barriers are:

- Inconsistent interconnection, net metering, and time-varying utility rate structures to capture the value of solar-generated electricity
- Lack of policies and incentives that capture the environmental, distributed generation, and risk mitigation benefits of PV
- Complex permitting procedures and fees
- The lack of flexible, sophisticated, and proven financing mechanisms
- The relatively high tax burdens on solar energy systems with "free fuel" but high capital costs[15]

---

[14]From www.solarpaces.org.
[15]Solar energy systems have "free fuel," but the cost of equipment to convert the "fuel" to electricity is high. Taxing solar energy equipment as real property presents a significant cost to generating electricity.

For CSP systems, the principal market barriers are:

- A lack of long-term, predictable economic policies supporting utility-scale deployment
- A lack of sufficient transmission capacity linking CSP systems to load-use regions such as cities
- The difficulty in leasing and permitting land favorable to utility-scale projects

Meeting these market challenges will decrease the time required for solar technologies to reach grid parity and accelerate PV's potential to mitigate carbon emissions.

10.4 High penetration limits

Tens of terawatts of carbon-free electricity may be required to significantly reduce CO_2 emissions from the electric sector [14, 24]. Although the solar resource is sufficiently large to meet this capacity many times over, the technical barriers of land, water, and resource use need to be analyzed along with the inherent challenge of integrating an intermittent energy source into the electricity grid.

The global land area suitable for CSP and PV is enormous and will not limit solar deployment in high-penetration scenarios. For example, one estimate of the total roof area suitable for PV in the United States is roughly 6 billion m^2, even after eliminating 35–80 percent of roof space to account for panel shading and suboptimal roof orientations [8]. With current PV performance, this area has the potential for over 600 GW of capacity, which could generate over 800 TWh per year, providing over 20 percent of 2006 electricity demand. If roof area grows at the same rate as electricity consumption, this PV potential (in terms of fractional energy supply) will remain constant. However, PV efficiency is expected to increase over time, resulting in an increase in the potential PV capacity.

Beyond rooftops, there are many opportunities for installing PV on low- (or zero-) opportunity-cost land such as parking structures, awnings, airports (which have the added advantage of being secure environments), and along freeways and farmland set-asides. The land area required to supply all end-use electricity in the United States using photovoltaics is about 0.6 percent of the U.S. land area (181 m^2 per person), or about 22 percent of the "urban area" footprint [9]. A similar analysis for Europe estimates that between 19 and 48 percent of total electricity consumption could be met with PVs placed on building rooftops and facades [17].

The global land area suitable for CSP is similarly large. A study of the CSP potential in the southwestern United States found 11,165 GW of potential CSP sites, which could generate six times the electricity used by the United States in 2007 [20]. This study included only the land area with at least 6.75 $kWh/m^2/$ day of direct normal insolation, a terrain slope of less than 1 percent, and at least 1×1 km^2 of continuous area, while excluding urban area and environmentally sensitive land. After narrowing the selection criteria to land with adjacent transmission lines, the study found a potential for 7000 GW of installed CSP [20]. This land area is shown in Figure 10.19, along with the locations of major transmission lines.

Another recent study suggested building 20 5 GW capacity high-voltage DC lines connecting Europe to the high solar resource regions in north Africa, which could enable Europe to import about 15 percent of its electricity demand from low-cost CSPs at roughly 5 €-cents/kWh [28]. Even though extrapolating these studies to estimate the global CSP potential would require further study, it is clear that the land area suitable for CSP is enormous, and

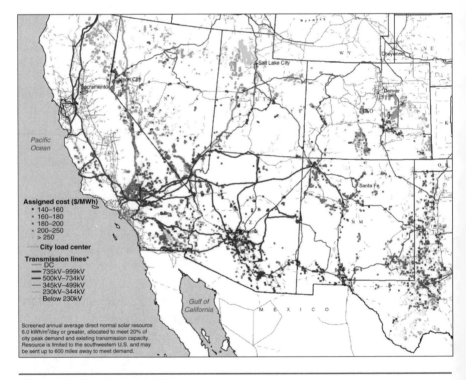

Figure 10.19 Optimal CSP sites.
Source: Mehos and Kearney [20] and POWERmap (2007; powermap.platts.com).

wide-scale CSP deployment would be limited more by access to transmission lines than the availability of suitable land area.

If PV is deployed at the terawatt scale, some technologies may be limited by the availability of feedstock materials. The silicon feedstock is virtually unlimited, but crystalline silicon panels may be limited to a capacity of 8–12 TW by the availability of silver, which is used as an electrode [10].[16] Amorphous silicon thin films may be limited by the use of indium to roughly 0.5 TW but have a potential greater than 40 TW if they use different electrode materials such as ZnO. There are varied estimates of the potential for CdTe, ranging from 0.1 TW [10] to 30 TW [35],[17] based on the limited supply of tellurium. Similarly, the potential of CIGS ranges from 0.1 TW [10] to 30 TW [35] where capacity is limited by the supply of indium. Dye-sensitized PV and multijunction CPV units could account for tens of TW of PV capacity with adjustments to electrode and substrate materials.

Solar CSP facilities are primarily constructed of steel, glass, and concrete. Though these materials are not subject to rigid supply limits, they will be affected by changes in commodity prices. However, the availability of water resources may limit the scale of CSP deployment in arid regions. CSP plants typically use 0.63 gallons (2.4 L) per 1 kWh [26]. This represents a large and possibly prohibitive freshwater requirement for traditional cooling towers at arid CSP sites. Water loss can be reduced by 90 percent using dry-cooling towers where fans and radiators facilitate heat loss rather than evaporative cooling. Building and operating dry towers will lead to additional capital costs (e.g., building radiators and fans), and the dry cooling process may decrease the efficiency of the CSP process if the heat engine is operated at a higher ambient temperature. Closed-cycle heat engines such as dish/Stirling engines do not require water cooling and will not be affected by limited water resources.

For photovoltaic systems, the challenge of integrating intermittent solar electricity into the existing electrical grid presents a barrier to large-scale penetration of the electricity market. There are two fundamental limitations that must be addressed: first, the inherent mismatch of PV supply and electricity

---

[16]The study uses known global material reserves from the USGS, standard PV efficiencies for each technology, solar insolation of 4.0 $kWh/m^2/day$, and no losses for recycling materials from old panels and assumes that 25 percent of each material will be deployed in PV panels at any given time. The study quotes PV energy in TWy, which we use to calculate PV capacity by assuming a 20 percent capacity factor: TW capacity = 5*TWy.

[17]The study also uses similar parameters but assumes that 100 percent of the material is used in thin-film panels. However, this factor of four does not account for the discrepancy in estimates, and projecting maximum PV capacity remains an active area of research.

demand, and second, the inability of conventional thermal generators to adjust output in response to the hourly variations in PV output.

The flexibility of electric power systems to accept increased fluctuations in demand resulting from PV output is limited. Historical U.S. market data indicates that when the net demand for electricity drops to below 30–40 percent of the annual peak, wholesale electricity prices often drop below the variable costs of producing electricity [5], implying that generators are willing to sell electricity at a loss to keep plants running. This reflects the variety of technical and operational constraints on conventional thermal generators, including the long ramp times of nuclear plants, the stability limits on coal-fired steam plants, the required number of plants operating as "spinning reserve," and so on. The flexibility of electric power systems varies by region and country, where systems dominated by nuclear power (such as in France) will likely have less flexibility, whereas systems that rely largely on hydroelectric generation (such as in Norway) will probably have greater flexibility. This minimum load constraint ultimately determines the amount of PV-generated electricity that could be used at high penetration levels.

Figure 10.20 illustrates the limits of PV contribution into the grid and potential options to increase the usefulness of PV-generated electricity. In this simulation of the Texas electricity grid, the Electric Reliability Council of Texas (ERCOT), the hourly supply of a large distributed PV generator is superimposed on the normal load, along with the resulting net load that would have to be met by conventional generators. On this day, PV produces a large amount of energy, resulting in a drop in the residual load. Also shown on the graph is the normal system minimum load below which conventional generators may be unable to cycle. In this case, minimum load constraints of base-load generators could restrict the amount of PV accepted by the system,

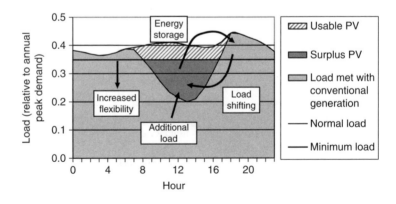

Figure 10.20   Options for using surplus solar PV generation.
Source: Adapted from Denholm and Margolis [6].

resulting in curtailed energy. This minimum-load constraint is well understood and is reflected in operational constraints that result in curtailed wind generation in locations such as Denmark that currently use wind generation to meet a large fraction of demand. This limited coincidence between energy supply and demand represents the ultimate economic and technical limits in incorporating solar PV in the grid.

There are a number of options to increase the penetration of nondispatchable solar energy sources such as PV, as illustrated in Figure 10.20 [6]. These include increasing the flexibility of the grid, reducing dependence on inflexible "base-load" generators, and increasing the ability to transmit energy over long distances. Other possibilities include deploying a variety of enabling technologies such as "smart" appliances that can shift demand to periods of low energy prices, dispatchable load from plug-in hybrid or other electric vehicles, or stationary energy storage.

It should be noted that CSP with thermal storage largely avoids these problems via the ability to dispatch the solar resource as needed. CSP could also act as an enabling technology for other solar energy technologies and other variable resources such as wind, acting as a load-following resource and greatly increasing the flexibility of electric power systems. Ultimately, a combination of enabling technologies will likely be needed to support extremely large-scale deployment of solar PV, along with other nondispatchable energy sources. This will require substantial restructuring of the electricity system—from a centrally controlled to a highly distributed and interactive system with increased interaction between utilities and end users and greater use of solar resource forecasting.

10.5 Potential for mitigating carbon emissions

Many studies have forecast the future capacity of solar-generating technologies for both the U.S. and global markets and the role of solar energy technologies in reducing global carbon emissions from the electric sector. These forecasts often assume different prices for solar generation technologies and different electricity rates and rate structures.

As a baseline, 700 GW of continuous solar generation would be required by 2050 to reduce carbon emissions by 1 GT C/year or one "wedge," as defined by Socolow and Pacala (2006). Assuming a mix of CSP and PV with an overall capacity factor of 33 percent,[18] this corresponds to 2100 GW of solar-generation capacity by 2050. In a U.S. study to 2050, forecast growth in PV and CSP

---

[18]CSP capacity factors range from 30 percent to greater than 40 percent with six hours of storage. Future low-cost storage could result in capacity factors approaching 70–80 percent.

led to cumulative installed capacities ranging from 210–750 GW of PV and 11–100 GW of CSP based on different assumptions of capital costs, carbon prices, and incentive levels [18].

Though it seems likely that solar technologies could lead to significant reductions in carbon emissions, the challenges of integrating the intermittent PV generation into the electricity grid will begin to present barriers at about the one "wedge" level, requiring substantial deployment of enabling technologies, as illustrated in Figure 10.20. If a significant fraction of solar-generated electricity is supplied by dispatchable CSP, solar electricity could mitigate carbon emissions at well beyond the "one-wedge" level.

In considering the carbon reduction potential of PV, CSP, and other solar energy technologies, the net energy yield and the "embodied" carbon emissions in these technologies must be considered. In the carbon-intensive U.S. electricity grid, the energy used to produce PV and CSP generates carbon emissions ranging from 20–55 g CO_2-eq./kWh for PV [11] and 13 g CO_2-eq./kWh for CSP [21], whereas coal and natural gas emit 970 and 450 g CO_2-eq./kWh, respectively [31]. This represents an 88–98 percent reduction in carbon emissions for each kWh of solar electricity that replaces electricity generated by natural gas or coal. As a result, the carbon reduction factor for solar technologies is large and will likely increase over time as the efficiency of manufacturing processes and solar technologies increases and as the carbon intensity of the electricity grid decreases.

10.6 Conclusion

Solar-generated electricity is the fastest-growing renewable energy technology, but it is starting from a small base of approximately 9.0 GW of installed capacity at the beginning of 2008, accounting for less than 0.1 percent of global electricity use. However, grid-connected solar could be cheaper than the electricity generated from conventional resources within the next decade in many states in the United States and countries in Europe, and achieving TW-scale deployment is possible with existing technology.

The large-scale penetration of solar electricity will not be limited by the solar resource or land availability. The short-term barriers to solar electricity deployment are the high initial cost of solar energy systems and a lack of clear, long-term incentives in many regions. In the long term, the barriers are the availability feedstock materials for some types of PV technologies, the challenge of integrating intermittent PV electricity into the grid, and the need for new transmission for utility-scale CSP and PV systems. Enabling technologies such as improving the flexibility of conventional generators, load shifting, and low-cost energy storage may allow solar-generation technologies to supply a significant fraction of the electricity market and substantially reduce carbon emissions from the electric sector.

REFERENCES

[1] Burch J, Hillman T, Salasovich J. Cold-Climate Domestic Hot Water Systems: Cost/Benefit Analysis and Opportunities for Improvement. NREL Report No. CP-550-37105; 2005.

[2] Christensen C, Barker G. Annual System Efficiencies for Solar Water Heating. In: Campbell-Howe T, Cortez T, editors. Proceedings of the 1998 American Solar Energy Society Annual Conference, June 14–7, 1998, Albuquerque, New Mexico. Boulder, CO: American Solar Energy Society p. 291–6. NREL Report No. 25569; 1998.

[3] CEAC. Western Governors' Association's Clean and Diversified Energy Advisory Committee (CDEAC). Clean and Diversified Energy Initiative 2006.

[4] CSP Global Market Initiative (CSP GMI). International Energy Agency (IEA), SolarPACES; 2004.

[5] Denholm P, Margolis RM. Evaluating the Limits of Solar Photovoltaics (PV) in Traditional Electric Power Systems. Energy Policy 2007a;35:2852–61.

[6] Denholm P, Margolis RM. Evaluating the limits of solar photovoltaics (PV) in electric power systems utilizing energy storage and other enabling technologies. Energy Policy 2007b;35: 4424–33.

[7] Denholm P. Technical Potential of Solar Water Heating to Reduce Fossil Fuel Use and Greenhouse Gas Emissions in the United States. NREL Report No. TP-640-41157; 2007c.

[8] Denholm P, Margolis R. Supply Curves for Rooftop Solar PV Generated Electricity for the U.S. NREL Report No. TP-670-44073; 2008a.

[9] Denholm P, Margolis R. Land-use requirements and the per-capita solar footprint for photovoltaic generation in the United States. Energy Policy 2008b;36:3531–43.

[10] Feltrin A, Freundlich A. Material considerations for terawatt level deployment of photovoltaics. Renewable Energy 2008;33:180–5.

[11] Fthenakis V, Kim HC, Alsema E. Emissions from Photovoltaic Life Cycles. Environ Sci Technol 2008;42(6):2168–74.

[12] Grama S, Wayman E, Bradford T. Concentrating Solar Power: Technology, Cost and Markets. Prometheus Institute for Sustainable Development and Greentech Media; 2008.

[13] Guinness O. The Future of Germany's Solar PV and Wind Energy Markets. Greentech Media, Inc.; 2008.

[14] Hoffert MI, Caldeira K, Jain AK, Haites EF, Harvey LD, Potter SD, et al. Energy implications of future stabilization of atmospheric CO_2 content. Nature 1998;395:881–4.

[15] International Energy Agency (IEA). Trends in Photovoltaic Applications: Survey report of selected IEA countries between 1992–2006. Photovoltaic Power Systems Programme, IEA-PVPS T1-16: 2007.

[16] International Energy Agency (IEA). Trends in Photovoltaic Applications: Survey report of selected IEA countries between 1992–2007. Photovoltaic Power Systems Programme, IEA-PVPS T1-17: 2008.

[17] International Energy Agency (IEA). Potential for Building Integrated Photovoltaics. Photovoltaic Power Systems Programme, IEA-PVPS T7-4: 2001.

[18] Margolis RM, Wood F. 2004. The Role for Solar in the Long-Term Outlook of Electric Power Generation in the U.S., Paper presented to at the 24th USAEE/IAEE North America Conference, Washington, DC, July 8-10, 2004.

[19] Marketbuzz 2008: Annual World Solar Photovoltaic Industry Report, 2008. Solarbuzz, http://www.solarbuzz.com/

[20] Mehos MS, Kearney DW. Potential Carbon Emissions Reductions from Concentrating Solar Power by 2030. In: Kutscher CF, editor. Tackling Climate Change in the U.S.: Potential Carbon Emissions Reductions from Energy Efficiency and Renewable Energy by 2030, Boulder, CO: American Solar Energy Society; p. 79–89; NREL Report No. CH-550-41270; 2007.

[21] Pehnt M. Dynamic life cycle assessment (LCA) of renewable energy technologies. Renewable Energy 2006;31:55–71.

[22] REN21. Renewables 2007: Global Status Report. Paris, France: REN21 Secretariat; 2008.

[23] Shaheen SE, Ginley DS, Jabbour GE. Organic-Based Photovoltaics: Toward Low-Cost Power Generation. MRS Bulletin 2005;30(1):10–5.

[24] Socolow R, Pacala S. A Plan to Keep Carbon in Check. Sci Am 2006;295(3):50–7.

[25] Solar Thermal Markets in Europe. Trends and Market Statistics 2007. European Solar Thermal Industry Federation (ESTIF); 2008.

[26] SolarPACES. International Energy Agency (IEA). www.solarpaces.org; 2008.

[27] Stoddard L, Abiecunas J, O'Connell R. Economic, Energy and Environmental Benefits of Concentrating Solar Power in California. National Renewable Energy Lab Subcontract Report; 2006 NREL/SR-550-39291; 2006.

[28] TRANS-CSP. Trans-Mediterranean Interconnection for Concentrated Solar Power. German Aerospace Center (DLR), Institute of Technical Thermodynamics; 2006.

[29] U.S. Department of Energy (DOE). Solar Energy Technologies Program: Multi Year Program Plan 2007–2011. 2007.

[30] U.S. Department of Energy (DOE). Solar Energy Technologies Program: Multi-Year Program Plan 2008–2012. 2008.

[31] U.S. Energy Information Administration (EIA), 2008b. Annual Energy Outlook 2008 with projections to 2030. DOE/EIA-0383 (2008).

[32] U.S. Energy Information Administration (EIA), 2008a. Residential Buildings Energy Consumption Survey, 2005, http://www.eia.doe.gov/emeu/recs/

[33] Weiss W, Bergmann I, Ganinger G. Solar Heat Worldwide: Markets and Contribution to the Energy Supply 2006. International Energy Agency Solar Heating and Cooling Programme, International Energy Agency (IEA); 2008.

[34] Wiser R, Barbose G, Peterman C. 2009. Tracking the Sun: The Installed Cost of Photovoltaics in the U.S. from 1998–2007, Lawrence Berkeley National Lab, LBNL-1516E.

[35] Zweibel K. The Terawatt Challenge for Thin-Film PV. National Renewable Energy Lab Technical Report. NREL/TP-520-38350; 2005.

Geothermal Power: The Baseload Renewable

Kenneth H. Williamson

Geothermal Consultant, Santa Rosa, California

Abstract

In a carbon-unconstrained world, geothermal power generation has grown at the rate of 3 percent annually and now provides 0.4 percent of the world's electricity, with an installed capacity of 10 GW. Two areas of technology breakthrough could accelerate geothermal growth, and a recent surge in investment has stimulated the industry, creating promising signs that a breakthrough is imminent. It is likely that a renewable energy source that can provide baseload, carbon-free power will command premium pricing in a carbon-constrained world, and geothermal could emerge as a significant player in the electricity market.

11.1 Introduction

We know that 10^{31} Joules of heat energy are stored within the Earth. Humans currently consume energy at the rate of 1.5×10^{13} Joules per second, so if it were possible for humans to utilize all the heat stored within the Earth, it could provide a global energy supply for 20 billion years! The Earth's heat was created from ongoing radioactive decay and from the energy of accretion generated when the Earth was formed. Heat flows to the Earth's surface at the rate of 1.4×10^{21} Joules per year.[1] For comparison, roughly 2000 times this rate reaches the Earth's surface in the form of solar energy. Geothermal power systems mine the heat stored below the Earth's surface.

The flow of heat from the interior creates a temperature gradient within the Earth so that temperatures in the Earth's crust increase with depth at a rate usually ranging from 10–80°C/km. In some localities, such as volcanically active areas, gradients may exceed 200°C/km in the upper crust. Geothermal power projects drill wells where temperature gradients are high and extract heat by producing hot water or steam from a permeable reservoir through the wells and pipelines into turbines on the surface, where thermal energy is converted to electricity. The cooled water and condensed steam are then injected back into the reservoir for reheating.

Geothermal contributed 0.4 percent to the worldwide electricity supply in 2003 and represented 0.2 percent of installed generating capacity, at about 10 GW. Geothermal electricity generation in the United States[2] in 2006 was 15 TWh, with 88 percent from California and 9 percent from Nevada. By comparison, wind power generated 32 TWh in the same year, having tripled since 2002, whereas geothermal generation had negligible growth in the same period in the United States. The net geothermal power plant capacity in the United States was 3 GW in 2008 (Table 11.1).

To date the geothermal industry has relied on naturally occurring porosity and permeability within hot rock to provide sufficient heat transfer to circulating fluids and has focused on parts of the Earth where geothermal gradients are anomalously high, to keep drilling costs down and geothermal power

---

[1]Pollack (1993).
[2]EIA (2008).

Table 11.1   U.S. geothermal installed capacity by state, 2008[3]

State	Installed Capacity (MW)
California	2555*
Nevada	318
Utah	36
Hawaii	35
Idaho	13
Alaska	0.4
U.S. total	2957

California capacity includes 460 MW on standby at The Geysers due to reservoir pressure decline.

prices competitive. This was the only way geothermal power could compete with fossil-fueled plants in a carbon-*un*constrained world.

The process of developing geothermal energy from naturally occurring reservoirs worldwide is described in the next section, followed by a discussion of the drivers that have shaped the geothermal industry in the United States. The enormous potential offered by enhanced geothermal systems (EGSs), the technical challenges that need to be overcome for EGS to compete economically, and a brief review of progress to date are then discussed in the following section.

11.2 Geothermal power from naturally occurring reservoirs

The worldwide electrical power output from geothermal sources increased from 2.6 TWh/yr in 1960 to 57 TWh/yr in 2005 as the installed geothermal plant capacity increased from 386 MW in 1960 to 8912 MW in 2005. Another 800 MW were added in the three years from 2005 to 2007,[4] half of which were in Iceland and Indonesia. The countries generating more than 100 MW of power from geothermal sources are listed in Table 11.2. Large-scale geothermal projects are located in regions with recent volcanic activity associated with tectonic plate boundaries.

Several of the world's largest geothermal reservoirs wells produce dry steam, which is fed into low-pressure steam turbines to generate power.

---

[3]www.geo-energy.org
[4]Bertani (2007).

Table 11.2 Countries generating more than 100 MW of geothermal power in 2007[5]

Country	Geothermal Installed Capacity (MW) in 2007	Geothermal Generation in 2004 (GWh/yr)
United States	2687	14,811
Philippines	1970	9419
Mexico	953	6282
Indonesia	992	6085
Italy	810	5340
Japan	535	3467
New Zealand	472	2774
Iceland	421	1406
Costa Rica	163	1145
El Salvador	204	967
Kenya	129	1088

However, in the most common configuration, water from high-temperature ($>230°C$) reservoirs is partially flashed to steam before the steam is passed through low-pressure steam turbines (Figure 11.1). Reservoir temperature has a strong influence on the conversion efficiency of heat to electricity, which increases from less than 5 percent at $100°C$ to more than 25 percent at $300°C$. In lower-temperature reservoirs or in some cases to utilize the heat from separated brine, power is generated using a heat exchanger and secondary working fluid to drive a turbine. This accounted for about 8 percent of worldwide geothermal generation in 2005.[6]

More than 97 percent of current geothermal production is from reservoirs heated by volcanic magma bodies. Driven by heat loss from underlying magma, hot fluids rise along pre-existing zones of high permeability. Buoyant up-flowing fluids enhance the permeability of the rocks through which they flow by chemical leaching and by explosive boiling. If the system becomes large enough and has high enough permeability, it has the potential to be a commercial-grade geothermal reservoir. Geothermal reservoirs in the depth range of 0.5 to 4 km are being developed economically using current drilling technology.

---

[5]U.S. net generation and capacity from EIA (2008), other countries from Bertani (2005, 2007). Generation numbers are from 2003 or 2004, depending on availability at the time of publication.
[6]Bertani (2005).

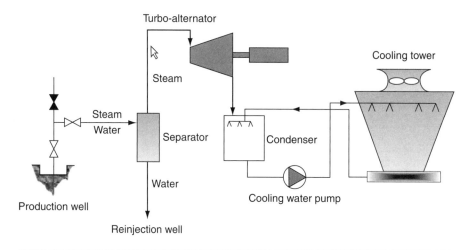

Figure 11.1   Process diagram for a geothermal flash power plant.[7]

Geothermal reservoirs may also develop outside regions of recent volcanic activity, where deeply penetrating faults allow groundwater to circulate to depths of several kilometers, become heated by the geothermal gradient, and then rise again through buoyancy. This up-flowing hot water may then accumulate in a shallow aquifer and create an economic geothermal resource. Such fault-based systems are typically smaller (usually tens of MW) than magma-based systems (often hundreds of MW) and constitute a small fraction of the world's installed capacity.

Compared to oil, gas, or coal, geothermal power projects are generally classified as clean, renewable energy sources with low or zero greenhouse gas emissions and low to moderate environmental impact in terms of air quality, land use and stability, and water quality. Geothermal fluids do, however, contain gases, mainly carbon dioxide, which can either be released to the atmosphere in the case of dry steam and flash plants or injected back into the reservoir in the case of binary plants. When the gases are discharged to the atmosphere, greenhouse gas emissions are typically an order of magnitude less than that of a coal plant. Figure 11.2 shows comparison data for CO_2 emissions for various power sources in the United States. Binary plants may be designed to have zero emissions when produced fluids are circulated through a heat exchanger and injected directly back into the reservoir.

Geothermal projects extract, on average, 8 MW for each square kilometer of reservoir area,[8] although with the use of directional drilling and large

---

[7]Fridleifsson et al. (2008).
[8]Bertani (2005).

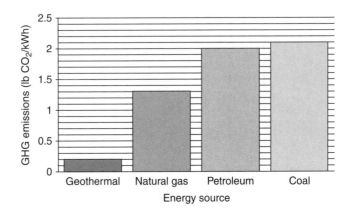

Figure 11.2 Geothermal and fossil-fuel CO_2 emissions in the United States.[9]

generation centers, the actual land usage dedicated to facilities is much lower. For example, the Salak geothermal field in Indonesia used about 10 percent (<2 km^2) of the total reservoir area for well pads, roads, pipelines, and power plants on a 330 MW project.[10]

Conventional geothermal reservoirs occur rarely in nature because they require the coincidence of high subsurface temperatures within drillable depths and highly permeable rock formations. Large volumes of fluid must be produced per well for the process to be economically feasible, and permeability must be so high that normally the rock formation must have been subjected to fracturing from tectonic activity. Most geothermal reservoirs were found by drilling near hot springs, in the same way that early oil exploration focused on oil seeps.

The ultimate growth potential for conventional geothermal systems is hard to estimate, because of the relatively primitive technology currently available for exploration. No doubt many geothermal systems have no telltale hot springs leaking to the surface, as was found to be the case for hydrocarbon accumulations and oil seeps. Recent estimates of the ultimate global potential of conventional geothermal resources have been roughly an order of magnitude above the current installed capacity of 10 GW. A group of U.S. experts[11] estimated that 70 GW could be produced worldwide using current technology and speculated that enhanced technology would raise their estimate to 140 GW. An alternative approach,[12] based on the correlation between geothermal fields and the distribution of magmatic heat in active volcanic belts, gave a

[9]Bloomfield et al. (2003).
[10]Williamson et al. (2001).
[11]Gawell et al. (1999).
[12]Stefansson (2005).

most likely value of 210 GW. This second approach took into account the likely occurrence of geothermal systems with no surface expression of hot springs.

The greatest potential for development of conventional resources is in the volcanic chains of the Pacific Rim: North, Central, and South America, the Philippines, Indonesia, Japan, Russia (Kamchatka Peninsula), and New Zealand. Iceland also has large expansion potential, as do the countries straddling the East African Rift valley. The existence of geothermal resources throughout the Pacific Rim has been known for many years, but the pace of development in the past has been constrained by government regulation and competition from fossil fuels.

Geothermal production in the Philippines began in 1977 and as of 2003 accounted for 19 percent of the country's generation.[13] It is now second to the United States in terms of geothermal production and continues to expand. The third largest, Mexico, had 953 MW of geothermal power installed in 2003, 95 percent of which was in two fields.[14] Geothermal power represented 2.2 percent of the country's installed capacity.

Indonesia was aggressively developing its abundant geothermal resources in the 1990s, when growth was interrupted by the Asian financial crisis. The government is trying to restimulate investment in the sector and a new 110 MW plant was added in 2007.

Italy was the first country to generate geothermal power, starting in 1904 and reaching 790 MW installed in 2004. Government policies supporting the development of renewable energy resources, and a "green power" price of 14.3 €cents/kWh stimulated the building of 10 new power plants within existing geothermal fields during 2000–2004.[15]

Geothermal expansion in Japan was promoted for many years by the New Energy and Industrial Technology Development Organization (NEDO) through its exploration, research, and field development programs totaling $1.5 billion over 14 years, and 535 MW geothermal of capacity has been installed.[16] The deregulation of the power market and the ending of NEDO subsidies caused geothermal development to stall. Environmental regulations restrict access to many potential geothermal sites in Japan.

New Zealand has seen a resurgence in investment in geothermal plants, with 225 MW under construction in 2008.[17] Iceland has also experienced rapid

---

[13]Benito et al. (2005).
[14]Gutiérrez-Negrín et al. (2005).
[15]Cappetti et al. (2005).
[16]Kawazoe et al. (2005).
[17]White (2007).

growth in geothermal, with 220 MW coming online in 2006–2007[18] and an additional 160 MW under construction in 2007.[19]

The rough estimates of ultimate potential described above suggest that conventional exploration and development of the world's naturally occurring geothermal systems could ultimately generate enough electric power to meet a few percent of world demand, providing the power could be transmitted economically from geothermal sites to major load centers. A much greater potential exists if manmade geothermal reservoirs can be developed economically.

11.3 Geothermal in the United States

There are three geothermal fields in the United States that generate more than 100 MW: the Geysers in northern California, Coso in eastern California, and the Salton Sea in southern California. Naturally occurring geothermal systems have only been found in the Western states, in association with volcanic activity and with the Basin and Range province, where the Earth's crust is thin and the region is heavily faulted.

Early large-scale commercial development of geothermal power occurred at the Geysers in northern California in the 1960s and 1970s by Union Oil Company and Pacific Gas and Electric. Union drilled wells and delivered steam under contract to Pacific Gas and Electric, which built and operated power plants. Other companies followed, and the Geysers capacity was expanded to 2000 MW by 1989, but overdevelopment resulted in an accelerated decline that was ultimately stabilized by enhanced water injection.[20] In 2008 the field was capable of generating around 900 MW.

In the late 1970s and 1980s the U.S. government spent more than $800 million on geothermal research over 10 years, and an industry-coupled program spent $32 million in 4 years. During the same period, the California PUC approved Standard Offer #4 contracts (ISO4) in 1983, and 29 were issued to geothermal projects totaling 815 MW. The Coso and Salton Sea fields were largely developed with ISO4 contracts. During the first 10 years of an ISO4 contract, energy price was relatively high, e.g. 14¢/kWh in a 27 MW plant in California during 1998.[21] In the 11th year of ISO4, contract holders experienced a sharp drop in revenues, so geothermal exploration activity dropped to

---

[18]Bertram (2009).
[19]Bertani (2007).
[20]Goyal et al. (2007).
[21]EIA (1997).

a low level in the 1990s and the boom fueled by government cost-sharing and ISO4 contracts came to an end.

However, electricity prices have now increased, and renewable portfolio standards (RPS) have been adopted by 24 states,[22] including California (20 percent by 2010), Nevada (20 percent by 2015), Hawaii (20 percent by 2020), Utah (20 percent by 2025), Oregon (25 percent by 2025), and Washington State (15 percent by 2020). Most RPS phase in over years, and the dates shown in parentheses refer to the year that the full requirement takes effect. Geothermal projects are eligible as a renewable energy source under RPS rules and are also eligible for the federal production tax credit (PTC). The PTC is valued at 1.9¢/kWh for the first 10 years of operation but has been allowed to expire several times before being reinstated by Congress, creating a disadvantage for geothermal developers because of the multiyear development cycle.

The cost of geothermal power is strongly dependent on capital cost and drilling, and power plants comprise more than 75 percent of the total.[23] The capital cost of conventional geothermal projects was recently estimated to be approximately $3500 per kW installed.[24] Dry steam projects are usually cheapest to develop because fewer injection wells are required and brine-handling facilities are unnecessary. Lower temperature and deeper liquid reservoirs are the most expensive.

In the United States, it is likely that the Cascade volcanic belt that extends from California to Washington hosts numerous geothermal systems, but restrictions on land use have impeded exploration and development to date. Also, the local hydrology tends to suppress hot springs in this environment, so prospects are hard to find with conventional exploration tools. Rapid reconnaissance tools are needed to survey large areas and identify prospects where no hot springs are leaking to the surface. Government-supported research in this area is ongoing but at a low level.

Naturally occurring geothermal reservoirs are situated wherever the Earth's tectonic and magmatic forces have placed them, without regard for their proximity to electricity load centers. In the United States, some Western state agencies are in the process of identifying where geothermal resources are likely to exist and how to facilitate transmission access to developers.[25] The Federal Energy Regulatory Commission (FERC) recently approved a new mechanism

---

[22]DOE (2007).
[23]Williamson et al. (2001).
[24]Sanyal (2007).
[25]RETAAC (2007) and CGEC (2005).

to facilitate financing new transmission lines needed to connect geothermal and other renewable energy projects to the power grid.[26]

The most spectacular geothermal resources often occur in areas of great natural beauty, such as Yellowstone National Park or Lassen Volcanic National Park, which are off limits to development, and most new geothermal prospects in the United States are on land controlled by the federal government through the Bureau of Land Management (BLM). In areas that are open to development, regulatory delay can be a factor in the success of a geothermal project, with the permitting and environmental review process in California taking 1 to 3 years, and even longer in areas with environmental sensitivity.[27] An extreme case is a geothermal project poised for development at Medicine Lake in northern California, which had its BLM and U.S. Forest Service permits nullified by the U.S. Ninth Circuit Court of Appeals following a lawsuit claiming that the original environmental impact statement did not recognize the historic and cultural value of the area, 25 years after exploration began.

The BLM began an initiative in 2007 to attempt to expedite the leasing of BLM and Forest Service lands with high geothermal potential.[28] The BLM issued 291 leases between 2001 and 2007, compared to 25 leases from 1996–2001, and currently administers about 420 geothermal leases.

Rising power prices coupled with renewable incentives have created a resurgence of interest in geothermal in the United States, with the Geothermal Energy Association[29] reporting on about 100 new geothermal power projects totaling more than 3000 MW in the western United States in 2008, of which 10 new projects totaling 424 MW were actually under construction. The U.S. Geological Survey[30] estimates that approximately 6500 MW of additional geothermal power could be developed in the United States from already identified sources and that 30,000 MW may be available from undiscovered sources. Geothermal power may further increase in value to power utilities as CO_2 emission legislation, such as California Assembly Bill 32 (AB32), is implemented[31,32] and renewable sources that can supply baseload power are sought.

---

[26]FERC (2007).

[27]CEC (2007).

[28]BLM (2008).

[29]www.geo-energy.org

[30]Williams (2008).

[31]AB32, signed into law in 2006, aims to reduce greenhouse gas emissions to 1990 levels by 2020—a reduction of about 25 percent—and then an 80 percent reduction below 1990 levels by 2050.

[32]Stern et al. (2009).

11.4 Engineered geothermal systems: manmade reservoirs

The requirement that reservoir rocks have high permeability severely limits the ultimate potential of geothermal energy derived from conventional sources. Temperatures typically increase with depth at a rate of 10–80°C/km (and more than 200°C/km in some volcanic areas) in the upper crust of the Earth, depending on the geological setting and rock type, but permeability generally decreases with depth. Drilling technology allows wells to be drilled routinely to 5 km, and the current practical limit is around 10 km, so that temperatures within the range required for power generation can be accessed by drilling over a substantial fraction of the Earth's surface. Estimated temperatures 10 km beneath the surface of the United States are shown in Figure 11.3. It has been estimated that the thermal energy stored below the land surface of the United States, to a depth of 10 km, is more than 100 times the annual energy demand of the country.

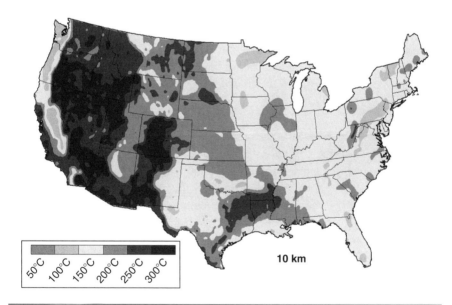

Figure 11.3 Estimated temperatures at a depth of 10 km in the United States.[33]

[33]MIT (2006).

However, it is necessary to circulate fluid through much larger surface area than is available on the circumference of a wellbore to mine heat from rock in commercial quantities, requiring a technology capable of creating multiple pathways with large surface area throughout a region of hot rock.

Hydraulic fracture stimulation is the key technology for creating such an artificial reservoir. In petroleum reservoirs, permeability enhancement by hydraulic stimulation is achieved by creating tensile cracks. Experiments in hard rock geothermal environments have shown that shear failure along existing fracture planes is likely the dominant mechanism of fracturing. Shear failure occurs when the conditions of normal and shear stress on a fracture plane are such that the frictional resistance to sliding along the plane is overcome. Movement along the fracture plane creates new porosity and permeability because the fracture surface is rough.

Experiments attempting to perform hydraulic fracture stimulation have been conducted in the United States, United Kingdom, France, Germany, Sweden, Switzerland, Australia, and Japan over the past 30 years. Government-sponsored projects in the United States, United Kingdom, and Japan were stopped before demonstrating commercial levels of production because of insufficient funding. A European Union–funded project at Soultz-sous-Forêts in France (Figure 11.4) was successful at achieving subcommercial flow rates between two wells by successfully stimulating a 200°C granite reservoir at 5 km depth, and a pilot plant is being built on the site. These manmade geothermal systems have been termed *hot dry rock* or *engineered* (or *enhanced*) *geothermal systems* (EGS).

Technical challenges to commercial development of manmade geothermal reservoirs emerged from the government-funded experiments. One or all of the following occurred during testing at the sites in the United States, United Kingdom, and Japan: low productivity, high water loss, and rapid thermal drawdown. The three factors are linked, since if the permeability of the artificial fracture network is enhanced too much by stimulation in an effort to increase well productivity, the circulating fluid may be channeled through "short circuits" in the rock, or excessive water loss into surrounding rock may occur. To overcome these challenges, the primary need is to gain more control over the process of creating, monitoring, and maintaining the permeable system of reopened cracks that is key to the extraction of heat from hot rock.

This will require a better understanding of the ways in which cracks form and propagate in different stress regimes and rock types. New tools need to be developed that allow specific zones in a hot borehole to be isolated for both fracture creation and short-circuit repair. This will allow multiple fracture zones to be created, enhance the water circulation rate while increasing the heat transfer area, and reduce the cost of development.

Figure 11.4 Experimental EGS project at Soultz-sous-Forêts, France.[35] Water injected at high pressure opens up cracks in the rock, allowing water to circulate between wells and carry heat to the surface. (a) Cross section showing wellcourses. GPK2 and GPK4 are producers, GPK3 an injector. (b) Microseismic events during hydraulic fracturing in well GPK4.

Water might not be the ideal circulating fluid in an engineered geothermal system. Preliminary modeling suggests that the use of supercritical carbon dioxide improves heat extraction and reduces parasitic load.[34] Such a system could be designed to simultaneously provide underground storage of a fraction of the CO_2 being circulated and extract heat for power generation with the remainder.

Local geological conditions have a strong influence on the potential for creating manmade geothermal reservoirs. An improved characterization of the geology, thermal and stress regimes, and the condition and orientation of existing joints and fractures is necessary to plan for an EGS. Zones of enhanced natural permeability, related to lithology or to geological structure, will normally be sought as targets.

Downhole pumps can be used to increase the productivity of geothermal wells, but the temperature and depth limitations on existing pumps need to be extended to optimize the output from manmade geothermal systems. This technical barrier will be overcome once the economic drivers are in place to create demand for EGS and motivate pump manufacturers to upgrade their systems.

The cost of geothermal power from EGS will be strongly dependent on drilling costs. The costs for a 6 km deep geothermal well[36] are likely to be split evenly among rock reduction and removal, permanent well stabilization, and miscellaneous items such as management, well control, logging, and wellheads. Thus it will be necessary to create improvements in all three areas to reduce costs significantly. Geothermal wells need to be designed with larger-diameter deep casing than typical oil and gas wells because larger volumes of fluid need to be produced. Larger casing sizes require larger bits, wellheads, and rigs and greater volumes of mud and cement. Dramatic reductions in drilling costs have been achieved in the oil and gas industry. For example, Unocal Corporation experienced an 80 percent lowering of natural gas well-drilling costs by process and technology improvement over a period of 15 years in the Gulf of Thailand.[37] An important factor was a continuous drilling campaign and a constant focus on incremental improvement. Breakthroughs that lower drilling costs in the geothermal industry are most likely to occur once large-scale developments are under way.

Well cost depends strongly on depth, but the efficiency of converting heat to electricity in geothermal plants increases with temperature, e.g., doubling when

---

[34]Pruess (2006).
[35]Gerard et al. (2006).
[36]Mansure et al. (2005).
[37]Pinto et al. (2004).

fluid temperatures are increased from 100–200°C. Drilling deeper wells to access higher temperatures typically improves project economics.

Building on what had been learned at the European Union research project at Soultz-sous-Forêts, France, a private venture in Australia obtained funding to lease and develop geothermal power from artificially stimulated reservoirs in 250°C granite at 4.4 km depth.[38] The geological environment in the Cooper Basin is particularly favorable because anomalously high concentrations of heat-producing radioactive elements within a large granite body are overlain by a thick blanket of sediments that have a low thermal conductivity. This condition created a high temperature gradient within the Earth and enabled temperatures of 250°C to be reached by drilling to less than 5 km depth. As of mid-2008, the project had successfully completed two wells and created a reservoir connecting the wells by hydraulic stimulation.

The results of the Soultz-sous-Forêts project also stimulated EGS projects in Germany, where a renewable energy law in 2004 introduced a tariff of €0.15/kWh for small geothermal plants. As of mid-2008, 150 geothermal projects were being planned, most with plans to utilize fluids in the temperature range 100–150°C at depths of 2–4 km.

The U.S. Department of Energy Geothermal Technologies Program has renewed its focus on EGS, indicating in 2008 the intention to provide cost-share funding of up to $40 million for EGS technology development and $50 million for EGS demonstration projects.[39] A major venture capital firm in the United States[40] launched a U.S. company dedicated to EGS in 2007. The U.S. Geological Survey[41] released a provisional assessment in 2008 that approximately 500 GW of geothermal power could be generated from EGS in the United States.

An extensive study organized by the Massachusetts Institute of Technology and commissioned by the U.S. Department of Energy estimated that about 100 GW, or roughly 10 percent of current U.S. electricity generation, could be met at competitive prices from EGS within 50 years,[42] providing there is a reasonable investment in research and development. The study concluded that the key to achieving commercial viability is having the ability to create sufficient connectivity between injection and production wells to allow for high mass throughput and correspondingly high energy extraction rates. It was proposed that experiments be conducted initially in areas where temperature

[38]Wyborn (2005) and Geodynamics (2008).
[39]DOE (2008).
[40]Perkins, Kleiner, Caufield, and Byers, www.kpcb.com
[41]Williams (2008).
[42]MIT (2006).

gradients are high, such as the margins of conventional geothermal fields or in oil fields where deep wells have encountered hot water, so that drilling costs can be minimized while the technologies for stimulation are being developed. The U.S. Department of Energy recently announced the formation of an industry-government-academia multidisciplinary team to perform stimulation experiments in a well on the margins of the Desert Peak geothermal field in Nevada.[43]

For large-scale developments in areas of high geothermal gradient,[44] the cost of EGS projects was recently projected to be at least $4000 per kW installed.[45] At $4000/kW it was estimated that the levelized power price would have to exceed 7¢/kWh for the project to be economically feasible. A recent geothermal power purchase agreement in northern California[46] for a conventional project had a levelized price of approximately 10¢/kWh.

Thus the conditions necessary for a breakthrough in EGS are in place, and companies prepared to address the technical challenges are attracting funds from capital markets, venture capitalists, and government cost-share programs. Spectacular growth may be just a few years away.

11.5 Conclusion

Rising fuel prices and incentives for greenhouse-gas-free energy have created renewed interest in geothermal power, with projects becoming economically viable at smaller scales and at lower temperatures, and after years of stagnation the geothermal industry is growing rapidly. Geothermal power has some very desirable features in a carbon-constrained world: low to nil greenhouse gas emissions, baseload generation, renewable classification, and predictable pricing. It does not rely on imported fuels and will not be impacted by changing weather patterns in the future.

It is likely that as the need for renewable energy continues to increase in response to greenhouse gas reduction initiatives, the unique value that geothermal can add as a baseload power source will be sought after by the electric power market. Whereas naturally occurring geothermal resources may ultimately supply a few percent of the electricity demand in the United States, engineered geothermal systems have the potential to provide baseload power

---

[43]DOE-EERE (2008).
[44]*High geothermal gradient* means high temperatures at shallow depth, resulting in lower drilling costs.
[45]Sanyal (2007).
[46]NCPA (2008).

in much larger quantities, up to 10 percent or more of the U.S. demand at current levels. The significant investments in EGS being made now in Australia, Germany, and the United States will demonstrate within a few years whether the necessary technology breakthroughs can be achieved.

If they can, a vast new energy source will be ready to serve the carbon-constrained electricity markets of the future.

REFERENCES

[1] Benito FA, Ogena MS, Stimac JA. Geothermal Energy Development in the Philippines: Country Update. Proceedings World Geothermal Congress 2005. Antalya, Turkey: International Geothermal Association; 2005.

[2] Bertani R. World Geothermal Generation 2001–2005: State of the Art. Proceedings World Geothermal Congress 2005, Antalya, Turkey, April 24–29, 2005. International Geothermal Association; 2005.

[3] Bertani R. World geothermal generation in 2007. In: Geo-heat Center Quarterly Bulletin, vol. 28, no. 3. Klamath Falls: Oregon Institute of Technology; 2007.

[4] Bertram G. Implications of New Zealand's 90% Renewable Portfolio Standard for Electricity. In: Sioshansi FP, editor. Carbon Constrained: Future of Electricity. Elsevier; 2009.

[5] BLM. Geothermal Resources Leasing Programmatic EIS. U.S. Department of Interior, Bureau of Land Management. www.blm.gov/wo/st/en/prog/energy/geothermal/geothermal_nationwide.html; 2008.

[6] Bloomfield K, Moore J, Neilson R. Geothermal energy reduces greenhouse gases. Geothermal Resources Council Bulletin 2003;77–9.

[7] California State of Assembly Bill No. 32. California EPA Air Resources Board. 2006. www.arb.ca.gov/cc/docs/ab32text.pdf.

[8] Cappetti G, Ceppatelli L. Geothermal Power Generation in Italy 2000–2004 Update Report. Proceedings World Geothermal Congress. Antalya, Turkey: International Geothermal Association; 2005.

[9] CEC. Geothermal Permitting Guide. California Geothermal Energy Collaborative: Expanding California's Confirmed Geothermal Resources Base. 2007. www.energy.ca.gov/2007publications/CEC-500-2007-027/CEC-500-2007-027.PDF.

[10] CGEC. Transmission planning issues of interest to geothermal development. California Geothermal Energy Collaborative 2005–2008 .

[11] DOE. Geothermal Technologies Program: Open Solicitations. U.S. Department of Energy Office of Energy Efficiency and Renewable Energy. www1.eere.energy.gov/geothermal/current_solicitations.html; April 2008.

[12] DOE. States with Renewable Portfolio Standards. DOE, EERE State Activities and Partnerships. June 2007. www.eere.energy.gov/states/maps/renewable_portfolio_statcs.cfm.

[13] DOE-EERE. DOE Supports Creation of an Enhanced Geothermal System. Desert Peak, NV: Energy Efficiency and Renewable Energy. DOE-EERE, 2 13, 2008. 7 9, 2008. www1.eere.energy.gov/news/progress_alerts/progress_alert.asp?aid=261

[14] EIA. Electricity Net Generation From Renewable Energy by Energy Use Sector and Energy Source. Energy Information Administration. www.eia.doe.gov/cneaf/alternate/page/renew_energy_consump/table3.html; 2008.

[15] EIA. Examples of Contract Arrangements at The Geysers. Energy Information Administration. www.eia.doe.gov/cneaf/solar.renewables/renewable.energy.annual/appe.html; 1997.

[16] FERC Federal Energy Regulatory Commission 119 Ferc ¶ 61,061. Federal Energy Regulatory Commission. www.ferc.gov/whats-new/comm-meet/2007/041907/E-5.pdf; April 19, 2007.

[17] Fleischmann D. An Assessment of Geothermal Resource Development Needs in the Western United States. Geothermal Energy Association. www.geo-energy.org/publications/reports/An %20Assessment%20of%20Geothermal%20Resource%20Development%20Needs%20 January%202007.pdf; 2007.

[18] Fridleifsson IB, et al. The possible role and contribution of geothermal energy to the mitigation of climate change. IPCC Scoping Meeting on Renewable Energy Sources, 20–25 January 2008. Luebeck, Germany: IPCC; 2008. p. 59–80.

[19] Gawell K, Reed M, Wright PM. Preliminary report: Geothermal energy, the potential for clean power from the earth. Washington, DC: Geothermal Energy Association; 1999.

[20] Geodynamics Geodynamics Ltd. Power from the earth. www.geodynamics.com.au/IRM/content/home.html; 2008.

[21] Gerard A, et al. The deep EGS (Enhanced Geothermal System) project at Soultz-sous-Forets (Alsace, France). Geothermics 2006;35:473–83.

[22] Goyal KP, Pingol AS. Geysers Performance Update Through 2006. Geothermal Resources Council Transactions. GRC, 2007. 31.

[23] Gutiérrez-Negrín Luis CA, Quijano-León José L. Update of Geothermics in Mexico. Proceedings World Geothermal Congress. Antalya, Turkey: International Geothermal Association; 2005.

[24] Mansure AJ, Bauer AJ, Livesay BJ. Geothermal well cost analyses 2005 [Conference] Geothermal Resource Council Transactions. GRC 2005;29:515–9.

[25] MIT. The Future of Geothermal Energy: Impact of Enhanced Geothermal Systems (EGS) on the United States in the 21st Century: U.S. Department of Energy report. Massachusetts Institute of Technology; 2006.

[26] NCPA. NCPA Boosts California's Geothermal Power With Major Agreement. Northern California Power Agency (NCPA). www.ncpa.com/images/stories/News/080520westerngeopurchaseagree.pdf; May 20, 2008.

[27] Pinto JC, et al. Novel PDC bit achieves ultrafast drilling in the Gulf of Thailand. www.bakerhughesdirect.com. 2004. www.bakerhughesdirect.com/cgi/hello.cgi/BHI/public/bakerhughes/resources/indepth/10.no1.2004/indepth_vol10_no1.pdf.

[28] Pollack HN, Hunter SJ, Johnson JR. Heat flow from the earth's interior: Analysis of the global data set. Rev Geophys 1993;31:267–80.

[29] Pruess K. Enhanced geothermal systems (EGS) using CO_2 as working fluid—A novel approach for generating renewable energy with simultaneous sequestration of carbon. Geothermics 2006;35:351–67.

[30] RETAAC. Governor Jim Gibbons' Nevada Renewable Energy Transmission Access Advisory Committee Phase I Report. Nevada State Government. http://gov.state.nv.us/Energy/FinalReport/RETAAC%20Phase%20I%20Report.pdf; December 31, 2007.

[31] Sanyal S. Cost of electricity from enhanced geothermal systems. In: Thirty-Second Workshop on Geothermal Reservoir Engineering, January 22–24, 2007, vol. 32. Stanford, California: Stanford University; 2007. p. 1–11.

[32] Seiki K, Noriyuki S. Geothermal Power Generation and Direct Use in Japan. Proceedings World Geothermal Congress. Antalya, Turkey: International Geothermal Association; 2005.

[33] Stefansson V. World Geothermal Assessment. In: Proceedings World Geothermal Congress 2005. Antalya, Turkey: International Geothermal Association; 2005. p. 1–6.

[34] Stern G, Harris F. Navigating the California Regulatory Currents: The Global Warming Solutions Act and Its influence on Southern California Edison. In: Sioshansi F, editor. Electricity Generation in a Carbon-Constrained World. Elsevier; 2009.

[35] Wardlow CL. The history and future of geothermal energy as an independent power producer. Geothermal Resources Council Transactions. GRC 1994;18.

[36] White BR. Growth in the New Zealand Geothermal Industry. New Zealand Geothermal Association. www.nzgeothermal.org.nz/publications/Reports/GrowthInNZGeoIndustry.pdf; 11 2007. 9.

[37] Williams CF, Reed MJ, Mariner RH, DeAngelo J, Galanis Jr SP. Assessment of moderate- and high-temperature geothermal resources of the United States: Fact Sheet 2008–3082. U.S. Geological Survey; 2008.

[38] Williamson KH, et al. Geothermal Power Technology. Proc IEEE 2001;89(12):1783–92.

[39] Wyborn D. Development of Australia's First Hot Fractured Rock (HFR) Underground Heat Exchanger. In: Proceedings World Geothermal Congress. 2005. p. 1–7.

Hydroelectricity: Future Potential and Barriers

João Lizardo de Araujo

Centro de Pesquisas de Energia Elétrica – CEPEL

Luiz Pinguelli Rosa

Instituto Alberto Luiz Coimbra de Pós-Graduação e Pesquisa de Engenharia – COPPE-UFRJ

Neilton Fidelis da Silva

Instituto Federal de Educação, Ciência e Tecnologia do Rio Grande do Norte – IFERN, Instituto Virtual Internacional de Mudanças Globais – IVIG/COPPE-UFRJ

Abstract

There is considerable untapped hydroelectric potential around the world, especially in developing countries. Electricity generated from hydro resources is characterized by extremely low operating costs and essentially zero fuel costs, which do not escalate over time, as is the case with fossil-fueled resources. With negligible greenhouse gas emissions relative to fossil-fueled generation, hydroelectric power offers an attractive solution to address global climate change. However, a number of barriers, including ecological impact and front-loaded investment in developing transmission and reservoir infrastructure, must be overcome for the full potential of the resource to be realized.

12.1 Introduction

Since the dawn of civilization, hydropower has been an important source of energy for mankind. The earliest known application of water power dates back to at least 3000 years ago. Romans used water to run their flour mills as early as 21 centuries ago, as did other ancient civilizations. The earliest mills featured a vertical wheel on a horizontal shaft, with gears to transmit the movement of passing or falling water to a vertical shaft. Such water wheels are the ancestors of today's modern hydroelectric turbines used to convert the potential energy of water stored in an elevated reservoir into kinetic energy—which turns a turbine to generate electricity.

Until the invention of the steam engine—which led to the Industrial Revolution—humans relied principally on water, wood, biomass, and limited use of wind, as well as domesticated animals in agriculture and for transportation. This meant limited mobility and limited flexibility. Since hydropower could not be transmitted with primitive mechanical devices of the day, cities and factories had to be located close to rivers. Water transportation via rivers and canals was the main source of shipping heavy goods around.

The steam engine, the emergence of railroads, and later the internal combustion engine changed all this. Raw material could be transported across land and oceans and manufactured products distributed to where they were needed. The increased mobility was initially fueled principally by coal and subsequently by oil. The fossil-fuel era, accompanied by an ever-rising carbon footprint, effectively started in the 1800s.

Current concerns about the environmental impact of burning fossil fuels, particularly in electricity generation, are a relatively recent phenomenon [15, 16, 17, 19]. Between 1970 and 2004, for example, the largest contribution to anthropogenic CO_2 emissions—representing a 145 percent increase—is attributed to electricity generation [11].

Hydroelectric power generation, which has played a dominant role in many parts of the world in the past century, releases relatively small amounts of methane and CO_2 due to decomposition of organic material in reservoirs. But these emissions are miniscule compared to those from fossil thermal power plants of equal size [18, 23]. This, plus the fact that a considerable amount of hydro resources remain untapped around the world, especially in a number of developing countries, offers attractive opportunities to address global climate concerns.

This chapter focuses on an assessment of the potential contribution of hydro resources to meet a significant portion of the global demand for electricity in the foreseeable future. Section 12.2 provides an overview of the current use of hydroelectricity worldwide and the potential of remaining untapped resources. Section 12.3 describes the existing barriers and constraints that limit the use of these remaining resources. Section 12.4 introduces a number of suggestions to overcome these barriers. Section 12.5 shows that the increased use of hydropower is among the options to decarbonize electricity generation. This section is followed by the chapter's main conclusions.

▌ 12.2 Current use and potential of hydropower

Biomass, once the dominant source of energy for mankind, has been gradually replaced by coal and, more recently, oil (Figure 12.1). Today oil accounts for roughly 35 percent of global energy demand, followed by coal and natural gas, each contributing around 20 percent. Together, these three fossil fuels represent three quarters of world energy consumption. Oil is the dominant fuel in

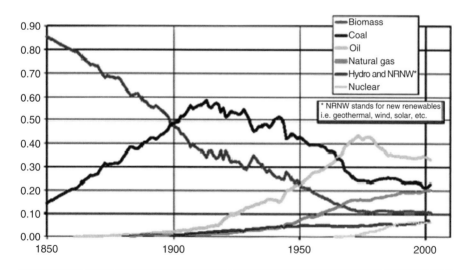

Figure 12.1 Evolution of primary energy sources, 1850–2000.
Note: For hydro, the equivalent thermal energy for electricity generation is used.
Source: Institut Français du Pétrole (IFP) [8].

transportation, coal is principally used in power generation, and natural gas is used for many purposes, including heating and for power generation. Clearly, the growing global reliance on fossil fuels is among the reasons for increased concentration of carbon in the atmosphere.

On a global scale, the contribution of hydroelectric power became noticeable in the 1990s with the development of large-scale dams, reaching roughly 8 percent of global primary energy supply by 2000, if we compute the use of the equivalent thermal power to produce the same electric energy output as shown in Figure 12.1 [8]. However, considering hydro's contribution based on mechanical energy, hydropower accounts for only 2.2 percent of total primary energy [7], as shown in Figure 12.2.

The contribution of hydro is more pronounced in electric power generation, especially since the first oil shock of 1973. As shown in Figure 12.3, following the rise of oil prices, there has been a sharp decrease in the share of oil used for electricity generation, with noticeable increases in the use of natural gas and nuclear. Coal's share has remained virtually unchanged.

Despite a doubling of hydro generation from 1300 TWh/year to 2900, the relative share of hydropower declined from 21–16 percent during this period [7]. Yet the potential scale for increasing hydro generation remains large. To put matters in perspective, the net global water cycle produces 42,600 km^3 per year of water resources, a significant portion of which can ultimately be captured and put to good use—not necessarily as hydroelectric power [24].

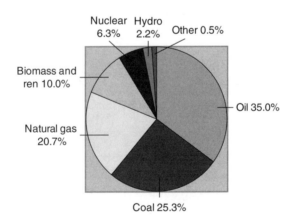

Figure 12.2 World primary energy sources in 2005.
Note: For hydro, the equivalent mechanical energy is used.
Source: World Energy Council [26].

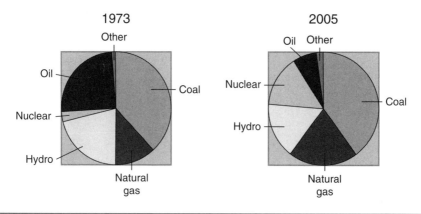

Figure 12.3 Global share of primary energy sources in electric power generation, 1973 and 2005 (%). Source: International Energy Agency [7].

Hydro resources are widely spread across the continents, but some countries are especially blessed with large and commercially viable resources (Table 12.1). Norway, Iceland, and Brazil, for example, can virtually run their entire electric grid on hydro resources during periods when rain and snowfall are plentiful.

Table 12.1 Top ten countries with largest water resources

	Thousands Km³/Year	M³/Year/Inhabitant*
Brazil	8.2	48.3
Russia	4.5	30.9
Canada	2.9	94.3
Indonesia	2.8	13.3
China	2.8	2.2
United States	2.0	7.4
Peru	1.9	74.5
India	1.9	1.8
Congo	1.3	25.1
Venezuela	1.2	51.0

Source: D'Araujo [3]; Food and Agricultural Organization [5].
Per capita data is for 2001.

To appreciate the difference between what is theoretically available and what may be commercially viable, one must define three distinct hydrological metrics:[1]

■ *Physical hydro potential* (PHP) refers to the total kinetic energy stored in the volume of water that could theoretically be tapped.

■ *Technical hydro potential* (THP) refers to the upper limit of what may physically be exploited from the PHP—even if it is not necessarily viable commercially.

■ *Economic hydro potential* (EHP) refers to that portion of THP that might be commercially viable—also considering environmental factors and other constraints.

Viewed from this perspective, the global theoretical hydroelectric potential (PHP), which is based on physical volumes of water from the natural water cycle, is estimated to amount to 41,203 TWh/year, roughly 40 percent of it in Asia [26]. The technical hydroelectric potential (THP), which is based on assumptions about what may be technically feasible, is estimated at 16,494 TWh/year, roughly 40 percent of the PHP. Finally, the economically exploitable hydroelectric potential (EHP), which takes into account not only economics and financial viability but also environmental constraints, is estimated at 8000 TWh/year [6], about half of THP and less than 20 percent of PHP, as shown in Figure 12.4.

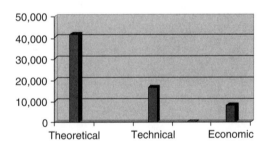

Figure 12.4 Global theoretical, technical, and economic hydroelectric potential (TWh/year).
Source: World Energy Council [26].

---

[1]A similar concept is used in defining how much energy conservation is commercially viable. Refer to Chapter 8 by Prindle et al. in this volume.

As with any estimates of this nature, it must be noted that these figures are somewhat imprecise, especially when it comes to deciding how much of the physical potential is technically feasible and what portion of that might be commercially viable. These measurement issues are more problematic in developing countries where reliable data are scarce. Table 12.2, however, is an attempt at providing the best that can be done with the known data and assumptions about technical and commercial viability at current prices.

As indicated in Table 12.2, global installed hydro capacity in 2005 was 778 GW, with 124 GW under construction—representing a 16 percent increase. Several large hydro projects currently under construction include the Three Gorges Dam in China (partially completed) and two Madeira River plants in Brazil, totaling 6300 MW, now starting.

The scale of hydro plants currently under construction represents a significant percentage of the installed capacity in many cases (Table 12.4)—for example, 50.0 percent in China, 41.3 percent in India, 31.5 percent in Brazil, 25.0 percent in Turkey, 15.3 percent in Venezuela, 12.2 percent in Russia, and 10.1 percent in Austria. Although there are other projects under construction in several countries, these represent a small percentage of the installed capacity. For instance, Canada has 1.4 GW under construction, which, though significant, represents only 2 percent of the country's installed capacity.

Europe and Asia have the largest amount of installed hydro capacity, followed by North and South America. A comparison of the hydro energy generated in 2005 with the theoretical and technical potentials shows the magnitude of as-yet untapped resources that may be utilized in the future (Figure 12.5).

As shown in Figure 12.5, the two most developed regions, North America and Europe, currently use more than 20 percent of the estimated technical hydro potential (THP), while—ironically—the less developed regions such as Africa use only 4 percent of the estimated THP. The percentages are even lower when the theoretical hydro potential (PHP) is considered.

Due to the lack of precise hydrological and feasibility studies, however, the THP values for the less developed regions have a higher degree of uncertainty. In such cases, it may be reasonable to consider the PHP as the upper limit for the THP. Based on this assumption, there is enormous untapped technical hydro potential in the world, especially in Africa, where the needs are acute.

Table 12.3 lists countries with economic hydro potential (EHP) exceeding 100 TWh/year. Indonesia, Pakistan, Madagascar, Chile, Argentina, Bolivia, and Greenland have technical hydro potential (THP) above 100 TWh/year, but the economic hydro potential (EHP) is lower than 100 TWh/year due to economic, financial, or environmental reasons.

According to a survey by World Energy Council [26], 18 countries have an EHP exceeding 100 TWh/year, or a total of 7167 TWh/year. Together, these

Table 12.2 Hydroelectric potential, technical potential, energy generated, installed capacity, and projects under construction by region, 2005

Region	Theoretical Potential (TWh/y)	Technical Potential (TWh/y)	Energy in 2005 (TWh)	Installed Capacity, 2005 (GW)	In Construction in 2005 (GW)
Asia	16,285	5523	718.1	222.6	85.7
South America	7121	3036	596.5	123.7	9.0
North America	8054	3012	675.5	164.1	2.9
Europe	4945	2714	705.4	225.2	10.0
Africa	3884	1852	83.7	21.6	5.6
Oceania	495	189	40.4	13.4	0.1
Middle East	418	168	16.8	7.1	10.5
World	41,202	16,494	2836.7	778.0	124.0

Source: World Energy Council [26].

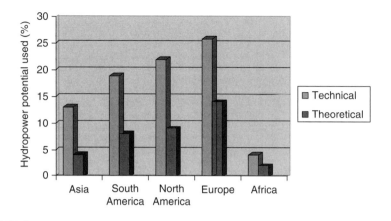

Figure 12.5 Percentage of the technical and theoretical hydropower potential that is currently utilized in each region.
Source: World Energy Council [26].

Table 12.3 Theoretical, technical, and economic hydro potential for countries with EHP >100 TWh/year (in TWh/year)

Potential →	Theoretical	Technical	Economic
China	6083	2474	1753
United States	4485	1752	501
Russian Federation	2295	1670	852
Brazil	3040	1488	811
Canada	2216	981	536
Congo	1397	774	419
India	2638	660	600
Peru	1577	395	260
Ethiopia	650	260	160
Tajikistan	527	264	264
Venezuela	320	246	130
Turkey	433	216	130
Colombia	1000	200	140
Norway	560	200	187
Japan	718	136	114
Ecuador	167	134	106
Cameroon	294	115	103
Paraguay	130	106	101

Source: Survey of Energy Resources, World Energy Council [26].

hydro-rich countries account for nearly 90 percent of the estimated global EHP (Table 12.3). Thirteen of the 18 are developing countries, with limited financial resources to develop their hydro resources—and some of these same countries are suffering from severe and chronic electricity shortages.

Not surprisingly, countries with large land mass and watersheds, including China, Russia, Brazil, India, Canada, and the United States, appear on the list. Together, these six countries have the potential to develop 5053 TWh/year of what are believed to be economically and commercially feasible hydro resources, or 63 percent of the global total.

There are, of course, many ways to look at global hydro players. Figure 12.6 shows the countries with the highest share of hydropower in their overall electric energy generation, with Norway on top. Another measure would be to look at current installed capacity and generation figures. Table 12.4 shows the top league, defined as countries with more than 10 GW of installed capacity.

Not surprisingly, the same countries appear on top of the list in both Tables 12.3 and 12.4. However, some European countries, such as France, Spain, Italy, and Sweden, were not included in Table 12.3 because their EHP is lower than 100 TWh/year, even though their installed capacity exceeds 10 GW.

A comparison of data from Tables 12.4 and 12.3 shows 2005 generation data among a selected group of countries that already tap a significant percentage of EHP. In the cases of Norway, Japan, and Canada, for example, more than 60 percent of the economically viable hydro resources have already been tapped, whereas countries such as China and India have a long way to go (Figure 12.7).

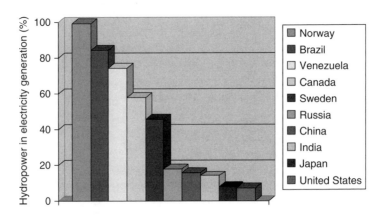

Figure 12.6 Top countries with the highest percentage of hydropower in their electricity generation (%).
Source: International Energy Agency [7].

Table 12.4   Countries with installed hydro capacity exceeding 10 GW (2005 data)

	Energy (TWh)	Installed Capacity (MW)	In Construction (MW)
Canada	358.605	71.978	1.460
Brazil	337.457	71.060	23.100
China	337.000	100.000	50.000
United States	269.587	77.354	8
Russian Federation	165.000	45.700	5.648
Norway	136.400	27.698	300
India	97.403	31.982	13.245
Japan	80.715	27.759	745
France	56.245	25.526	0
Spain	23.215	18.674	51
Italy	36.067	17.326	45
Venezuela	77.229	14.413	2.250
Sweden	72.100	16.100	0
Turkey	35.065	12.788	3.197
Austria	39.019	11.811	1.200
Switzerland	30.128	13.356	221
Mexico	27.967	10.285	754

Source: World Energy Council [26].

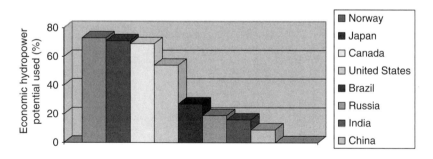

Figure 12.7   Economic hydropower potential that is currently utilized in selected countries (%).
Source: World Energy Council [26]; National Energy Balance [1] for Brazil estimate.

This is an important observation: There is a huge gap between what is actually on the ground relative to what is believed to be technically available and commercially viable in a number of countries and regions, especially in a number of developing countries. For instance, in the case of Congo, which currently uses only 2.5 percent of its economic hydro potential, there is significant room for growth. In fact, most developing countries use less than 30 percent of what they appear to be endowed with from a hydro resource point of view. In a world pressed with growing demand for electricity and confronted with high fossil-fuel prices and concerns about climate change, this offers hope.

In summary, it is estimated that only 35 percent of the global EHP, 17 percent of the THP, and a mere 7 percent of the PHP are currently utilized. The message to take away from the preceding discussion is that no matter how one looks at it, significant amounts of hydro resources remain to be developed if the constraints—especially environmental and ecological ones—are carefully addressed, which is the subject of the following section.

12.3 Barriers to future development of hydro resources

As already discussed, dams have been built for thousands of years for flood control, for water supply, and, more recently, for power generation. In the recent past, nearly 47,600 large dams have been constructed globally, roughly half of them primarily or exclusively for irrigation purposes. Figure 12.8 and Table 12.5 provide an overview of the largest dams in the world and their locations.

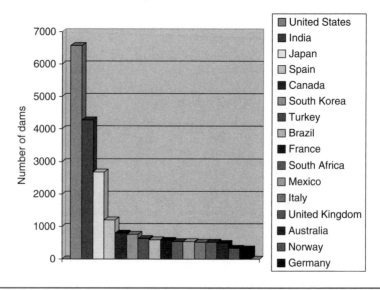

Figure 12.8 Countries with a large number of big dams.
Source: World Commission on Dams [25], excluding China, which has over 22,000 dams.

Table 12.5 Number of large dams, average areas, and volumes of reservoirs

	World	Europe	Asia	North America	South America	Africa	Australasia
Number	47,655	5480	31,340	8010	979	1269	577
Area (km^2)	23	7	44	13	30	43	17
Vol. (km^3)	269	70	268	998	101	883	205

Source: World Commission on Dams [25].

The largest number of dams can be found in Asia, a region that also has the largest average reservoir area, whereas the largest average volumes are located in South America, followed by North America, China, the United States, India, Japan, and Spain, which each have more than 1000 large dams. The World Commission on Dams (WCD) estimates that construction of dams and reservoirs has affected the lives of millions of people. For a great many people, dams, reservoirs, and clean and cheap hydroelectricity have been a blessing, but some have been adversely affected by loss of land and destruction of traditional ways of life.

It is, therefore, rather ironic that at a time when the world, especially many developing countries, needs more environmentally friendly energy, hydroelectric development faces considerable barriers. These barriers may be grouped into three major categories:

■ Considerable front-loaded investments required to develop reservoirs, construct dams, and install generating turbines

■ Investment in transmission lines and related infrastructure to bring the generated power to major load centers

■ Address considerable ecological concerns associated with building large dams

The first two are not fundamentally different to barriers facing nuclear or wind, and the issues are similar to those addressed by authors of two chapters on these topics that appear in this volume. But the considerable environmental and ecological opposition confronting hydro is serious, complicated, and perhaps unique. It is also ironic that a resource that is *renewable* and, at least on the surface, appears environmentally benign has come to be opposed by many environmentalists.[2]

The first barrier is the considerable investment required to develop the infrastructure that is needed up front before any revenues can be generated. On the positive side, once dams and reservoirs are built, the water—in addition to its capacity to generate power—is a valuable commodity.[3] Many dams and reservoirs serve multiple purposes, including flood control and irrigation—making them more valuable than a single-purpose power plant, which is typical of nearly all other means of electric power generation.

---

[2] In California, as in many other places, large hydro resources are distinguished from small hydro; the former is considered environmentally undesirable, the latter acceptable. For example, in meeting the state's renewable portfolio standards, large hydro does not qualify.

[3] Many dams and reservoirs were in fact originally designed for water storage and irrigation purposes or for flood control, with electricity generation seen as a secondary benefit.

The front-loaded investment associated with hydro resources makes them less attractive to private investors, who generally prefer less up-front investment and shorter construction times. Like all renewables, hydro plants, however, have low operating costs and no fuel costs. Moreover, fuel costs do not escalate with energy costs in typical thermal plants. If hydro reservoirs are of significant size, they can be used as peaking, intermediate, or base load, greatly increasing their economic advantage and their operational flexibility.

The second barrier to hydroelectric development is the need for investing in long distance transmission lines to bring the power to major load centers, in some cases requiring transmission lines stretching 1000 kilometers or more.[4] Unlike nuclear or natural gas plants, hydro resources cannot be sited where the power is needed. This constraint has many similarities to other renewable resources, notably wind, where the best resources are generally far from major cities.

The third, and arguably the most serious, barrier is related to environmental, ecological, and social impacts attributable to *large* dams and reservoirs, which often evoke unfavorable popular reactions worldwide. This problem, to a large extent, can be blamed on insensitive historical practices on the part of hydro developers and central planners who paid little or no attention to what might be submerged following the construction of large reservoirs and even less to preservation of unique and scenic rivers, wildlife, and indigenous populations.

In a typical case, engineers and hydrologists optimized the location of the dams and set the height and size of the reservoirs to maximize output with little or no attention to environmental, biological, archeological, or social issues. The legacy of these unfortunate practices seriously haunts the prospects for future hydro development in many parts of the world.

Although there is no scientific definition for what can be considered a large dam,[5] a working definition may be anything that is over 15 m high or over 5 m high and with a reservoir volume of more than 3 million cubic meters, according to the World Commission on Dams [25].[6]

With the realization that more sensitivity—and sensibility—must be exercised in building hydro infrastructure, new guidelines have emerged on handling the multiple facets of hydro resources with their considerable

---

[4]Examples include long lines from northern Quebec to Montreal and New York or from British Columbia and the Pacific Northwest to feed California.

[5]The State of California, for example, defines large hydro as anything with capacity larger than 5 MW.

[6]The WCD, established in 1997 to address considerable controversies surrounding large dams, released a report with recommendations in 2000.

physical footprint. More recently, it has become customary for developers to carefully consider the impact of water reservoirs on physical, biological, and human systems, in addition to performing an environmental impact assessment, which is now virtually mandatory around the world.

The pioneering studies (e.g., Penner and Icerman [13]) that have contributed to this new awakening typically categorized the impacts of hydro resources to be considered into three broad areas:

- Physical systems, including atmospheric, microclimate, hydrological, water quantity, discharge, current, ground water evaporation, water quality, turbidity, sedimentation, earthquakes, and so on

- Biological systems, including aquatic upstream and downstream ecosystems, nutrients, phytoplankton, zooplankton, plants, fish, disease vectors, terrestrial ecosystems, submerged land, soil, plants, animals, and man

- Social systems, including economic production, fishing, agriculture, manufacturing, commerce, tourism and recreation, and cultural and archeological issues

With new guidelines in place, perhaps a new chapter is opening in the history of hydro development—one in which a reasonable balance between what is technically possible and what is socially and environmentally acceptable can be reached. Adding to the urgency are rising concerns about climate change. As discussed in Section 12.5, hydro resources, like nuclear energy, are not entirely carbon-free, but their greenhouse gas emissions are minimal compared to those of thermal power plants.

12.4 Overcoming barriers to future hydro development

The three principal barriers to hydro development can, and should, be addressed, especially with the growing interest in decarbonizing electricity generation. These are briefly addressed here.

Overcoming hydro's front-loaded investment is not unlike challenges confronting nuclear plant developers. But just as is the case with nuclear plants, hydro facilities, once completed, enjoy relatively low operating costs and virtually zero fuel costs. In an environment where fossil-fuel prices keep rising and exhibit volatility, hydro—as with all renewables—is immune to future cost escalations. And should an implicit or explicit cost on carbon come to pass, hydro investments will become even more attractive. Another advantage of hydro is that it can complement other intermittent renewable resources, such as wind and solar, potentially serving as a storage medium.

Table 12.6 Comparative costs of hydro, nuclear, and thermal plants for Brazil

Technology	Investment Costs (US$)	Energy Costs (US$)
Hydroelectricity	1600/kW	50/MWh
Thermoelectricity	1200/kW	80/MWh
Nuclear	3500/kW	92/MWh

Source: Energy Research Enterprise (EPE) [4].

Since much of the unexploited hydro potential is in developing countries, cost considerations become paramount. Depending on the location, investment requirements of hydro resources vary, and the cost of transmission lines must, of course, also be factored in. Table 12.6, which is applicable to Brazil, for example, suggests that hydro generation can compete favorably with fossil and nuclear plants [4].

One way to address the significant up-front investment of hydro generation is to involve both national and multilateral banks and funding agencies, such as the National Bank for Economic and Social Development (BNDES) in Brazil and the World Bank, as well as private banks.

Regarding long-distance transmission lines, there are a number of technical developments that promise to make transmission across long distances more efficient while reducing costs.[7]

As for the environmental impacts and controversies surrounding large reservoirs, it must be pointed out that roughly 50 percent of the world's existing dams are primarily or exclusively built to serve as water reservoirs and/or for flood control and irrigation purposes. The percentage of dams designed for hydroelectric power is probably around 25 percent of the total, with another 25 percent designed for multiple uses, including water supply, flood control, and other uses such as recreation. Moreover, many existing and proposed dams with dual purposes or for electricity generation are not necessarily large.

Regardless of the intended purpose of these large reservoirs, specific care must be taken to reduce and mitigate any adverse environmental impacts,

---

[7]To facilitate long-distance transmission lines (LDTLs), two new concepts deserve special attention: (a) the grid shock absorber concept, using segmentation of alternating current system in clusters and direct current interconnection (e.g., refer to Clark and Epp, 2006), and (b) a little longer than half wavelength for very long transmission lines (e.g., Portela, Silva and Alvim, 2007). Both concepts increase the operational and reliability aspects of LDTLs while reducing the costs.

especially the size of the areas flooded by future hydro reservoirs. One example of the compromises involved can be found in the case of the Belo Monte project in Brazil, where the original size of the reservoir was significantly reduced to minimize the environmental impact of the dam. The adjustments in this case resulted in reductions in the project's hydropower potential—the price to be paid for minimizing environmental impacts.[8]

Ultimately, building a major hydro project is not fundamentally different from building any other major energy project. Governments and regulators must make sure that affected stakeholders are informed and the process is transparent and democratically managed. Figure 12.9, for example, illustrates how the pros and cons of a proposed hydro project can be viewed in a holistic manner to its successful acceptance and approval. In the end, environmental regulations must be strictly followed, and the government must successfully convince the stakeholders that the benefits of a proposed scheme significantly outweigh the costs and the risks.

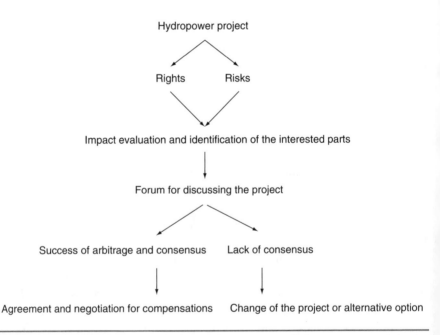

Figure 12.9   Basic framework for seeking approval for a proposed hydro project.

[8]Addressing the concerns of environmentalists frequently leads to development of so-called *run-of-the-river plants*, which do not offer multiyear storage capacity and have no capability for water-flow regulation. This limitation has been compensated in the Brazilian interconnected grid up to now by relying on the already existing large reservoirs in the system.

12.5 GHG emissions associated with hydropower

Hydroelectric reservoirs have been shown to emit methane (CH_4) and CO_2. These emissions depend on several factors, including the age and depth of the reservoirs, land vegetation prior to flooding, and climate. Just as in the case of natural lakes and rivers, manmade reservoirs produce biogenic gases through decomposition of organic matter and other processes.

The emissions of methane from reservoirs are always unfavorable to hydropower, whereas emissions of CO_2 can be attributed to the natural carbon cycle between the atmosphere and the reservoir.[9]

The *2006 IPCC Report* noted that current data on GHG emissions from hydroelectric reservoirs does not give an accurate account and concluded that there are currently no reliable data from a number of countries with a significant number of reservoirs [10]. The sampling did not consider the net flux of emissions from manmade reservoirs because it did not account for the natural fluxes of gases, mainly CO_2, which can come from the atmosphere. This makes it harder to separate anthropogenic emissions from emissions that would have occurred in the absence of the manmade reservoir.

The same *IPCC Report* points out that reservoirs in warm-temperate, moist, tropical-wet, and tropical-dry areas produce higher emissions. Recent measurements from 10 hydro plants in Brazil,[10] which amount to a total of 28,537 MW, compare emissions to those from thermal power plants burning natural gas to produce the same electric energy [23].

Gross GHG emissions from each of the selected reservoirs were assessed through sampling, with subsequent extrapolation of the findings to obtain a value for the total reservoir area. A methodology for obtaining a representative average gas flux, taking spatial and temporal variations into account, was developed. A wide variation in the intensity of the emissions was noted, indicating the influence of many factors, including external organic matter washed in from the soils and slopes of the watershed basin [20].

Other researchers who have measured emissions in the case of the Balbina hydropower plant in Central Amazon [12] have concluded that downstream

---

[9]GHG emissions may be categorized as diffusive emissions, bubble emissions, and degassing. Most of the emissions that have been measured around the world refer to diffusive emissions. In the case of Brazilian reservoirs, however, both diffusive and bubble emissions have been measured (e.g., Santos et al., 2006) and initial studies have been looking into downstream and degassing emissions.

[10]The hydro plants studied are Balbina, Tucuruí, and Samuel, in the Amazon Rainforest; Xingó, Serra da Mesa, Três Marias, and Miranda, in the savannah; and Barra Bonita, Itaipu, and Segredo, in the Atlantic forest (south).

methane emissions are of the same order of magnitude as emissions from the water surface of the reservoir.[11]

The results of the studies to date are highly sensitive and require careful analysis and interpretation [21]. However, in what follows it is safe to assume that the 10 hydropower plants studied have, as an upper limit, downstream emissions roughly equal to their surface emissions.[12]

Though research on improved measurement techniques continues to progress, it is fair to say that GHG emissions from hydro reservoirs per MWh are, on the average, lower than those of natural gas-fueled power plants in 7 of the 10 hydro plants with installed capacity of 27,675 MW that have been studied. Even if the surface emissions are multiplied by two, assuming the upper limit of downstream emissions, the resulting average emissions from the 10 reservoirs, expressed in mass of carbon equivalent[13] per MWh (tCeq./MWh), are lower than those from natural gas-fueled power plants with simple cycles or combined cycles.[14] Table 12.7 reinforces this proposition showing comparison GHG emissions from hydro and natural gas plants.

Table 12.7 Comparison of GHG emissions from hydro and natural gas plants

Type of plant	Emissions (tCeq./MWh)
Natural gas power plant emissions[15]	
With simple cycle (A)	0.2330
With combined cycle (B)	0.1399
Average emission of 10 hydropower plants in Brazil	
Measured surface emissions (C)	0.0127
Including estimated upper limit of downstream emission (D)	0.0254
Natural gas power plant emission/hydropower plant emission	

[11]The Balbina plant has already been studied (Santos, 2000; Rosa et al., 2004c). The power density of Balbina (0.09 W/m^2) is very low: 100 and 500 times lower than that of Itaipu, in the south, and Xingó, in the northeast, respectively. Balbina is the worst case of GHG emissions among all Brazilian hydropower plants.

[12]This figure is probably an overestimation, because the methane flux from water passing through the turbines is not directly measured; instead, it is calculated by subtracting two measurements that feature overlapping experimental errors.

[13]Carbon equivalent is the mass of the gas multiplied by the ratio between the mass of carbon atoms and the mass of the gas molecule, the result multiplied by the global warming potential (GWP) for 100 years, adopted by the U.N. Climate Convention (IPPC, 1995).

[14]The only three hydropower plants with emissions per MWh higher than those of natural gas-fueled power plants (Portela, Silva and Alvim, 2007) have very low power density (defined as power generated divided by reservoir area), less than 0.4 W/m^2. These reservoirs represent only 3 percent of total installed capacity considered in the study.

[15]CO_2 emitted in the natural gas combustion process and CH_4 fugitive emissions from natural gas were considered for natural gas-fired plants.

Average hydropower plant emissions are lower than natural gas power plant emissions by the following percentages:

C is 5.4% of A

D is 10.9% of A

C is 9.1% of B

D is 18.2% of B

Based on research results to date, it is safe to say that GHG emissions from hydro plants are negligible compared to natural gas plants on a per MWh basis. As shown, under a *worst-case scenario*, emissions from hydro plants are 10.9 percent of those from a simple cycle and 18.2 percent of a combined cycle. The comparison to coal would be considerably more dramatic.[16]

12.6 Conclusion

Hydroelectric power is a renewable and extremely low-carbon energy resource with significant untapped potential. Moreover, many developing countries are blessed with large potential that can be developed while avoiding serious environmental implications.

Hydro resources, like all other energy resources, have their limitations and face barriers, in this case front-loaded infrastructure investments in reservoirs and transmission lines, but the benefits in terms of extremely low operating and essentially zero fuel costs are overwhelming. At a time when humankind is looking for clean and affordable energy with a small carbon footprint, renewable, clean, and plentiful hydroelectric generation cannot be ignored.

Painful lessons learned from insensitive developments of the past that did not consider the environmental impact of large dams and reservoirs must guide governments and policymakers in terms of the ways in which hydro resources are developed and managed in the future.

Dedication

L. P. Rosa and N. F. Silva would like to dedicate this chapter to the memory of João Lizardo de Araujo, an important Brazilian researcher in energy planning and systems engineering, a concerned citizen, and a good friend and colleague who contributed to this paper before his untimely departure.

---

[16]It must also be noted that based on these observations, hydroelectric plants with power densities of less than 1 W/m^2 should be avoided.

REFERENCES

[1] BEN. National Energy Balance. Brazil: Ministry of Mines and Energy; 2007.

[2] Clark H, Epp K. Technical Assessment of the Grid Shock-Absorber Concept. Montreal: FACTS User's Group & Task Force Meeting; 2006.

[3] D'Araujo R. Seminar on the Future of Northeast Region. Salvador, Brazil: National Confederation of Engineering and Architecture (CONFEA); 2008.

[4] EPE. Energy Research Enterprise. Brazil: Ministry of Mines and Energy; 2008.

[5] FAO. Report. UN: Food and Agriculture Organization; 2003.

[6] HDWAIG. Hydropower & Dams. World Atlas and Industry Guide; 2000.

[7] IEA. Report. International Energy Agency; 2006.

[8] IFP. Rapport. Institut Français du Pétrole; 2003.

[9] IPCC. Second Assessment Report.

[10] IPCC. Guidelines for National Greenhouse Gas Inventories, Volume 4: Agriculture, Forestry and Other Land Use, Appendix 3, CH_4 Emissions from Flooded Land: Basis for Future Methodological Development. 2006.

[11] IPCC. Fourth Assessment Report. 2007.

[12] Kemenes A, Forsberg BR, Melack JM. Methane Release Below a Tropical Hydroelectric Dam. Geophys Res Lett 2007.

[13] Penner SS, Icerman L. Non-Nuclear Energy Technologies, vol. II. Addison-Wesley Publishing Co.; 1975.

[14] Portela C, Silva J, Alvim M. Non Conventional AC Solutions Adequate for Very Long Distance Transmission, IEC/CIGRE. Beijing: UHV Symposium; 2007.

[15] Rosa LP, Ribeiro SK. The Present, Past and Future Contributions to Global Warming of CO_2 Emissions from Fuels. Climatic Change 2001;48:289.

[16] Rosa LP, Ribeiro SK, Muylaert MS, Campos CP. Comments on the Brazilian Proposal and Contributions to Global Temperature Increase with Different Climate Responses. Energy Policy 2004a;32:1499.

[17] Rosa LP, Muylaert MS, Cohen C, Pereira AS. Equity, Responsibility and Climate Change. Climate Research 2004b;28:89–92.

[18] Rosa LP, Santos MA, Matvienko B, Santos EO, Sikar E. GHG Emissions from Hydroelectric Reservoirs in Tropical Regions. Climatic Change 2004c;66:9–21.

[19] Rosa LP, Campos CP, Muylaert MS. Historical CO_2 Emission and Concentrations Due to Land Use Change of Croplands and Pastures by Country. Sci Total Environ 2005;346:149–55.

[20] Rosa LP, Santos MA, Matvienko B, Sikar E, Santos EO. Scientific Errors in the Fearnside Comments on GHG Emissions from Hydroelectric Dams and Response to his Political Claiming. Climatic Change 2006;75:91–102.

[21] Rosa LP, Santos MA. GHG Emissions from Ten Hydroelectric Power Plants in Brazil and Comparison with Natural Gas-Fueled Power Plants. to be published; 2008.

[22] Santos MA. Energy Planning. D.Sc. thesis, COPPE, Federal University of Rio de Janeiro; 2000.

[23] Santos MA, Rosa LP, Sikar B, Sikar E, Santos EO. Gross greenhouse gas fluxes from hydropower reservoir compared to thermo-power plants. Energy Policy 2006;34:481.

[24] UNESCO. World Water Research at Beginning of 21st Century—HP UNESCO. 2003.

[25] WCD. Report. World Commission on Large Dams; 2000.

[26] WEC. Survey of Energy Resources. World Energy Council; 2007.

Case Studies

Part 3

Ontario: The Road to Off-Coal Is Paved with Speed Bumps*

Roy Hrab and Peter Fraser

Ontario Energy Board, Toronto, Ontario, Canada

Chapter 13

Abstract

In 2003, the government of Ontario pledged to retire all coal-fired genera-tion by 2007, primarily for environmental and health reasons. At the time, coal-fired generation accounted for 25 percent of the province's installed

---

*The views expressed here are those of the authors and do not reflect those of the Ontario Energy Board.

Doi: 10.1016/B978-1-85617-655-2.00013-4.

generation capacity. Removing and replacing such a significant portion of the province's capacity involves many financial and technical challenges. These include procuring adequate alternative supply, the financial costs of a zero coal supply mix, reliability issues, and understanding the impact of the policy on the province's environment, health, electricity system, and economy. Progress has been made, and there are no signs that the policy will be abandoned; however, the complexities encountered resulted in the deadline being pushed back to 2014.

13.1 Introduction

With a population of 13 million, Ontario is Canada's most populous province, accounting for 39 percent of the country's total population and roughly the same percentage of its GDP. In 2006, Ontario was the second largest producer of greenhouse gas (GHG) emissions in Canada (behind the province of Alberta) at 190 megatonnes (Mt; CO_2 equivalent), roughly 27 percent of the country's GHG emissions.[1] Electricity generation accounted for 28.5 Mt of Ontario's emissions, of which 24 Mt originated from coal-fired facilities.[2]

In 2003, the government of Ontario pledged to retire all coal-fired generation by 2007 for environmental and health reasons. At the time, 25 percent of the province's installed generation capacity was from coal-fired generation. Removing and replacing such a significant portion of the province's capacity involves many financial and technical challenges.

Coal-fired generation in the province plays a critical role in Ontario's current electricity supply mix. With baseload power generation supplied principally by nuclear power plants and with peaking hydroelectric generation with limited storage capability, Ontario's coal-fired generation performs as a flexible "swing" resource, producing 20 percent of the electricity generated within the province in 2007. Coal generation is relied on to deal with unplanned supply availability and to follow load. The existing Ontario grid has been designed to utilize these resources.[3]

The province's electricity system, operated by the Independent Electricity System Operator (IESO), is interconnected to two neighboring Canadian provinces, Manitoba and Quebec, and three U.S. states, Michigan, Minnesota, and New York. Much of Ontario's electricity trade occurs with New York

---

[1]Canada (2008).
[2]Ibid.
[3]OPA (2007a). For example, the 4000 MW Nanticoke station has a critical role to play in providing reactive power to Ontario's main load center in the Greater Toronto area.

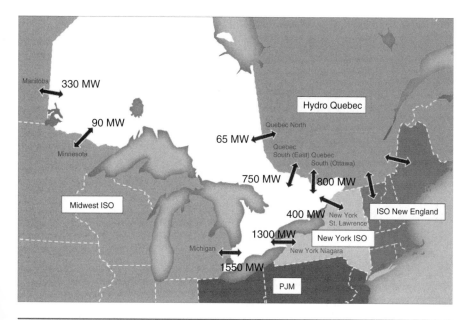

Figure 13.1 Interconnections between IESO (Ontario) and neighboring systems (with summer import limits).
Source: IESO.

through the New York Independent System Operator and with Minnesota and Michigan through the Midwest Independent System Operator (Figure 13.1).

In this regard, Ontario power production is competing with power generators in these states and with neighboring U.S. states that rely heavily on coal for power generation, including Ohio (86 percent), Indiana (95 percent), Michigan (60 percent), and Pennsylvania (56 percent).[4]

In recent years, Ontario has moved to become a significant exporter of electricity to the United States. According to one estimate by the IESO, coal generation supplies approximately half of the electricity exported from Ontario.

Replacing these important resources while maintaining reliability was always anticipated to create significant technical challenges. It was because of those challenges that the deadline was pushed back, first to 2009 and then to 2014. However, despite these obstacles, the government remains committed to eliminating coal-fired generation from the province's supply mix, making the policy the primary component of its climate change action plan. The replacement process is well under way, including procuring adequate alternative supply and investing in necessary transmission upgrades, to meet the

---

[4]Energy Information Administration (2007).

2014 deadline. The process involves addressing important public policy issues, including the impact on cost, system reliability, and estimating the benefits for the province's (and the planet's) environment, health, electricity system, and economy.

This chapter reviews the institutional and market structure in which Ontario's coal-replacement policy is occurring, the evolution of the policy over time, the replacement plan, stakeholder assessments of the policy and plan, and various challenges involved in implementing the strategy. The case of Ontario provides valuable lessons to other jurisdictions contemplating similar action.

▌ 13.2 The setting in Ontario

Beginning in the mid-1990s, the Ontario government began to implement a number of restructuring initiatives aimed at dismantling a vertically integrated, government-owned monopoly (Ontario Hydro), developing a competitive generation sector and liberalizing prices.

The structure of the sector at the time was dominated by provincially and municipally government-owned companies. The provincial utility, Ontario Hydro, generated most of the province's power and owned the main transmission system. Most of the province's 4.5 million electricity consumers were served by about 300 (now consolidated to about 80) municipal distribution companies. Today, the provincially owned Ontario Power Generation has about 70 percent of the province's installed capacity, with the remainder privately owned and/or operated. The province's electricity grid is almost completely owned by Hydro One, a firm also wholly owned by the provincial government.

Two agencies were mandated to oversee the electricity market: the pre-existing Ontario Energy Board (OEB) and the newly created Independent Electricity Market Operator (IMO), subsequently renamed the Independent Electricity System Operator (IESO) in 2004. At the time of restructuring, the OEB was given the responsibility to regulate the monopoly segments of the electricity market by setting transmission and distribution rates. The IESO's responsibilities included operating the wholesale spot market and dispatch function. However, in the fall of 2002, less than a year after prices were liberalized, voter response to unexpected electricity price increases led the government to intervene by freezing retail electricity rates for residential and small consumers.[5]

---

[5]For a detailed account of the short-lived experiment to privatize and deregulate Ontario's electricity sector, see Trebilcock and Hrab (2006).

In October 2003, the provincial government changed from Conservative to Liberal. Among the promises made by the incoming government was a pledge to retire all the province's coal-fired generation plants by 2007, for environmental and health reasons. At the time, coal units accounted for 7560 MW, or approximately 25 percent of the province's total installed generation capacity. Coal generation was to be replaced through conservation measures and the procurement of natural gas-fired and renewable generation.

The new government made three major reforms to the structure of the province's electricity sector through legislation (Electricity Restructuring Act, 2004):

1. It established a new body called the Ontario Power Authority (OPA). The OPA was given the responsibility and ability to ensure, plan, and procure new generation and transmission capacity for the province through designing an Integrated Power System Plan (IPSP) to be reviewed by the OEB. Further, a Conservation Bureau was created within the OPA to develop conservation programs (Figure 13.2).

2. The OEB's regulatory role was expanded. The OEB was given the responsibility of regulating retail electricity rates for residential and other

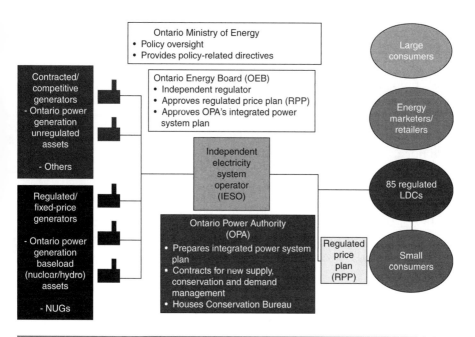

Figure 13.2 Ontario's hybrid electricity market structure, 2008.
Source: Ontario Energy Association.

designated consumers through the Regulated Price Plan. The OEB would also regulate the prices of certain of Ontario Power Generation's assets.

3. The government gave itself the authority to direct the future power generation mix of the province, including "goals relating to ... the phasing out of coal-fired generation facilities."

It is within this "hybrid" structure of significant nonmarket pricing, a mixture of public and private ownership, and centralized system planning by a government agency under government direction that Ontario's coal phase-out initiative is occurring.

13.3 The evolution of the coal-replacement policy

Ontario's oldest coal plant, the 1140 MW Lakeview facility, was closed in April 2005.[6] The remaining coal plants were initially to be closed by the end of 2007. At the time of the Lakeview shutdown, the government released a study it had commissioned to support the decision to eliminate coal-fired generation.[7] The study was a cost/benefit analysis that quantified the financial, health, and environmental costs of various generation portfolio options. Assuming that replacement capacity would be available by the end of 2007, it concluded that replacing coal units with nuclear and natural gas-fired generation provided greater net benefits to the province than adding emissions control equipment to all the coal-fired units (Table 13.1). The study estimated that the air pollution emitted from the province's existing coal units resulted in 668 premature deaths a year and that these would be largely avoided by a coal phase-out (Table 13.2).

Two months later, the government announced that although 9145 MW of additional generation capacity and demand-side projects had either been finalized or were under negotiation,[8] the deadline for shutting down coal-fired generation capacity needed to be extended from 2007 to 2009. The delay was needed to maintain network reliability.[9]

In June 2006, following public consultations, the government issued its Supply Mix Directive to the OPA.[10] The Directive set out generation, conservation, and other objectives that the OPA would be required to fulfill in its IPSP (Figure 13.3).

---

[6]The Lakeview shutdown date had been established by the previous Conservative government.
[7]DSS Management Consultants Inc. and RWDI Air Inc. (2005).
[8]Ontario (2005a).
[9]Ibid.
[10]Ontario (2006a).

Table 13.1   Estimated annualized financial costs and health and environmental damages of supply mix scenarios

	Scenario			
	Status Quo	All Gas	Nuclear and Gas	Stringent Emission Controls
Financial costs	$985	$2076	$1529	$1367
Health damages	$3020	$388	$365	$1079
Environmental damages	$371	$141	$48	$356
Total cost of generation	$4377	$2605	$1942	$2802
Net benefits	—	$1772	$2435	$1575

All values are expressed as annualized costs/damages/benefits in 2004 CAD Millions (based on purchasing power parity used in the report of 1 CAD = US$0.80).
Source: DSS Management Consultants Inc. and RWDI Air Inc. (2005) [6].

Table 13.2   Estimated health impact of supply mix scenarios

	Scenario			
	Status Quo	All Gas	Nuclear and Gas	Stringent Emission Controls
Premature deaths (total)	668	11	5	183
Premature deaths (acute)	103	2	1	28
Hospital admissions	928	24	12	263
Emergency room visits	1100	28	15	312
Minor illnesses	333,660	5410	2460	91,360

Source: DSS Management Consultants Inc. and RWDI Air Inc. (2005) [6].

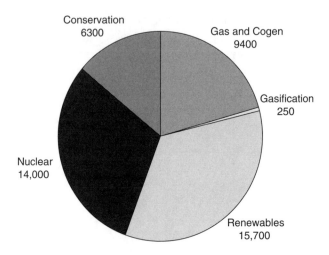

Figure 13.3   The government of Ontario's planned electricity supply mix, 2025 (MW, installed capacity).
Source: Ontario (2006a,b) [16, 17].

At this time, the government also announced that it could not meet its timeline to close all coal-fired generation plants by 2009. Instead, as part of the Supply Mix Directive, the government instructed the OPA to:

Plan for coal-fired generation in Ontario to be replaced by cleaner sources in the earliest practical time frame that ensures adequate generating capacity and electric system reliability in Ontario. The OPA should work closely with the IESO to propose a schedule for the replacement of coal-fired generation, taking into account feasible in-service dates for replacement generation and necessary transmission infrastructure.[11]

In addition, the OPA was also "asked to recommend cost-effective measures to reduce air emissions from coal-fired generation."[12]

In November 2006, the OPA released a preliminary plan, indicating that the coal-fired generation could be completely phased out by the end of 2014 with the possibility of shutting the units down early, pending the results of demand growth, conservation initiatives, and replacement generation resources going in-service as scheduled.[13]

---

[11]Ibid.
[12]Ontario (2006b).
[13]OPA (2006).

Table 13.3 Capital cost of environmental control technology options

Alternative		Capital Cost, Including Financing ($M)
Alt. 1	Existing ECT	0
Alt. 2	Baghouses on Nanticoke Units 5–8 and SCRs on Units 5 and 6	761
Alt. 3	FGDs on Nanticoke Units 5–8 and SCRs on Units 5 and 6	1580
Alt. 4	FGDs on Nanticoke Units 7 and 8	666

Source: OPA (2007b) [25].

Further, in April 2007, the OPA released a discussion paper presenting an evaluation of four emission control technology (ECT) options. These options involved the installation of baghouses, selective catalytic reduction (SCR), and/or flue gas desulphurizer (FGD) technologies (Table 13.3).[14] The proposed ECT options could have been in-service during 2010–2011. It should be noted that Lambton Units 3 and 4 have FGDs and SCRs to reduce SO_2 and NO_x emissions, resulting in these units having the lowest SO_2 and NO_x emissions of Ontario's coal-fired units.

In response to the OPA report, the government stated it would not install any additional ECTs because of the cost and the expected timing of shutting down the coal plants and that ECTs did not reduce CO_2 emissions.[15] This statement represented a significant shift in the primary goals of the coal-replacement policy from air quality concerns to climate change. Indeed, the coal-phase plan became a central component in the government's plan to address climate change with the expectation that CO_2 emissions from all Ontario power plants will drop to less than 7 Mt by 2015 (Figure 13.4).[16]

In July 2007, the government proposed a new regulation under Ontario's Environmental Protection Act, requiring the complete cessation of burning coal at OPG's four remaining coal-fired generation stations after December 31, 2014. The regulation went into force in August 2007.[17]

---

[14]OPA (2007b).
[15]Canadian Press (2007).
[16]Ontario (2007a,b).
[17]Reg. 496/07, Environmental Protection Act (Ontario).

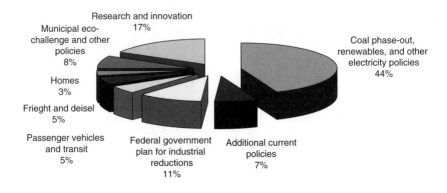

Figure 13.4 Ontario's emission reductions to be achieved by 2014.
Source: Ontario (2007b) [19].

Last, in May 2008, the government announced that it was further amending
its coal phase-out regulation to limit CO_2 emissions from the four OPG coal
plants to 11.5 Mt annually from 2011 through 2014.[18]

▌ 13.4 The coal-replacement plan

In August 2007, the OPA submitted its IPSP to the OEB for review and
approval. The IPSP contained the OPA's two plans to replace coal-fired generation.[19] One plan is focused on Ontario's Northwest system, which is
isolated by a significant degree from the rest of the province's electricity system, involving the Thunder Bay (306 MW) and Atikokan (211 MW) units.
The second plan pertains to the rest of the province and involves the Lambton
(1972 MW installed capacity) and Nanticoke (3945 MW) facilities (see
Table 13.4).

The Northwest replacement plan relies on conservation, renewable
resources, and interconnections. To ensure supply adequacy, it is expected that
half of the capacity of Thunder Bay and Atikokan will be taken out of service
in 2011 and the remainder in 2014.

The plan for the rest of the province relies on conservation, renewables,
nuclear, and gas-fired generation to replace coal. The OPA's analysis demonstrated that the installation of 1400 MW of gas-fired generation along with
transmission system enhancements are required to meet local area reliability

[18]Ontario (2008).
[19]OPA (2007a), Exhibit D, Tab 7, Schedule 1.

Table 13.4   Energy production from coal-fired facilities, 2003–2007 (TWh)

Station	2003	2004	2005	2006	2007
Lambton	10.6	7.7	9.4	6.9	8.9
Nanticoke	20.4	14.5	17.7	16.2	18.1
Thunder Bay	1.5	1.0	1.0	1.0	0.6
Atikokan	1.0	1.0	1.0	0.7	0.6
Lakeview	2.8	2.3	0.7	0.0	0.0
Total	36.3	26.4	29.7	24.7	28.2
% of actual Ontario annual energy	23.9	17.2	18.9	16.3	18.0

Note: The Lakeview station was shut down in 2005 and taken out of service.
Source: OPG, IESO.

needs in the absence of coal. Between 2008 and 2027, the OPA estimates capital expenditures of $3.6 billion (in 2007 dollars) on gas-fired generation.[20] This is in addition to approximately $2.7 billion invested in 2000 MW of gas-fired capacity ordered between 2005 and 2007 and approximately $140 million of transmission enhancements to ensure that adequate capacity and system reliability have been identified by the OPA.[21]

In its replacement plan, the OPA considered but ruled out converting the Nanticoke and Lambton facilities to gas-fired facilities. The conversion of Lambton to gas was ruled prohibitively expensive because of the cost of associated transmission enhancements. The Atikokan facility will potentially be converted to a biomass facility.

In aggregate, the IPSP foresees 17,244 MW of renewable, nuclear, conservation and demand management, and gas-fired resources being added to the province's supply mix by 2014, allowing for the replacement of coal-fired resources. It is expected that all these replacement resources will be acquired by the OPA through a combination of competitive and noncompetitive procurements. If all committed and planned resources come into service on time and are performing reliably, the OPA suggests that coal could be replaced by 2012. However, to mitigate the risks associated with new supply resources and to meet NPCC reserve planning requirements, the OPA finds that prudent planning practice suggests that coal is a valuable insurance resource

---

[20]OPA (2007a), Exhibit G, Tab 2, Schedule 1.
[21]OPA (2006).

that should not be completely phased out until 2014; however, the OPA also suggests that the phase-out could occur sooner, depending on system conditions.

The overall impact of the IPSP on a consumer's all-in electricity bill is expected to rise from \$88/MWh in 2006 to \$108/MWh in 2015 (in 2007 dollars).[22] However, the OPA states that consumers who take aggressive conservation and demand management actions will not experience higher electricity bills and, in fact, can lower their electricity bills.[23]

13.5 Stakeholder assessments of the phase-out strategy

As the government's coal-replacement strategy has evolved over time, so have the assessments of the plan by various stakeholder groups. The groups representing large power users and the labor unions have been highly critical of the policy. Interestingly, although environmental groups have been supportive of the policy, they have harshly criticized the supply portfolio recommended by the OPA to replace coal-fired generation.

The coal replacement program has been criticized most vigorously by large consumers of energy. In 2006, following the release of the OPA's *Supply Mix Advice Report*, the Association of Major Power Consumers in Ontario (AMPCO), whose membership comprises 40 of Ontario's largest companies that represent consumption of up to 15 percent of the province's electricity production, released a series of papers arguing that the costs of coal replacement would negatively impact economic activity through higher electricity prices.

Specifically, the AMPCO commissioned research[24] concluding that the coal shutdown would:

- Increase electricity rates 55 percent by 2025, reduce Ontario's real GDP by \$16 billion per year, and reduce annual employment by approximately 100,000 jobs by 2025

- Result in increases in coal-fired generation and corresponding emissions from neighboring jurisdictions supplying imported energy to Ontario

AMPCO argued that the installation of emission control technology on the coal units would be more cost effective.

---

[22]OPA (2007a), Exhibit G, Tab 2, Schedule 1, p. 30.
[23]Ibid. p. 31.
[24]See: ICF Consulting Canada Inc. (2005); The Centre for Spatial Economics (2006); Navigant Consulting Ltd. (2006).

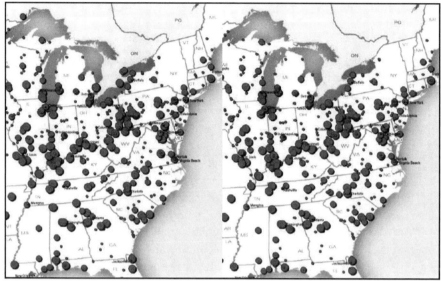

Coal-fired generation in Ontario airshed   Ontario plants removed

Figure 13.5   Coal-fired generation in Ontario's airshed.
Source: Society of Energy Professionals (2004) [29].

The two unions representing workers at Ontario Power Generation have
also been critical of the decision. The largest, the Power Workers Union
(PWU), argued that the cost of transmission enhancements alone to facilitate
coal replacement would exceed $3 billion.[25] The PWU launched a website
(www.abetterenergyplan.ca) and an advertising campaign in attempts to sway
public views on the issue. The other major union, the Society for Energy Professionals, argued that the coal replacement program will have minimal health
benefits, maintaining that the primary sources of air pollution and GHGs arise
from transborder pollution from the United States (Figure 13.5) and from
transportation emissions.[26] In support, they point to the government's own
statistics that show that Ontario's transportation emissions account for over
30 percent of Ontario's GHG emissions.[27] Further, they cite government analysis showing that transboundary air pollution from the United States causes
2751 premature deaths and an estimated $5.2 billion in damage to Ontario's

---

[25]Global Energy Advisors (2006).
[26]Society of Energy Professionals (2004).
[27]Ontario (2007b).

people, economy, and environment,[28] far in excess of the damages attributed to Ontario's coal-fired generation.[29]

Environmental groups have been generally supportive of the policy to eliminate coal-fired generation. Their position on the implementation of environmental control technologies (ECTs) on the existing plants is of particular interest. Though environmental groups have criticized the delays in the phase-out date, they have in general supported the government's decision not to require installation of additional ECTs, despite the short-term benefits to air quality and to public health. Instead, environmental groups described installation of such controls as a "band-aid" and apparently are concerned that such measures would lead to a postponement of final closure.

That said, environmental groups have remained critical about the resource choices to replace the coal-fired plants. For example, the Ontario Clean Air Alliance (OCAA) has argued that the OPA's plan to build 1350 MW of single-cycle gas turbines to replace coal is economically inefficient and also emits greenhouse gas at a higher rate than alternative technologies.[30] The OCAA argues that if OPA planned to more aggressively pursue conservation and demand management, renewable generation, followed by higher-efficiency gas-fired generation (i.e., combined heat and power and combined cycle generation), all coal plants could be phased out by 2010.

Further, other environmental groups that support the coal phase-out have criticized the continued reliance on nuclear power to meet Ontario's electricity needs. For example, the Pembina Institute argues that nuclear generation proposes numerous environmental risks related to the wastes created by uranium mining and refining as well as the storage of nuclear waste.[31] Moreover, they argue that the historical poor performance and cost overruns associated with Ontario's existing nuclear generation fleet make nuclear power a poor resource choice. Similar to the OCAA, Pembina recommends a more aggressive use of conservation and demand management as well as renewable resources to replace coal.

The result of the various lobbying efforts by stakeholder groups and the government's own communication strategy appears to have resulted in a fractured view in overall public perception of the plan. For example, the results of a February 2006 survey found that 47 percent of Ontarians supported the

---

[28]Ontario (2005b).

[29]Other critics of the policy have stated that the air quality in Ontario is better in the 2000s than it was in the 1970s and that the contribution of coal-fired power plants to pollution, smog, and mercury emissions is quite small. See McKitrick, Green, and Schwartz (2005).

[30]OCAA (2007).

[31]Pembina Institute (2006).

government coal shutdown decision, 39 percent opposed, and 14 percent were neutral; the remaining 10 percent of those polled were either unsure or refused to answer.[32]

13.6 Challenges with the phase-out strategy

The plan to eliminate coal-fired generation in Ontario has shown three distinct types of challenges:

- First, there are technical challenges to the elimination of this resource related to its importance in Ontario system reliability, both its adequacy and system security.

- The second challenge is cost; a significant amount of investment and time will be required before the phase-out can be completed.

- The final challenge relates to the net environmental impact, particularly in the context of an interconnected North American electricity system.

This section discusses each of these challenges in turn.

13.6.1 TECHNICAL CHALLENGES

The replacement of coal-fired generation in Ontario poses, first of all, a number of technical issues. These issues arise because of importance of the plants to the existing grid layout and to its operation. The existing grid in Ontario has evolved to accommodate a small number of large central generating stations. Nanticoke, at 4000 MW, is second only to the Bruce Nuclear complex (Table 13.5).

Not surprisingly, the high-voltage grid in Ontario is configured to accommodate such large stations. The deployment of replacement capacity in different locations will change the grid's topology. Grid planning has been necessary to ensure that the grid can accommodate the replacement capacity. It has also necessitated a number of upgrades to the grid. In particular, the role played by Nanticoke in supplying reactive power to the grid will require the use of static var compensators at the Nanticoke site after the units are closed.[33]

---

[32]SES Research (2006). In a random telephone survey of 500 Ontarians, the question asked by pollsters was: "The government of Ontario is in the process of closing down all of Ontario's coal burning power plants. Do you strongly support, somewhat support, neither support nor oppose, somewhat oppose, or strongly oppose this decision?"

[33]OPA (2007a), Exhibit E, Tab 2, Schedule 6, p. 3.

Table 13.5 Ontario's largest generating stations

Facility	Fuel	Installed Capacity (MW)
Bruce	Nuclear	5000
Nanticoke	Coal	3960
Darlington	Nuclear	3512
Pickering	Nuclear	3094
Sir Adam Beck	Hydroelectric	2287
Lennox	Natural gas	2120
Lambton	Coal	1976

Source: OPG.

A larger technical challenge exists because of the role played by the existing coal-fired generation in following Ontario system load. Ontario's generating mix relies heavily on nuclear power plants, which do not have load-following capability. Although there is significant hydropower, much of it is from generators with limited water storage capability, limiting the ability of these resources to follow load. The existing system is thus dependent on coal-fired resources to fill this role and manage the swings in electricity demand.

Coal-fired resources are expected to be replaced principally by gas-fired generation. Although these resources have a number of desirable flexibility characteristics, the relatively high minimum electric output of the gas-fired generation facilities implies less flexibility than the existing coal units. Coal replacement will also be occurring at a time when the share of wind, an intermittent generation resource, will be significantly increased, from about 470 MW in 2008 to over 3000 MW by 2015. This will further raise the requirement for generation with load-following capability.

13.6.2 DIRECT COST OF COAL REPLACEMENT

That the elimination of coal will result in the procurement of higher-cost resources is well established. For example, between May 2004 and April 2005, coal-fired generation set the wholesale price 49 percent of the time, at an average price of 4.48¢/kWh; hydroelectric set the price 35 percent of the time at 4.84¢/kWh; and gas and oil set the price 16 percent of the time at 7.42¢/kWh.[34] The removal

---

[34]IESO (2005).

of coal will put gas-fired generation on the margin more frequently, raising both electricity costs and wholesale market prices.

In addition, a substantial increase in renewable resource generation is also being procured with long-term contracts at prices higher than coal, since current prices do not include environmental costs such as CO_2. The higher costs of renewable electricity will be passed through to consumers as higher prices.

However, coal replacement is one of a number of factors that has driven substantial investment in both supply and conservation resources in Ontario. Along with resources required to meet load growth, much of the province's generation capacity must be replaced or refurbished over the next 20 years. Therefore, it is difficult to quantify the expenditures that are strictly related to coal replacement, and the IPSP does not provide such allocation of costs.

As a general estimate, however, it can be argued that much of gas-fired generation capacity that has been (and will be) procured and certain specific transmission expenditures are aimed at coal replacement. Furthermore, this replacement generation capacity has been (and will be) obtained through long-term contracts with generators, leaving the risks of capacity underutilization with electricity ratepayers. As of May 2008, the OPA had executed long-term contracts with 13 gas-fired generation projects, including combined heat and power projects, scheduled to begin operations between 2008 and 2010, totaling 4267 MW of capacity[35] and costing approximately \$4.5 billion.[36] In the IPSP, the OPA plans on procuring an additional 2786 MW of gas-fired capacity, scheduled to begin operations between 2011 and 2014[37] and costing about \$3 billion.[38] Last, the OPA has identified roughly \$140 million in coal-replacement transmission expenditures.[39]

13.6.3 Environmental Challenges and the Market Context

Closing coal-fired power plants is expected to reduce emissions that affect air quality and the concentration of greenhouse gases in the atmosphere. The Ontario government's initial reliance on the air-quality objective made the coal closure more urgent—particularly when the Ministry of Energy's own study suggested that such emissions were responsible for hundreds of deaths each year. When the aggressive deadline proved impractical, the primary objective

---

[35]North Side Energy, LLC. (2008).
[36]Authors' calculations based on Ministry of Energy news releases.
[37]North Side Energy, LLC. (2008).
[38]OPA (2007a). Exhibit G, Tab 2, Schedule 1, p. 4.
[39]OPA (2006).

was shifted to that of limiting greenhouse gas emissions. Although climate change is also seen as an urgent problem, it is widely perceived as a problem that will take many years to address successfully and consistently with the practical timeframe for the phase-out.

With hindsight, had a more realistic deadline for coal phase-out been determined at the outset, the installation of additional environmental control technologies (ECTs) would have been attractive as a complementary part of the coal phase-out strategy. The ECTs would have been available much more quickly than replacement capacity, and their operation would have resulted in significant environmental and health benefits to offset their cost.

The ultimate environmental impact and the ultimate cost to electricity rate-payers of the current policy are difficult to assess because the Ontario electricity market is interconnected with its neighboring provinces as well as with neighboring U.S. electricity markets. Much of Ontario's electricity trade occurs with New York and Michigan. In this regard, Ontario power production is competing with power generators in these states and with neighboring U.S. states such as Ohio and Indiana that rely heavily on coal for power generation. Currently, there are no restrictions on CO_2 emissions in most of these jurisdictions.

For this reason, because coal-fired generation is restricted and eliminated within Ontario, coal-fired generation in the neighboring jurisdictions can be expected to increase, both to replace exports of Ontario energy and through increased imports to Ontario. The increase in imports would be expected to mitigate price rises within the Ontario market by reducing gas-fired power production within Ontario but also, when these imports are from coal-fired generation, mitigate the global environmental benefit from the coal phase-out. The government of Ontario is relying on what it calls its "leadership position" on this issue to persuade neighboring jurisdictions to adopt a similar policy.[40]

13.7 Conclusion

This chapter has touched a number of particular complexities associated with the Ontario government's policy to eliminate coal-fired generation. Many of these details are specific to the Ontario context, such as its particular market and grid structure, the predominance of public ownership, and relative importance of central planning in determining the future electricity supply mix. At the same time, the specific and concrete nature of Ontario's strategy can be contrasted with those other jurisdictions pursuing carbon abatement, such

---

[40]Ontario (2007c).

as the United Kingdom's decision to invest in the development of carbon capture and storage technologies[41] (the Government of Ontario has ruled out investing in these technologies) and California's broader, though less tangible, market-based approach through making use of a cap-and-trade program.[42]

There are, however, at least three clear general lessons to other jurisdictions considering a similar policy:

1. *Set realistic deadlines that recognize the technical challenges associated with coal replacement.* The network properties of an electricity system, the lengthy process to approve projects for replacement capacity and transmission, and the different operational characteristics of alternative forms of generation mean that one must take a careful approach. A belated appreciation of these difficulties has led to several delays in the phase-out date. More realistic replacement dates would have also changed perception of the relative benefits of interim measures such as emissions controls that would have improved air quality and public health.

2. *Recognize that there are bound to be significant direct costs associated with replacement.* This includes both acquiring replacement generation and necessary transmission upgrades. Clearly, the more reliant a jurisdiction is on coal generation for supply, the greater the direct costs; for example, the cost of reducing reliance on coal-fired electricity generation in New Zealand[43] will be significantly lower than in Ontario. These costs and the resultant impact on prices pose significant challenges to policymakers since increases in electricity prices have met with strong political resistance in many jurisdictions over the past several years.

3. *Work with neighboring jurisdictions to ensure that the full environmental benefits of coal replacement are obtained.* A reduction of coal-fired generation in Ontario may lead to replacement of at least some of this production with coal-fired generation among its neighbors. When replacement generation has a higher fuel cost (i.e., natural gas), coal-fired generation from neighboring jurisdictions may be able to compete, limiting the global environmental benefits. This is particularly the case if there are no (or modest) restrictions on emissions in the interconnected jurisdictions.

Ontario's experience illustrates that the elimination of a large quantity of coal-fired generation from an electricity network is technically feasible.

---

[41]See Cornwall (2009).
[42]See Stern and Harris (2009).
[43]See Bertram (2009).

Moreover, the progress to date gives every indication that the province will achieve its 2014 off-coal deadline. However, the economic and environmental impacts of the policy are outcomes that will not be observable for several years. These impacts are difficult to judge because of the interconnected nature of power systems, where reduced production in one jurisdiction can be replaced by increased production elsewhere.

The effectiveness of Ontario's policy to phase out coal-fired generation and limit global greenhouse gas emissions will depend on the climate change policies pursued by Canada's federal government and the federal and state governments in the United States. These policies, to a large extent, remain ill defined, limiting the environmental value of Ontario's strategy. A coordinated and cooperative approach to climate change on both sides of the border and on a global basis is needed to maximize these benefits.

REFERENCES

[1] Bertram G. Kicking the Fossil-Fuel Habit: New Zealand's Ninety Percent Renewable Target for Electricity. In: Sioshansi F, editor. Electricity Generation in a Carbon Constrained World. Oxford: Elsevier; 2009.

[2] Canada. National Inventory Report: Greenhouse Gas Sources and Sink in Canada. 2008.

[3] Canadian Press. Ontario rejects pollution scrubbers. 2007.

[4] Centre for Spatial Economics. The Impact on the Ontario Economy of an Alternative Electricity Supply Plan. Report prepared for the Association of Major Power Consumers of Ontario; 2006.

[5] Cornwall N. 2009, Carrots and Sticks: Will the British Electricity Industry Measure Up to the Carbon Challenge? In: Sioshansi F, editor. Electricity Generation in a Carbon Constrained World. Oxford: Elsevier.

[6] DSS Management Consultants Inc. and RWDI Air Inc. Cost Benefit Analysis: Replacing Ontario's Coal-Fired Electricity Generation. Report prepared for the Ontario Ministry of Energy; 2005.

[7] Energy Information Administration. State Electricity Profiles. 2007.

[8] Global Energy Advisors. Transmission Analysis for Ontario: Transmission Enhancements for Off-Coal, Nuclear, and Renewable Plan. Report prepared for the Power Workers' Union; 2006.

[9] ICF Consulting Canada Inc. Final Report: Impacts of Government Energy Policy in Ontario. Report prepared for the Association of Major Power Consumers of Ontario; 2005.

[10] IESO. Your Road Map to Ontario Wholesale Electricity Prices. 2005.

[11] McKitrick R, Green K, Schwartz J. Pain Without Gain: Shutting Down Coal-Fired Power Plants Would Hurt Ontario. Fraser Institute; 2005.

[12] Navigant Consulting Ltd. Ontario Electricity Rates and Industrial Competitiveness. Report prepared for the Association of Major Power Consumers of Ontario; 2006.

[13] North Side Energy, LLC. Natural Gas-Fired Generation in the Integrated Power System Plan. Report prepared for the OPA; 2008.

[14] Ontario. McGuinty Government Unveils Bold Plan to Clean Up Ontario's Air. Ministry of Energy; 2005a.

[15] Ontario. Transboundary Air Pollution in Ontario. Ministry of the Environment; 2005b.

[16] Ontario. Supply Mix Directive. Ministry of Energy; 2006a.

[17] Ontario. McGuinty Government Delivers a Balanced Plan for Ontario's Electricity Future. Ministry of Energy; 2006b.

[18] Ontario. Ontario Greenhouse Gas Emission Targets: A Technical Brief. Ministry of the Environment; 2007a.

[19] Ontario. Go Green: Ontario's Action Plan on Climate Change. Ministry of the Environment; 2007b.

[20] Ontario. McGuinty Government's Approach to Climate Change Demonstrates Leadership. Ministry of the Environment; 2007c.

[21] Ontario. Moving Forward on Coal Replacement. Ministry of Energy; 2008.

[22] OCAA. The Ontario Power Authority's Coal Phase-Out Strategy: A Critical Review. 2007.

[23] OPA. Ontario's Integrated Power System Plan, Discussion Paper 7: Integrating the Elements— A Preliminary Plan. 2006.

[24] OPA. Ontario's Integrated Power System Plan and Procurement Process. 2007a.

[25] OPA. Emission Control Alternatives for Ontario Coal Generators. 2007b.

[26] Pembina Institute. Nuclear Power in Canada: An Examination of Risks, Impacts and Sustainability. 2006.

[27] Reg. 496/07, Environmental Protection Act (Ontario).

[28] SES Research. Coal Power Plants in Ontario. SES/Osprey Media Poll; 2006.

[29] Society of Energy Professionals. Energy Professionals on Coal Myths and Facts. 2004.

[30] Stern G, Harris F. 2009, California Dreaming: The Economics, Politics, and Mechanics of Meeting California's Carbon Mandate. In: Sioshansi F, editor. Electricity Generation in a Carbon Constrained World. Oxford: Elsevier.

[31] Trebilcock M, Hrab R. 2006, Electricity Restructuring in Canada. In: Sioshansi F, Pfaffenberger W, editors. Electricity Market Reform: An International Perspective. Oxford: Elsevier.

Kicking the Fossil-Fuel Habit: New Zealand's Ninety Percent Renewable Target for Electricity

Chapter 14

Geoffrey Bertram

School of Economics and Finance, Victoria University of Wellington, New Zealand

Doug Clover

School of Geography, Environment, and Earth Sciences, Victoria University of Wellington, New Zealand

Abstract

The New Zealand Government in 2007 set its sights on 90 percent renewable electricity by 2025, mainly via the expansion of large-scale, centrally dispatched geothermal and wind generation. The country's resource

Doi: 10.1016/B978-1-85617-655-2.00014-6.

endowments would make this transition feasible at low incremental cost relative to a business-as-usual trajectory, although the foreclosure of small-scale demand-side and distributed generation options by New Zealand's present electricity market design means that the new policy would mainly benefit the large incumbent generators. A renaissance of decentralized and demand-side energy solutions could potentially strand some of the new large-scale renewable projects as well as some legacy thermal capacity. New Zealand's resource endowment is unusually favorable for achieving a return to low-emission electricity generation without resorting to the nuclear option, compared to other countries covered in this book.

14.1 Background: NZ energy policy and its context

In October 2007 the New Zealand government declared that 90 percent of the country's electricity should be generated from renewable resources by 2025.[1] The policy measures announced to achieve this goal [32, 38, 39], and passed into law by Parliament in September 2008,[2] were the imposition of a carbon tax[3] on electricity generation provisionally beginning in 2010, and a 10-year restriction on construction of new baseload fossil-fueled electricity generation capacity "except where an exemption is appropriate (for example, to ensure security of supply)."[4] Shortly after passage of the legislation the government fell in the November 2008 general election, and both the Emissions Trading Scheme and the renewables target were put on hold by the incoming National Party administration.

The regulatory approach set out in the 2008 legislation required any new investment in thermal plant to secure an explicit exemption from the Minister of Energy and to carry the burden of an emissions tax on its operating costs. These measures fell well short of an outright ban, since future ministers would have political discretion at any time to invoke one of the numerous

---

[1] New Zealand Energy Strategy to 2050: Powering our future—Towards a sustainable low emissions energy system. Available at: www.med.govt.nz/upload/52164/nzes.pdf, p. 22. October 2007.

[2] Climate Change Response (Emissions Trading) Amendment Act 2008, No. 85; and Electricity (Renewable Preference) Amendment Act 2008, No. 86.

[3] Although described as an "emissions trading scheme," the New Zealand scheme is in fact a tax, with the tax rate determined by arbitrage with the world carbon market. See Bertram and Terry (2008), Chapter 4.

[4] Electricity (Renewable Preference) Amendment Act 2008, section 4, new s.62A of the Electricity Act 1992. The bill passed into law in September 2008.

loopholes built into the legislation[5] and allow a raft of new nonrenewable generation to be built, and the legislation itself could be repealed. Neither the emissions tax nor the requirement for new thermal plant to gain "exemption" have enjoyed bipartisan political support, which means that neither was entrenched.

New Zealand is nevertheless well endowed with resources to sustain increased renewables-based generation [44]. Over the next three decades New Zealand is likely to require 8000 MW of additional generation capacity (roughly a doubling of the existing total); against this, around 6500 MW of feasible large-scale (over 10 MW) renewables-based options have been identified with long-run marginal cost below NZ\$130/MWh (13 cents/kWh). Five thousand MW of this has cost below \$100/MWh. Building this 6500 MW of renewables as part of the 8000 MW expansion would raise the renewable share of capacity from its present 69 percent up to 75 percent. Achieving the 90 percent target would then require a further 15 percent shift in the makeup of the country's generation portfolio, with fossil-fired generation displaced by some combination of greater renewables penetration and changes in electricity demand.

The prospects of success seem good. On the supply side, technological progress is cutting the costs of wind, wave, and solar technologies, whereas fossil-fuel prices for electricity generators in New Zealand have been rising after four decades of access to cheap natural gas. The country's potential large-scale wind resource, including feasible projects costed at over \$130/MWh, is assessed at over 16,000 MW.[6]

On the demand side, including distributed small-scale generation, progress has been held back more by institutional barriers than by lack of options. The oligopolistic structure of the electricity market has effectively foreclosed entry by independent brokers and small generators; pro-competitive regulatory measures such as feed-in tariffs and net metering are yet to be introduced, two decades after market restructuring began. Over time these obstacles to technological progress and competitive entry are unlikely to be sustainable.

With relative prices and technological progress swinging the market balance in favor of renewables over the past 5 years, the dominant New Zealand generators have been racing to secure strategic footholds on key renewable

---

[5]The Electricity (Renewable Preference) Amendment Act 2008 s.4 automatically exempts all existing generation plants and allows new plants to be exempted by regulatory declaration. The new Electricity Act Section 62G allows exemptions to be granted for baseload plants that mitigate emergencies, provide reserve energy, supply isolated communities, function as cogeneration facilities, use a mix of renewable and fossil fuels or waste and fossil fuels, or replace existing plant with a more emission-efficient process.
[6]See Table 14.4.

resources by constructing large wind farms and geothermal plants.[7] As the next section describes, this move represents the reversal of a half-century-old trend away from renewables.

The chapter explores the feasibility of the renewables target and the 2008 policy framework. Section 14.2 sets out the record of New Zealand's 1970–2000 shift away from its historically high renewables share; Section 14.3 reviews some common issues with integrating renewables into an electricity system. Section 14.4 reflects on the achievement of 100 percent renewable electricity supply in Iceland and compares it with New Zealand, and Section 14.5 reviews the New Zealand government's modeling work on the future evolution of the generation portfolio in New Zealand and considers some implications of the supply-side bias built into New Zealand's electricity market design. Section 14.6 pulls together the main conclusions.

14.2 Historical development of the New Zealand system

14.2.1 THE RISE AND (RELATIVE) FALL OF HYDRO

Electricity reached New Zealand in the 1880s, when the country was still in its pioneering phase [31]. By the time of the First World War the country had a patchwork of local standalone supply systems and associated distribution networks, each with its own voltage and frequency standards. Starting in the 1920s an integrated supply network was established in each of the two main islands under government auspices, including the construction of large state-owned hydroelectric stations, which dominated supply by the mid-1960s.

Because of its mountainous topography, New Zealand was well endowed with opportunities to construct large-scale hydro. By the 1940s the share of fossil fuels in total capacity had fallen below 10 percent (Figure 14.2), with small oil-fired plants providing local peaking capacity and about 50 MW of coal-fired plants in Auckland and Wellington providing backup supply. Through the 1950s demand grew ahead of the pace of hydro construction and the gap was filled by investment in coal and geothermal plants (Figure 14.1). New hydro construction accelerated in the 1960s as a cable connecting the North and South Islands made possible the development of large hydro resources in the far south, to supply the northern market [43].

As Table 14.1 and Figure 14.1 show, the pace of hydro and geothermal construction slowed in the 1970s while that of fossil-fired thermal generation

---

[7]One would-be new entrant/entrepreneur is experimenting with very large-scale, subsea tidal generation: Neptune Power Ltd., "Response to the MED request for submissions to the Draft New Zealand Energy Strategy," March 2007, www.med.govt.nz/upload/47260/205.pdf; "Trial Approved for Strait Tidal Power," *Dominion Post* 2 May 2008, www.stuff.co.nz/4505727a11.html.

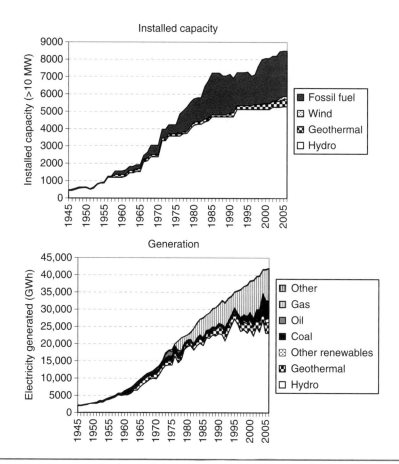

Figure 14.1 New Zealand electricity installed capacity and generation by fuel type, 1945–2006.
Source: 1945–1956 calculated from *Annual Reports* of the New Zealand Electricity Department, 1945–1956; 1956–1973 Ministry of Economic Development unpublished data; 1974–2006 from Energy Data File, June 2008, p. 100.

increased sharply. Over the two decades from 1965 to 1985 the fossil-fuel share of capacity rose from 11% to 33%. In 2004 it was still 32%.[8]

---

[8]As Bertram (2007), pp. 224–225, notes, the introduction of "commercial" incentives and behavior under the reforms of 1987–1992 led quickly to the decommissioning of reserve thermal capacity, which was costly to maintain but held prices down during the dry winter of 1992, thereby reducing generation profits. The demolition of this 620 MW of privately unprofitable plant temporarily cut the fossil-fuel share of capacity to 25 percent in the mid-1990s while sharply reducing the system's security margin and increasing the economy's exposure to blackouts in dry years.

Table 14.1   Fuel shares of NZ electricity generated for the grid, 1945–2007 (5-year averages, GWh per year)

	Geothermal	Hydro	Wind	Biomass	Coal and Oil	Gas	Cogen	Other	Total	Renewables (%)
1945–1949	0	2322	0	0	111	0	0	39	2472	94
1950–1954	0	3204	0	0	171	0	0	241	3616	89
1955–1959	27	4720	0	0	319	0	0	297	5363	89
1960–1964	716	6136	0	0	825	0	0	319	7995	86
1965–1969	1204	9240	0	0	770	0	0	346	11,561	90
1970–1974	1215	13,027	0	63	1986	42	0	299	16,632	86
1975–1979	1243	16,035	0	357	1498	2303	0	0	21,436	82
1980–1984	1194	19,300	0	403	613	3276	0	0	24,787	84
1985–1989	1314	21,633	0	442	697	5123	0	0	29,209	80
1990–1994	2176	23,067	1	488	758	6177	0	0	32,667	79
1995–1999	2244	24,791	17	494	1318	7227	0	0	36,089	76
2000–2004	2669	24,427	182	513	2598	9132	63	0	39,585	70
2005–2007	1896	13,901	431	415	2690	5849	26	0	25,208	66

Source: NZED, annual statistics in relation to electric power development and operation; Ministry of Economic Development Energy Data File, June 2008.

14.2.2 CHEAP GAS, RELATIVE COSTS, AND THE RISE OF NONRENEWABLES

New Zealand's transition from 90 percent renewable electricity in the early 1970s to 65 percent in 2006 (in terms of generation output) was a direct result of relative-cost trends. The availability of cheap natural gas from the giant offshore Maui field[9] and the rising cost of large hydro construction, as development of the most accessible and suitable river systems was completed and diminishing returns to hydro set in, produced a relative-price swing directly contrary to the international effect of the first two oil shocks.

New Zealand's Maui gasfield was developed under a long-term take-or-pay contract signed in 1973 with the government as buyer, at a delivered-gas price that was only incompletely inflation indexed. As a result, the real fuel cost of state-owned thermal generation fell steadily through the 1970s and 1980s (Figure 14.2). A fully indexed purchase-and-sale agreement between the Crown and ECNZ[10] was negotiated in 1989, but the fuel cost per kWh of generation continued to fall during the 1990s due to the rising efficiency of base-load thermal capacity and the scrapping of reserve thermal plant.

The oil shocks of 1973 and 1980 would probably have forced a reorientation back to renewables (especially geothermal) but for the fortuitous coincidence of major natural gas discoveries with no means of exporting the gas. The result was to delink thermal generation costs from world oil prices.

Figure 14.2 shows a sharp increase in fuel cost in the two years after the first oil shock in 1973, when existing thermal capacity was coal or oil fired, but over the following decade natural gas completely displaced oil and largely displaced coal, so that the second world oil price shock of 1979–1980 had no effect on the downward-trending fuel cost of generation. A large oil-fired plant at Marsden Point, which had accounted for over 6 percent of total supply in 1974, was downgraded to dry-year reserve status by 1980.[11]

Figure 14.3 shows the rapid post-1973 elimination of oil (and to a considerable extent, coal) from thermal electricity generation, a trend eventually reversed by a revival of coal use only from 2003 on as Maui output fell and the gas price rose.[12]

---

[9]Discovered 1969, onstream in 1979, peaked in 2001, now in decline.

[10]Electricity Corporation of New Zealand, the corporatized successor to NZED.

[11]The second major oil-fired plant at Marsden was completed in 1978 but never commissioned.

[12]New Zealand's coal reserves are large, and the lifetime cost of electricity from coal plants remains competitive in the absence of a carbon tax. However, the combination of the planned emissions trading scheme and 10-year moratorium on new baseload thermal plants will keep coal at the margin of the future electricity generation portfolio.

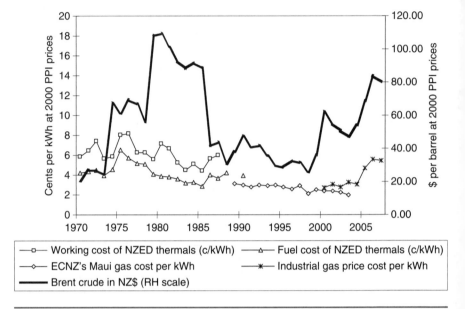

Figure 14.2 Real fuel cost of fossil-fired generation in New Zealand compared with world oil price trends, 1970–2005.
Source: Brent crude price from IMF, *International Financial Statistics*, converted to New Zealand dollars at current exchange rates and deflated by the New Zealand Producer Price Index (Inputs). NZED per-kWh fuel cost and thermal operating cost 1970–1991, calculated from NZED, *Annual Statistics in Relation to Electric Power Development and Operation*. ECNZ's fuel cost per kWh using Maui gas is the 1989 contract price of $2.225/GJ escalated to 2000 dollars using the PPI (Inputs), combined with thermal generation data from *Energy Data File*, June 2008, www.med.govt.nz/upload/59482/00_EDF-June2008.pdf, Table G2, p. 100, and gas used in generation from Ministry for the Environment, *Revised New Zealand Energy Greenhouse Gas Emissions 1990–2005*, December 2006, www.med.govt.nz/upload/38637/GHG%20report.pdf, Table 2.2.1, p. 33. Fuel cost per kWh 2000–2007 at the industry gas price: calculated using industry gas price from *Energy Data File*, June 2008, www.med.govt.nz/upload/59482/00_EDF-June2008.pdf, Table J4, p. 136; thermal generation data from *ibid.*, Table G2, p. 100; and gas used in generation from Ministry for the Environment, *Revised New Zealand Energy Greenhouse Gas Emissions 1990–2005*, Table 2.2.1, p. 33. PPI deflator from *Statistics New Zealand Long-Term Data Series*, www.stats.govt.nz/tables/ltds/ltds-prices.htm, Tables G3.1 and G3.2, updated 2004–2007 using the *INFOS* database.

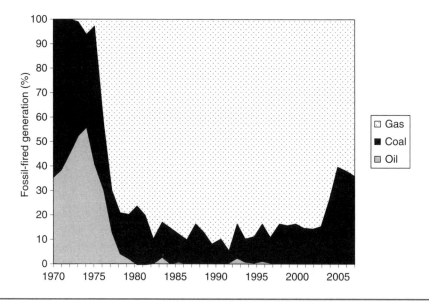

Figure 14.3   New Zealand's switch to gas in thermal generation, 1970–2005.
Source: NZED, annual statistics in relation to electric power development and
operation; Ministry of Economic Development Energy Data File, June 2008.

The switch to cheap gas and consequent rising reliance on fossil fuel, seen in
Figures 14.1–14.3, cannot be repeated today in the face of the rising world oil
price since 2003, because no new gasfield on the scale of Maui has been found
and because the emergence of a global LNG market means that the domestic
price of gas has become linked once again to the oil price.[13] In the coming two
decades, the cost of gas for New Zealand generators will move with (and to)
the world oil price, placing a squeeze on the profitability of thermal generation
relative to renewables. This squeeze will be exacerbated to the extent that a
carbon tax is actually imposed on thermal generation.

The change in the profitability of renewables relative to nonrenewables
since 2000 has been rapidly reflected in a surge of new investment in wind
and geothermal capacity. By October 2007, when the Labour government
announced its new strategy of aiming for 90 percent renewables and restrain-
ing construction of new thermal plants, market forces were already moving
strongly in that direction. Electricity sector modelers in the New Zealand

[13]New Zealand does not yet have any LNG terminal, but the world LNG price is already used
by the industry and the government as a pricing benchmark.

Electricity Commission and the Ministry of Economic Development estimated in late 2007 that a carbon tax of NZ$50/tonne[14] CO_2-equivalent would by itself make a 90 percent renewables share fully economic by 2030.[15]

14.3 Integrating renewables

With oil and gas prices trending upward and carbon taxes in prospect, fossil fuels will increasingly be confined to specialized roles in electricity generation. The two main ones in New Zealand are cogeneration (where electricity is a joint product from the burning of fuel for industrial process heat) and reliability support for the system: dry-year backup for hydro and reliable peaking capability to offset the intermittency of some renewable generation technologies. This section reviews the intermittency problem and some other issues with the displacement of fossil fuels by renewables.

14.3.1 INTERMITTENT RENEWABLES AND RELIABLE NONRENEWABLES

Primary energy sources are generally classified as renewable or nonrenewable on the basis of whether they draw on a depleting energy resource. Fossil fuel is nonrenewable, whereas hydro, wind, solar, and wave power are generally treated as renewable. On the borderline are nuclear,[16] which depletes its fuel stock but at a relatively slow rate, and geothermal energy (Williamson, Chapter 11 of this volume), which in most cases draws on an underground reservoir of heat sufficiently large to enable depletion to be ignored within the usual planning horizons for energy supply.[17] Here geothermal is treated as renewable. It is also a technology that is relatively benign in terms of carbon emissions—emissions are low, though not zero.

An important difference between renewables and nonrenewables is the degree of flexibility and controllability in the rate and timing of generation. A well-designed portfolio of fossil-fuel generating plants can be operated to

---

[14]Roughly US$30.

[15]Samuelson R, et al. Supplementary Data Files, "Emission Pricing on all Sectors," Figure 6b; 2007.

[16]Nuclear power is ruled out for New Zealand by a long-standing bipartisan political consensus.

[17]Note, however, the case of the geothermal project developed in New Zealand at Ohaaki, where a 104 MW plant was commissioned in 1989 but had been derated to 40 MW by 2005 due to unexpected depletion of the resource, accelerated by the cooling effect from reinjection of cooled fluids directly into the reservoir. See www.nzgeothermal.org.nz/geothermal_energy/electricity_generation .asp.

follow load with few constraints. Renewables-based generation, in contrast, is dependent on natural processes to supply the primary energy, which means that electricity systems with very high percentages of renewable generation must be designed with an eye to constraints that are outside the control of the system operator: wind and wave fluctuations, rainfall, the daily cycle of solar radiation, the regular but time-varying movement of tides. This intermittency must be offset in some way—by storage technologies that enable generation and consumption of electricity to be separated in real time, or by reliance on nonrenewable generators able to ramp up and down to fill gaps in renewable supply, or by a demand side that is able to respond in real time to price signals reflecting fluctuations in supply.

The operational difference between a fully renewable system and a fully nonrenewable one lies not in the baseload part of the spectrum but in the nature and extent of output variations in nonbaseload plants (Figure 14.4).

In a nonrenewables generation portfolio, the system operator is able to use peaking plant to follow load fluctuations, which means that the adequacy and reliability of supply are straightforwardly determined by human decisions on construction, maintenance, fuel procurement, and system dispatch. The "increasing variability" on the left side of Figure 14.4 is therefore a positive feature of the generation portfolio.

In a renewables portfolio, "increasing intermittency," on the right side of Figure 14.4, reflects output variations that are driven not by load following but by natural processes that are largely uncorrelated with demand peaks. The system operator therefore needs to have some controllable component of the overall system that can be called on to keep supply and demand continuously in balance. These issues are discussed in relation to wind power by Wiser and Hand (Chapter 9 in this volume).

Research by the New Zealand Electricity Commission [13, 14, 15, 16] suggests that there is no physical feasibility limit to integrating wind into the New Zealand system up to around 50 percent of total generation, but there are likely to be

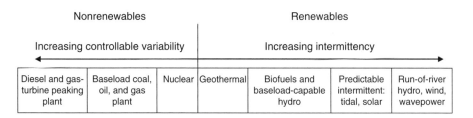

Figure 14.4   Schematic comparison of renewable and nonrenewable technologies.

rising costs of ancillary services to maintain reliability of supply, and these costs would be reflected in wholesale prices ([1], Section 7).

In most electricity systems, wind generation is treated as a nondispatchable source of variation in the residual demand faced by central generators [18, 19]. In contrast, New Zealand's large new wind farms are included in the system operator's central dispatch schedule on the basis of a 2-hour-ahead "persistence forecast" of their output and at a constrained must-run offer price of zero or NZ\$0.01/MWh ([1]; Electricity Governance Rule 3.6.33[18]). Dispatch is possible because virtually all the wind farms are owned by large generator-retailers with sufficiently diversified generation portfolios to allow intrafirm backup, usually from hydro, and because of the relatively high load factor of wind in New Zealand, generally 30–45 percent. The virtual absence of distributed wind generation, injecting power downstream of exit points from the grid, means that variability of residual load on the grid due to distributed wind has not yet been an issue in New Zealand.

In New Zealand, hydro generation has historically provided controllable variability. Hydro is a high-quality renewable, combining baseload and peaking capability, although it faces limitations imposed by New Zealand's rivers, which allow only limited storage and which are subject to minimum and maximum flow restrictions for environmental reasons. Development of hydro resources in New Zealand has, however, reached a mature stage, with few major rivers remaining undammed and rising costs of developing them for electricity—not only construction costs but also the rising opportunity value of wild and scenic rivers to the country's tourism industry, which is now the leading earner of foreign exchange.

The planned return to 90 percent renewables would therefore have to rely mainly on geothermal development combined with wind, wave, and tidal generation. To offset the intermittency of these last three technologies, a traditional solution would be to construct gas-fired or oil-fired peaking plant to cover for periods when demand is high and wind and wave are offline. However, if such new fossil-fired capacity is built and allowed to bid for dispatch, the market will be apt to "choose" a significant amount of electricity supply from these fossil-fired stations, which would rule out a 100 percent renewables system and could make even 90 percent problematic.

The combination of a commercially driven wholesale market for generation and a rising systemic requirement for backstop capacity that would operate for only part of the time raises issues of contract design and regulation that have not been resolved. In recent times a perceived shortfall of backstop capacity to

---

[18]The full set of Electricity Governance Rules is posted on the Web at www.electricitycommission .govt.nz/pdfs/rulesandregs/rules/rulespdf/complete-rules-5Jun08.pdf.

cover for dry years (when hydro generation is low) led to the government constructing a peaking station that is blocked by regulation from bidding into the market except at times of penal wholesale prices (over $200/MWh). If a rising renewables share is accompanied by increasing need for backstop reserve capacity, and if the backstop technology is fossil-fueled supply, restraining thermal generation below 10 percent of total generation is likely to require either a very high carbon tax or regulatory limits on the dispatch of thermal capacity once its construction cost has been sunk, or both. At this stage such policy issues have not been addressed, at least not publicly.

The problem of intermittency is obviously far less in an electricity system that is interconnected with other countries, as are the United Kingdom (with backup from the EU) and most states of the United States apart from Hawaii. In such cases, a target for the proportion of renewables in domestic generation may be met even when a substantial proportion of demand is served from externally located nonrenewables.

New Zealand, like Hawaii and Iceland, is an island system without interconnection to any other country, although the country's two main islands are interconnected and provide mutual support. Integrating intermittent renewables is in principle more challenging for island systems than for continental ones because of the lack of external backup. When the island market is small, it also suffers from inability to reap economies of scope and scale in maintaining reliability standards.

Much depends, of course, on precisely which mix of renewables is actually installed [23]. Diversification helps: a range of technologies spread over a range of locations can smooth out the consequences of intermittency at the level of the single generating unit [42]. Having wind farms dispersed across a wide geographical area should result in a more reliable flow of generation because wind speeds vary from place to place and fluctuations in wind speed are less likely to be correlated across widely dispersed sites. Intermittency patterns of wind, waves, tides, and rainfall can offset one another so that the probability of securing a reliable, hence easily dispatchable, flow of electricity rises as the number of interlinked technologies increases [22].

14.3.2 A MODEL OF THE TRADE-OFF

Conceptually, the intermittency problem can be captured by a diagram such as Figure 14.5. Here iso-reliability contours (indexed with 100 percent reliability as the initial target) are drawn sloping up on the assumption that as the share of renewables in the generation portfolio rises (horizontal axis), the cost of procuring the necessary capacity reserves to maintain any target level of reliability (vertical axis) rises at the margin (as is the case for, e.g., wind

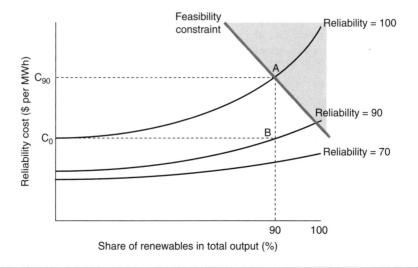

Figure 14.5 Framework for the integration of renewables into a hypothetical electricity system.
Source: Marconnet (2007), p. 78a [30].

penetration in the EU ([2], pp. 6–7). At Point A, to meet a 90 percent renewables target with 100 percent of target reliability, the cost C_{90} must be incurred, whereas the system with zero renewables is shown as having a full-reliability cost of C_0. The difference between these two represents the cost of moving toward more renewables without sacrificing quality of supply. Holding the electricity price at C_0 while pushing the renewables share up to 90 percent in this case would reduce reliability to R = 90 (Point B).

A hypothetical feasibility constraint is included in Figure 14.5 to take account of the possibility that, for a particular country, its resource endowment or particular characteristics of its electricity load may place some ceiling on the ability of the system to "buy" reliability as renewables increase their share. The position and slope of the constraint would be determined by both resource endowments and the state of technology. If it exists, the menu faced by policymakers seeking to maximize renewables subject to cost and feasibility constraints would be the set of corner solutions between the reliability contours and the feasibility constraint, including in this case Point A.

The position and slope of the contours in Figure 14.5 depend on the nature, diversity, and geographical dispersion of a country's renewable resources. An important modeling issue in the New Zealand case is the slope of these contours, which will dictate the long-run costs of moving to a high-renewables system relative to a status quo one.

The intermittency problem can be addressed on both demand and supply sides of the market. On the supply side, intermittency can be reduced greatly by technological progress in the design of wind and wave farms to render them more controllable and able to contribute directly to maintenance of frequency and voltage on the overall grid, and by installing substantial excess renewables capacity in diversified locations [24]. On the demand side, real-time pricing to final consumers and implementation of a range of energy-efficiency innovations can increase the flexibility of demand response to variable supply.

Two of the renewable supply technologies are not subject to intermittency: geothermal (Williamson, Chapter 11 in this volume) and hydro with storage. These are the key to the ability of Norway and Iceland to operate fully renewables-based generation portfolios, discussed in the next section.

14.4 Norway and Iceland as models

Within the OECD there are two very high-renewable electricity systems: Norway (99 percent renewables) and Iceland (100 percent). New Zealand ranks third behind these so long as nuclear is classified as nonrenewable (Figure 14.6).

Norway is not comparable with New Zealand since its hydro has massive storage capacity and is backed up by neighboring Sweden's large nuclear capacity, which gives Norway almost complete security of supply.

Iceland, however—an island system like New Zealand—is 100 percent renewable in terms of generation on the main island.[19] Iceland confronts no operational problems with integration of renewables, because its portfolio is dominated by two perfectly matched renewable technologies: hydro and geothermal. Geothermal provides reliable baseload and is fully dispatchable; hydro provides peaking capacity and is also dispatchable. In 2006 Iceland had five major geothermal plants producing 26 percent of total electricity consumption, while 0.1 percent came from fossil fuels and the remaining 73.4 percent was from hydro.[20]

Like New Zealand, Iceland embarked on large hydro construction in the 1920s and has ever since had a system based primarily on hydro. In the 1960s and 1970s, roughly 100 MW of oil-fired plant was built, bringing the total thermal capacity up to 125 MW, but following the oil shocks of the 1970s this capacity was stranded by a dramatic expansion of renewable capacity as part

---

[19]The offshore island of Grimsey has a diesel-powered generator.
[20]Geothermal Power in Iceland. Available at: http://en.wikipedia.org/wiki/Geothermal_power_in_Iceland, downloaded April 2008.

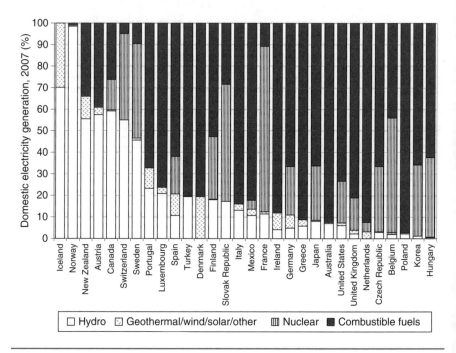

Figure 14.6 Electricity generation by primary energy source, OECD countries. Source: IEA, *Electricity Statistics,* www.iea.org/Textbase/stats/surveys/MES.XLS

of a policy of reducing dependence on oil and coal [21]. Between 1975 and 1985 installed hydro capacity doubled from 389 MW to 752 MW while geothermal capacity increased fifteen-fold, from 2.1 MW to 41.2 MW. After 1981 Iceland's fossil-fuel plants never supplied more than 9 GWh per year (around 0.1 percent of total supply), mainly to areas not connected to the grid. Geothermal now accounts for 25 percent of total installed capacity of 1698 MW, and hydro another 68 percent. The remaining 7 percent appears to be mainly residual thermal capacity, which provides a backstop for the system's reliability of supply and peaking ability but is hardly ever required.[21]

Table 14.2 gives comparative data for Iceland and New Zealand. Although with a population less than one tenth that of New Zealand, Iceland has per capita electricity generation more than three times as great. Both nations have over 60 percent of capacity accounted for by hydro, but Iceland's greater storage enables it to convert this to 73 percent of total supply, whereas New Zealand's hydro accounts for only 55 percent of supply.

---

[21]"Energy Statistics in Iceland," Orkustofnun (Iceland Energy Authority), www.statice.is/ Statistics/Manufacturing-and-energy/Energy

Table 14.2 New Zealand and Iceland compared, 1970 and 2006

	Iceland				New Zealand			
	1970	Share (%)	2006	Share (%)	1970	Share (%)	2006	Share (%)
Population (000)	204	—	304	—	2852	—	4173	—
Generation per capita, MWh	7.2	—	33	—	4.5	—	10	—
Electricity generated (GWh)	**1460**	—	**9925**	—	**12,926**	—	**42,056**	—
Hydro	1413	97	7289	73.4	9889	76.5	23,220	55
Geothermal	12	1	2631	26.5	1243	9.6	3210	8
Wind	0	0	0	0.0	0	0.0	617	2
Fossil fired	35	2	5	0.0	1471	11.4	14,322	34
Generation capacity, 2006 (MW)	**334**	—	**1698**	—	**3040**	—	**8517**	—
Hydro	244	17	1163	12	2373	18.4	5283	13
Geothermal	2.6	0.2	422	4	157	1.2	435	1
Wind	0	0	0	0	0	0	171	0.4
Fossil fired	88	6	113	1	510	3.9	2628	6

Source: Iceland data from Statistics Iceland Webpage, www.statice.is. New Zealand from Ministry of Economics Development, Energy Data File, www.med.govt.nz/templates/StandardSummary___15169.aspx, and population from Statistics New Zealand, www.stats.govt.nz/tables/ltds/default.htm.

The two obvious contrasts between the two countries are their different reactions to the 1970s oil shocks and the extent to which they have developed their geothermal resources. Looking at the historical evolution of the New Zealand generation portfolio (Figure 14.1), geothermal development stalled after the 1950s, despite the existence of a large-scale resource, and its share of total supply fell from around 12 percent in the mid-1960s to only 4 percent by 1990 (see Figure 14.7). Although New Zealand pioneered geothermal generation in the 1950s, the technology fell back to below 5 percent of capacity after 1970, whereas in Iceland, where the first geothermal plant appeared only in the 1970s, geothermal rose rapidly to a quarter of total generation capacity by 2006.

Confronted with the oil shocks of the 1970s, both countries delinked their electricity supply systems from world oil prices, but they did so by very different means. Iceland, whose thermal generation relied entirely on imported oil, delinked by building enough new hydro and geothermal capacity to effectively eliminate fossil fuels from its generation mix by 1983. New Zealand, as outlined earlier, delinked by switching to locally produced natural gas via a large-scale thermal generation construction program that raised the nonrenewables share of capacity to about one third by 2006 (Figure 14.8).

Iceland's strategy of delinking from oil prices by eliminating fossil fuels from its electricity sector means it now has a permanent buffer against volatile oil

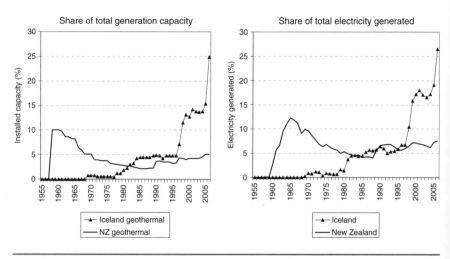

Figure 14.7   Geothermal shares of capacity and generation, New Zealand and Iceland, 1955–2005.
Source: Iceland, from Statistics Iceland website, www.statice.is/Statistics/ Manufacturing-and-energy/Energy; New Zealand, from Ministry of Economic Development *Energy Data File.*

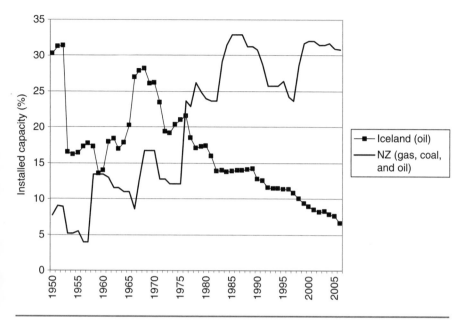

Figure 14.8   Nonrenewables share of installed capacity, 1950–2005.
Source: Iceland, from Statistics Iceland website, www.statice.is/Statistics/
Manufacturing-and-energy/Energy; New Zealand, from Ministry of Economic
Development Energy Data File.

markets, whereas New Zealand's strategy of a switch to cheap gas was effective
only so long as the Maui Contract dictated the local gas price. New Zealand is
now in the process of embarking on the Icelandic path, 40 years later.

14.5 Modeling the future NZ portfolio

Whether moving to 90 percent renewables is feasible for New Zealand at
acceptable cost is an issue best addressed by systematic modeling. This section
reviews recent work on the future evolution of electricity generation in New
Zealand under a variety of assumptions about policies and prices.

Since 2000 the New Zealand Ministry of Economic Development has conducted several rounds of scenario work using its SADEM model [33, 34, 35].
In addition, the Parliamentary Commissioner for the Environment has produced a scenario study focusing on renewables, distributed generation, and
demand-side response [41, 45], and Greenpeace has carried out a less formal
study [20] as part of a worldwide modeling exercise [5]. The leader in the field
at present is the Electricity Commission, the new sector regulator set up in
2003 (for background, see [6], p. 232).

14.5.1 THE ELECTRICITY COMMISSION'S GEM MODEL

The Electricity Commission has developed a Generation Expansion Model (GEM) to simulate alternative scenarios for the generation portfolio and select the most cost-effective one [8, 9]. The GEM determines the optimal commissioning dates of new generation plants and transmission equipment in response to an exogenously imposed forecast of demand for electricity. The GEM also simulates the optimal dispatch of both existing and new plants.

The model's objective function is to build and/or dispatch plants in a manner that minimizes total system costs while satisfying a number of constraints. The main constraints are to:[22]

- Satisfy a fixed load in each load block of each time period within each year
- Satisfy peak-load security constraints
- Provide the specified reserves cover
- Account for both capital costs incurred when building new plants and fixed and variable operating costs of built plants, including any specified carbon charge on the use of CO_2-emitting fuels
- Satisfy energy constraints arising from the limited availability of hydro inflows
- Satisfy HVDC constraints[23]

Underlying the "generation scenarios" part of the model [10, 11] is a database of possible new generation options, their associated capital and fuel costs, plant performance, depreciation, and load factors, based on Parsons Brinckerhoff Associates findings [40] and subsequent updates. The model also requires estimates of future hydro flows, the cost of carbon, and forecast loads during the various load blocks.[24] These technical supply-side data appear in

---

[22]This list is from the Electricity Commission's programmers' notes within the main GAMs batch file.

[23]$HVDC$ refers to the high-voltage direct current link between the two main islands of New Zealand.

[24]The load blocks used by the commission are:

b0n   A no-wind peak spike
b0w   A windy peak spike
b1n   A peaky no-wind block
b1w   A peaky windy block
b2n   A shoulder no-wind block
b2w   A shoulder windy block
b3   A mid-order block
b4   An off-peak shoulder block
b5   An off-peak block

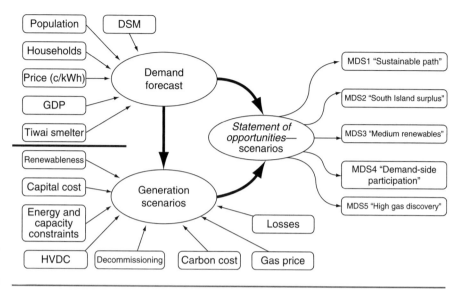

Figure 14.9 Schematic representation of the Electricity Commission's modeling. Source: Adapted from Hume and Archer, Figure 14.2, p. 2 [26].

the lower-left part of Figure 14.9 as inputs to the least-cost generation scenarios.

The other key input, also shown in Figure 14.9, is the demand forecast, which is based on modeling of three sectors—residential, commercial, and industrial—and "heavy industry" (the Tiwai Point aluminium smelter, which accounts for 17 percent of national load). Forecasts are done at both national and regional levels [27].

The national-level modeling of residential and commercial/industrial demand uses regression analysis, with GDP/capita, number of households, and electricity price as the explanatory variables. The commercial and industrial model has only two variables: GDP and "shortage."[25] Demand from heavy industry is assumed to be constant, unless the GEM scenario involves closure of the aluminum smelter. The forecasts currently assume that future rates of improvement in energy efficiency are the same as historical rates, with no feedback to the "DSM" input box in Figure 14.9.

Regional-level load forecasts cannot be undertaken with econometric methods due to lack of historical data. Therefore, the model's regional forecasts are based

---

[25]The shortage variable is a dummy that removes from the regression results years in which "shortages" have occurred. This is done to ensure that demand is not biased downward due to extraordinary circumstances; see Electricity Commission (2004).

on an allocation of national demand, using regional population forecasts for residential demand and regional GDP growth for commercial and industrial.

The forecasts are subjected to Monte Carlo analysis to provide an estimate of the forecast error, and before the figures are incorporated into the GEM they are passed through the Commission's hydrothermal dispatch model to estimate electricity demand per year, month, and island and to divide the load into blocks.[26]

With demand and generation opportunities thus exogenously determined,[27] the GEM uses programming techniques to design a least-cost generation portfolio to meet that demand. The model does not incorporate risk/return tradeoffs of the sort pioneered by Awerbuch [3] and Awerbuch and Berger [4], and it does not include in its output a future wholesale price path for each scenario, although such a path is implicit. Although the GEM does not calculate wholesale electricity prices, the Commission does use the model outputs to estimate the price levels necessary to achieve life-cycle revenue adequacy for the marginal generator(s) in each generation scenario. This does not, however, feed back to the demand block in Figure 14.9.

Figure 14.10 compares the Commission's demand forecasts with those of other modelers. Over the period to about 2040, the Commission's central projection is for demand to grow by 50–60 percent, an increase of 20,000–25,000 GWh over current annual generation. The projected annual growth rate of around 1.2 percent reflects linkage to expected GDP growth but with a steady exogenous improvement in efficiency. There are very wide uncertainty bands around this demand projection. At the lower end, both Webb and Clover [45] and MED [34] have estimated that major innovations on the demand side (high uptake of energy efficiency and distributed generation) could reduce required cumulative grid-connected generation growth to less than 40 percent. At the top end comes the high-demand scenario [45], in which increased electricity intensity of the economy drives projected demand up 70 percent over the three and a half decades.

---

[26]The EC uses PSR Inc.'s SDDP software package for this task (www.psr-inc.com.br/sddp.asp). The package is designed to calculate the least-cost stochastic operating policy of a hydrothermal system, taking into account the following aspects:
Operational details of hydro plants
Detailed thermal plant modeling
Representation of spot markets and supply contracts
Hydrological uncertainty
Transmission network performance
Load variation
[27]The scenario headed "demand-side participation" in Table 14.3 is based on ad hoc exogenous adjustments to the projected demand path rather than endogenous feedback from price within the model.

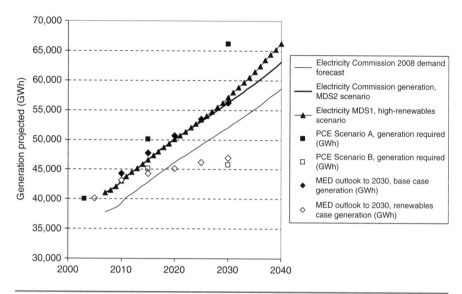

Figure 14.10   Projections of electricity demand and generation, 2000–2040. Source: Electricity Commission projections from the February 2008 demand forecast and June 2008 generation scenarios; MED scenarios from supporting data to Ministry of Economic Development [34]; Parliamentary Commissioner (PCE) projections from Webb and Clover [45].

The Electricity Commission's projected need for generation reaches 55,000 GWh by 2030 and 63,000 GWh by 2040, with the higher figure applying if there is a shift toward electricity away from other fuels (due, for example, to electrification of the transport vehicle fleet). Greenpeace ([20], p. 34, Figure 14.13, and p. 62, Appendix 2) similarly projects 59,000 GWh in 2040. Generation in Figure 14.10 must run above projected demand to allow for line losses and system constraints.

The least-cost capacity and generation to meet demand under the scenarios currently modeled by the Electricity Commission are summarized in Table 14.3 on the basis of results published in mid-2008 ([17], Chapter 6). The scenarios cover a range from the high-renewables "sustainable path" MDS1 to a low-renewables "high gas discovery" case, MDS5. Over the period to 2040, the renewables share exhibits a low of 61 percent and a high of 88 percent. This range reflects, at the low end, a minimum-renewables constraint imposed by already installed hydro, geothermal, and wind capacity and, at the high end, the need to allow for cogeneration and least-cost (thermal) backup supply. No scenario to date has incorporated the 90 percent renewables goal as a binding constraint, but it is clear that there are sharply rising costs to the system of driving fossil-fired generation below 10 percent of the total.

Table 14.3 NZ Electricity Commission scenarios, June 2008

		MDS1: Sustainable Path	MDS2: South Island Surplus	MDS3: Medium Renewables	MDS4: Demand-Side Participation	MDS5: High Gas Discovery
2007	Installed capacity (MW)	8553	8553	8553	8553	8553
	Modeled generation GWh	41,079	41,069	43,067	41,075	43,074
2025	Total MW	12,488	12,481	10,899	10,934	10,934
	Portion of which is renewable (MW)	9935	9161	7317	7164	7084
	Renewable share (%)	*79.6*	*73.4*	*67.1*	*65.5*	*64.8*
	Total GWh	53,393	53,133	51,513	53,288	55,051
	Portion of which is renewable (GWh)	46,832	42,729	37,496	35,868	35,737
	Renewable share (%)	*87.7*	*80.4*	*72.8*	*67.3*	*64.9*

2030	Total MW	13,532	13,286	11,239	11,916	11,459
	Portion of which is renewable (MW)	10,899	9676	7692	7244	7285
	Renewable share (%)	*80.5*	*72.8*	*68.4*	*60.8*	*63.6*
	Total GWh	57,147	56,187	53,035	56,991	58,103
	Portion of which is renewable (GWh)	50,239	44,705	38,349	34,957	37,566
	Renewable share (%)	*87.9*	*79.6*	*72.3*	*61.3*	*64.7*
2040	Total MW	15,988	14,328	12,559	13,081	13,247
	Portion of which is renewable (MW)	12,500	9676	8467	8209	7855
	Renewable share (%)	*78.2*	*67.5*	*67.4*	*62.8*	*59.3*
	Total GWh	66,223	63,066	59,917	65,826	65,029
	Portion of which is renewable (GWh)	55,662	45,106	42,116	39,875	39,854
	Renewable share (%)	*84.1*	*71.5*	*70.3*	*60.6*	*61.3*

Source: Draft Statement of Opportunities, background tables downloaded from www.electricitycommission.govt.nz/opdev/transmis/soo/08gen-scenarios#generation-scenario-outlines [17].

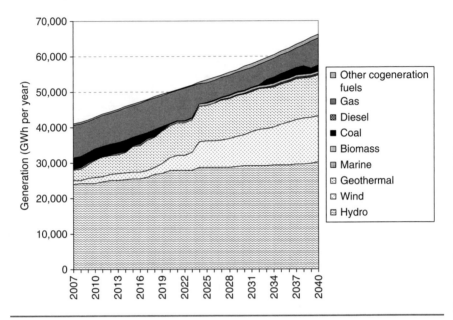

Figure 14.11   Generation by fuel, Electricity Commission scenario MDS1, 2007–2040.
Source: *Draft Statement of Opportunities*, background tables downloaded from www.electricitycommission.govt.nz/opdev/transmis/soo/08gen-scenarios# generation-scenario-outlines [17].

Figure 14.11 shows details of the Commission scenario that comes closest to the 90 percent target, namely scenario MDS1, Sustainable Path.[28] In this scenario the rapid expansion of wind and geothermal generation outpaces demand growth until the mid-2020s, when the renewables share reaches 88 percent. Renewables growth then slows while demand continues to rise, bringing coal back into the picture and reducing the renewables share back to 84 percent by 2040.

Inspection of the Commission's results highlights the importance of changes in, and the definition of, the denominator in calculating a "renewables share." Demand for electricity is affected by the same policy and relative-price forces

---

[28]The scenario "storybook" runs as follows: "New Zealand embarks on a path of sustainable electricity development and sector emissions reduction. Major existing thermal power stations close down and are replaced by renewable generation, including hydro, wind, and geothermal backed by thermal peakers for security of supply. Electric vehicle uptake is relatively rapid after 2020. New energy sources are brought onstream in the late 2020s and 2030s, including biomass, marine, and carbon capture and storage (CCS). Demand-side response (details not specified) helps to manage peak demand."

as those that drive the changing generation portfolio. Scenario MDS1 actually has higher demand in 2040 than the other scenarios in Table 14.2, partly because of the assumed shift to electric vehicles in the transport sector, with no change in the baseline energy-efficiency trend. In contrast, the High Gas Discovery scenario has lower electricity demand because of substitution of direct gas use for electricity. This simultaneous impact of modelers' assumptions on demand and supply makes 90 percent renewables a moving target. Unhelpfully vague specification of the target by the government to date has left this ambiguity unresolved.

14.5.2 Ministry of Economic Development Modeling Work

The Electricity Commission's published results do not enable construction of renewables/price reliability contours along the lines of Figure 14.5, but work by the Ministry of Economic Development [34] has produced wholesale price estimates for a range of 14 supply/demand scenarios out to 2030, with solutions at 5-year intervals. These scenarios were designed to test a range of alternative assumptions about technological progress, feasibility of adopting identified renewable resources for electricity generation, and adoption of energy-efficiency measures on the demand side of the market.

Figure 14.12 (with the same axes as Figure 14.5) plots the wholesale price of electricity in each of the 14 scenarios against the proportion of renewables in

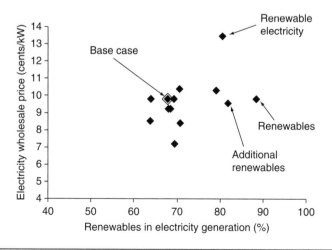

Figure 14.12   Renewables share and wholesale price: MED 2006 scenarios at 2025 [34].
Source: Calculated from Ministry of Economic Development [34].

total electricity generated. The business-as-usual base case has only 69 percent renewables in 2025, with a wholesale price of 9.8¢/kWh. Three of the alternative scenarios reach a renewables share of over 80 percent, and one has a share of 88 percent, with a price equal to the base case. If the points in Figure 14.12 are thought of as indicating where the cost/renewables contours run for New Zealand, then apart from one conspicuous outlier they suggest a remarkably flat curve up to the vicinity of 90 percent renewables. (The Ministry's modeling, however, may not fully incorporate the external cost of the ancillary backup services required to integrate a large volume of renewable generation into grid supply.)

Three of the MED scenarios in Figure 14.12 achieve over 80 percent of electricity generated from renewables: In Figure 14.12 they are labeled Renewables, Renewable Electricity, and Additional Renewable Electricity ([34], pp. 130–131, 99–102, and 102–104, respectively). The first and third of these have the same wholesale price as the 69 percent renewable base case, which seems to hint at opportunities to shift the generation portfolio toward 90 percent renewables by 2025, with little or no consequent increase in the wholesale electricity price—the renewability/price contours appear to be flat or only shallowly sloped across these scenarios.

The prominent high-cost outlier Renewable Electricity in Figure 14.12 is not a like-with-like comparison relative to the other observations and has to be interpreted with care. For this scenario, the modelers assumed that policymakers intervene directly to reduce the use of fossil fuels in electricity, with no action in other energy sectors—an approach similar in some respects to the now abandoned legislated moratorium. Under this assumption, no new coal-fired plant is built, the sole existing coal-fired plant is closed in 2014, and no new gas-fired plant is built, although existing gas-fired generation remains in operation. A steep rise in wholesale price is then required to bring in large volumes of new high-cost hydro and wind generation, and some high-cost geothermal,[29] to meet unrestrained demand growth. This scenario certainly raises the renewables share of generation but at relatively high cost.

The lower-cost Additional Renewable Electricity scenario assumes relaxation of planning and land-use constraints on the exploitation of renewable resources, allowing the model to build a large amount of moderate-cost

---

[29]The treatment of geothermal in the MED scenarios is problematic, since it is given no credit for its ability to provide reliable baseload. Instead, the modelers assumed that it would be crowded out of the dispatch order for much of the time by must-run hydro and wind, on the basis that the latter have lower short-run marginal costs (MED 2006, p. 100). In fact, it is likely that geothermal would be bid in at a zero offer price designed to undercut wind and hydro.

renewable generation that would (in the modelers' judgment) otherwise be ruled out by collateral damage to the environment. The renewables share then rises by 12 percentage points relative to the base case, with effectively no increase in wholesale price (marginal cost). Relative to the Renewable Electricity case, the model results suggest that overcoming resource consent hurdles could bring the wholesale price down by a full 4¢/kWh at the 2025 horizon, a reduction of 28 percent. Since the New Zealand government has a reserve power under planning law to "call in" selected projects seeking planning consent, there exists a straightforward policy instrument that could effectively eliminate the financial cost of a drive to renewables if the MED scenarios are taken as accurate.

The results from the Renewables scenario highlight the shortcomings of any policy that is limited simply to banning new fossil-fuel generation in electricity or overriding commercial merit-order dispatch, with no supportive price-based measures to promote renewables and energy-efficiency economywide, incentivize demand-side savings and response, and place prices on environmental externalities. In this third scenario, the MED modelers assumed that resource consents remain constrained as in the Renewable Electricity scenario, but they allowed for exogenous energy-efficiency improvements on the demand side and the installation of 750 MW of marine wave-power generation by 2025 at a cost of 10.2¢/kWh. The results are dramatic: Energy-efficiency gains reduce the amount of generation required in 2025 by over 7000 GWh (13 percent) so that even though total renewables generation is 2000–3000 GWh lower than in the other two renewable scenarios, the reduced demand enables fossil fuels to be squeezed to the margin of supply while keeping the wholesale price down, equal to the business-as-usual base case.

The demand side of the market thus emerges as crucial to securing a swing toward 90 percent renewables at low cost without sacrificing the competing environmental and social values protected by the planning laws. Even with demand reductions, however, the 2006 MED results suggested that costs turn up sharply at around 90 percent renewables, with an incompressible residual tranche of fossil-fired capacity.

In 2007, MED and the Electricity Commission combined their models to evaluate a further set of policy scenarios designed to nudge the economy toward renewables [36]. Options explored included carbon taxes ranging from $15–50/tonne, outright bans on fossil-fuel generation, and subsidies to renewables funded from consumers or from general taxation. Again, no scenario reached the 90 percent target (the highest was 88 percent). The 14 scenarios are plotted in Figure 14.13 in ascending order of wholesale electricity price. The height of each bar corresponds to the amount of generation required from grid-connected generation in each case. The Improved Energy

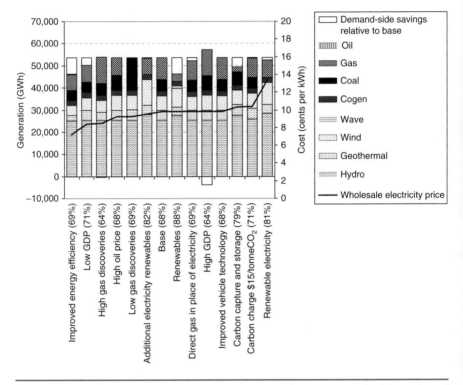

Figure 14.13 MED 2007 generation scenarios for 2025, ranked in order of wholesale electricity price.
Source: Calculated from Ministry of Economic Development [36].

Efficiency case at the left side of the diagram has both lowest generation and lowest price. The potential importance of demand-side savings in holding down the cost of a renewables-focused policy is clearly apparent, but this finding has not been picked up in the 2008 Electricity Commission work discussed earlier.

Figure 14.14 uses the results from MED [35] to indicate the location of the renewability/price contours. Renewables shares of generation ranging from 75 percent to nearly 90 percent turn out to be compatible with a wholesale electricity price only slightly above the 68 percent renewable base case. At the high-renewables end of the range, the difference between a scenario that achieves 88 percent renewable generation by subsidies to renewables and one that achieves the same target by a $50/tonne emissions charge on generators (Points A and B, respectively, in Figure 14.14) is 1.2¢/kWh, implying that the level of subsidy required to meet a 90 percent target could be less than 10 percent of the wholesale price.

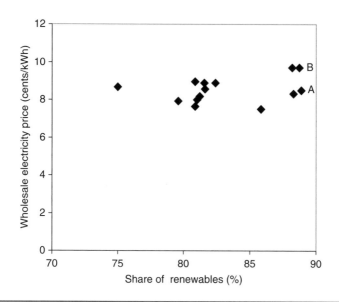

Figure 14.14   Renewables share and wholesale price, MED 2007 scenarios at 2025.
Source: Calculated from supporting data tables to MED 2007, downloaded from www.med.govt.nz/templates/MultipageDocumentTOC____31983.aspx [36]

In short, the evidence from recent modeling studies points to a nearly flat supply curve of renewable generation for New Zealand up very close to 90 percent. This in turn means that implementation of price-based instruments such as a carbon tax should be expected to elicit a high-elasticity response from the electricity supply side in terms of the composition of new investment, bringing the 90 percent target within easy reach.

14.5.3 THE LONG-RUN RENEWABLES SUPPLY CURVE

The Electricity Commission's preparation of its generation opportunities database turned up an unexpected wealth of opportunities—especially in wind resources, which are potentially in unlimited supply relative to national demand.[30] The Commission has identified new renewable projects totaling over 6400 MW at a long-run marginal cost of NZ$130/MWh or less, plus a further 13,000-plus MW of renewables that are either somewhat higher cost or cost-competitive but subject to other constraints in early development (see Table 14.4).

---

[30]Details of the database and the model are posted on the Commission website at www .electricitycommission.govt.nz/opdev/transmis/soo/08gen-scenarios/?searchterm = TTER and www.electricitycommission.govt.nz/opdev/transmis/soo

Table 14.4 Scope of feasible renewable projects

$/MWh	80	80–85	85	85–100	95	90–115	100–120	125	130	Total	Other	Total potential
Geothermal	250–300	–	400	–	–	–	–	–	–	650–700	56–106	756
Wind	–	800	–	–	3000	–	–	–	–	3800	12,590	16,390
Hydro	–	–	200	200	–	600	400	–	–	1400	537	1,937
Biomass cogen	–	–	–	–	–	–	–	–	150	150	–	150
Marine	–	–	–	–	–	–	–	400	–	400	–	300
Total MW	250–300	800	600	200	3000	600	400	400	150	6400–6450	13,183–13,233	19,533

Source: Electricity Commission ([16], pp. 65–80), and ([17], pp. 93–95); SSG, 2008.

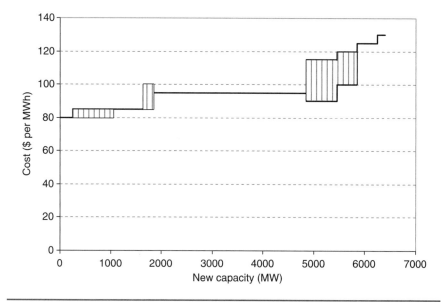

Figure 14.15   Estimated renewables supply curve from Electricity Commission database.
Source: Electricity Commission ([16], pp. 65–80), and ([17], pp. 93–95); SSG, 2008.

Figure 14.15 constructs an approximate supply curve from this data for the 6400 MW that has been provisionally costed by the Commission. Six thousand MW of new renewable capacity is estimated to have life-cycle (long-run) costs of NZ$120/MWh or less; 5000 MW of this is costed below $100/MWh. With very large volumes of wind potential still uncosted, the renewability supply curve appears likely to continue to flatten in the future.

Gas and coal plants are estimated to have long-run marginal costs competitive with most of the renewables in Figure 14.15 only if gas is priced at $7/GJ (below the LNG benchmark) and if there is no carbon charge. A carbon charge of NZ$30/tonne CO_2 would push thermal generation to or above the top of the range in the chart, making the full 6400 MW of listed renewable generation competitive on cost and relegating thermal to a support role as peaking plant and dry-year backup.

The conclusion is that New Zealand has sufficient hydro, geothermal, and wind resources to bring the 90 percent renewables target within easy reach at little if any cost penalty relative to fossil fuels, once carbon-emission externalities are priced in. The problem to be confronted in reaching the 90 percent target will not, therefore, be limited resource endowment. Rather, it will be institutional barriers and the overhang of legacy thermal capacity.

There is an apparently incompressible slice of nonrenewable generation associated with cogeneration, sunk-cost existing capacity, and reliability constraints in the absence of a responsive demand side, which does not fall below 10 percent in any of the Electricity Commission's scenarios to date. Ironically, across-the-board gains in energy efficiency and consequently lower demand growth could make it more, rather than less, difficult to achieve 90 percent renewables, because reduced need for new large-scale generation plants to meet demand growth means a larger share of legacy plants in the portfolio. To pursue the 90 percent target with radically reduced demand growth, policymakers would have to force the decommissioning of existing thermal capacity.

14.6 Evaluating the current policy

14.6.1 SUPPLY-SIDE BIAS?

The Electricity Commission is charged "to ensure electricity is produced and delivered to all consumers in an efficient, fair, reliable and environmentally sustainable manner," subject to a government policy that states:

> ... *[e]lectricity efficiency and demand-side management help reduce demand for electricity, thereby reducing pressure on prices, scarce resources and the environment. The Commission should ensure that it gives full consideration to the contribution of the demand side as well as the supply side in meeting the Government's electricity objectives.* ...[31]

The Commission has in practice been almost entirely preoccupied with the supply side (large-scale remote generators connected to the transmission grid), and this has strongly colored its modeling work. A recent Commission discussion of "nontransmission alternatives" ([16], pp. 39–41) contains no mention of downstream and demand-side options that might relieve grid constraints or strand grid-connected generators. This is particularly significant given the explicit instructions to the Commission in the most recent Government Policy Statement that modeling work should "enable identification of potential opportunities for ... transmission alternatives (notably investment in local generation, demand-side management and distribution network augmentation)."[32]

---

[31]Government Policy Statement on Electricity Governance, updated to May 2008, www.med.govt.nz/templates/MultipageDocumentPage___37639.aspx, paragraph 34.
[32]Government Policy Statement on Electricity Governance, updated to May 2008, www.med.govt.nz/templates/MultipageDocumentPage___37639.aspx, paragraph 89.

The dominance of incumbent-generator concerns in the work of the Electricity Commission has been reinforced by the reluctance of the New Zealand government to tackle barriers to entry facing distributed generation and decentralized demand-side response [6]. This means that the 90 percent renewables target has to date been conceived of by policymakers almost exclusively in terms of the construction of new large-scale grid-connected generating plants.

Insofar as price-responsive demand-side options can be brought into the market with real-time price incentives, there is good evidence from modeling work internationally that they are often more cost-effective than, for example, installation of quick-response supply-side options such as open-cycle gas turbines. The GreenNet modeling project carried out for the European Commission found, for example, that demand response could reduce the system cost of maintaining capacity margins in a high-wind-penetration scenario to as little as 25 percent of the cost of the thermal-generation equivalent (Figure 6.5, [2], p. 18; see also [25, 28]).

This suggests that small islanded systems should be especially eager to maximize demand-side flexibility and load management. Ironically, although demand-side measures were willingly developed in New Zealand half a century ago, they have been shut out of the new "deregulated" market by a complex rulebook drafted by and for the dominant large generation companies, combined with the absence of any pro-competitive regulations requiring retailers to post feed-in tariffs or make other provision for small independent suppliers to reach customers.

Looking back to Figure 14.10, there is a wide gap between the mainstream projected demand path and the low-demand scenarios of some analysts, suggesting that implementation of demand-side and distributed-generation options might cause substantial stranding of grid-connected generation investment. That prospect will provide a strong incentive for the incumbent generators and network operators to oppose policy initiatives to decentralize the market.

14.6.2 GAPS IN THE CURRENT POLICY FRAMEWORK

Having articulated its strategic goal of achieving 90 percent renewable generation, the New Zealand government had not, as of 2008, settled on a fully credible set of policy instruments to pursue that goal. In particular, market-based regulatory instruments have been missing. Neither the U.K. adoption of regulated renewable quotas for electricity retailers (Cornwall, Chapter 15 in this volume) nor the Australian tradeable renewable quotas scheme [37] has struck any chord with New Zealand policymakers. Nor is there serious discussion of demand-side measures such as real-time pricing and net metering, notwithstanding the potential importance of real-time demand-side response as a means of coping with intermittency of wind and wave generation [28].

For a New Zealand generator that anticipates that the previous government's 90 percent goal may be abandoned and a lower renewables share allowed in the future, it remains rational to proceed with the planning of fossil-fuel generation projects to the point of final decision on major expenditure. A considerable lead time for major projects is required because of the need to secure planning consents for land use and emissions and to complete design work and possibly install infrastructure for the new plant. The government's 2007 announcement of its moratorium on construction of new baseload thermal plants did not trigger abandonment of any existing plans to build new nonrenewable generators.

Two major generators (Contact Energy at Otahuhu and Genesis Energy at Huntly) have fossil-fired sites with planning consent already in place and are in a position to build at quite short notice. Genesis Energy, meantime, is pressing ahead to secure planning consents for a new 400 MW CCGT plant at Rodney, near Auckland.

In the face of this direct challenge to its credibility, the Labour government appeared weak. The State-Owned Enterprises Minister sent a letter to all state-owned generators in October 2007 [29], informing them of the moratorium and asking to be kept informed of their plans, but the letter made it clear that the minister would not use his powers to give direction under the State-Owned Enterprises Act of 1986, leaving the companies effectively free to proceed. (Contact Energy is privately owned and not subject even to this mild level of influence.) The test of whether the newly elected National government will grant an exemption for Genesis Energy's Rodney project is still to come.

14.7 Conclusion

New Zealand remains some distance from full policy commitment to a renewable future, but the direction in which market forces will push the country's electricity sector seems increasingly well defined and can be expected to deliver something close to the 90 percent target with minimal policy activism, provided the emissions tax proceeds.

This is a reversal of the dominant trend of the past half-century. At the time when many countries began to move away from dependence on fossil fuels under the spur of high oil prices in the 1970s, New Zealand embarked on a deliberate program of raising the fossil-fuel intensity of its economy to take advantage of its windfall of domestic natural gas. Only as gas prices began to rise from 2003 with depletion of the Maui field, accompanied by an upward trend in the electricity wholesale price, have the economics of geothermal become attractive again, while the rapidly falling cost of wind generation has triggered a wind-farm boom.

The New Zealand government's 2008 approach of placing a blanket restriction on the construction of new baseload fossil-fired capacity is likely to leave sufficient legacy thermal capacity in place to supply more than 10 percent of total generation in 2025, unless some further restriction is placed on the ability of thermal plants to bid for dispatch. To some extent the planned carbon tax will provide such a restriction, but the possibility of a need for more direct regulatory restraint on the operation of thermal plants cannot be ruled out if the 90 percent goal is seriously pursued.

Long suppressed by policymakers and the dominant generators, the potential for small-scale distributed generation and an active, responsive demand side might become a problem rather than a support for the 90 percent target if central generation is overbuilt and then stranded by an eventual demand-side renaissance. Policymakers would be well advised to take proper stock of their demand-side options earlier rather than later.

Turning to the wider global picture, New Zealand combines a number of characteristics that are not shared by the majority of the countries covered in this book. It is an island system without external backup, which means that its domestic electricity price is set in isolation from wider markets. It has a century-long history of a dominant role for renewables (hydro and geothermal) in its generation mix; the strong trend toward greater reliance on fossil fuels since 1970 now appears as an aberration that is already being reversed by relative-price trends in fuels and technology. The likely dominant renewable technologies for the next generation of investment—geothermal and wind—are well proven and mature, and the New Zealand resource endowment is known to be on a scale that makes a 90 percent renewables target entirely realistic. The cost of bringing in these renewables appears to be little if at all higher than the cost of fossil-fuel generation, especially in the context of a carbon charge and with the prospect of increased urgency of climate-change policy in the coming decades.

For other countries, high-renewables targets in electricity are probably feasible only at much higher cost. From Figure 14.6 it would appear that for most OECD countries, nuclear power offers a more likely path toward carbon-free generation than the renewables that are at the heart of New Zealand's determinedly nonnuclear future. In this respect one lesson to be learned from New Zealand (and Iceland) is that to escape from both nuclear and fossil fuels at reasonable cost requires an unusual combination of low population and an abundant natural resource endowment—or technological breakthroughs on a truly epochal scale.

For New Zealand, probably the most important lesson yet to be learned from the rest of the OECD is the importance of real-time demand-side response and distributed generation in a modern electricity system. Deregulation and corporatization of New Zealand's electricity sector since 1987 have left untouched the centralized engineering solutions that served the country well from the 1950s to the 1970s. The current market institutions built around that structure present obstacles to the widespread adoption of a 21st-century smart grid and small-scale-generation technology. A substantial regulatory agenda remains to be tackled.

REFERENCES

[1] Ancell G. Effect of Unpredictability of Wind Generation Output on Scheduling, Investigation 1 Part B. Wind Generation Investigation Project, Wellington: Electricity Commission. www .electricitycommission.govt.nz/pdfs/opdev/comqual/windgen/implications/Investigation-1-PtB .pdf; 2007.

[2] Auer H, et al. Action Plan: Guiding a Least-Cost Grid Integration of RES-Electricity in an Extended Europe. Vienna University of Technology Energy Economics Group; 2007.

[3] Awerbuch S. Portfolio-Based Electricity Generation Planning: Implications for Renewables and Energy Security, SPRU, University of Sussex. www.sussex.ac.uk/spru/documents/ portfolio-based_planning-dec-26-04-miti-final-for_distribution.doc; 2004.

[4] Awerbuch S, Berger M. Applying Portfolio Theory to EU Electricity Planning and Policy-Making. IEA Report EET/2003/03, Paris. www.iea.org/textbase/papers/2003/port.pdf; 2003.

[5] Ballesteros A, et al. Future Investment: A Sustainable Investment Plan for the Power Sector to Save the Climate, Netherlands: Greenpeace International and European Renewable Energy Council, www.energyblueprint.info; 2007.

[6] Bertram G. Restructuring the New Zealand Electricity Sector 1984–2005. In: Sioshansi FP, Pfaffenberger W, editors. Electricity Market Reform: An International Perspective. [Chapter 7]. Amsterdam: Elsevier; 2007.

[7] Bertram G, Terry S. The Carbon Challenge: Response, Responsibility, and the Emissions Trading Scheme, Wellington: Sustainability Council, www.sustainabilitynz.org/docs/The CarbonChallenge.pdf; 2008.

[8] Bishop P. GEM: A Brief Description, Wellington: Electricity Commission, www.electricity-commission.govt.nz/pdfs/opdev/modelling/GEM/docs/GEM2.pdf; 2007.

[9] Bishop P, Bull B. GEM: An Explanation of the Equations in Version 1.2.0, Wellington: Electricity Commission, www.electricitycommission.govt.nz/pdfs/opdev/modelling/GEM/docs/ equations-explained.pdf; 2007.

[10] Bull B, Hemery E. Draft 2008 SOO Generation Scenarios, Electricity Commission, www .electricitycommission.govt.nz/pdfs/opdev/modelling/GPAs/presentations-29Feb08/Draft08-SOO-scenarios.pdf; 2008.

[11] Bull B, Smith B. 2008 Grid Planning Assumptions: Consultation Material on Draft Generation Scenarios, www.electricitycommission.govt.nz/pdfs/opdev/modelling/pdfconsultation/ GPA/ consultation.pdf; February 2008.

[12] Electricity Commission. Electricity Demand Forecast Model Review (DRAFT). Wellington, www.electricitycommission.govt.nz/pdfs/opdev/modelling/pdfsmodelling/DemandForecast-ModelReview2004.pdf; 2004.

[13] Electricity Commission. Consultation Paper: Wind Generation Investigation Project, www.electricitycommission.govt.nz/pdfs/opdev/comqual/windgen/Wind-Project-Jun-05.pdf; June 2005.

[14] Electricity Commission. Summary Report: Effect of Large Scale Wind Generation on the Operation of the New Zealand Power System and Electricity Market, www.electricitycommission.govt.nz/pdfs/opdev/comqual/windgen/implications/Summary-Report-July07.pdf; June 2007a.

[15] Electricity Commission. Wind Generation Investigation Project: Discussion Paper on Initial Options Assessment, www.electricitycommission.govt.nz/consultation/optionsanalysis/?searchterm = wind%20discussion%20paper; October 2007b.

[16] Electricity Commission. Final report on the Transmission To Enable Renewables Project (Phase 1), www.electricitycommission.govt.nz/pdfs/opdev/transmis/renewables/TTER-Final-report.pdf; July 2008a.

[17] Electricity Commission. 2008 Statement of Opportunities: Draft for Consultation, www.electricitycommission.govt.nz/pdfs/opdev/transmis/soo/pdfssoo/2008-draft/Draft%202008%20SOO.pdf; July 2008b.

[18] Ford R, Milborrow D. Integrating Renewables, British Wind Energy Association 2005; http://www.bwea.com/pdf/RAEIntegrationfinal.pdf; 2005.

[19] Fox B, Flynn D, Bryans L, Jenkins N, O'Malley M, Watson R, et al. Wind Power Integration: Connection and System Operational Aspects. London: Institute of Engineering and Technology Power and Energy Series 50; 2007.

[20] Freeman P, Atkinson V, Teske S. New Zealand Energy Revolution: How to Prevent Climate Chaos. Auckland: Greenpeace NZ; 2007.

[21] Grimsson OR. A Clean Energy Future for the United States: The Case of Geothermal Power. Testimony before the U.S. Senate Committee on Energy and Natural Resources, 26 September, http://energy.senate.gov/public/_files/testimony.pdf

[22] Grubb M. The Integration of Renewable Electricity Sources. Energy Policy 1991; (September):670–88.

[23] Grubb M, Jasmab T, Pollitt MG, editors. Delivering a Low-Carbon Electricity System: Technologies, Economics and Policies, Occasional Paper 68 forthcoming. Department of Applied Economics, University of Cambridge; forthcoming June 2008.

[24] Grubb M, Butler L, Twomey P. Diversity and security in UK electricity generation: The influence of low-carbon objectives. Energy Policy 2006;34:4050–62.

[25] Huber C, Faber T, Resch G, Auer H. The Integrated Dynamic Formal Framework of Green-Net, Energy Economics Group Work Package 8. Institute of Power Systems and Energy Economics, Vienna University of Technology; 2004.

[26] Hume D, Archer M. Power Systems Analysis for the 2008 Statement of Opportunities, Wellington: System Studies Group NZ Ltd.; www.electricitycommission.govt.nz/pdfs/opdev/transmis/soo/pdfssoo/2008/SSG-report.pdf; 2008.

[27] Kirtlan B. Electricity Demand Forecast Review February 2008. Electricity Commission, www.electricitycommission.govt.nz/pdfs/opdev/modelling/pdfconsultation/GPA/Demand-Forecast-Review.pdf

[28] Klobasa M, Ragwitz M. Demand Response: A New Option for Wind Integration, Karlsruhe: Fraunhofer Institute for Systems and Innovation Research, http://colorsofthecity.org/all-files2/94_Ewec2006fullpaper.pdf; 2006.

[29] Mallard T. Letter dated 11 October 2007 to chairs of state-owned generation companies.

[30] Marconnet M. Integrating Renewable Energy in Pacific Island Countries. MCA thesis, Victoria University of Wellington; 2007.

[31] Martin J. People, Politics and Power Stations: Electric Power Generation in New Zealand, 1880–1990. Wellington: Bridget Williams Books; 1991.

[32] Ministry for the Environment. Proposed National Policy Statement on Renewable Electricity Generation. Wellington, www.mfe.govt.nz/publications/rma/nps-renewable-electricity-generation/proposed-nps-for-renewable-electricity-generation.pdf; 2008.

[33] Ministry of Economic Development (MED). New Zealand Energy Outlook to 2025, Wellington: MED, www.med.govt.nz/upload/27823/outlook-2003.pdf; 2003.

[34] Ministry of Economic Development (MED). New Zealand Energy Outlook to 2030, Wellington: MED, www.med.govt.nz/upload/38641/eo-2006-final.pdf; 2006.

[35] Ministry of Economic Development. Energy Data File June 2007, www.med.govt.nz/upload/48437/000-200707.pdf; 2007a.

[36] Ministry of Economic Development. Benefit-Cost Analysis of the New Zealand Energy Strategy, www.med.govt.nz/templates/MultipageDocumentTOC____31983.aspx; 2007b.

[37] Office of the Renewable Energy Regulator. Fact Sheet: Mandatory Renewable Energy Target Overview Version 2, Canberra, www.orer.gov.au/publications/pubs/mret-overview-feb08.pdf; 2008.

[38] Parker D. 90% Renewable Energy Target is Achievable, Press release, www.beehive.govt.nz/release/90+renewable+energy+target+achievable; 6 March 2008.

[39] Parker D. Wind Energy in a Sustainable New Zealand. press release, 8 April 2008.

[40] Parsons Brinckerhoff Associates. Electricity Generation Database: Statement of Opportunities Update 2006. www.electricitycommission.govt.nz/pdfs/opdev/modelling/GPAs/SOO-Update-Final.pdf; October 2006.

[41] Potter N, McAuley I, Clover D. Future Currents: Electricity Scenarios for New Zealand 2005–2050, Wellington: Parliamentary Commissioner for the Environment, www.pce.govt.nz/reports/allreports/1_877274_55_0.shtml; 2005.

[42] Sinden G, Keirstead J, Milborrow D. Diversified Renewable Energy Resources, London: Carbon Trust and Environmental Change Institute, www.carbontrust.co.uk/NR/rdonlyres/DF0A5EAA-5F96-4469-AE61-2DBBE6EFEFA9/0/DiversifiedRenewableEnergyResources.pdf; 2006.

[43] Taylor P. White Diamonds North: 25 Years Operation of the Cook Strait Cable, 1965–1990. Transpower: Wellington; 1990.

[44] Watson JDF, et al. In: 2020: Energy Opportunities: Report of the Energy Panel of the Royal Society of New Zealand. Royal Society of New Zealand Miscellaneous Series No. 69, 2 vols., downloadable from www.rsnz.org/advisory/energy/; August 2006

[45] Webb M, Clover D. Future Currents: Electricity Scenarios for New Zealand 2005–2050: Technical Report, Wellington: Office of the Parliamentary Commissioner for the Environment, www.pce.govt.nz/reports/allreports/1_877274_57_7.pdf; 2005.

Carrots and Sticks: Will the British Electricity Industry Measure Up to the Carbon Challenge?

Nigel Cornwall

Cornwall Energy Associates

Abstract

The United Kingdom is leading the international debate on climate change and has translated words into action with innovative domestic carbon abatement policies—several targeted at the electricity sector, a significant polluter, which is playing a key role in delivering the government's environmental objectives. The United Kingdom was one of the first countries to develop a quota-based obligation for renewable energy and is now deepening programs on energy efficiency. It is trying to kick-start development of carbon capture and storage and complete a U-turn on nuclear policy. However, as this chapter shows, the sector has generally been slow to respond to these initiatives, meaning targets are being missed.

15.1 Introduction

"The U.K. government has been a pioneer in many aspects of energy policy," noted the IEA in 2006.[1] It was the first country to formally adopt a major long-term carbon emission cut of at least 60 percent by 2050. In the middle of 2008 it was in the throes of entrenching this goal in statute and making it binding on the government. The IEA concluded, "The Government deserves credit for the fresh approach and new ideas it has brought to the energy sector," and most of these developments are being driven by carbon abatement objectives that are constraining or priming market participant choices.

We think the IEA's assessment is rose-tinted for the reasons we set out in this chapter, and there is a widening gap between theory and practice. However, the various low-carbon policy initiatives taken thus far are certainly worthy of investigation, not least to understand why participant response has generally been slow.

The chapter briefly looks back to 1990, when the electricity sector was reorganized, and this is coincidentally the year against which carbon reduction targets are referenced. We also look forward to 2020, the focus point for most current targets and strategies.

The chapter is structured as follows:

■ Section 15.2 sets the scene for the electricity sector and some of the key changes that have occurred over the past 15 years. It covers the emergence of the climate-change program and the various energy reviews.

■ Section 15.3 considers incentives and obligations in the electricity sector to date, progress against them and participant responses.

■ Section 15.4 sets out conclusions on progress, likely directions, and future challenges.

---

[1]*UK Country review*, IEA (May 2006).

15.2 Context

This section provides background on U.K. energy supply, recent trends, and the changing carbon balance. It then describes the evolving policy framework.

15.2.1 ENERGY SETTING

The major trend of the past 20 years has been the rise in natural gas at the expense of coal and oil, a development enabled in 1989.[2] Exploitation of substantial gas reserves in the U.K. Continental Shelf (UKCS), mostly in the North Sea, led to greatly expanded gas use, which led to high investment in gas-fired combined cycle gas turbine (CCGT) electricity generation and two "dashes for gas." In parallel, coal use has steadily fallen. Since 1973 domestic production has dropped 80 percent. The phasing out of government subsidies after the 1984 miner's strike had a profound effect, and in the 1990s coal consumption declined further as natural gas increased. Oil production peaked in 1999, falling thereafter through 2004 by 30 percent as UKCS fields matured. The production profile for gas is similar, with a peak in 2000 before a fall of 11 percent by 2004.

The United Kingdom's commitment to nuclear energy has been variable. From the mid-1950s there were three main waves of development, each coincident with higher oil prices. The last nuclear station, Sizewell B, was commissioned in 1994 but was followed by a loss of policy interest in new nuclear, given abundant cheap supplies of gas. However, following an abortive revival by the nuclear lobby in 2002–2003, the government reaffirmed its support of further nuclear development in 2007.

Although new supply technologies are now emerging, their contribution to total supply is currently limited. In 2007 the total renewable energy contribution (including heat and transport) was 1.6 percent, or 18.1 TWh.

The government estimates that oil and gas will continue to be the dominant fuels up to 2020. In 2007 it estimated that gas consumption would increase to 40 percent overall, with renewables expanding substantially but still only reaching 5 percent of TPES in 2020 without further policy interventions.

With relatively healthy indigenous supply, the United Kingdom has been a net importer of fossil fuels in only 8 of the past 27 years. However, with North Sea production now falling, it is likely the United Kingdom will be increasingly

---

[2]Prior to the Energy Act 1989, gas was classified as a premium fuel and could only be used in power stations in limited circumstances.

dependent on fossil-fuel imports at a time of growing political concern about both carbon abatement and supply security. Further, the recent increase in net imports occurred more rapidly than expected and coincided with commodity price volatility, which has caused political tremors.

15.2.2 ELECTRICITY GENERATION

These changes have had a profound effect on generation in Britain, as Figure 15.1 illustrates.

In 1990 the electricity system was heavily reliant on coal, though successive programs since the mid-1950s had seen development of a bedrock nuclear contribution for some diversity. From virtually none in 1990, CCGT capacity rose to 12 GW in 1996 and to nearly 20 GW in 2000. The proliferation in investment in gas-fired generation technologies was such that the government imposed a moratorium in 1997, introducing a tougher consenting policy on new gas build because it was worried about the indigenous coal industry.

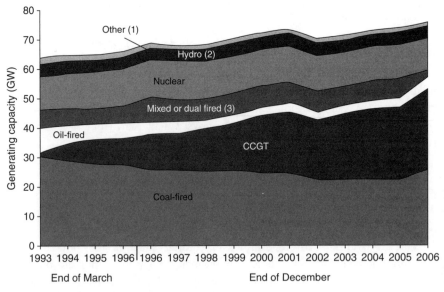

(1) Gas turbines, oil engines, and renewables other than hydro
(2) Natural flow and pumped storage
(3) Includes gas-fired stations that are not CCGTs

Figure 15.1 Generating capacity, 1993–2006.
Source: Digest of U.K. Energy Statistics (DUKES) (2007).

Although gas generation grew, that of other fossil fuels fell dramatically, with 18 GW of coal-fired capacity closed during the 1990s, mostly from less efficient plant but not at the end of their design lives. As of the winter of 2007–2008, Britain had a total of 75 GW of generating capacity.

Industry structure has been very fluid. The 1990s was a period of regulatory-driven restructuring, with the three large generators encouraged to divest plant and retailers encouraged to enter generation. At the end of the decade, cross-investment began to occur en masse. Though many new entrants emerged in generation, the incumbents began a renewed drive for scale that continues today. This reintegration was reinforced by the bilateral Neta trading model introduced in 2001, which encourages integration.

Today the market is dominated by six large vertically integrated players—termed the Big Six[3]—who seek to balance their large domestic customer portfolios with in-house generation. Generation and supply positions are shown in Figure 15.2, and do not include the EDF acquisition as this is still subject to regulatory and shareholder clearances at the time of this writing.

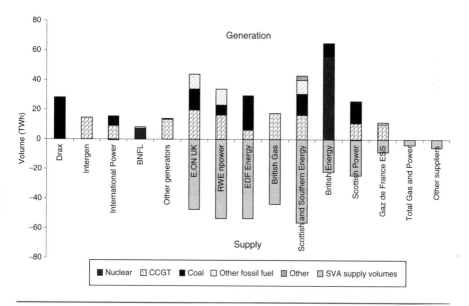

Figure 15.2   Participants in the GB electricity market, April 2008.
Source: Cornwall Energy from Berr data (April 2008).

---

[3]They are Centrica, EDF Energy, E.ON U.K. RWE Npower, Scottish Power, and Scottish and Southern Energy.

The shape of the United Kingdom's generation mix is set to undergo further major change, though in all scenarios dependence on gas as a fuel increases significantly as domestic production declines.

15.2.3 U.K. Emissions Profile

Government statistics show a steady fall in emissions from 1979 to the mid-1990s. This was due to the displacement of coal in electricity generation by gas and nuclear as well as a fall in industry emissions. Subsequently there have been annual variations in emissions but no clear trend.

In 2007 U.K. emissions of the basket of six greenhouse gases (GHGs) covered by the Kyoto Protocol were provisionally estimated at 639.4Mt CO_2e, 2 percent lower than 652.3Mt CO_2e in 2006. CO_2 accounted for 85 percent, or 554 mn tonnes, of total U.K. GHG emissions in 2006, the latest year for which final results are available. In 2007 U.K. net emissions of CO_2 were provisionally estimated at 544 mn tonnes, 2 percent lower than the 2006 figure. This corresponds to a figure of 148 mn tonnes of carbon (MtC).[4] Energy supply was the major source of emissions in 2006, representing almost 40 percent of the total.

The most recent GHG projections were published by the (then) Department of Trade and Industry (DTI), in May 2007. U.K. carbon emissions were forecast at 119.2–128.9 MtC in 2020, equating to a 20–26 percent reduction on 1990 levels. Additional policies are acknowledged by the government as needed to increase prospects for meeting targets, and the electricity sector will remain a key focus.

15.2.4 Policy Initiatives

The United Kingdom places itself at the forefront of diplomatic initiatives to combat climate change. It is difficult for the authorities to sustain this claim, given faltering progress against targets at home. Nevertheless the government has endeavored to translate sentiment into tangible programs domestically, to date with mixed success.

The first real moves started with the previous Conservative government's support for the U.N. Framework Convention on Climate Change, agreed at the 1992 Rio Earth Summit and ratified by the United Kingdom in December 1993. Under this treaty the United Kingdom made a voluntary commitment to

---

[4]1 MtC corresponds to 3.67 $MtCO_2$.

reduce GHG emissions to 1990 levels by 2000. That government published the first U.K. Climate Change Programme in January 1994, which set out high-level measures for achieving this target.[5]

Shortly after coming to office in 1997, Tony Blair's Labour government ratified the Kyoto Protocol and agreed to a target of a 12.5 percent GHG reduction as part of European "burden sharing." It later adopted a more stretching domestic target of a 20 percent CO_2 emissions reduction by 2010.

15.2.5 CLIMATE CHANGE PROGRAM

The Labour government set out the current Climate Change Programme (CCP) in 2000,[6] reviewed it in 2004, and published an update in 2006 (see Table 15.1).[7]

Table 15.1 Key policy developments, 2000–2008

Year	Development
2000	Climate Change Programme
2003	Aviation White Paper
	Energy Review and Energy White Paper
2004	Energy Efficiency Action Plan
	Future of Transport White Paper
2006	Climate Change Programme
	Energy Review
	Stern Review
2007	Draft Climate Change Bill
	Energy White Paper
	Planning White Paper
2008	Nuclear White Paper
	Energy Bill
	Planning Bill

---

[5] No longer available.

[6] *Action in the U.K.: The U.K. Climate Change Programme*, Defra (2000).

[7] *The U.K. Climate Change Programme*, Defra (2006).

15.2.5.1 2000 program

The CCP set out how the government would meet its targets.[8] It made clear that a 60 percent cut in emissions was likely to be the minimum required in the longer term.

Key measures fell under six headings, three impacting on the electricity sector:

■ *Stimulate new, more efficient sources of power generation.* There were two main initiatives here:

- Electricity retailers were to be obliged to increase the proportion of renewables supply to 10 percent by 2010 under a new Renewables Obligation (RO)
- A target to double the United Kingdom's combined heat and power (CHP) capacity to 10 GW by 2010

■ *Improve business use of energy.* This included:

- The Climate Change Levy (CCL), an energy tax for business, together with improvement targets for energy-intensive sectors administered through Climate Change Agreements (CCAs) and support for energy efficiency measures in the business sector
- Establishment of a new government-funded Carbon Trust to accelerate the take-up of cost-effective, low-carbon measures by business
- Exemption of good-quality CHP and renewable sources of electricity supplied to businesses from the CCL

■ *Promote better domestic energy efficiency.* Actions included:

- A new Energy Efficiency Commitment (EEC), through which electricity and gas retailers were required to help their domestic customers demonstrate measures that reduced energy supply

In aggregate the government estimated that the CCP was likely to achieve reductions of CO_2 of 23 percent below 1990 levels by 2010, with quantified measures delivering a 19 percent reduction.

15.2.5.2 2006 revised program

A government review begun in September 2004[9] evaluated existing measures and appraised new policies. The main elements of the original program with estimated carbon savings in 2010 as updated during the review in 2005 are shown in Table 15.2, together with new measures from 2006.

---

[8]*Climate Change: The U.K. Programme*, Department of the Environment, Transport and the Regions (DETR), November 2000.
[9]*Review of the Climate Change Programme, Terms of Reference.*

Table 15.2 Key elements of the United Kingdom's Climate Change Programme (CO_2 saved in 2010)

	CCP2005		CCP2006	
Sector	Main Policy	Mn Tonnes	Additional Policy	Mn Tonnes
Energy supply	RO	2.5	EU ETS Phase 2	3.0–8.0
	—	—	Biomass heat support	0.1
	Subtotal	2.5	—	3.1–8.1
Business	CCA	2.9	—	—
	Subtotal	5.0	—	0.2
Domestic	EEC	1.6	EEC	0.3–0.6
	Subtotal	3.6	—	1.2
Other	—	6.2	—	2.1
Total	—	17.3	—	6.6–11.6

Source: Cornwall Energy from Defra data (2005).

Thus in 2005 the rebased estimates of measures set out in the original CCP showed an expected 2.5 MtC saving from the RO but with a further 2.9 MtC expected from the CCAs and 1.6 MtC from EEC out of total program reductions of 17.3 MtC. The new measures added 6.6–11.6 MtC to expected savings, including more stringent EEC targets, adding 0.3–0.6 MtC. However, the main change was the addition of the EU ETS, implemented on January 1, 2005, which more than doubled the expected contribution of the energy supply sector by 2010.

Despite the new measures it became clear that the 20 percent domestic target set for 2010 was unlikely to be achieved. The government still says it can be, but most commentators disagree. Details of U.K. emissions since 1990 are set out in Figure 15.3.

15.2.5.3 The Stern Review

The direction of government thinking was reinforced by the publication of the influential Stern Review.[10] This report estimated that the cost of inaction on climate change significantly outweighed that of coordinated global action.

---

[10] *The Economics of Climate Change: The Stern Review*, Nicholas Stern, October 2006. Sir Nicholas Stern was head of the Treasury Economics Service and climate change adviser to the government.

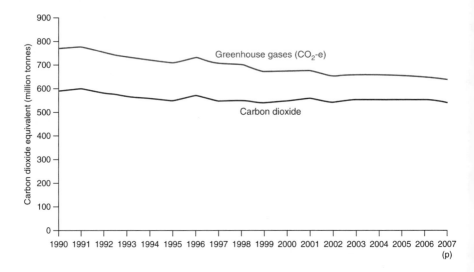

Figure 15.3 U.K. GHG emissions, 1990–2007.
Source: Second Annual Report to Parliament, Defra (July 2008).

Without effort to tackle climate change, Stern predicted, the loss of GDP from climate change could cost the global economy significantly more than that to stabilize atmospheric GHG concentrations at 550 pgm CO_2.

The government published a Climate Change Bill in March 2007. It proposed binding legal commitments to reduce CO_2 emissions by 60 percent below 1990 levels by 2050 and by 26–32 percent by 2020. Government will be held to account through five-year carbon budgets.[11] Failure to comply will make the government open to judicial review. An independent Committee on Climate Change has been created, and it recommended in October 2008 that the 60 percent target for 2050 should be increased to 80 percent, and the higher objective was formalized through the bill, which has since been enacted.

---

[11]The proposal is for five-year carbon budgets that would allow banking and borrowing from one five-year period to the next. It would also allow a proportion of carbon credits to be purchased from abroad to be included in the budget. Three successive budgets would be presented at the same time—covering a total of 15 years—with the aim of providing certainty for business when making long-term decisions. Powers to amend the budgets would be available using secondary legislation but only as a result of significant changes in circumstances and after advice from the proposed Committee on Climate Change.

15.2.6 Energy White Papers and Reviews

Environmental and energy policies have increasingly overlapped. Interwoven with the development of the CCP have been three energy white papers and two energy reviews. Dieter Helm has termed this "the new energy paradigm."[12] He acknowledged the "repoliticization of energy" and the "return of energy policy." The radical changes are well illustrated by reference to the white papers.

15.2.6.1 1998 White Paper

The first energy white paper in October 1998 largely dealt with legacy issues from the plight of the British coal industry. The white paper's main practical outcome was the stricter consents policy, but this was removed in 2001.

15.2.6.2 2003 White Paper

An energy review was initiated in 2001 in response to rising oil prices. The subsequent 2003 energy white paper recognized that the United Kingdom needed a new energy policy to address:

- The threat of climate change
- The implications of becoming a net energy importer
- The need to replace energy infrastructure

Its key theme was how to put the country on a path to cut carbon emissions by the target advocated by an independent Royal Commission of 60 percent by about 2050, with "real progress by 2020." It set out how 15–25 MtC of cuts could be achieved by 2020.[13]

The program headings are shown in Table 15.3. In terms of those impacting the electricity sector:

- Energy efficiency in households: The government undertook to consult on new targets for the EEC, which were nearly doubled.
- Energy efficiency in other sectors: Existing measures such as the CCL and CCAs were retained.

---

[12]*The New Energy Paradigm*, edited by Dieter Helm (OUP, 2007).
[13]*2003 Energy White Paper*, p. 7.

Table 15.3   2003 white paper targeted reductions for 2020

	Estimated MtC Reductions
Energy efficiency in households	4–6
Energy efficiency in industry, commerce, and the public sector	4–6
Transport: continuing voluntary agreements on vehicles; use of biofuels for road transport	2–4
Increasing renewables	3–5
EU carbon trading scheme	2–4

Source: Energy White Paper, Berr (2003).

■ Increasing renewables: The main plank was to double renewables' share of electricity from the 2010 target to 20 percent by 2020. However, firm targets were set only out to 2015–2016.

■ EU ETS: The government committed to entry in 2005.

Regarding nuclear power, the white paper said current economics made it "unattractive for new carbon-free generating capacity, and there are also important issues of nuclear waste to be resolved."

The overall tone was of priming markets. It extolled them and diversity of imports, emphasizing that these would deliver carbon targets and security of supply. The main proposals were programs to stimulate renewables and energy efficiency, but most headlines were dominated by its refusal to back new nuclear. Its most significant element was endorsement of carbon trading and commitment to the EU ETS. The reprioritization of policies represented a modernization of thinking.

15.2.6.3 2006 Energy Review

As the effects of import dependency kicked in, it was quickly apparent that the *2003 Energy White Paper* would have to be rewritten. That Britain became a net importer within little more than a year was something of a shock, especially as it coincided with soaring global fuel prices.

In July 2006 the government released *The Energy Challenge*, the first report from another energy review launched in late 2005. This represented a fundamental change of tone and the shift in political and market sentiment. The assumption at the heart of the *2003 Energy White Paper*—that strategic security could be achieved through mainly developing import capacity—was viewed as flawed.

15.2.6.4 White Paper 2007

Trade Secretary Alistair Darling announced the energy white paper, which was strongly influenced by the Stern Report, in May 2007.[14] The dominating theme—the need for increased diversity of supply—was reflected in proposals to build on measures outlined in 2003 on renewables and energy efficiency targets. Far more attention also went into areas such as carbon capture and storage (CCS). It also previewed the consultation on the merits of new nuclear. However, the focus was firmly on creating the right environment in which long-term investment could occur on scale, and an important adjunct was an overhaul of the planning regime in another white paper issued in mid-2007.[15]

The 2007 white paper set out specific programs to:

- "Strengthen" the RO so that the renewables contribution to electricity supply could rise up to 20 percent

- Double energy suppliers' obligation under the EEC through a new Carbon Emission Reduction Target (Cert)

- Launch a competition to build the world's first end-to-end CCS plant

These proposals were set out in the Energy Bill, which in October 2008 was still passing through Parliament.

15.2.7 EU CLIMATE AND ENERGY PACKAGE

Although the latest white papers—energy, nuclear, and planning—represent a seismic shift, further policy changes are already being discussed. The EU Climate and Energy package, published in January 2008, set out proposals to achieve a reduction in EU-wide greenhouse gas emissions of 20 percent by 2020, compared to 1990. Other key elements are to strengthen the EU ETS, proposed targets for each member state to reduce emissions in sectors not covered by the EU ETS, and plans to promote and regulate CCS technologies. The package also stipulates that the renewable energy—not generation—

---

[14]House of Commons Debate, May 23, 2007, c1281.

[15]www.communities.gov.uk/publications/planningandbuilding/planningsustainablefuture. The Planning Bill seeking to implement its proposals has since been introduced in Parliament.

contribution across the community should be 20 percent by 2020. The proposed U.K. target is 15 percent renewable energy by 2020.

The government had at October 2008 only just set about translating this goal into a renewable energy strategy. Its initial thinking was that the renewables generation contribution would need to almost double.

15.3 Incentives, obligations, and responses to climate change

The CCP and the Energy White Papers have provided the policy framework to develop a complex series of incentives and obligations on electricity participants to reduce GHG emissions. This section explains in more detail the main mechanisms implemented and the industry's response to them to date. It looks at supply side incentives first, including the RO, then at demand-side initiatives, including the EEC/Cert arrangements. It then also looks at new environmental regulations and the new European trading system for carbon permits. It finally considers the investment plans of some of the main players, to see how the various low-carbon incentives are impacting wider strategies.

15.3.1 RENEWABLES DEVELOPMENT

15.3.1.1 Renewables obligation

The RO is a quota-based system designed to stimulate new renewables generation. It replaces the Non Fossil Fuel Obligation[16] from April 2002. It aims to boost demand for renewables by mandating suppliers to obtain a given proportion of electricity from eligible renewable sources or face a financial penalty. Certificates (Rocs) are awarded to producers of qualifying generation and have a value that is then realized from suppliers.

Suppliers can meet this obligation by: constructing renewable generation of their own and producing their own Rocs; by purchasing electricity with Rocs (or Rocs on their own) from merchant operators; or by paying a fee to "buy itself out" of the obligation in any given year.

A key British design feature, in contrast to feed-in tariffs in Germany, renewable portfolio standards in the United States, or capacity tender mechanisms in Ireland, is that the Roc's value is variable, depending on the relative positions of demand and supply. This is because all buy-out payments are

---

[16]The NFFO was a tendering system. It involved a competitive bidding process through which developers competed for purchase agreements. There were five tender rounds in England and Wales, each targeting different technologies.

recycled to Roc holders. The Roc value to the holder is therefore the combination of the avoided buy-out fee plus its expectation of the value of the recycle payment it can retain, and the combined value is reflected in the commercial terms paid to generators. The higher the Roc payment, the more economically viable more marginal technologies should become.

Current obligation levels are shown in Figure 15.4, with the obligation remaining at 15.4 percent from 2016–2017 to the end of the life of the RO, currently set at 2026–2027. Roc values have fluctuated between £45 and £55/MWh.

The government conducted a review in 2005–2006, after which it made several decisions, which have yet to be implemented. The earliest proposed date for implementation is April 1, 2009,[17] which includes:

■ Extension of the obligation level to 20 percent

■ "Banding" the RO to group technologies needing similar levels of support, through additional Roc awards to encourage post-demonstration

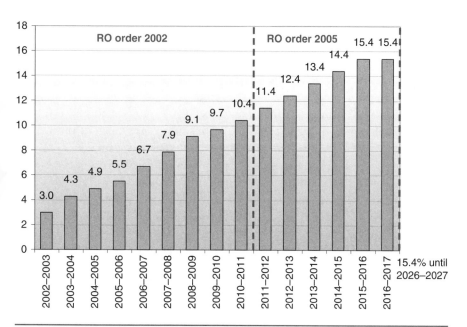

Figure 15.4   Suppliers' annual RO, 2002–2017.
Source: Renewables obligation orders (various).

---

[17]A statutory consultation was initiated in June 2008, with the draft order expected to be brought forward before the end of the year.

technologies (offshore wind and dedicated biomass) or those considered uneconomic at present (tidal and wave), and reduced awards for established technologies (cofiring and landfill gas)

The bands will be subject to review every five years. The proposals are underpinned by a principle of "grandfathering," so developers would not see a loss of incentive through a lower Roc award if a technology were later banded down.

Suppliers pass the associated Roc or buy-out costs to customers. Although this is a commercial choice, the supplier would usually look to recover at least the buy-out price. As such, the level of cost recovery will increase as the obligation level rises and as the buy-out fee (which is indexed to consumer inflation) increases. The RO currently costs the consumer over £1 bn/year.

The RO also ensures a payment stream to renewable generators in addition to monies from power sales. This is protected by the buy-out price, usually paid to the generator in return for the Rocs. Off-take contracts usually specify that a portion of the recycle payment is also paid to the generator.

15.3.1.2 New renewables investment

British renewables capacity was 4.7 percent in 2007 and by March 2008 had grown to 3.8 GW (see Figure 15.5 for country distribution) but is now set for rapid growth. In March 2008 there was 8.7 GW consented and a further 9.4 GW in planning, though access limitations mean that much of this cannot get onto the system in the short to medium term. There is a broader spread of investment in the renewables sector than for conventional technologies with a number of specialist operators, several new to the United Kingdom. Box 15.1 shows main investments and strategies of the Big Six, which have been active if cautious players. The RO is still young and the main participants are still defining strategies.

Combined with the surge in power prices, revenues available to developers have increased dramatically over recent years (Figure 15.6). However, despite official hype, new build is well below target. Despite considerable potential, wind penetration at the end of 2007 was less than 2 percent of U.K. consumption (see Wiser and Hand, Figure 3). Most commentators forecast a shortfall of 20–25 percent against the 2010 target. The government's own calculations suggest the best that can be realized from the "strengthened RO" would be a 13.4 percent share of electricity supply against a target of 15.4 percent in 2015–2016, and even this is, in our view, extremely optimistic.

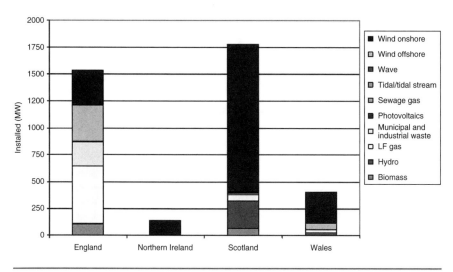

Figure 15.5   Renewables technologies by country and type (installed MW), February 2008.
Source: Berr statistics (March 2008).

Box 15.1   BIG SIX INVOLVEMENT IN RENEWABLES

Centrica

- Currently developing its own resources, primarily offshore wind; dates from July 2003. Subsequently acquired 72 MW Braes of Doune and 26 MW Glens of Foudland onshore developments.
- One offshore wind development—Barrow Offshore Wind (90 MW JV with Dong) operational. Lynn and Inner Dowsing (97 MW each) under construction.
- Lincs, Race Bank, and Docking Shoal offshore developments in planning (1.3 GW combined capacity).

E.ON U.K.

- Fifteen onshore wind farms in the United Kingdom (133 MW), 54 MW Rheidol hydro plant, 64 MW of offshore wind (Scroby Sands, 60 MW; Blyth, 4 MW).
- Cofiring at Ironbridge and Radcliffe on Soar coal stations.

Continued

Box 15.1   BIG SIX INVOLVEMENT IN RENEWABLES—Cont'd

■ Recently opened 44 MW biomass plant at Lockerbie, with consent for another 25 MW plant in Sheffield. Developing a further 90 MW of offshore wind at Scarweather Sands and Solway, plus the 1 GW London Array as a JV with Dong Energy.

EDF Energy

■ Smallest renewables portfolio of the Big Six. Owns and operates two small onshore wind farms (capacity <10 MW).
■ Developing 90 MW Teesside offshore wind farm.
■ Cofiring at Cottam and West Burton.

Npower Renewables (RWE)

■ Operates 18 onshore wind sites (400 MW capacity), North Hoyle offshore, hydro (15 stations/60 MW).
■ Cofires (35 MW) at Didcot A, Aberthaw, and Tilbury.

Scottish and Southern Energy

■ Claims 3.5 GW of renewables generation in operation, construction, or consented, most recently the 456 MW Clyde onshore wind farm acquired with Airtricity. Some 1.5 GW is wind, all onshore; 2 GW operational.
■ Airtricity acquired for $1.5 bn in January 2008, including 308 MW of operating onshore wind capacity in the British Isles, plus a development stream of 104 MW consented capacity, a 50 percent share in the 504 MW Greater Gabbard offshore wind development, and 1.5 GW of other potential wind projects.
■ Cofiring at Fiddlers Ferry and Ferrybridge.

Scottish Power Renewables

■ Part of Iberdrola Renovables since 2007.
■ Claims to be "the UK's largest developer of onshore wind farms with over 30 operational, under construction or in planning."
■ Undertaking large-scale feasibility study into wave and tidal power potential in Scottish waters.
■ Cofiring at Longannet and natural flow hydro (around) 150 MW remains part of Scottish Power.
■ Has 531 MW wind capacity and 271 MW under construction (including 235 MW at Whitelee).

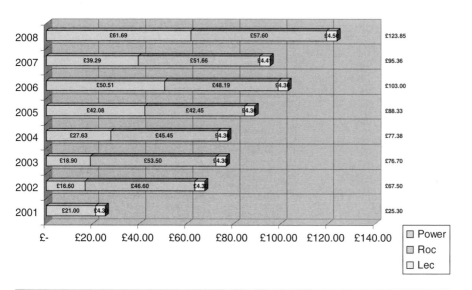

Year	Power	Roc	Lec	Total
2008	£61.69	£57.60	£4.56	£123.85
2007	£39.29	£51.66	£4.4	£95.36
2006	£50.51	£48.19	£4.3	£103.00
2005	£42.08	£42.45	£4.3	£88.33
2004	£27.63	£45.45	£4.3	£77.38
2003	£18.90	£53.50	£4.3	£76.70
2002	£16.60	£46.60	£4.3	£67.50
2001	£21.00	£4.3		£25.30

£-   £20.00   £40.00   £60.00   £80.00   £100.00   £120.00   £140.00

□ Power
□ Roc
□ Lec

Figure 15.6   Market value of renewable power, 2001–2008. *Note:* Power prices are calendar averages. Roc and Lec years commence April 1. Source: Cornwall Energy from Berr and Ofgem data (July 2008).

15.3.1.3 Assessment of renewables record

Four factors should be highlighted to explain the bottleneck in projects:

■ The RO itself has already undergone several substantive changes, which arguably has undermined the confidence of investors. But until the new RO order comes into force, there is no guarantee the banding proposals will be implemented, which is delaying possible projects.

■ Planning consent for some projects, particularly onshore wind, often meets local opposition. By summer 2007 there were 227 renewable energy developments with submitted planning applications for a total installed capacity of 12 GW, of which 8.5 GW was for onshore wind. Success rates have been modest, as low as 20 percent in some areas. The planning bill before Parliament should remove some of these barriers for the largest schemes, but local issues will remain.

■ Developers often experience difficulties in acquiring connection to the transmission system, which has resulted in a "queue" of approximately 12 GW in Scotland alone. The industry and government are working to improve this situation to identify and prioritize the schemes most viable to connect. Five years after the government committed to addressing this issue, firm proposals are awaited.

■ Other factors that have a bearing on investment decisions are the global availability of equipment, which is pushing up costs, and competition from other, possibly more attractive markets. For instance, Shell pulled out of the proposed 1 GW Thames Array wind farm in May 2008 to focus elsewhere, citing higher expected returns.

In autumn 2008 renewable targets and incentives were once again being reviewed as part of the renewable energy strategy development process, with formal proposals expected in the first quarter of 2009. The government remains committed to the RO but is considering a feed-in arrangement for microtechnologies. To meet the 15 percent renewable energy target for 2020, it is estimated that renewables will need to contribute between 32 and 35 percent of electricity supply, with a further 30 GW of renewable capacity needed requiring £100 billion in new investment. To reach these implied levels of deployment requires a quantum leap in just over 12 years. It is no surprise that there is profound skepticism about the prospects for meeting the new targets.

15.3.2 CHP

15.3.2.1 CHP incentives

All political parties have promoted large-scale gas-based CHP for several years. Labour first proposed a target of 10 GWe for good-quality CHP capacity by 2010, before it was elected. But the CHP strategy was repeatedly delayed, allegedly because of differences of view between parts of government, before being adopted in April 2004.

Government policy "recognizes that CHP makes a significant contribution to the U.K.'s sustainable energy goals, bringing environmental, economic, social and energy security benefits." With an assessed potential in excess of 30 GW,[18] the possible benefits are huge. However, the main elements of the strategy are partial and include:

■ Exemption from the CCL, presently worth 0.46 p/kWh of electricity and 0.16 pkWh of gas

■ Eligibility for enhanced capital allowances for good-quality CHP,[19] introduced in 2001

---

[18] *The Government's Strategy for CHP to 2010*, Defra (2004) (though the document noted that in reality 20 GW is likely to represent a very challenging medium- to longer-term cap).

[19] Since 2001, CHPQA has provided the United Kingdom's methodology for assessing the quality of CHP schemes and their qualification as good-quality CHP for all or part of their inputs.

- Business rates exemption for CHP plant and machinery
- Extension of the eligibility for Rocs to include mixed-waste plants that use CHP
- Separate allocation to good-quality CHP in Phase 2 of the EU ETS

In addition, in December 2006 new power station consent guidelines were published to encourage CHP. Proposals for new stations of over 50 MW must show that they have fully explored opportunities for use of heat before being consented.

The CCL exemption is worth £60–80 million a year (total value: £300–400 million since 2001), and the total enhanced capital allowance tax reduction benefit for investment in CHP since 2001 is estimated at around £59 million, for a total investment on the order of £1.2 billion.[20]

When it adopted the 10 GW target, the government said it could be met by these fiscal support mechanisms, without introducing an RO-like CHP obligation, which it claimed would be too costly for the consumer.

15.3.2.2 Developments in CHP

Policies have and still do make little if any real difference to large-scale development. The CCL exemption worked well initially and led to the development of ConocoPhillips' Immingham 734 MW project and improved load factors at existing plants. With the exception of RWE Npower, most of the Big Six have at least temporarily withdrawn from new CHP developments. There has been the realization virtually everywhere outside government that the 2010 target will be missed.

Only two of the many power station proposals in planning or construction utilize CHP. Besides Phase 2 at Immingham, the other is E.ON U.K.'s new Isle of Grain LNG terminal, with good heat-load. Two further schemes have just come forward.[21] A recent study said U.K. support was the least generous of countries considered and that investment had moved overseas.[22]

---

[20] These estimates were calculated by Defra in late 2007: www.defra.gov.uk/environment/climatechange/uk/energy/chp/pdf/progress-report.pdf.

[21] In August 2008 Thor Cogeneration was granted government approval for a 1020 MW scheme at Seal Sands, and an 800 MW project led by ConocoPhillips at Teesside received local planning.

[22] Benchmarking study of the United Kingdom against other countries in terms of CHP policies and impact, Delta (November 13, 2007). This is available from CHP Association at www.chpa.org.uk.

15.3.2.3 Assessment of CHP policy

The strategy has been dependent on a single gesture: the time-limited exemption of CHP from the CCL. Further, in recent years the industry has faced serious economic difficulties, mainly due to high gas prices.

The United Kingdom has scarcely tapped the potential for CHP. A study in June 2008 identified potential for 14 GW of further industrial CHP facilities, saving up to 4.6 percent of U.K. carbon emissions.[23] However, of all EU members the United Kingdom's CHP capacity is the fourth lowest, largely because of fitful and insufficient incentives reflecting the government's entrenched view that the technology is competitive. CHP use has remained at about 6 percent of electricity supplied for the past 10 years. In contrast, some EU member states have achieved levels as high as 40 percent.

The difference in approach from several utilities to CCS and CHP is striking. Largely silent on large-scale CHP, they are pushing the government to believe that commercial cleaner coal development with CCS is just a round the corner; it just requires more incentives to unlock this capability. In the meantime, extensive CCGT investment is proceeding, with a significant abatement opportunity from CHP technology being largely disregarded. With a large amount of new gas capacity to come, this error will be compounded. This position raises questions as to why in the late 1990s a tougher consent policy was feasible at a time when gas prices were low and stable, yet it is not considered desirable now, when high commodity prices and import dependency are facts of life.

15.3.3 NUCLEAR

15.3.3.1 Nuclear policy

The last nuclear station in the United Kingdom, Sizewell B, was commissioned in 1995. Rumors circulating in 2001 suggested that it was only a last-minute intervention from the Prime Minister's office that led Labour to omit reference in its election manifesto to opposing future nuclear build. Although the 2002 energy review revisited nuclear options, the *2003 Energy White Paper* concluded: "[I]ts current economics make it an unattractive option for new, carbon-free generating capacity and there are also important issues of nuclear waste to be resolved."

In January 2006, when the government began its latest energy review, it indicated a preference to explore new nuclear build for reasons of security of

---

[23]www.greenpeace.org.uk/blog/climate/a-surprising-solution-to-our-energy-needs-20080619.

supply. However, in February 2007 Greenpeace won a High Court ruling that threw out its in principle statement of support. The judge said the review was "seriously flawed" because key details of the economics of the argument were published only after it was completed. A fresh consultation followed, and the *2007 Energy White Paper* contained a preliminary view that "it is in the public interest to give the private sector the option of investing in new nuclear power stations." Alongside the *Energy White Paper* the government published a consultation document, *The Future of Nuclear Power*,[24] and a nuclear white paper in January 2008 announced government support for new nuclear stations,[25] provided they were built without public subsidy.

Consequently, the Energy Bill proposed legislation to ensure operators of new stations had secure financing arrangements in place to meet the full share of decommissioning and waste management costs. The government also committed to reduce regulatory uncertainty—for instance, by improving the planning system for major generating stations and pushing for a strengthening of the ETS so that investors had confidence in continuing carbon prices. The government's view of the likely way forward is shown in Figure 15.7.

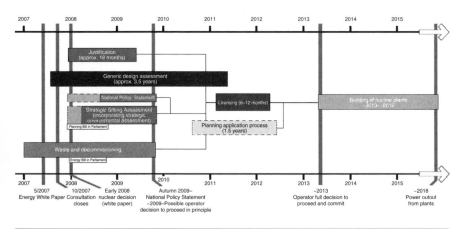

Figure 15.7   Indicative pathway to new nuclear, 2007–2018.
Source: *Meeting the Energy Challenge: A White Paper on Nuclear Power*, Berr (January 2008).

---

[24]www.berr.gov.uk/energy/whitepaper/consultations/nuclearpower2007/page39554.html.
[25]However, the Scottish National Party–led Scottish government has made clear that it opposes new nuclear power stations being built in Scotland.

15.3.3.2 Nuclear developments

At present, the operating lives of existing nuclear plants are scheduled to end as shown in Table 15.4.

Operating lives can be extended if approved by the Nuclear Installations Inspectorate. Seven stations have already been granted extensions. There are no proposals to extend the lives of either of the remaining Magnox nuclear power stations. British Energy expects to decide shortly on whether to apply for life extensions for Hunterston B and Hinkley Point B.

Decisions on new nuclear build remain some way off, not least because the necessary planning legislation has yet to be passed. The licensing to the United Kingdom of generic reactor designs is progressing, and much will depend on which achieve approval. Four designs have applied, though one has since dropped out.

Another factor will be the timing of the completion of the deal, announced in September 2008, of EDF's acquisition of British Energy, targeted for late 2008, and whether any remaining regulatory hurdles have to be overcome. Other U.K. players, including Centrica, RWE, and E.ON, have said they want

Table 15.4 U.K. nuclear plants and closure dates

Station	Estimated Closure	Extension Granted	Type of Reactor	Installed Capacity (GW)
Oldbury	2008	—	Magnox	0.47
Wylfa	2010	—	Magnox	0.98
Hinkley Point B	2011	10	AGR	1.26
Hunterston B	2011	10	AGR	1.21
Hartlepool	2014	5	AGR	1.21
Heysham 1	2014	5	AGR	1.20
Dungeness B	2018	10	AGR	1.08
Heysham 2	2023	10	AGR	1.20
Torness	2023	10	AGR	1.20
Sizewell B	2035	—	PWR	1.19
Total	—	—	—	**11.00**

Source: Cornwall Energy from Berr data.

to invest in new nuclear but most likely as equity partners in specific projects.[26] However, a tie up with British Energy's new owners will probably be essential for any new developments, given its ownership of most available sites.

15.3.3.3 Assessment of nuclear prospects

The nuclear debate has concluded, but decisions on new builds remain some way off. The government's U-turn has required careful political management and experienced a few problems. The legislation that will be a precondition of new builds is fully defined and the legislative process nearly complete. With licensing of new reactor designs in the United Kingdom under way, a sale of the government's British Energy stake advancing, and joint-venture partners emerging, the government's patience is set to pay off. However, until a British Energy deal is concluded and the legislative changes implemented, no formal proposals for new investment will emerge.

It is widely assumed within the industry that a minimum of four new stations will be built but that these are unlikely to be operational before 2020. This means, among other things, that new nuclear will be unable to contribute to existing targets, though expected decisions on life extensions of existing plants would.

15.3.4 COAL

Despite the downturn in its fortunes, coal retains a high level of political support, and the government sees an enduring role for a continuing coal-generation sector, provided clean coal technology (CCT) applications can be demonstrated. It has also promoted technological advances, including CCS. It summarized its policy in a 2006 review: "it is right to make the best use of U.K. energy resources, including coal reserves, where it is economically viable and environmentally acceptable to do so."

15.3.4.1 Clean coal incentives: CCS

CCS first came to prominence in the United Kingdom in the energy review of 2002. The government investigated its potential and published a review[27] and a strategy for carbon abatement technologies.[28] A summary of technology options is also provided in Chapter 6, by Lackner, Park, and Miller, in this volume.

---

[26]Centrica has indicated an interest in investing in British Energy should the EDF transaction go through.

[27]*Review of the Feasibility of Carbon Capture and Storage in the UK*, Berr (2003).

[28]*A Strategy for Developing Carbon Abatement Technologies for Fossil Fuel Use*, DTI (2005).

At present there is no requirement that new coal plants should be carbon-capture ready. Likewise, there are no CCS incentives as yet due to the early state of development of the technology. However, it is regarded as the only option available to make significant cuts in global emissions from fossil-fuel generation.

Because the CCS processes have not been demonstrated on a commercial scale at a power plant, in 2007 the government launched a competition to develop the United Kingdom's first full-scale demonstration project, due to be commissioned in 2011–2014. The competition will focus on post-combustion technology.[29] In mid-2008 the government announced a short list of companies, and it is aiming to announce the winner later in 2008. This process is intended to allow the United Kingdom to play a full part in wider programs endorsed by the EU, which wants up to 12 demonstration projects over the next decade.[30] Commercial deployment is targeted for 2020.

The government is undertaking other activities to develop the technology, including investing in R&D. It has also committed £550 million to a newly formed Energy Technologies Institute, which is considering CCS as one of its technology themes. The Energy Bill aims to develop a comprehensive legal basis for licensing CO_2 storage.

The government has said that, once technically proven, CCS could be an important part of an emissions reduction effort, but "within the framework of the EU ETS." This suggests that there are no further incentives beyond the startup phase competition payments.[31] It follows that CCS would need to receive a preferential allocation under the EU ETS post-2012 for it to be seen by the market as attractive.

15.3.4.2 Clean coal developments

New advanced and clean technologies and potential schemes have been openly debated by the Big Six, and each has a preferred scheme it has publicly backed. However, decisions on investments have been in stasis as the policy debate has

---

[29]http://nds.coi.gov.uk/environment/fullDetail.asp?.
ReleaseID=321108&NewsAreaID=2&NavigatedFromDepartment=True.
[30]At the EU Spring Council in March 2007, EU leaders called for the European Commission to develop a mechanism to stimulate the construction and operation by 2015 of up to 12 demonstration plants and for member states and the Commission to work toward the necessary technical, economic, and regulatory framework to bring environmentally safe CCS to deployment in new fossil-fuel power plants, if possible by 2020.
[31]"If UK-based companies deployed CCS where not justified by the price of carbon, this would have the effect of substituting CCS for more cost effective emissions reductions elsewhere. As a result there would be no net change in the overall level of CO_2 emissions for which the EU is responsible." *Towards Carbon Capture and Storage*, Berr (June 2008).

dragged on and as the competition reaches its conclusion. However, there is only one scheme that has as yet entered planning: a controversial plant brought forward by E.ON U.K. at Kingsnorth, where debate is focused on whether approval should be conditional on the plant being carbon-capture ready.

The risk of higher emissions from new coal has prompted proposals to stop new plants being built. This is a contentious area, not least because preventing new, more efficient plants could increase emissions as older, less efficient facilities are kept open. Environmental groups want to limit new development by applying an emissions performance level but which the government and energy companies oppose, saying that this could undermine the ETS. Some groups propose an EU-wide moratorium on all investment on new coal, at least until policy is clearer.[32]

15.3.4.3 Assessment of coal policies

The debate on clean coal started late and is unresolved. Virtually all the Big Six talk up CCS and clean coal but have not invested, and Kingsnorth has triggered significant controversy. As investment in renewables slips and opted-out LCPD plants approach closure (see Section 3.5), the prospect of a capacity gap around 2015 is being highlighted by the industry and the government.

Despite strong policy rhetoric, the proposed incentives for clean and advanced coal are modest, and there is disappointment in the industry that the government has opted to support only one CCS scheme, based on post-combustion technology. Consequently there is uncertainty about how much new coal may be built in the United Kingdom. This is due not only to volatile fuel prices but also the development of carbon trading and the associated prices (see Section 3.6). It is possible that if decisions on carbon capture readiness (CCR) are not taken quickly, as much as 9 GW of further coal generation could materialize, despite the environmental cost. If so, this would destroy any prospect of meeting 2020 targets in the Climate Change Bill.

15.3.5 ENVIRONMENTAL SCHEMES: LCPD

Government policy has also embraced two European measures that are intended to ensure that environmental costs are fully targeted on coal generation: the Large Combustion Plants Directive (LCPD) and the EU ETS.

---

[32]www.ippr.org/members/download.asp?f = /ecomm/files/after_the_coal_rush2.pdf&a=register#register.

15.3.5.1 LCPD policy

The LCPD, which restricts emissions of SO_2 and NO_x, requires large generators to meet stringent air-quality standards after January 1, 2008. To date owners of two thirds of existing U.K. coal plants (20 GW capacity) are investing in new equipment, primarily flue gas desulfurization (FGD) equipment to ensure compliance. Those who "opt out" (that is, have decided they prefer not to comply) will have to close by the end of 2015 or after 20,000 hours of operation from January 1, 2008, whichever is the earlier.

15.3.5.2 LCPD closures

Some 12 GW of coal and oil-fired plants opted out. The timing of these closures is a commercial matter for owners, who would be expected to take into account carbon prices and allocations, environmental restrictions, and the age of the plant. Hence, it is impossible to predict the timing of the impact of the LCPD on capacity. However, all opted-out coal plants will probably have used allowances by the end of 2015.

The remaining 20 GW are investing to achieve compliance, largely through the installation of FGD.

The directive is likely to affect oil-fired plants in a different way since, as peaking plants, they are less likely to run out of hours.

15.3.5.3 Assessment of LCPD

It is early days. The directive has had a significant effect, with the industry investing heavily in new FGD. Opted-out plants are now adjusting to the new environment, but there are suggestions that this process has been operationally fraught, with system operating costs increasing dramatically.

15.3.6 EU ETS

The government views emissions trading as a cornerstone of its carbon abatement strategy, especially the EU ETS, introduced in 2005.

15.3.6.1 Emissions trading policy

This EU-wide scheme trades allowances to cover GHG emissions from permitted installations. It has two fundamental components: a cap on emissions and a system for trading the "right to emit." Further details on scheme mechanics can be found in Chapter 2, by Dijkema, Chappin, and de Vries, in this volume. Some 11,000 installations with combustion plants above 20 MW across Europe, including all coal-fired power stations and most gas-fired stations[8] in the United Kingdom, are required to operate within this cap.

15.3.6.2 EU ETS development

The first phase ran from January 1, 2005, to December 31, 2007, with the total allowances allocated to the United Kingdom 736.3 $MtCO_2$, around 65 $MtCO_2$ (around 8 percent) below projected emissions of installations covered by the scheme. The U.K. installations were responsible for 46 percent of CO_2 emissions in 2002.

The government then allocated allowances across participating sectors that it decided should receive allowances equivalent to projected emissions, with power stations receiving the rest. So the electricity industry was responsible for delivering the savings the United Kingdom expected in Phase 1. This was because the sector was considered to face limited international competition and to have a relatively large scope for low-cost abatement opportunities. Once the sector's allocation was determined, the allowances were distributed using a baseline period from 1998 to 2003. All allocations in Phase 1 were free.

In Phase 1 the United Kingdom had a shortfall of 88 million allowances, meaning permits had to be purchased from elsewhere in the EU from countries that adopted more lax allocations.[33]

Phase 2 of the scheme, from January 1, 2008, to December 31, 2012, was similar, but overall the cap represents an annual reduction of 29.3 $MtCO_2$ against business-as-usual emissions. The government has introduced important enhancements, including auctioning 7 percent of allowances.

Formal consultation on Phase 3, from 2013 to 2020, began in May 2008. Again, most targeted savings are to come from the power sector. The Commission wants to include other gases and involve all major emitters. Allowances will be reduced 21 percent in 2020 over 2005 levels, setting the average annual cap at 1.8 billion tonnes between 2012 and 2020, with a constant annual reduction of 1.74 percent over the period. There will be full auctioning for power from 2013, but for other sectors only from 2020. The arrangements are summarized in Figure 15.8.

15.3.6.3 Assessment of EU ETS

Analysis undertaken by consultant OXERA shows the power sector has seen increased profitability because—despite extensive free allowances—power prices have reflected the full marginal permit cost. In Phase 1 earnings rose

---

[33]www.defra.gov.uk/news/2008/080520a.htm.

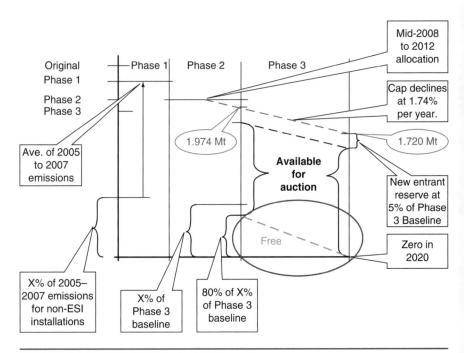

Figure 15.8 Overview of EU ETS, Phases 1–3.
Source: Cornwall Energy or Cornwall Energy based on CHPA

47 percent, spread unevenly across the sector, with the most carbon-intensive generators benefiting least.

In addition, passing costs onto customers is estimated to have added 4–10 percent to electricity prices, triggering calls for a windfall tax. During 2008 an intense campaign was waged by trade unions and consumer groups in the United Kingdom about carbon rentals earned by large generators at a time of rising consumer prices. The matter seems to have been resolved with a package of measures announced in September extending energy efficiency obligations on suppliers but also with the creation of a new Community Energy Programme costing £350 million, which was to be half-funded by larger generators implicitly from their carbon windfalls.

EU ETS Phase 1 was regarded as a learning phase. The view of commentators has tended to be that carbon prices, which are shown in Figure 15.9, have not as yet been high enough to encourage significant, if any, fuel switching. Investment decisions by participants are still heavily skewed by largely free carbon allocations. Current projections suggest prices will continue to be firm. However, decisions on new nuclear and CCS (and marginal renewables) will need to be taken before there is a proven track record of carbon prices, probably on the back of strong balance sheets from large utilities. This is good for carbon but possibly bad for competition.

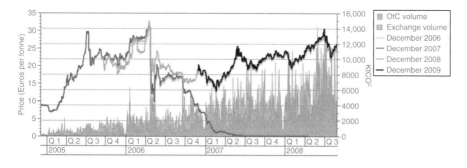

Figure 15.9 Carbon prices under the EU emissions trading scheme, 2005–2008. Source: Point Carbon (August 2008).

As Bouttes, Trochet, and Dassa conclude in Chapter 16 in this volume, the EU ETS scheme design could and should be improved to remedy structural weaknesses. But they also concur that the trading scheme "can and must play a fundamental role towards implementing" technology change.

15.3.7 ENERGY EFFICIENCY IN THE DOMESTIC SECTOR

The U.K. government has been aggressive in stimulating energy efficiency. The main mechanisms implemented to date involving electricity are:

- Legal obligations on retailers to invest in energy and carbon abatement measures domestically
- A tax on business use of energy administered through retailers

15.3.7.1 Incentives in the domestic markets

Energy efficiency commitments

Six countries currently have energy efficiency obligations on energy companies in Europe: Belgium (Flanders region), France, Italy, Ireland, and Denmark as well as the United Kingdom. Similar activities are under development in the Netherlands, Poland, and Portugal. The World Energy Council[34] has estimated that the United Kingdom typically spends about a factor of three more per year through the supplier obligation than the next highest spender (France). The next country, Italy, spends a sixth of the amount, with other countries spending considerably less.

---

[34]*Energy Efficiency Policies Around the World: Review and Evaluation* (January 2008).

Between 2002 and 2008 electricity and gas retailers were mandated to seek reductions in energy use across their domestic customer bases through EEC. The programs were administered by Ofgem, whose remit was to determine the target for each supplier, deciding whether a proposed activity could be considered qualifying and setting the level of improvement to be attributed to a qualifying action.

EEC1 ran from 2002 to 2005. Defra set a savings target of 62 TWh over the lifetime of installed measures in domestic households. At least 50 percent had to be met in relation to a *priority group*, defined as those on income benefits to help alleviate fuel poverty. All suppliers with at least 50,000 gas or electricity domestic consumers were subject to a target. The formula set progressively higher targets for suppliers with larger consumer numbers. The targets were set in terms of fuel-standardized, lifetime-discounted energy savings in homes using gas, electricity, coal, oil, or liquefied petroleum gas. Suppliers were not restricted to their customer bases in pursuing measures and could choose which ones to use.

In the *2003 Energy White Paper*, the government wanted to double the level of activity, and the overall target set for EEC2, from April 2005 through March 2008, was 130 TWh, again with half sourced from the priority group.

Carbon emissions reduction target

The target set for the Cert, which began on April 1, 2008, and runs until March 31, 2011, was more ambitious again. Instead of fuel standardized lifetime discounted energy savings, it was set as an overall lifetime CO_2 saving, at 154 $MtCO_2$, effectively doubling activity under EEC2. Ofgem must be satisfied that any resulting reductions in emissions will be more than would have happened without Cert. It is expected to deliver 1.1 MtC savings a year in 2010 and stimulate £2.8 billion of investment by suppliers in carbon reduction measures.[35] Suppliers must now achieve at least 40 percent of carbon savings, reduced from 50 percent in EEC2, from a priority group. Cert has extended qualifying actions to include micro-generation.

In September 2008 the government announced a 25 percent extension in Cert targets in response to widespread concern about high prices and their impact on vulnerable customers at an estimated additional cost of £560 million to retailers.

Post-2011 supplier obligations

In the 2006 energy review report, the government reaffirmed its commitment to continuing to deliver carbon savings from the domestic sector using some form of supplier obligation until at least 2020. It proposed a minimum of

---

[35]www.opsi.gov.uk/si/si2008/draft/em/ukdsiem_9780110805306_en.pdf.

3–4 MtC a year by 2020 for the obligation. In June 2007 the government called for evidence[36] on the form of the supplier obligation post 2011. It is considering an outcome-based obligation with trading, where the overall objective would be based on a cap-and-trade mechanism. The alternative is an extension of Cert, where suppliers would meet targets, set in terms of carbon savings but with trading of energy-saving measures ("white certificates").

15.3.7.2 Developments in domestic efficiency

All but two suppliers (who ceased trading) comfortably met their EEC1 targets. Overall suppliers installed energy efficiency measures, which would result in an energy saving of 86.8 TWh. The main measure offered was insulation; the distribution of energy efficient light-bulbs achieved one quarter of total savings with appliances, mainly energy-efficient white goods, contributing 11 percent and heating measures 9 percent. Suppliers were allowed to transfer excess savings to EEC2.

A further 185 TWh was saved under EEC2 through implementation of similar measures, with 55 TWh to be carried over into Cert. Expected carbon savings in 2010 were put by the government at 0.3 MtC/y for EEC1 and 0.5 MtC/y for EEC2 in the *2007 Energy White Paper*. EEC1 stimulated about £600 million investment in energy efficiency and delivered net benefits to householders valued at over £3 billion. The government said when EEC was appraised under the 2006 CCP review, it was confirmed as one of the most cost effective of any of the major government policies, generating net benefits of £270/tonne of carbon saved.[37] Reporting under Cert has only just commenced.

15.3.7.3 Assessment of domestic efficiency

The mechanism of an energy efficiency obligation administered through large domestic retailers has proved very successful in the United Kingdom and surpassed expectations. Early indications are Cert will build on EEC's success. By targeting priority customers, the industry and the government have also been able to show they are tackling fuel poverty.

All major retailers have developed diverse energy services business, providing these players with a very strong foothold in these new markets. Because the costs are recovered across domestic customers as a whole and because these players control almost all this market, they are all able to recover the costs they incur.

---

[36]www.defra.gov.uk/environment/climatechange/uk/household/supplier/pdf/evidence-call.pdf.
[37]Ibid.

15.3.8 Energy Efficiency in the Business Sector

15.3.8.1 CCL and CCAs

Introduced in April 2001, the CCL is an energy tax that adds 15 percent to typical energy bills of U.K. businesses (not domestic customers). It was designed to be revenue neutral, and businesses paying it can claim a reduction in National Insurance contributions. To protect the competitiveness of the most energy-intensive industry and to improve business energy efficiency, CCAs were also introduced to provide up to an 80 percent discount from the levy, subject to achievement of agreed energy efficiency targets. The levy is applied as a specific rate per nominal unit of energy. There is a separate rate for each category of taxable commodity, as shown in Table 15.5. Until 2007–2008 the rates were flat, but since then they have been indexed to inflation. The aim was to achieve annual savings of 5.4 MtC.

Businesses paying the tax can sign agreements with suppliers to buy renewable electricity or good-quality CHP, both of which are levy exempt to encourage less-polluting energy sources. Levy Exemption Certificates (Lecs) are issued to qualifying generators, one per MWh of generation, and evidence CCL exempt electricity supply generated from qualifying sources. The renewable or qualifying CHP generator negotiates with suppliers to obtain value for Lecs. In turn they are redeemed by suppliers to HM Customs and Excise, to demonstrate the amount of nonclimate change electricity able to be levied that had been supplied to nondomestic customers. In effect, the customer then receives exemption from the levy on these volumes. The CCL is legislated to run to 2013, though most current agreements only run to 2010.

Table 15.5   U.K. climate change levy rates, 2008–2009

Commodity	Normal Rate
Electricity	0.456 pence per kilowatt hour
Gas supplied by a gas utility or any gas supplied in a gaseous state that is of a kind supplied by a gas utility	0.159 pence per kilowatt hour
Any petroleum gas or other gaseous hydrocarbon supplied in a liquid state	1.018 pence per kilogram
Any other taxable commodity (e.g., coal)	1.242 pence per kilogram

Source: Cornwall Energy.

15.3.8.2 Developments under CCL

At the time the revised U.K. CCP was published in March 2006, the CCL and CCAs were projected to be two of the three most important contributions toward the government's 2010 target of a 20 percent reduction in U.K. CO_2 from 1990 levels. Since then, projections of their impacts have been reduced slightly, but the CCL still features as the second biggest measure, with a projected annual saving of 12.8 $MtCO_2$ in 2010. Between its announcement in 1999 and the end of 2005, the levy alone is estimated to have saved a cumulative total of some 60.5 $MtCO_2$. The government continues to project that the levy and agreements together will be responsible for around a third of all the carbon savings achieved by policy by 2010.

15.3.8.3 Assessment of business energy efficiency incentives

The CCL/CCA arrangements are now in their eighth year. In 2007 the National Audit Office,[38] the government's value-for-money watchdog, gave a mixed assessment of the arrangements. It concluded that initially the levy "contributed to a significant refocusing of attention on energy use in the years after 1999." This has driven energy efficiencies and emissions reductions relative to business as usual in both energy-intensive and less intensive industries. But, it continued, the extent to which the levy has continued to drive further energy efficiencies in more recent years "is harder to discern," especially because other policies had come into play. Results of a survey carried out by the NAO suggested it "is no longer seen as a major driver of new energy efficiencies," and the cumulative carbon savings achieved by the levy could not be measured.

As for CCAs, "the negotiation of agreements and the development of monitoring regimes to measure progress against Agreement targets raised awareness of the potential for energy efficiencies. These efficiencies were then made." It went on to note that "Not all Agreement targets were stringent, but early overachievement against them was the result of genuinely significant improvements in efficiency as much as weak targets."

However, consultant Cambridge Econometrics has estimated that CCL reduced overall energy demand in the economy by 0.2 percent in 2000, 1 percent in 2001, and 1.8 percent in 2002, rising to 2.9 percent by 2010, compared with a situation in which the CCL package had not been implemented.[39]

All major suppliers offer levy-exempt power into the business market. Appetite for Lecs has increased to a point at which suppliers are now paying

[38] *The Climate Change Levy and Climate Change Agreements, National Audit Office* (August 2007).
[39] www.camecon.com/aboutce/our_work_on_climate_change_levy.htm

generators significantly more than the 85 percent of the value conventional at market start. A growing market exists for the import of Lecs from mainland Europe.[40]

A vigorous intermediary sector has developed on the back of these arrangements, assisting larger companies that are part of exemption arrangements under CCAs to reduce their consumption, negotiating bespoke commercial arrangements with retailers, and carrying out administrative support functions such as data analysis and bill checking.

15.3.9 COST OF CARBON ABATEMENT

Not all measures introduced by the government to date have been summarized here. The assessed costs of these and other measures are widely divergent. The government's own assessment from the 2007 white paper is shown in Figure 15.10.

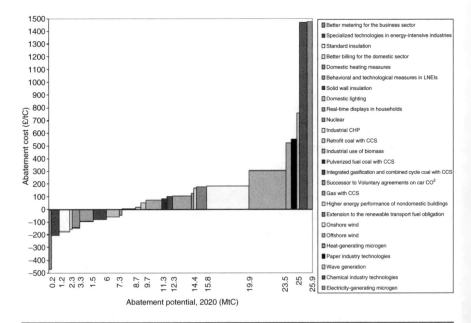

Figure 15.10 Government assessment of costs of carbon abatement measures. Source: *Energy White Paper* (July 2007), Berr.

---

[40]The rules are different from Rocs, which must be generated in the United Kingdom.

The most beneficial investments tend to be on the demand-side. They include new metering applications for the business sector, but the government is struggling to adopt a coherent strategy to access these. Better billing arrangements for domestic consumers have been defined but not yet implemented.

On the supply side, all options have a positive net cost and, although industrial CHP and new nuclear are assessed as lowest cost, incentives are being targeted at higher-cost renewables, especially offshore wind. The government wants to achieve retrofitted coal with CCS, but the technology remains unproven. At the same time there are programs to bring currently high-cost technologies closer to commercialization, including marine (wave and tidal). The government is also rolling out programs to encourage local generation, including micro-gen, despite high costs for reducing carbon and modest abatement contributions over the foreseeable future.

15.3.10 INVESTMENT OUTLOOK IN THE ELECTRICITY SECTOR

15.3.10.1 Drivers

The government's view of likely supply from individual technologies from the *2007 Energy White Paper* is shown in Table 15.6. To varying degrees it assumes early and deeper penetration of the various measures and polices outlined previously.

In all scenarios, new gas will proliferate (output up 15–65 percent) by 2020. Coal will fall between 38 and 54 percent, and nuclear 56–66 percent. Renewables generation is expected to grow 270–400 percent. Broadly speaking, the more slowly coal and nuclear disappear and/or the faster renewables are deployed, the less will be the growth in gas.

But there are important variables that will have impact on the path forward:

- The wider economic environment and its impact on sectoral demand
- Individual participants' investment plans and success in planning, connection, and operation
- How quickly new commercial renewables technologies are deployed
- Operating decisions by coal operators following the LCPDs
- The rate of retirement of existing nuclear stations and any further life extensions
- The timing of CCS technologies from demonstration to commercialization and whether the government mandates CCR

Table 15.6 U.K. future generation mix estimates by fuel, 2005–2020

TWh	2005	Baseline Projections		Low Policy Estimates		Central Policy Estimates		High Policy Estimates	
		2010	2020	2010	2020	2010	2020	2010	2020
Coal	125	121	119	113	67	113	71	113	77
Oil	2	2	1	2	1	2	1	2	1
Gas	135	129	202	136	223	129	195	123	156
Nuclear	75	68	25	68	25	68	25	68	33
Renewables	17	31	48	29	46	33	57	36	67
Imports	11	11	16	11	16	11	16	11	16
Storage	3	3	3	3	3	3	3	3	3
Total	**368**	**365**	**415**	**362**	**381**	**359**	**367**	**357**	**352**

Source: Energy White Paper, Berr (May 2007).

- The rate of depletion of indigenous supplies
- Last but certainly not least, levels of carbon prices under the EU ETS

Each will impact preferred individual technologies as well as their timing.

15.3.10.2 New baseload investment plans

By autumn 2007, some 14 GW capacity of major conventional generating plant was rated as under active development by Berr. Most of that (12.4 GW) was gas-fired, apart from E.ON U.K.'s proposed 1.6 GW coal-fired station on the Isle of Grain.[41] An additional 2 GW of interconnector capacity to the Netherlands was also being considered; 5.9 GW of renewable generation had secured planning but was not operational, most wind powered, onshore but also offshore.

Investors in the United Kingdom continue to favor gas and onshore wind. Renewed interest in coal-fired plants in 2005–2006 waned following increases in world coal prices and policy uncertainty, which appears to have reinforced gas as the preferred scale investment technology. Note that this is in the combined cycle mode rather than combined heat utilization, which some argue is a lost opportunity.

Most new baseload investment comes from the large integrated players, who also hold a 99 percent share of the domestic retail market. These developments have been announced despite persistently low spark spreads[42] but are required to provide replacement capacity for coal and nuclear stations due to retire over the next decade.

15.4 Conclusion

U.K. policy development has been strongly influenced by climate-change concerns, and environment and energy policies are now closely interwoven. The CCL was one of the first energy taxes to be imposed on business by a leading economy. The government has adopted and revised a comprehensive

---

[41]The latest official summary is given in Berr's *Energy Markets Outlook* report (October 2007). A more recent industry view for transmission-connected plants is provided in National Grid's *Seven Year Statement* and is current as of May 2008.

[42]The *spark spread* is the theoretical gross income of a gas-fired power plant from selling a unit of electricity, having bought the fuel required to produce this unit of electricity. All other costs (operation and maintenance, capital, and other financial costs) must be covered from the spark spread.

climate change program and undertaken two major energy policy reviews in the past six years. It is legislating for challenging emissions reduction targets, and the *2006 Stern Review* has been very influential. As a consequence, the electricity sector is now being transformed by complex incentives and rules.

Because of the electricity sector's high emissions and since it is not exposed to international competition, it has been singled out to help deliver government targets. In 2003, when the 2010 domestic CO_2 target looked achievable, the government's focus was on increased renewables generation. It increased targets and outlined programs to tackle bottlenecks to new capacity. It also doubled energy efficiency targets in the domestic sector. Although it did not pursue new nuclear, it did not rule it out. But with the benefit of hindsight, the proposals in that white paper look inadequate and naïve.

A characteristic of the United Kingdom is the number and scope of mechanisms and targets now in place. As targets have slipped, the government has broadened policy, introduced new measures, or increased existing obligations and widened programs into new market segments. This trend was evidenced again by the *2007 Energy White Paper*, which in the round will lead to a further diversification and deepening of policy initiatives, with nuclear power and clean coal being added to the armory.

The government also wants to entrench the climate-change program within a statutory framework, with binding legal targets and budgets. The importance of this approach cannot be overemphasized. The detail of this institutional framework is yet to be concluded, but the approach will be closely monitored by other governments.

Despite these good intentions, performance against targets has been at best mixed. In October 2008 delivery of the framework was in hand but a work in progress. A lot of emphasis is placed over time on the efficacy of the EU ETS but not before 2013, yet auctioning has not yet begun, and carbon prices are at levels as yet unlikely to bring about major behavioral change.

It is easy to claim that the approach is failing because targets are being missed, and problems with delivery of measures have been greater than expected. A 2008 report by a right-wing think tank[43] was damning. It claimed: "Of 138 high level targets surveyed, 60% of targets have been missed; are unlikely to be achieved or are worded so vaguely as to make meaningful analysis impossible. ... Targets on climate change are also a serious concern, with almost two thirds of targets looking unlikely to be met." But this judgment is unfair.

---

[43]*Green Dreams: A Decade of Missed Targets.*

Overall the electricity situation can be best described as in flux. The main lesson from the U.K. experience for other jurisdictions is that climate-change mitigation is a complex matter, requiring multidimensional solutions that take some time to fit together.

Responses by participants in retail have been relatively swift, with effective responses to imposed measures and obligations such as EEC and Cert, and they show that demand-side measures, properly incentivized, can bring real benefits.

But investments in generation more generally are in a holding pattern. This is best illustrated by the following:

- Clean coal investments, which the industry continues to talk up but which remain in abeyance
- Large-scale renewable investment, which, despite strong incentives (and the promise of stronger ones for some technologies), are falling well short of increases in targets
- Nuclear plans, which remain conceptual

Ultimately many of the negative comments arise from transitioning issues, especially the gap in time between setting policy and delivering legislation. This hiatus could well endure for some time, certainly a year. Policy is attempting to create a benign investment framework so that market participants can evaluate options that suit them but also in a way that is responsive to a robust carbon price.

This philosophy was summed up well in the preamble to the June 2008 draft renewable energy strategy consultation: "By taking a leading role in developing and demonstrating CCS technologies, alongside measures to facilitate the deployment of nuclear and promote the deployment of renewables, we are ensuring that there are a range of options available to power companies when they are looking at managing their emissions. They can then decide which low-carbon technologies to invest in within this framework."[44] But even though the framework is forming, it still depends on the development of a much more robust carbon price, and many believe that will not be before 2012.

Against this background and the government's insistence on market mechanisms, the U.K. outlook remains uncertain.

---

[44] *Towards Carbon Capture and Storage*, Berr (June 2008).

References

[1] Action in the U.K.: The U.K. Climate Change Programme, Department of the Environment, Food and Rural Affairs (DEFRA) (2000).

[2] Climate Change: The U.K. Programme, Department of the Environment, Transport and the Regions, November 2000.

[3] Digest of U.K. Energy Statistics, (DUKES) (2007 and 2008).

[4] Energy Efficiency Policies Around the World: Review and Evaluation, World Energy Council (January 2008).

[5] Energy White Paper: Our Energy Future - Creating a Low Carbon Economy, Department of Trade and Industry (DTI) (2003).

[6] Green Dreams: A Decade of Missed Targets, the Policy Exchange (May 2008).

[7] Meeting the Energy Challenge: A White Paper, BERR (July 2007).

[8] Meeting the Energy Challenge: A White Paper on Nuclear Power, BERR (January 2008).

[9] Review of the Feasibility of Carbon Capture and Storage in the UK, Department of Business, Enterprise and Regulatory Reform (BERR) (2003).

[10] Strategy for Developing Carbon Abatement Technologies for Fossil Fuel Use, DTI (2005).

[11] The Climate Change Levy and Climate Change Agreements, National Audit Office (August 2007).

[12] The Economics of Climate Change: The Stern Review, Nicholas Stern (October 2006).

[13] The Government's Strategy for CHP to 2010, Defra (2004).

[14] The New Energy Paradigm, edited by Dieter Helm (OUP, 2007).

[15] The U.K. Climate Change Programme, Defra (2006).

[16] Towards Carbon Capture and Storage, BERR (June 2008).

[17] UK Country Review, International Energy Agency (May 2006).

CO_2 Regulations: The View of a European Electricity Company[1]

Jean-Paul Bouttes

Executive Vice President, Corporate Strategy, Prospective and International Affairs, Electricité de France (EDF)

Jean-Michel Trochet

Senior Economist, Prospective and International Affairs, EDF

François Dassa

Head of International Corporate Relations, Prospective and International Affairs, EDF

[1]The authors would like to thank the Institut Français des Relations Internationales (Ifri) for its permission to reprint, with minor amendments, a paper originally published as Bouttes JP, Dassa F, Trochet JM. Assessment of EU CO_2 Regulations. In: Lesourne J, Keppler JH, editor. Abatement of CO_2 Emissions in the European Union, European Governance and the Geopolitics of Energy, Ifri research; 2007.

Abstract

As a European leader in the energy sector, EDF Group has long been involved in the climate-change debate. Like other major electric companies in Europe, EDF's investment strategy needs to be assessed within at least a European level, notably within the frameworks of the electricity internal market and the CO_2 market. Based on the electricity sector experience, we know that investing in available CO_2-free electricity technologies can enable us to achieve ambitious climate targets within the coming decades. We show that cap and trade as well as other public policy instruments should be appropriately designed to send the right signals for such investments.

16.1 Introduction

As a European leader in the energy sector,[2] EDF Group has long been involved in all aspects of the climate-change debate, from engagement in scientific research to plans for concrete action. The expertise offered by EDF played a central and practical role when a cap-and-trade mechanism for curbing greenhouse gases was still a matter of debate and experimentation in Europe. Supported by its existing generation assets and by new investment in nuclear and renewable sources,[3] one of EDF's main goals is to remain the lowest emitter of greenhouse gases of the seven major European electricity companies. For EDF, energy efficiency is also at the heart of its sales and marketing strategy, with a comprehensive range of energy-saving products and services.[4]

Climate policy and CO_2 regulation in France are key to shaping the investment strategies of EDF and its competitors. Yet French climate policy is intimately linked to EU climate commitments; the French electricity sector is a part of electricity and CO_2 markets that are integrated at the European

---

[2] In 2007, 98 percent of EDF revenue was generated in Europe (54 percent in France, 44 percent elsewhere in Europe).

[3] EDF is Europe's leading producer of renewable energy and the world's leading nuclear producer; its average specific emissions in Europe have been maintained at under 120 g/kWh over the past 3 years.

[4] EDF policy is formally and extensively presented in "EDF Group, 2007 Document de reference" (English version), §6.4.3.2.1, "Fighting Climate Change," available at http://investisseurs.edf .com/accueil-com-fr/edf-investisseurs/publications/. See Box 16.1 in this chapter.

level through both the electricity internal market and the EU European Trading Scheme (EU ETS). It turns out, in fact, that investment strategies of the electricity companies already represent broader European strategies. This means that any relevant assessment of CO$_2$ regulation needs to be done at the European level, even if some subsidiary aspects are more appropriate to a national level. Thus, the focus of this chapter addresses the issue of much-needed analysis at the European level.

The European energy landscape is fundamentally different from in the 1990s, when discussions on climate policy were in the early stages. In the 1990s, gas prices were low, overcapacity was prevalent in many European countries, and we were not yet thinking in terms of stringent CO$_2$ targets for 2050. Today there is no overcapacity; new power plants are needed in almost all the EU countries.

Box 16.1   EDF GROUP'S STRATEGY: FIGHTING CLIMATE CHANGE

The following is an extract from unofficial translation of the French document de reference registered with the Autorité des marchés financiers on April 14, 2008. "Fighting climate change," pp. 99–100.

As an energy corporation, EDF is involved in challenges at a global level. It is an active participant in both international climate-change negotiations and the implementation of the Kyoto Protocol and the European ETS Directive (emission quotas, flexibility mechanisms) Group is Europe's biggest energy producer, but due to the proportion generated by its nuclear and hydropower stations in its generation mix, it also has one of the lowest rates of CO$_2$ emissions. In France, 95 percent of electricity generation emits no CO$_2$, keeping its specific emission rate to less than 50 g CO$_2$/kWh, compared with the European average of around 400 g CO$_2$/kWh. The EDF Group's specific emissions rate at a global level was 120 g CO$_2$/kWh in 2007 (outside Edison and Dalkia; EDF's estimations).

The Group's ambition is to remain the electricity producer with the lowest CO$_2$ emissions per kWh among the seven biggest European producers based on its policy of optimizing the operation of its assets, its investment policy, its creation and promotion of commercial proposals, and advice on energy use to all end users.

EDF has a number of tools for reducing its greenhouse gas emissions, including, in the short term, optimizing its current plants, taking due account of operating

Continued

Box 16.1 EDF GROUP'S STRATEGY: FIGHTING CLIMATE CHANGE—Cont'd

methodologies (by factoring in carbon costs when ranking generating methods), and improving efficiency; it is also active on the emissions trading and electricity markets. In the longer term, the most important currently available tools are modifications to generating equipment (modernizing power stations, conserving hydroelectric potential, drawing on sources of renewable energy, and minimizing the most polluting methods); developing services to help customers manage their energy demands; exploiting renewable energy sources; and using the project mechanisms proposed in the Kyoto Protocol. To facilitate the performance of the commitments of the Kyoto Protocol, it is possible to use "flexibility" mechanisms in addition to policies and measures implemented at the national level .... EDF is now studying the possible circumstances in which investment projects could factor in some use of these mechanisms.

The development of renewable sources of energy is at the heart of the EDF Group's strategy, with the aim of developing a sustainable and profitable presence in the renewable energy sector in both France and Europe through the industrial management of mature plants and technological innovation in areas that are still under development.

These developments include centralized power generation projects (e.g., the construction of a hydroelectric power project at Nam Theun 2 in Laos): semicentralized projects and decentralized projects, which comprise a wide range of initiatives for energy management (included in the building, in particular: thermal and PV solar energy, heart pumps, wood-fired central heating). The EDF Group is also focusing on PV energy throughout the value chain: investments in silicon technology; manufacture of panels by the subsidiary Tenesol; development of ground-based solar power systems; new PV products, and so on. Thereby, the EDF Group intends to treble, whether alone or in partnerships, its installed capacities in renewable sources (except hydraulic) by 2012.

A substantial proportion of the R&D budget is devoted to non-CO_2-emitting technologies. Environmental R&D projects focus on the whole range of electricity issues, both upstream and downstream, including:

■ Research on CO_2 capture and storage techniques
■ Nuclear technology: fourth-generation reactors that will eventually replace EPRs; geological storage of radioactive waste
■ Decentralized generating technologies: micro-cogeneration, fuel cells, offshore wind farms, new solar energy technologies, biomass gasification

> **Box 16.1   EDF GROUP'S STRATEGY: FIGHTING CLIMATE CHANGE—Cont'd**
>
> ■ Intelligent management of networks and metering systems, combining centralized and decentralized production
> ■ Efficient use of electricity, e.g., more efficient heat pumps, electric cars, rechargeable hybrid vehicles
>
> **Demand-side management (DSM)**
>
> EDF is heavily committed to DSM for all types of customers by offering them plans to suit their needs. The aim is to extend this effort by adding a CO_2 emission reduction component that will enable customers to make their own contributions to fighting climate change. The Group's marketing proposals focus on energy saving and energy efficiency. In France its plans are increasingly predicated on eco-energy efficiency and reducing CO_2 emissions. For example, for buildings the Group offers DSM services (insulation, renovation) and thoroughly integrated renewable energy sources (geothermal using heat pumps, photovoltaic and solar thermal energy, and biomass). EDF's commercial plans also allow customers to choose low- or non-CO_2-emitting options.
>
> EDF is also implementing this commitment in-house, working to reduce all its emissions by monitoring office construction and vehicle fleets and offering a DSM program for employees. . . .
>
> EDF is committed to fighting climate change and is developing electricity offers based on energy savings through its Bleu Ciel d'EDF product range, a package of assorted renewable sources (including PV, heat pumps, solar water heating, etc.), teleservices and innovative services, and insulation/renovation services. It issues energy-saving certificates based on customer services, which focus on energy efficiency.

According to the IEA's *2006 World Energy Outlook*,[5] the EU will need to build 862 GW of generation capacity between 2005 and 2030, mostly to replace its existing fleet (723 GW in 2004). Given expected final prices of gas at 6 or 7 dollars/MMbtu (compared with 2.5 dollars/MMbtu in the 1990s[6]), combined cycle gas turbines are no longer necessarily the most competitive technology for baseload facilities. Gas will, of course, remain part of

---

[5]IEA (2006b). Corresponding data is not given by the IEA more recent World Energy Outlook (2007), but orders of magnitude are of the same.
[6]IEA (2006b).

the solution, but it is no longer the miracle technology for European supply security and emissions reduction, as the scenario for 2030 published in the *2006 European Green Paper* clearly shows.[7]

In the *Green Paper* scenario, the outlook for the coming decades is characterized by the need to meet the challenge of supply security and the pressing need to reduce CO_2 emissions while upholding the conditions that make the European economy competitive.[8] We must seize the opportunity offered by the need to renew power fleets to massively reduce emissions in the electricity sector. This we can do by turning to the best energy-saving technologies and by building a cleaner energy mix that combines the best available nuclear, renewable, coal, and gas technologies. In so doing, Europe as a whole could by 2030 make significant headway on leveling off its carbon emissions, with emissions in the electricity sector actually significantly dropping below the 1990s level. This would be consistent with a strong decarbonization target for 2050 (level of emissions divided by 4 compared to the present level) if the use of technologies reaching maturity—notably carbon capture and storage (CCS), fourth-generation nuclear, and next-generation photovoltaics—is factored in for 2030–2050.

The European Union Greenhouse Gas Emission Trading Scheme (EU ETS) can play a key role in achieving this ambitious target if it succeeds in effectively orienting the massive investments needed in the coming years toward least-emitting technologies.[9] As a matter of fact, unlike other industrial sectors, the electricity sector's potential for emissions reduction relies essentially on building new low-emitting plants.

In spite of a positive start in terms of concrete implementation and of its ability to reflect market players' beliefs about short-term conditions of supply and demand, the EU ETS has proven less than convincing in terms of environmental results. This has raised numerous questions—if not controversies—about its ability to actually drive investments along a sustainable path. In the first part, we examine the market design improvements that would enable ETS to actually provide the right incentives for investments in least-emitting technologies. But a "good" ETS design alone is not enough. In the second part, we address other requirements and look at the various complementary instruments needed for a sound public policy. After a comparative review of

---

[7]EC (2006).

[8]In this scenario, despite ambitious targets for the share of renewables in the electricity mix, Europe's dependence on imported gas will double by 2030 and CO_2 emissions will increase 10 percent in the electricity sector.

[9]For a precise overview of the two initial phases of the EU ETS, see Cornwall (2009), §3.4.1.3, and notably Figure 3.6, in this issue.

cap and trade vs. taxation, we provide insights on three key issues—namely, security of supply, research and development (R&D), and control-and-command regulation in energy policy to illustrate this argument in detail. To further enhance the EU ETS, we need to examine certain international issues, which are reviewed in the third part. This international perspective will draw the case for a genuine European sectoral approach for electricity in a context of growing concern over the impacts of the ETS on Europe's competitiveness. Finally, the fourth part contains some concluding remarks.

16.2 Improving ETS market design for massive deployment of mature low-emitting technologies

The major issue with ETS is that it should efficiently provide the incentive, as of today, for investment in clean technologies. This is not the case as yet. Improvement of ETS market design calls for a three-step analysis: (1) recognizing where the real potential lies in the electricity sector; (2) identifying the major shortcomings of the current mechanism; and (3) learning from proven successful design.

(1) In the electricity sector, the possibilities for existing generation facilities to reduce their emissions are limited. Potential efficiency gains are minimal. So is potential for developing interconnections between low- and high-emitting areas, that is, between zones where generation is mainly nuclear and hydro based and zones where coal is dominant.

Thus, most of the potential reduction in emission volumes lies in the construction of low-carbon generation facilities. Such plants take 2 to 6 years to build and have a lifetime ranging from 20 years to more than 50 or 60 years for best available technologies (i.e., supercritical coal, combined-cycle gas, third-generation nuclear, wind power, etc.).

Given the time constants and the wide range of potential carbon emissions for these technologies (1 tonne CO$_2$/MWh on average for existing coal plants in Europe; 0.8 for new supercritical coal plants, 0.4 for combined-cycle gas turbines, known as CCGTs, and close to 0 for nuclear and wind power), cap-and-trade systems should be designed to provide a clear incentive to invest in least-emitting projects.

(2) From this point of view, the ETS design has, in effect, proven particularly ineffective, providing scarce incentive to base investment decisions on carbon emissions:

■ Free allowances for new projects and cancellation of initial allowances following decommissioning are a counterincentive to factoring the implied carbon cost into decision making. This gives an added edge to emitting technologies over least-emitting technologies.

■ The periodic renegotiation of quotas every five years, together with "periodic grandfathering," can actually encourage investment in higher-emitting technologies to benefit from free allocations based on emissions from the previous period.

■ The lack of market design beyond 2012 makes it difficult to (1) anticipate any medium- to long-term CO_2 price, precisely during the first years of operation for the plants whose construction is being decided today, and therefore (2) justify any extra cost involved when investing in low-carbon or carbon-free technologies.

■ Finally, unauthorized allowance banking between periods artificially increases price volatility, rendering the short- to medium-term price trend chaotic. As shown in Figure 16.1, there was a total disconnection between CO_2 allowance prices (with the same long-term environmental impact) over a period of only months overlapping two trade periods (2007 and 2008).

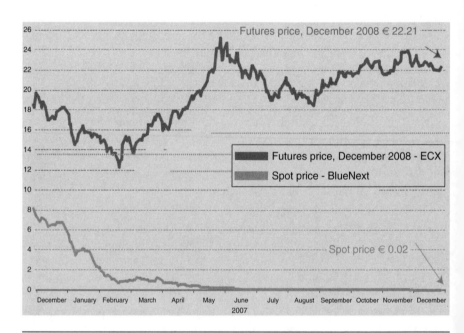

Figure 16.1 Carbon prices in Europe, 2007. Spot price of first period (2005–2007), futures price December 2008 of second period (2008–2012). Source: Blue Next, ECX, in *Tendances Carbone*, no. 21, January 2008 [4].

Here we must point out that the normal trend of the 2007 price, either toward zero, which is the case, or toward the penalty, merely reflects the mechanical decline of emissions uncertainty corresponding to the gradual collation of information on the supply/demand balance. So, the price level at the end of this period was consistent with a slight overallocation, without implying any significant overallocation at the start of the period.

Figure 16.1 clearly shows the total disconnection of CO$_2$ allowance prices between the two periods. Today, the price of the first period is of course no longer relevant. For more recent price data on the current period, see Cornwell [5], Chapter 15 in this volume.

(3) The fact that emissions trading figures among the flexibility mechanisms provided for by the Kyoto Protocol and that Europe has embraced a cap-and-trade system owes much to the success of the United States SO$_2$ market. This system's successful achievement of an environmental target at least cost can in large part be attributed to the fact that the designers and legislators behind the 1990 Clean Air Act, which instituted the SO$_2$ market, held closely to the recommendations of economic theory:

- The SO$_2$ market design, and namely the rules for initial allocation, were set for a period of 30 years—consistent with the lifetime of the equipment—without periodic renegotiation during the intermediate trading periods.

- This has made it possible to draw a clear distinction between plants built before the law's enactment and those built after it. For the former, lawmakers designed a system for free initial allocations based on emissions prior to the Act (grandfathering, including trade-offs among polluting and polluted states based on a rationale of "political economy"). Attribution was once and for all for a set duration beyond which there were no free allocations. As for new facilities, there was no question of allocating free allowances. Any new project must pay its emissions in full through auctions, and decommissioning has no effect on the duration of free allowances allocated at the time of the law's enactment.

- Allowance banking over several years is authorized to encourage the leveling of price variations over time and to lessen price volatility.

- The target for overall emissions volumes can be periodically adjusted to factor in new data on, for instance, environmental impacts. But should this occur, allocation is based on a straightforward rule known to all players from the start, and allowance recipients are notified several years in advance to avoid a process of renegotiation with uncertain results.

16.3 Choosing complementary tools for public policymaking

Improving ETS market design is no doubt indispensable, but on its own it is not enough. Efficient public policy requires additional, complementary tools that are adapted to the policy's multiple objectives. The issue here is twofold: to assess, on one hand, the value of cap and trade vs. taxes (Section 16.3.1) and to determine how, on the other hand, cap and trade must be used with other, complementary instruments to most effectively meet multiple objectives. Following the recommendations of institutional economics, we present the case for targeted instruments rather than a broad but somehow vague use of cap and trade. Insights on three key issues—namely, security of supply (Section 16.3.2), R&D (Section 16.3.3), and the complementary role of control and command regulation (Section 16.3.4) will enable us to illustrate this approach in detail.

16.3.1 TAX VS. CAP AND TRADE: THE CASE FOR HYBRID INSTRUMENTS

In the final run-up to the Kyoto Summit, both cap and trade and taxation were the focus of debate. Each of these instruments has its advantages, which must be compared in order to choose effectively.

In his seminal article Prices Versus Quantities [24], Weitzman indicates that when the cost of environmental damage and the cost of mitigating such damage are uncertain, the relative efficiency of "price" (taxation) or "quantity" (cap and trade with fixed global cap) types of regulations may vary in favor of one or the other.

With taxation, players reduce their CO_2 emissions up to the point at which their marginal costs equal the value of the tax. When costs are uncertain, the overall level of reduction is uncertain, whereas the marginal cost of reduction remains certain.

Conversely, cap and trade with a fixed global cap leads to an uncertain marginal cost and therefore to an uncertain overall cost of reduction while reduced volumes remain certain. The cost/benefit analysis for the community therefore differs depending on which instrument is used.

Weitzman showed (1) the superiority of price mechanisms when the curve of marginal environmental damage is relatively flat (i.e., when in the case of quantitative variability the marginal cost of damage varies little or shows little uncertainty), and (2) the superiority of quantity mechanisms when the curve of marginal costs of reduction is relatively flat (i.e., when in the case of variability of quantities emitted the marginal cost of reduction varies little or shows little uncertainty).

Presumptions regarding the risk of major damages in the long run due to abrupt climate change argue in favor of quantified targets in terms of cumulative annual CO$_2$ emissions over the next decades. In the absence of well-assessed probabilities regarding extreme climatic events, this can also be justified by the precautionary principle, as in the *Stern Review*.[10]

On the other hand, uncertain emissions reduction costs resulting, in the short term, from the impact of fluctuations on the use of existing emitting facilities[11] and, in the medium term, from the uncertain cost of clean technologies, rather argue in favor of a price mechanism in the short and medium term. This would make it possible, thanks to a favorable technological context, to control the marginal cost of emissions reduction while dramatically curbing the overall volume of emissions.

In fact, the current perspectives on existing and future clean technologies allow us to draw up, for the electricity sector, a trajectory for significant emissions reduction over 40 to 50 years at reasonable cost.

This prospect was put forth in a recent study by the International Energy Agency[12] (among others) proposing long-term scenarios (2050) based on a detailed assessment of the various technologies, present and future.

This benchmark study shows that if (1) we choose now to massively deploy the best clean technologies available, thereby seizing the natural opportunity offered by the need to renew today's emitting plants, and (2) we prepare to deploy the technologies that will be ready in 30 years, time, we could, by 2050, reach a worldwide electricity mix with an emissions level half that of the 1990s. Furthermore, this target could be reached at controlled cost, that is, a cost of avoided CO$_2$ not surpassing on average 25 dollars/tCO$_2$.[13] These results are similar to those of a study on the electricity sector by [2].

As an illustration, let us start with the realistic hypothesis that by 2030 coal plants using carbon capture and storage (CCS) will have reached a level of industrial and economic maturity that enables them to be massively deployed. Let us assume that by this time, CCS, used in conjunction with the massive use

---

[10]Stern (2006).

[11]Interannual volatility of CO$_2$ emissions in the electricity sector results from economic fluctuation and the variability of climate (temperature and hydro flows), fossil-fired generation being used to meet high levels of electricity consumption, and low hydro generation. This variability over an average of 10–\20 years has no impact on climate since all that counts are cumulative emissions. A 5-year, quantified, nonadjustable target may be costly to the economy, without benefit for the environment.

[12]IEA (2006a).

[13]Tech+ scenario in IEA (2006a).

of other existing technologies (demand-side management, nuclear, renewables) will represent the marginal means of CO_2 emissions reduction at a cost of 50 euros/tCO_2 with respect to standard coal technologies.[14] Optimizing the trajectory of CO_2 emissions relative to a cumulative emissions target over 50 years, we can apply Hotelling's rule[15] and justify the marginal cost of avoided CO_2 (and therefore a price) that increases over time, on the order of 12 to 25 euros/tCO_2 in the short run and 25 to 35 euros/tCO_2 in 2020, depending of the discount rate level.

This reasoning provides rationale for the use of a hybrid "price/quantity" instrument, combining the ETS with a safety valve or price floor, to make available technologies work toward achieving a quantified emissions goal in the long run.

Over the past few years, a number of authors[16] have suggested using a safety valve as an instrument of climate policy based on Weitzman's original reasoning. As Pizer explained (2002):

A hybrid policy gives producers the choice of either obtaining a permit in the marketplace or purchasing a permit from the government at a specified trigger price. Such a policy operates like a permit scheme with uncertain costs and fixed emissions as long as the marginal cost, reflected by the permit price, remains below the trigger. When the trigger price is reached, however, control costs are capped and emissions become uncertain, as in a tax scheme. By setting the trigger price high enough or the number of permits low enough, the hybrid policy can mimic either a pure quantity or pure price mechanism, respectively. Since it encompasses both tax and permit mechanisms as special cases, the hybrid policy will always perform at least as well as either pure policy.

The main advantage to safety valves and price floors is that they encourage convergence of investor expectations regarding the final price of CO_2. This said, they warrant further, in-depth discussion before they are implemented.

---

[14]As mentioned in IEA (2006a) and a recent MIT report (MIT, 2007), lower costs are still a target.
[15]According to Hotelling's rule, efficient exploitation of a nonrenewable resource (which is approximately our case, with a stabilized concentration of CO_2 in the atmosphere targeted in the long run), the percentage change in price per unit of time is equal to the discount rate over the period of emissions (in excess of a reference level supposed to be low enough).
[16]At both an academic level (Cournède and Gastaldo, 2002; Ellerman, Joskow, and Harrison, 2003) and a governmental level (the U.S. National Commission on Energy Policy, 2004; RGGI, 2005). See also Baron and Philibert (2005) and Philibert and Reynaud (2004, IEA).

This is a major issue that Helm and Hepburn's proposal for "carbon contracts" in the United Kingdom also aims to resolve.[17] Carbon contracts in some ways resemble the hybrid "price/quantity" mechanism described earlier. Long-term CO$_2$ contracts are auctioned off by government in exchange for a commitment to reduce emissions over a 20- to 30-year period. The successful bidder gains an income based on the volume of emissions avoided, at a price guaranteed for the long term. As added incentive to investors, risk allocation is improved because the risk involved with long-term commitment to climate policy is borne by the government, and investors need only assume the industrial and commercial risks tied to best technologies. This incidentally provides incentive to the government to clarify its own long-term commitments.

16.3.2 CAP AND TRADE AND SECURITY OF SUPPLY

The debate raised by Y. Smeers over free allocation for new projects is key to the success of the ETS.[18] It points to the multiplicity of public goals, particularly with regard to the combined objectives of climate, security of supply, and economic competitiveness. The focus may therefore shift to European coal for reasons of security of fuel supply, to reduce external dependency on gas, for instance. Similarly, backing for investment in generation capacity is advocated by some as a way to ensure security of power supply (minimize the risk of loss of load).

Once again, before deciding which of the various instruments to put in place to achieve these complementary objectives, we would no doubt do well to weigh their respective advantages and disadvantages and to analyze their interrelation, to avoid disappointment. Institutional economic analysis provides possible criteria for assessment and decision making.[19]

First, common wisdom, backed by advocates theory,[20] suggests that multiple objectives call for multiple instruments:

- Security of fuel supply would tend to require measures other than cap and trade, such as diversifying gas supply, rethinking relations with Russia and other producer countries, for instance, or possibly taxing imported gas, the justification being the "internalization" of a possible "externality" of security of supply.

---

[17]See Helm and Hepburn (2005).
[18]See Smeers (2007).
[19]See, for example, Dewatripont and Tirole (1999) and Maskin and Tirole (2004).
[20]See Dewatripont and Tirole (1999).

■ Security of electricity supply, on the other hand, depends on the ability of the generation fleet to meet demand in extreme peak situations. This requires incentive to invest in peak facilities and to ensure that all possible existing generation facilities are available during periods of high demand. This implies that electricity prices be able to cover fixed costs of investment in peak facilities.

A number of mechanisms have been considered or tried on wholesale markets in Europe and the United States. Capacity markets, for instance, involve remunerating installed generation capacity based on availability during peak demand, whatever the technology used to meet that demand. Financing of the mechanism can be secured through demand-side contracts for firm power shaving.[21]

Free allocation of CO_2 permits could possibly provide incentive to invest in generation facilities. Nevertheless, effective incentives to reduce carbon emissions would only hold on the condition that there be a single allocation benchmark for all kilowatt (kW) invested, regardless of the technology: A coal-generated kW would receive the same CO_2 allocation as a wind (or gas, or nuclear) generated kW. But this benchmark allocation only applies to the supply side of electricity. On the demand side, it remains less effective than the combined use of a capacity market or of a sufficiently high kWh price during extreme peak periods, with full payment of emissions by new projects, which make it possible to influence demand through electricity prices and consumer incentives to save energy—and therefore CO_2—during periods of peak demand.

Developments such as these involving the price of electricity bring us to the debate over "windfall profits," from which electricity producers are said to have benefited.

We must bear in mind that for the users of wholesale markets (producers, traders, and wholesale customers), the reasoning in terms of opportunity cost is irrefutable: Integration of the CO_2 reduction marginal cost into the marginal cost of electricity and into the market price is mechanical and consistent with competitive equilibrium. This provides added incentive for energy efficiency, which is where much of the potential for reduction resides.

This situation can nevertheless be handled intelligently in that, like supply, key demand-side solutions also constitute long-term investment choices. Because short-term elasticity is low, high short-term CO_2 price has little effect. It is better to start with a moderate price to smooth over the transition while sending clear signals for the future and consequently for investment decisions.

---

[21]For more details on such mechanisms, see Bouttes and Trochet (2004).

This same reasoning leads us to endorse a rapid, Europe-wide phase-out of free allowances for existing facilities, in a context of gradual increase in electricity prices. The measure is easy enough to calibrate and could at last forego discussion over "windfall profits."[22]

16.3.3 R&D POLICY INSTRUMENTS FOR IMMATURE TECHNOLOGIES

A cap-and-trade program with the right market design would provide investors with institutional visibility and long-term price signals (20–30 years). This would enable them to steer investment toward low- or nonemitting best technologies. But it is not enough to provide the incentive needed to accelerate mass deployment of technologies for which industrial and commercial maturity is still far off. Other instruments can and must be used to complete the picture. The choice of tools could, in our opinion, rely on a distinction between technologies based on a double segmentation:

■ The first segmentation breaks technologies down into three categories according to how mature they are: (1) technologies nearing a competitive phase for which the main issue is to lower costs through progress on the learning curve, such as wind power; (2) technologies for which costs are still far from competitiveness and that require further advancement, such as CCS or fourth-generation nuclear; and (3) technologies that, to mature, will require technological breakthroughs and are therefore still in the R&D phase, such as hydrogen fuel cells.

■ The second segmentation distinguishes between two types of equipment: (1) large-scale facilities (100 MW or more, requiring several hundred million euros in investment), which involve sizing government financing according to the degree of maturation of the various technologies, fostering public/private partnerships and providing regulatory conditions conducive

---

[22]Incidentally, to avoid common misunderstandings, it is important that we clarify the meaning of key expressions and use them with rigor and precision:

■ Considering that most new plants in Europe are built by existing gas or electricity companies (whether inside or outside their previous area of service), *new projects* would be more suitable than the term *new entrants,* used currently in the ETS Directive.

■ *Grandfathering* should be used in the American sense, that is, strictly limited to allowances allocated prior to the Directive.

■ We should also be aware of the proper use of *windfall profits.* Since new plants have to be developed anyway, the competitive market price should reflect the total cost of an efficient new plant, including investment and maintenance costs and not only variable fuel costs.

to the deployment of the facilities, and (2) small-scale equipment intended for more diffuse uses and often installed directly in consumer homes, such as heat pumps and solar water heating. This last category will require public authorities to lower the transaction costs that tend to hinder the deployment of technologies that have reached or will soon reach maturity.

The junction between these two segmentations determines how specific targeting of public policies will be adapted for each technology.

Let us take three examples from the electricity sector:

■ For CO_2 capture and storage,[23] the right public policy instruments would help accelerate the construction of industrial demonstrators, thus enabling the validation of the main technological options under competition and to determine an adequate regulatory framework on CCS. Insofar as the range is broad enough,[24] with the potential and degree of maturity for each of them still heterogeneous, it seems indispensable to size public support based on thorough analysis of each available technology. Finally, public policy in this area must closely follow the dynamic rise of demonstrators to prepare the transition from today's prototypes (a few dozen MW) to industrial deployment (several hundred MW and several million tonnes per year of CO_2 to be stored).

■ Other mechanisms are needed for photovoltaics, which are very different in terms of expected maturity and, as with diffuse technologies, industrial characteristics. Their current level of cost in Europe, roughly 10 times higher than the price of competitive technologies, suggests that we should intensify efforts on R&D rather than on encouraging immediate massive deployment. In other words, lower costs for photovoltaics seem more likely to come from technological breakthroughs than from mere progress on the learning curve. However, there are other arguments in favor of PVs' rapid deployment. These include the development of a new high-technology sector and thus job creation as well as increased consumer awareness of diffuse technologies. These are legitimate arguments that only serve to illustrate our point that the multiple goals of public policy should be clarified and quantified.

---

[23]For a more in-depth analysis of the technological aspects of CCS and the public policy for implementation it implies, we refer the reader to the MIT report on coal (MIT, 2007).

[24]Three major technologies for capture—post-combustion, oxycombustion, and precombustion—and two main geological solutions for storage: deep aquifers and depleted oil and gas fields, to which must be added the various generation technologies associated (coal and gas).

■ As for wind power, the degree of industrial and commercial maturity reached by onshore technologies suggests that we are already most of the way down the learning curve. Thus we can expect that the market price of CO_2 in an ETS with improved market design will soon suffice to incite investment in this area.[25]

16.3.4 COMPLEMENTARITY OF CONTROL AND COMMAND INSTRUMENTS FOR A CONSISTENT ENERGY POLICY

Public policy must take into account complementary tools not only to meet its varied objectives. Cap and trade alone, even with improved market design, will hardly be able to fulfill the specific goal of the ETS, which is to orient investment at least cost. Other measures are a very real necessity for the success of the energy policy, of which the ETS is but a part:

■ On the supply side, the right incentive for the massive development of new clean-generation technologies will have little effect if it is not possible to actually build them, and rapidly. The price mechanism is dependent on clear procedures for licensing and site authorization, which is able to address local opposition (Nimby syndrome). This is true whether it is for boosting wind power capacity, bolstering transmission and distribution systems, or building new coal or nuclear plants, methane terminals for liquefied natural gas (LNG), and, later, CCS.

■ On the demand side, procedures for standardization and labeling of electrical appliances must go hand in hand with CO_2 and electricity price signals; home insulation and heat pumps should be encouraged, using tax breaks that compensate for an "imperfect" housing sector and provide benefits to both homeowners and tenants.

Obviously, the analysis and complementarity of these various instruments, especially price and control-and-command mechanisms, are applicable not only in the electricity sector but also to mature technologies in other energy-consuming sectors, such as transport. London's urban toll mechanism, carried out in parallel with an efficient public transport program, is one such example. The toll alone, without the transport program, would have little impact, given low consumer elasticity in the absence of a substitute for taking

---

[25]Levelized costs of wind power provided by Wiser and Hand (2008), Chapter 9 in this volume, are consistent with this opinion.

the automobile. Conversely, the development of public transport will be less efficient without a CO_2 price signal that penalizes using gasoline for automobiles.

16.4 ETS from an international perspective

In dealing with CO_2 regulation in Europe, we must consider how this public policy instrument fits in with the current international framework and the negotiations under way. The international perspective should enable us to shed light on two decisive components for the future of the ETS: the opportunity offered to build a genuine sectoral approach for electricity in Europe on one hand and issues of competition between sectors and between countries on the other.

16.4.1 TOWARD A SECTORAL APPROACH IN EUROPE

The history of the ETS is closely tied to the international negotiations on climate change that led to the Kyoto Protocol—so closely tied, in fact, that it tends to efface the distinction, useful though it may be, between negotiations held at the international level and those at the European level.

Each of these two levels has its own dynamics and its own capacity for commitment and control that are, by nature, very different. Measures taken by the international community and those of Europe should, in our view, be considered as much more complementary than they have been to date. Although an international agreement is indeed of utmost importance, we cannot allow the ETS to remain a mere offshoot, an approach that has, symptomatically, led to aligning the ETS Directive deadline with the Kyoto Protocol.

Europe has the capacity to commit to the long-term, stable design needed for an efficient ETS. Europe needn't, therefore, adhere to the reasoning behind periodic renegotiation at the international level. But if the European position is to be more than a scale model of international architecture, it will no doubt be necessary to reconsider the subsidiarity that continues to characterize the EU ETS.

Indeed, in its current form, the ETS appears to be more of a patchwork of national mechanisms than a unified system.[26] Yet it is both ineffective and costly (in environmental terms) to leave the definition of allocation rules up to the subsidiarity of each country, a process in which, as Gresham might have

---

[26] The expression dates back to 1858, from H. D. Macleod. See http://eh.net/encyclopedia/article/selgin.gresham.law

said, "bad rules drive the good rules out of circulation." This is a recurring issue that we can illustrate with an example from electricity. Imagine that a country decides to allocate free allowances to new plants while its neighbors decide the contrary. The consequences are fairly mechanical: Countries where new projects do not pay their emissions, and therefore with massive investment in plants that are CO$_2$-inefficient but also inexpensive, will compete with neighboring countries that have plants that respect the environment but are less competitive economically. Ultimately, the "virtuous" countries will align their own rules with the "bad rules" of their neighbor.

There is at present both an urgent need and an opportunity for a truly European public policy framework. A European sectoral approach for electricity would be based on an ETS with a cap defined at the European level for the whole sector and a single declining rate. In such a European scheme, all existing emitting plants would receive the same percentage of their past emissions, year by year, and all new plants would pay for their emissions, irrespective of the country in which they are located.

Such Europe-wide mechanisms would dramatically foster incentive to invest in lowest-emitting technologies, at the same time helping to resolve the state aid issues that are currently growing within the EU.[27]

16.4.2 COMPETITION BETWEEN SUBSTITUTABLE SECTORS AND BETWEEN COUNTRIES

When broaching a possible sectoral approach in Europe, two open questions arise: ETS scope and international competition:

■ At the European level, decisions have to be made as to which sectors it is adequate to include within the EU ETS. There should be discussion on broadening, as much as possible, the range of sectors included within the European trading scheme, beginning with the current scope, which appears too restricted. Some sectors with a significant potential for emission reductions are currently outside the ETS scope.[28] These are highly substitutable emitting sectors—that is to say, sectors that can be substituted with electricity at a cost comparable or inferior to the price of CO$_2$ avoided for electricity. Heating is a good example: Substantial CO$_2$ abatement can be achieved by end users through the substitution of fossil-fuel technologies

---

[27]As a "second-best solution," one could also envisage a common technology benchmark across Europe, but this would leave open some equity issues due to past investments.
[28]See, in particular, the Eurelectric report, *The Role of Electricity* (2007).

with heat pumps, an example to which Japan can testify with its very active policy in this area.

■ The transport sector could also be taken into consideration. For automobiles, future substitution of fossil-fuel vehicles by plug-in hybrids could be greatly encouraged through a CO_2 price signal applied to this sector. Yet this example might constitute a limit-case: As mentioned earlier, the extension of the ETS perimeter should carefully take into account the state of maturity of substitute technologies. Indeed, it would be counterproductive to include sectors with too-high substitution costs; this would increase the price of CO_2 across the whole market, with very little gain in terms of emission reductions.

■ Finally, the impact of the ETS or other European climate mechanisms on Europe's competitiveness with respect to other parts of the world warrants a closer look. This is a real economic and environmental challenge, given the risk of CO_2 leakage—that is, the risk that GHG-intensive firms relocate to countries lacking stringent CO_2 emission regulations. The issue is open for assessment and recommendations.[29]

Implementing a cross-border adjustment mechanism at the European level[30] could be one possibility. This mechanism requires that the carbon content of products imported by sectors under the ETS be treated as though the products were of European origin. Goods coming from a country where there is no system equivalent to the ETS would be subject to the purchase of allowances, just as though they were produced within the EU.

An ETS safety valve could be another possible measure for addressing leakage. Setting the price cap sufficiently low in the short term and at a reasonably increasing level in the medium term might mitigate leakage in a context where competing countries would progressively join global efforts against GHG emissions.

Both potential instruments must be discussed in depth to assess their possible impact. An assessment must take into account a changing international landscape that includes countries such as the United States, which might very well implement their own CO_2 market within the next 5–10 years, and developing countries such as China and India, where such implementation will probably take longer but where other mechanisms are nevertheless likely to emerge in the short run through Sustainable Development Policies & Measures (SDPAM) or sectoral commitments.

---

[29]As far as the leakage issue may be related in part with industrial sector exposed to international competition, see also the comments from Cornwall (2009) in Chapter 15 of this volume.

[30]See Ismer and Neuhoff (2004) and Godard and Avner (2007).

16.5 Conclusion

The credibility of European leadership in its efforts to mitigate climate change rests with achieving tangible CO_2 emission reductions. In the electricity sector, such results are attainable—at controlled cost—insofar as competitive CO_2-free technologies already exist and new technologies are visible on the horizon, especially those able to augment potential reductions in the most emitting facilities (coal and gas). These technologies were only briefly reviewed here but are extensively analyzed by specific chapters in this book, including chapters on energy efficiency, hydroelectricity, nuclear, clean coal, wind, and geothermal on the supply side.

The ETS can and must play a fundamental role toward implementing this technological scenario. To this end, throughout this chapter we have pointed to four main conditions:

- Reform the current ETS design in favor of a system in which all new projects pay for their allowances, where rules are stable and designed for the long run with the possibility of "banking" over time
- Build a truly European ETS, letting go of the national positions currently in force in favor of European market rules
- Ensure that the technologies of tomorrow will be ready on schedule thanks to R&D policies adapted to each
- Enable investment to effectively move forward by improving siting procedures

To these we might add another, equally crucial condition: the need to act fast. . . .

REFERENCES

[1] Baron R, Philibert C. Act Locally, Trade Globally: Emissions Trading for Climate Policy. Paris: OECD/IEA; 2005.
[2] Bouttes JP, Leban R, Trochet JM. A Low Carbon Electricity Scenario. Chaire Développement Durable de l'Ecole Polytechnique; 2006 Cahier No. DDX-06-10.
[3] Bouttes JP, Trochet JM. La conception des règles des marchés de l'électricité ouverts à la concurrence. Economie Publique, Etudes et Recherches 2004:14.
[4] Caisse des Dépôts. Tendances carbone January 2008:21.
[5] Cornwall N. Chapter 15 in this volume. 2009.
[6] Cournède B, Gastaldo S. Combinaison des instruments prix et quantités dans le cas de l'effet de serre. Economie et Prevision 2002:156.
[7] Dewatripont M, Tirole J. Advocates. Journal of Political Economy February 1999;107(1).

[8] EC. Annex to the Green Paper: A European Strategy for Sustainable, Competitive and Secure Energy. What Is at Stake: Background Document. Commission Staff Working Document. COM (2006) 105 final.

[9] Ellerman D, Joskow P, Harrison D. In: Emissions Trading in the US: Experience, Lessons and Considerations for Greenhouse Gases. Prepared for the Pew Center on Global Climate Change; May 2003.

[10] EU. Directive 2003/87/EC of the European Parliament and of the Council of 13 October, establishing a scheme for greenhouse gas emission allowance trading within the Community. 2003.

[11] Eurelectric. The Role of Electricity: A New Path to Secure and Competitive Energy in a Carbon-Constrained World. 2007.

[12] Godard O, Avner P. Post-Kyoto Climate Policy and Conditions for a Successful Unilateral Commitment of the EU. Ecole Polytechnique, research workshop on the impacts of climate change policies on industrial competitiveness. June 12, 2007.

[13] Helm D, Hepburn C. Carbon Contracts and Energy Policy: An Outline Proposal, Available at www.dieterhelm.co.uk/publications/CarbonContractOct05.pdf; 2005.

[14] IEA. Energy technology perspectives: scenarios and strategies to 2050. 2006a.

[15] IEA. World Energy Outlook 2006. 2006b.

[16] Ismer R, Neuhoff K. Border Tax Adjustments: A Feasible Way to Address Nonparticipation in Emission Trading, Cambridge Working Papers in Economics 0409, MIT-CEEPR; 2004.

[17] Maskin E, Tirole J. The Politician and the Judge: Accountability in Government. American Economic Review. 2004.

[18] MIT. The Future of Coal: Options for a Carbon Constrained World. Cambridge, MA: Massachusetts Institute of Technology; 2007.

[19] Philibert C, Reynaud J. Emissions Trading: Taking Stock and Looking Forward. Paris: OECD and IEA Information Paper; 2004.

[20] Pizer W. Choosing price or quantity controls for greenhouse gases. Climate Issues Brief, no. 17. July, Resources for the Future. Article reproduced in National Commission on Energy (2004), Policy Ending the Energy Stalemate, Technical Appendix. Washington, DC; 1999.

[21] RGGI. Regional Greenhouse Gas Initiative: Memorandum of Understanding. Signed by the states of Connecticut, Delaware, Maine, New Hampshire, New Jersey, New York, and Vermont, 2005.

[22] Smeers Y. Description and Assessment of EU CO_2 Regulations. In: Lesourne J, Keppler JH, editors. Abatement of CO_2 Emissions in the European Union, "Les Etudes de l'Ifri." Paris: Ifri; 2007.

[23] Stern N. Stern Review on the Economics of Climate Change. London: Report to the Prime Minister and the Chancellor of the Exchequer; 2006.

[24] Weitzman ML. Prices versus quantities. Review of Economic Studies 1974;41:477–91.

[25] Wiser R, Hand M. Chapter 9 in this volume.

Low-Carbon Electricity Development in China: Opportunities and Challenges

Joanna I. Lewis,* Qimin Chai,** and
Xiliang Zhang**

*Georgetown University
**Tsinghua University

Abstract

China's role in an international climate-change solution cannot be over-stated. Now the world's largest emitter of greenhouse gases, China has become the focus of scrutiny as climate-related impacts are revealed and global policy solutions are debated. At the center of this scrutiny is China's rapidly growing power sector: the largest single source of carbon dioxide emissions in the world. Coal has played and continues to play a crucial role in powering China's economic development, particularly for electricity generation. Diversification away from coal and toward a range of low-carbon power sources will need to be at the core of any climate change mitigation strategy that moves China to a lower-carbon development pathway.

17.1 Introduction

China's role in an international climate-change solution cannot be over-stated. Now the world's largest emitter of greenhouse gases, China has become the focus of scrutiny as climate-related impacts are revealed and global policy solutions are debated. At the center of this scrutiny is China's rapidly growing power sector: the largest single source of carbon dioxide emissions from any country in the world. It is for this reason that this chapter focuses on China's power sector in the context of this book on electricity and climate change.

China's rapid economic growth over the past few decades, and in particular over the past few years, has resulted in rapid expansion of its energy infrastructure and the electric power sector in particular. Coal has played and will continue to play a crucial role in powering China's economic development, particularly for electricity generation. Coal currently fuels about 80 percent of China's power generation, and this share has not fallen below 70 percent in 30 years. In 2006 and 2007 alone, China added about 170 GW of new coal-fired power plants—more than it added in the previous 6 years combined. To put this in an international perspective, China currently has more coal plants in use than the United States, India, and the United Kingdom combined.

Diversification away from coal and toward a range of low-carbon power sources will need to be at the core of any climate-change mitigation strategy that moves China to a lower-carbon development pathway. This diversification would also come with many other environmental benefits, since coal is also to blame for many of China's localized air- and water-quality issues, and the power sector is China's largest source of pollution. China's climate-change strategy to date, however, has not included a targeted program to move away from coal. Though some progress has been made in installing higher-efficiency coal plants, currently there are no plants in China for capturing and storing carbon dioxide emissions.

This chapter explores the current state of China's electric power sector and options for "low-carbon" power generation. Section 17.2 begins with an overview of the power sector in China including the current fuel mix, the structure of the power industry, and projected investment and emissions trends. Section 17.3 examines the current role of low-carbon power sector options for China, and Section 17.4 presents a modeling analysis of the feasibility of moving toward a low-carbon power sector in China. In the final section, these options are discussed in the context of China's current energy policy priorities, the politics of climate change, and the outlook for future technology developments that could shape the structure of China's power sector.

17.2 China's power sector

17.2.1 OVERVIEW

China's vast electricity sector faces problems surrounding regional resource disparity and the growing complexity of coordinating supply with demand. Although China currently ranks second in the world, after the United States, in terms of both installed generating capacity and annual power generation, per capita electricity consumption is only about one sixth to one quarter that of developed countries. The Chinese government estimates that economic growth and rising energy consumption will necessitate a substantial expansion of installed power-generating capacity in coming years; experts project that China will account for the largest increase in national power demand through 2030 [6]. This not only has tremendous local and global environmental implications; it also poses a huge technical challenge for a sector plagued by inefficient state-owned enterprises, aging capacity, and transportation bottlenecks.

The world's third largest country in terms of area, China has a rich supply of coal and hydropower potential for electricity generation. With 11 percent of the world's proven coal reserves, China has been the world's largest coal producer and consumer since 1988 [7]. As a result, coal dominates the country's power generation, accounting for 78.5 percent in 2005 (Figure 17.1). There is a substantial mismatch, however, between the geographic distribution of these rich resources and China's major centers of population, industry, and economic growth. Both the highest quality and highest concentration of coal reserves are generally found in the north, whereas hydropower potential is concentrated in the southwest. The energy-hungry and economically dynamic areas of south-central and eastern China have only about 9 percent of national coal reserves [18]. The transport of power over thousands of kilometers from the economically underdeveloped west and north to the major load centers of the east and south adds significantly to the cost of electricity supply and is a major constraint on the development of the power sector [34].

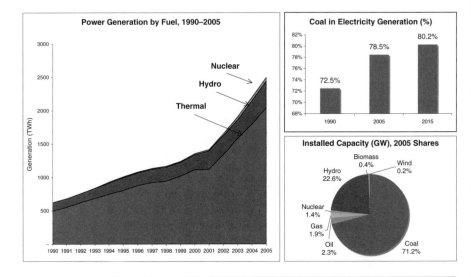

Figure 17.1 China's power generation by historic and future trends, 1990–2015. *Note:* Non-hydro renewable electricity (including wind and biomass) represented less than 1 percent of power generation in 2005.
Source: Sinton et al. (2004) [26], NBS and NDRC (2006) [20], IEA (2007b) [12], and EIA (2007) [6].

Recently, the Chinese government has begun to stress fuel diversification for electricity generation to mitigate some of the environmental and health impacts of burning coal. Plans emphasize the expanded use of natural gas and hydropower and include several recent transnational gas pipelines and large dams, such as the 18,200 MW Three Gorges Dam. Because only about 10 percent of China's total hydropower potential had been tapped as of 1995 [1], the Open Up the West (*Xibu dakaifa*) program, launched in 1999 to promote Western development, specifically calls for the development of hydropower in the western provinces. Aggressive policies also aim to promote nonhydro renewables such as wind and biomass energy.

Total electric power capacity additions in the past few years have been record breaking. In 2007 91 GW were added, and 105 GW were added in 2006—the largest year-on-year increase ever recorded in China or in any nation in the world. At the end of 2007, China's power sector had a total installed generation capacity of about 713 GW. Almost all recent capacity additions have been coal-fired—as much as 90 percent in 2006. Over half (51 percent) of China's total coal consumption is for power generation [11]. As a result, the share of electricity coming from coal has actually increased in recent years and is projected to continue to increase, despite the aforementioned efforts to diversify supply. Though nuclear and hydro generation has also increased, their share in the generation mix is still quite small. Nonhydro renewables still contribute less than 1 percent of power generation in China. As of 2007, there were about 554 GW of installed coal-fired thermal power capacity, 145 GW of hydropower capacity, 9 GW of nuclear capacity, and about 4 GW of wind power capacity.

China's electricity demand is currently driven by the industrial sector, which accounts for 68 percent of electricity consumption. Residential and commercial use is responsible for about a quarter of China's total electricity consumption today but is expected to represent an increasing share of demand in coming years.

Although about 60 percent of China's population lives in rural areas, China has a legacy of successful rural electrification programs. About 90 percent of China's rural inhabitants and 99.4 percent of China's counties (not including Tibet) are electrified, making China's electrification rates much higher than the average for the other developing countries in the region. The estimated number of citizens without access to electricity is between 50 and 100 million [1, 29]. It is estimated that about 10 million people still use solid fuels (coal, biomass) for cooking and heating in rural areas in China, though these are increasingly being replaced by electricity as incomes rise.

China's power sector is facing reliability issues tied to an aging grid infrastructure and "boom/bust" supply cycles marked by alternating periods of highly disruptive supply shortage and inefficient overcapacity [23, 34]. Power shortages are not just caused by a lack of sufficient supply but are also blamed on high fuel prices, bureaucratic inefficiencies and regional electricity transmission, and distribution and fuel transport issues. As a result, it will be difficult for China to achieve its electricity capacity goals without significant additional power sector reforms and policy changes.

Some of these changes could come in the form of pricing reforms. Historically (until at least 2002), coal was sold to generators at prices below market values, with prices determined annually at an Annual Coal Procurement Conference run by central government authorities [23]. This changed with a 2004 National Development and Reform Commission (NDRC) policy that linked coal prices for electricity generation to wholesale power prices and allowed generators to pass most of any increase in coal prices to consumers. However, few power markets in China are truly governed by market-based price signals, and many challenges remain.

Although there is some discrepancy in the data on the cost of generation, and exact power prices vary according to user, it is evident that coal is still the least expensive source of electricity in China, with costs ranging from as low as 2.8 cents per kWh up to just over 4 cents per kWh. The cost of power from advanced, higher-efficiency coal plants is only slightly higher. Natural gas is more expensive (4.5 cents to 8 cents per kWh); large hydropower ranges from 3 to 6 cents per kWh; and wind is around 6 to 9 cents per kWh. Nuclear power in China is estimated to cost around 4.5 cents per kWh [12].

17.2.2 INDUSTRY STRUCTURE

China has six regional power grids that are owned and operated by six regional state-owned grid companies: Northern China, Northeastern China, Eastern China, Central China, China Southern, and Northwestern China. All except the China Southern Power Grid Company are subsidiaries of the State Grid Corporation, which owns about 80 percent of the country's grid system; the rest is owned by the Southern Grid Company.[1] A series of power sector reforms in 2002 separated power generators from distributors and established the provincial and municipal grid utilities as the sole purchasers of power from generators. The grid companies then "resell" power to customers and distribution companies. Most power generation is sold in the form of long-term contracts with terms set by NDRC.[2] China's electricity grids are currently organized as illustrated in Figure 17.2.

There are five state-owned power-generating companies (see Table 17.1), which were established when the assets of the State Power Corporation were divided up in the 2002 reforms. The "big five" generating companies—China Huaneng Corporation, China Huadian Corporation, China Power Investment Corporation, China Guodian Corporation, and China Datang Corporation—each owns about 30 GW of generating capacity (35 percent of China's total capacity) distributed across the country. To some extent, these companies or their subsidiaries have been opened to partial privatization through listings on one or more international stock exchanges, facilitating foreign investment and ownership [23]. Local government corporations and private companies own other generation assets. Very little of the capacity is wholly foreign owned; Sino-foreign joint ownership is more common.

---

[1]The State Grid Company serves 1 billion people and has over 1.5 million employees [23].
[2]Separation of generation assets from the wires is not yet fully complete, since both the State Grid and Southern Grid companies retain ownership of some generation facilities. The Southern Grid Company owns only a very small amount, but the State Grid possesses over 30 GW, consisting mostly of hydro and coal-fired units. Furthermore, many provincial grid companies also own generation (RAP, 2008).

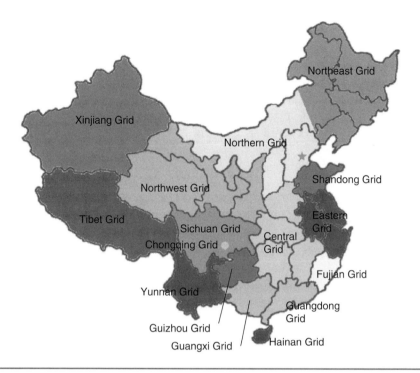

Figure 17.2   China's electric power grid system.
Source: Lewis (2005) [15].

Table 17.1   Electricity investment by ownership type, 2002

Ownership Type	Investment (Million Current Yuan)	Share (%)
State owned	208,218	74.2
Urban collective	1643	0.6
Joint venture	284	0.1
Shareholding corporations	62,058	22.1
Foreign invested	4210	1.5
Hong Kong, Macao, and Taiwan invested	4384	1.6
Total	**280,797**	**100.0**

Source: Sinton et al. (2004) [26].

17.2.3 PROJECTED GROWTH AND INVESTMENT NEEDS

Investment in China's power sector has been astonishing. Fifty billion dollars were invested in 2005 alone, a 31 percent increase over the previous year and a 231 percent increase from 10 years earlier.[3] The IEA projects that China will invest $2.7 trillion in the power sector (including generation, transmission, and distribution) between 2006 and 2030 as China's installed capacity reaches 1775 GW by 2030 [12]. Unless government policies direct this investment toward new technologies, this investment will go on the same carbon-intensive development path as it has for the past 30 years. That would amount to a colossal lost opportunity because the consequences of the investment decisions made in China today and tomorrow will be with us for decades to come.

The challenge of decarbonizing investment, however, will be significant. It is estimated that an additional $30 billion per year would be needed in developing countries to "green" the $160 billion of investment already needed in the power sector annually. It is estimated that global additional investment and financial flows of $200–210 billion will be necessary in 2030 to return global greenhouse gas (GHG) emissions to current levels [27]. The IEA estimates that to reduce global GHG emissions by 50 to 85 percent by 2050 from 2005 levels, as the IPCC recommends, we will need to invest around $45 trillion globally—and much of this investment will need to take place in China. This investment would include additional R&D, larger investments in deployment of technologies that are not yet market competitive, and commercial investment in low-carbon options stimulated by CO_2-redution incentives [14].

Despite the introduction of economy-wide market reforms in China that have resulted in competition, more realistic price signals, and increased foreign trade and investment, government involvement in power sector investment remains strong. However, public funding will be insufficient to meet the scale of the investment challenge China will face in transitioning to a low-carbon power sector. It will be critical for government funds to be used to leverage larger private investments by creating incentives for deviating from business-as-usual technologies.

International funding support could also play a critical role. Although the Chinese government is still concerned with allowing foreign control of crucial economic sectors, including power generation capacity, it is increasingly

---

[3]Estimate based on investment in fixed assets of state-owned units in energy industry (Sinton et al., 2004, Table 2-1).

willing to support the involvement of international financial institutions and foreign direct investment in China's electricity sector. Institutions such as the World Bank are only beginning to explore how their resources can be used to leverage low-carbon investments in the developing world through, for example, their Clean Energy Investment Framework [31].

Access to advanced technology will be equally important. China has placed a particularly heavy emphasis on technology transfer in the international climate negotiations, and at home China has adopted many innovative domestic policies that have helped promote technology transfer in commercially available low-carbon energy technologies. Many of these policies are in place to encourage technology transfer in the form of licensing intellectual property rights by foreign companies to Chinese companies. In discussions with China, the United States and the European Union have emphasized the importance of the removal of tariff and nontariff barriers to the trade of environmental goods and services so that broader trade in these technologies can occur unimpeded. The World Bank estimates that global trade in climate-friendly technology, currently about $70 billion per year, will need to expand substantially for the world to achieve cost-effective reduction in emissions.

17.3 Power sector emissions

17.3.1 CONVENTIONAL AIR POLLUTANTS

As China's electricity sector grows at rapid rates and continues to rely on coal, pollution levels are also increasing. Coal is one of the most polluting energy sources, with coal combustion for power releasing sulfur dioxide, nitrogen oxides, and particulate matter. Regulation of emissions from power plants can help reduce emissions, and China has relatively stringent environmental regulations in place. However, these regulations that target power plant emissions are not always effectively enforced, particularly because running pollution control equipment on a power plant can decrease the net power output, resulting in lost revenue from power sales to the plant owner. As of 2005, only 45 GW of the then 389 GW of installed thermal capacity had flue gas desulfurization units in place though this number has reportedly increased in recent years [23].

China's power sector is the single largest source of criteria air pollutants, emitting about 44 percent of China's SO_2 emissions and 80 percent of NO_x emissions. Consequently, China is home to 5 of the 10 most polluted cities in the world. Acid rain falls on one third of China's territory, and one third of the urban population breathes heavily polluted air. Studies have estimated that poor air quality imposes a welfare cost to China of between 3 and 8 percent of GDP [32].

17.3.2 CARBON DIOXIDE EMISSIONS

China's power sector is the largest source of CO_2 emissions in China, responsible for about one half of energy-related CO_2 emissions [13]. Mitigating the CO_2 emissions in China's power sector will consequently be a crucial part of any global climate-change solution.

It is now almost universally accepted that China is the largest national emitter of CO_2 on an annual basis. In 2007, China's emissions were up 8 percent from the previous year, which would mean that Chinese emissions surpassed U.S. emissions that year by 14 percent [19]. It is important to note that China's per capita emissions are still low (below the world average and only about one fifth those of the United States), as are its cumulative emissions over the past century [4].[4] But China's economic growth is more emissions intensive than that of not only the developed countries but also most other developing countries, and it is well above (26 percent) the world average. Currently, China emits 35 percent more CO_2 per dollar of output than the United States and 100 percent more than the European Union [6].

Looking ahead, and recognizing that projections of China's emissions have been notoriously inaccurate, the trends in China's emissions growth are clear. As illustrated in Figure 17.3, the EIA projects that China will emit over 11 billion tons of CO_2 in 2030, 26 percent of the global total in that year and 41 percent higher than in the United States [6].[5]

Coal also has a higher carbon content than any other fuel—over 700 Kg-C/TCE, compared with about 400 Kg-C/TCE for natural gas—resulting in higher carbon dioxide emissions per kWh generated than any other power source. Consequently, about 98 percent of China's power sector CO_2 emissions come from coal use [13].

---

[4]Greenhouse gases stay in the atmosphere for years and CO_2 for about a century; therefore it's the buildup of gases over time, rather than annual emissions, that is important from a scientific perspective. The United States is responsible for about 30 percent of energy-related CO_2 emissions into the atmosphere between 1900 and today and China for only about 10 percent.

[5]The IEA's projections are quite similar (IEA, 2007b). However, both organizations' projections of China's emissions have been very far off in the recent past. Between 2004 to 2007, the EIA revised its projections upward by 3.5 billion tons C: the equivalent of the total emissions coming from Central and South America, the Middle East, and Africa combined.

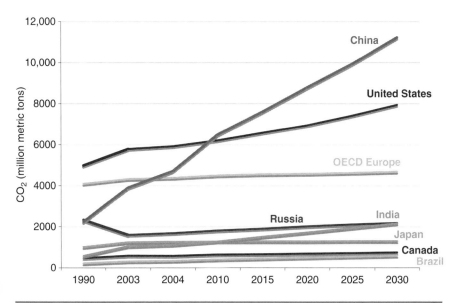

Figure 17.3 Selected countries' historic and projected carbon dioxide emissions, 1990–2030.
Source: EIA (2007, Table A10, "World Carbon Dioxide Emissions by Region, Reference Case, 1990–2030").

▌ 17.4 Decarbonizing China's power sector

This section reviews the supply-side technology options being promoted through current Chinese government policies to promote low-carbon technologies in China. It also discusses demand-side policies that can help curb the need for as rapid an expansion in the power sector, which will be crucial given the scale of growth expected in coming years.

17.4.1 RENEWABLES

China is a world leader in renewable energy development (see Figure 17.4). One of only a handful of countries with a national law to promote the development and dissemination of renewable energy technologies, China passed its Renewable Energy Law in 2005, and it went into effect in January 2006 [21]. The law offers financial incentives for renewable energy development, including favorable tariffs on renewable power generation, mandated purchase of renewable power, and mandated grid interconnection.

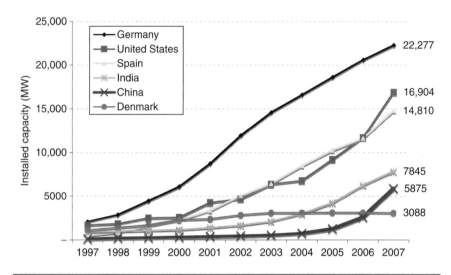

Figure 17.4 Cumulative wind capacity: leading countries 1997–2007.
Source: BTM, 2007; *Windpower Monthly*, 2008 [3, 30].

Targets that have been announced in conjunction with the renewable energy law and subsequent government documents include 15 percent of primary energy from renewables by 2020 (including large hydropower), up from about 7 percent today, and 20 percent of electricity capacity by 2020, which includes 30 GW of wind power, 20 GW of biomass power, and 300 GW of hydropower capacity. Policies to promote many renewable energy technologies in China also aim to encourage local technology industry development; China is already producing commercial large wind turbines that sell for approximately 30 percent less than similar European and North American technology [33, 36] and is the largest manufacturer of solar photovoltaic panels and solar hot water-heating systems in the world. Thirty-five million homes in China get their hot water from solar water heaters—more than the rest of the world combined [24].

China has been pursuing large dams for hydroelectric generation for decades and is currently the largest producer of hydroelectricity in the world. More hydropower dams are expected to be built in coming decades in China, but the share of hydropower in China's total power mix, currently 16 percent, is expected to decline. The government has a target of 300 GW of hydropower capacity by 2030.

Even though China's renewable energy development has been aggressive, nonhydro renewables still contribute less than 1 percent of China's electricity generation. In the near term, only wind energy is likely to be able to supply a more significant share of China's electricity. However, even if China's aggressive wind target of 30 GW in 2020 is far exceeded, as is expected, the contribution to power generation would likely remain well below 5 percent. Solar technologies are being explored for scale-up in China, particularly concentrating solar power (CSP) for large-scale solar electric generation. Even though China is one of the largest manufacturing bases for manufacturing solar photovoltaic panels in the world, the cost of deploying panels domestically is still thought to be too high to justify government support programs.

17.4.2 NUCLEAR

Nuclear power represented only about 2.4 percent of total electricity generation in 2005 in China. Despite ambitious plans to scale up nuclear capacity in the coming decades, this share is not expected to increase significantly. Approximately 21 new nuclear plants are either under construction or are about to be constructed and have already been approved by the State Council. China has set extremely aggressive goals for its nuclear expansion, including a target of 40 GW of nuclear capacity by 2020 and 20 percent of its electricity generated by nuclear plants by 2030.

Historically, the building of new plants has been strained by long construction times and bottlenecks in nuclear component technologies. Consequently, many think the 40 GW target is too ambitious and might not be achieved. For example, the IEA projects that China is likely to have 21 GW of nuclear capacity in place by 2020 and 31 GW in 2030 [12].

As in many other power generation technologies, China aspires to become technologically self-sufficient in manufacturing nuclear power plants, including in reactor design and construction. To date, China has utilized French and Russian pressurized water reactors as well as Canadian pressurized heavy water reactors [12]. Westinghouse, in the United States, has recently signed a multibillion-dollar deal with China for four nuclear reactors totaling 1100 MW, to be built by 2013 [2].

17.4.3 CLEAN FOSSIL POWER

Advanced coal technologies are in various stages of commercialization worldwide, and many are being developed within China. China does have a few state-of-the-art ultrasupercritical power plants and coal gasification plants and an increasing number of supercritical plants going online (20 percent of new builds in 2006), but the majority of new plants being deployed in China are subcritical plants. In addition, China is beginning to explore the use of integrated gasification combined cycle (IGCC) coal plants with the initiation of several government-sponsored demonstration projects.

Though not yet commercially available for coal-fueled power plants, technologies do exist that can prevent most of the CO_2 emissions from large point sources that combust or gasify coal from entering the atmosphere. The technologies referred to as carbon capture and storage (CCS) involve separating CO_2 from other exhaust gases, compressing the CO_2 to transport it through pipelines, and storing it in deep underground geological formations. Currently there is no large-scale commercial power plant equipped with CCS technology in China. China is, however, planning several CCS demonstration projects, including GreenGen (a 400 MW-scale IGCC plant with CCS being added in Phase Three, by 2020) and the Near-Zero Emission Coal (NZEC) partnership among China, the European Union, and the United Kingdom (goal of a coal plant with CCS by 2020). China's Huaneng power company, in cooperation with the Australian Commonwealth Scientific and Industrial Research Organisation (CSIRO), launched a post-combustion carbon-capture project (without storage).

In the near term, the priority to reduce emissions from coal is to ensure that new plants being deployed are high efficiency and that existing plants are running as efficiently as possible. In China, the average efficiency of coal power plants is rapidly catching up with that of developed countries as new, larger units come online and small, less efficient units are shut down. It is estimated that the average efficiency of China's coal-fired fleet was 32 percent (LHV, gross output) in 2005 (see Table 17.2) and is expected to approach 40 percent by 2030 as more large supercritical units come on line and older subcritical units are phased out [13].

17.4.4 ENERGY EFFICIENCY

The efficiency with which China uses energy is in part reflected in its energy-intensity metrics. China quadrupled its GDP between 1980 and 2000 while merely doubling the amount of energy it consumed over that period, marking a dramatic achievement in energy-intensity gains not paralleled in any other country at a similar stage of industrialization. This allowed China's energy intensity (ratio of energy consumption to GDP) and consequently the

Table 17.2   Comparison of coal-based power technology in China

Technology	Availability	Cost ($ per kW)	Efficiency (%)	Use in China
Subcritical	Now	500–600	30–36	Most of current generation fleet
Supercritical	Now	600–900	41	About half of current orders
Ultrasupercritical	Now, but needs further R&D to increase efficiency	600–900	43	Two 1000 MW plants in operation
Integrated gasification combined cycle (IGCC)	Now, but faces high costs and needs more R&D	1100–1400	45–55	Twelve units awaiting NDRC approval

Source: International Energy Agency, 2007b [12], p. 345.

emissions intensity (ratio of CO_2-equivalent emissions to GDP) of its economy to decline rapidly. Without this reduction in the energy intensity of the economy, China would have used more than three times the energy that it did during this period.

The trend of decreasing energy intensity reversed in China between 2002 and 2005, however, and energy growth surpassed economic growth for the first time in decades. This reversal has had dramatic emissions implications, with China's greenhouse gas emissions growing very rapidly since 2002. By 2006, China's energy demand had grown more in just 4 years than it had during the previous 25 years, with heavy industry largely to blame. Industry consumes about 70 percent of China's energy, and China's industrial base supplies much of the world. For example, China today produces about 35 percent of the world's steel and 28 percent of aluminum, up from 12 percent and 8 percent, respectively, a decade ago [8]. The steel sector alone consumes more energy and emits more CO_2 than all Chinese households; chemical production uses more energy than all the personal cars clogging the country's new roads; and aluminum smelters surpass the entire commercial sector in terms of electricity consumed [9].[6] China's role as the world's industrial base must be considered accordingly in evaluating its responsibility in contributing to global greenhouse gas emissions.

Out of concern for the reversing energy-intensity trends, Beijing's eleventh 5-year plan includes an aggressive goal to reduce national energy intensity 20 percent below 2005 levels by 2010. Implementation of this centrally administered government target has proven challenging, however, particularly at the local level. Supplementary programs have been established to encourage specific actors to help meet this national intensity goal, including a program established in 2006 to improve energy efficiency in China's largest enterprises [22]. Another government effort targets the elimination of a number of small, inefficient power plants, totaling around 8 percent of China's generating capacity, by 2010. Similar plant closings are planned across the industrial sector, including inefficient cement, aluminum, ferroalloy, coking, calcium carbide, and steel plants.

In addition, the 1997 Energy Conservation Law initiated a range of programs to increase energy efficiency in buildings, industry, and consumer goods. China has efficiency standards and labeling programs in place for many key energy-consuming appliances and is adopting energy standards for buildings in regions with high heating and cooling demands.

The recent surge in energy consumption by heavy industry in China has caused the government to implement measures to discourage growth in energy-intensive industries compared with sectors that are less energy intensive. In November 2006, the Ministry of Finance increased export taxes on energy-intensive industries. This includes a 15 percent export tax on copper, nickel, aluminum, and other metals; a 10 percent tax on steel primary products; and a 5 percent tax on petroleum, coal, and coke. Simultaneously, import tariffs on 26 energy and resource products, including coal, petroleum, aluminum, and other mineral resources, will be cut from their current levels of 3–6 percent to 0–3 percent [35]. Whereas the increased export tariffs are meant to discourage relocation of energy-intensive industries to China for export markets, the reduced import tariffs are meant to promote the utilization of energy-intensive products produced elsewhere.

17.4.5 Modeling a Low-Carbon Electricity Future for China

To examine how carbon emission constraints might play out in China, researchers at Tsinghua University have modeled a low-carbon scenario for

---

[6]In the United States, by contrast, only 25 percent of its emissions are industrial. Three quarters of U.S. emissions come from transport, commercial, and residential energy use.

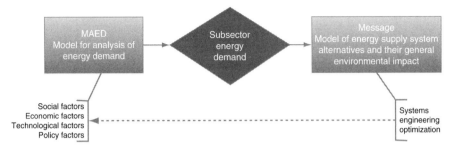

Figure 17.5 Modeling schematic.

China using the MESSAGE and the MAED model.[7] Social, economic, technological, and policy assumptions for each scenario are fed into the MAED model, which then analyzes the resulting energy demand (see Figure 17.5). This demand is then input into the MESSAGE model, which optimizes the total systems costs under the constraints imposed and configures the evolution of the energy system from the base year to the end of the time horizon. Outputs of the modeling exercise include the projected installed capacities of technologies, energy outputs and inputs, energy requirements at various stages of the energy systems, costs, and resulting emissions.

To identify the contributions and the challenges of establishing a sustainable energy supply system, Tsinghua developed two scenarios:

■ *Reference scenario.* A baseline scenario of alternative energy development, which assumes a continuation of the current energy technology development trends and market conditions and the absence of any additional policies and measures. The reference scenario reflects the expected "business-as-usual" pathway.

■ *CO₂ emission constraint scenario.* An alternative scenario for achieving the target of zero CO_2 emissions growth after 2025 through the aggressive development of alternative low-carbon energy, along with assumed improvements in market efficiency through policies and measures and a clearing of any relevant market barriers.

---

[7]Model for Analysis of Energy Demand (MAED) is a simulation model for evaluating the energy demand implications (in the medium and long term) of a scenario describing a hypothesized evolution of the economic activities and of the lifestyle of the population. The model provides a flexible simulation framework for exploring the influence of social, economic, technological, and policy changes on the long-term evolution of energy demand. Model of Energy Supply System Alternative and their General Environment Impact (MESSAGE) is a systems engineering optimization model used for medium- to long-term energy system planning, energy policy analysis, and scenario development.

17.4.6 REFERENCE SCENARIO

Energy demand and economic and social development are interrelated, yet there exist countries in different development stages with very different total energy demand and structural characteristics. According to China's current stage of development, the outlook for its energy demand will be affected by factors that will both support and inhibit its demand growth. In the reference scenario, it is assumed that energy demand growth will continue—along with industrialization, urbanization, and mobility—as driven by the size and health of its economy, the size of its population, and potential space for a sizable increase in energy consumption per capita. It is also assumed, however, that growth may be inhibited by aggressive policies aimed at building a resource-conserving and environmentally friendly society, including the Energy Conservation Law, the Mid- and Long-Term Energy Conservation Plan, and the 11th Five-Year Plan's goal to reduce the energy intensity of the economy by 20 percent by the year 2010. In addition, many macroeconomic policies aim to optimize China's industrial structure and improve technical efficiency. As a result, although China's medium- and long-term energy demand is expected to continue to grow, over time this growth is expected to gradually slow.

As illustrated in Table 17.3, it is assumed that final energy demand continues to rise in the long run, but with a slowdown in growth with a curve fit to a relatively progressive convex function.

In the reference scenario, China's final energy demand mix changes greatly in comparison with China's current primary energy mix (see Figure 17.6). Final electricity demand grows fastest, followed by final gas demand. Final coal demand rises only slightly in the short term and then declines in the medium and long term. From 2020 to 2025, final oil demand will exceed final coal demand as the number of vehicles owned and traffic volume both rise rapidly. Around 2030, final electricity demand will exceed final oil demand. From 2030 to 2050, final electricity demand accounts for the largest share of energy demand, followed by oil, coal, gas, and heat. The resulting shift in the mix of final energy demand over time reflects the continuous optimization of China's energy consumption structure.

The left side of Figure 17.7 shows the reference scenario of China's primary energy supply. The share of low-carbon sources (wind energy, hydropower, biomass energy, nuclear energy, and other alternative energy) in the primary energy mix continues to increase after 2010, ultimately reaching a share of 24 percent in 2050. The left side of Figure 17.8 shows the reference scenario of China's power installation capacity. The share of low-carbon power capacity increases over time, ultimately reaching 52 percent in 2050. The share of

Table 17.3 Modeling assumptions: socioeconomic parameters, 2005–2050

	2005	2010	2015	2020	2025	2030	2035	2040	2045	2050
GDP growth rate (%)	—	10	—	8.4	—	6	—	—	—	3.5
Population (millions)	—	1360	1410	1450	1480	1500	1510	1510	1490	1485
Industrial structure energy consumption										
Primary (%)	12.6	9.5	8.2	6.8	6.0	5.0	4.6	4.0	3.8	3.5
Secondary (%)	47.5	49.0	49.0	48.5	48.0	46.0	45.0	42.5	42.3	39.0
Tertiary (%)	39.9	41.5	42.8	44.7	46.0	49.0	50.4	53.5	53.9	57.5
Urbanization rate	43	46	51	55	59	62	65	67	68	70

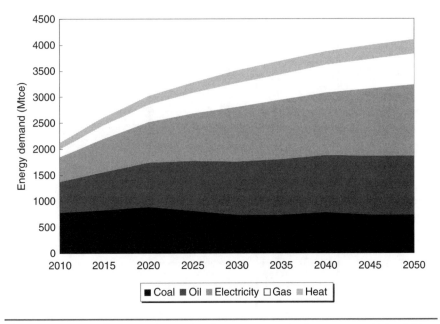

Figure 17.6 Final energy demand, reference scenario, by source, 2010–2050.
Source: Tsinghua University MESSAGE model output.

alternative electricity production also increases substantially, representing 44 percent in 2050. In addition, IGCC power plants are introduced beginning in 2025, after which time they will then play an increasingly important role.

China's future oil demand is projected to experience sustained growth, reaching over 800 Mtoe in 2050. The China Ministry of Land and Resources projects that China's domestic fuel output will not exceed 200 Mtoe after 2020 and that an inadequate supply of global market resources exists to fuel domestic consumption [5]. Consequently, China's oil import dependence will exceed 70 percent after 2030, resulting in energy supply security challenges. To help counter this phenomenon, the share of coal-based fuels, biofuels, natural gas, and other alternative fuels in the transportation fuel mix continues to increase after 2010, reaching 16 percent in 2050.

Primary alternative energy, alternative power, and alternative power technologies will play an increasing role over time in the reference scenario, but their scope is expected to be relatively limited. Only through substantial policy changes that will incent the further development of alternative energy will China's energy supply system be transformed in the coming decades to a sustainable, low-carbon energy system. If alternative energy technology development remains at the level specified in the reference scenario, China's CO_2

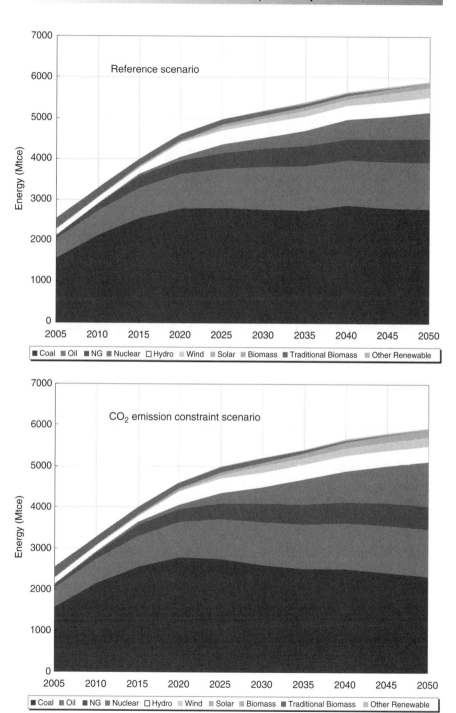

Figure 17.7 Primary energy mix scenario comparison, 2005–2050.
Source: Tsinghua University MESSAGE model output.

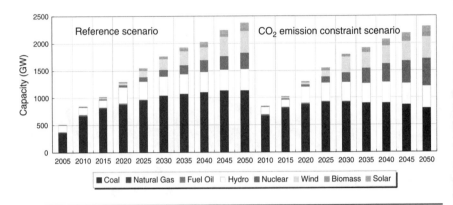

Figure 17.8 Installed power capacity by fuel, scenario comparison, 2005–2050. Source: Tsinghua University MESSAGE model output.

emissions will continue to rise, reaching 10 billion tons of CO_2 in 2050 (see Figure 17.8).

17.4.7 CO_2 Emission Constraint Scenario

Scientists increasingly project a smaller and smaller space for remaining GHG emissions growth in the atmosphere if dangerous climate change is to be avoided [10]. As a result, China is likely to face a future scenario in which its emissions growth is constrained. Without the rapid, expansive development of low-carbon energy sources, China will face a future scenario in which its CO_2 emissions will rapidly reach any prescribed atmospheric CO_2 concentration limits. Consequently, this scenario models how China's energy system might evolve in the face of CO_2 emissions constraints, set for this exercise at a leveling of emissions by the year 2025.

The right side of Figure 17.7 shows the CO_2 emission constraint scenario of China's primary energy supply through 2050. In this scenario, the total supply of low-carbon energy sources (wind, hydro, biomass, nuclear, and other alternative energy sources) reaches 1.89 Gtce in 2050. Compared with the reference scenario, the role of alternative energy has been noticeably enhanced, reaching a share of 25 percent in 2030 and 35 percent in 2050 to achieve the target of zero CO_2 emissions growth.

The right side of Figure 17.8 shows how the CO_2 emission constraint scenario shapes China's power installation capacity. The share of low-carbon power sources (wind power, hydropower, biomass power, nuclear power, and other alternative power) in the capacity mix significantly increases after 2020 compared with the reference scenario, with shares reaching 48 percent in 2030 and 63 percent in 2050. This is primarily achieved with the scaling of wind and nuclear power. The share of alternative electricity in total power production reaches 60 percent in 2050. In addition, IGCC power plants are introduced after 2025 and then begin to play an increasingly important role.

A key challenge in achieving the target of CO_2 emission zero growth after 2025 is that energy supply system costs increase significantly, causing an increase in energy prices. As illustrated in Figure 17.9, energy service supply costs in the CO_2 emission constraint scenario in 2030 increase 34 percent over the reference scenario and in 2050 are 11 percent over the reference scenario. Because the price of energy services depends on the cost of energy services, this increase in cost could have serious implications for the competitiveness of Chinese enterprises and on people's living standards.

According to the modeling results, the zero CO_2 emissions growth target after 2025 can be obtained by increasing the share of alternative energy in the primary energy mix to 35 percent, up from 25 percent in the reference scenario. This would include an increase in nuclear, hydro, and wind power development, supplemented by increased biomass and solar PV as well as IGCC. In addition, CO_2 emissions are reduced by 6.4 percent below reference scenario

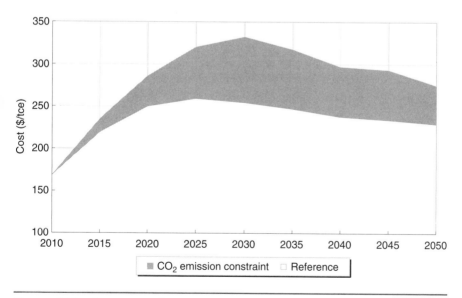

Figure 17.9 Energy supply cost scenario comparison, 2010–2050.
Source: Tsinghua University MESSAGE model output.

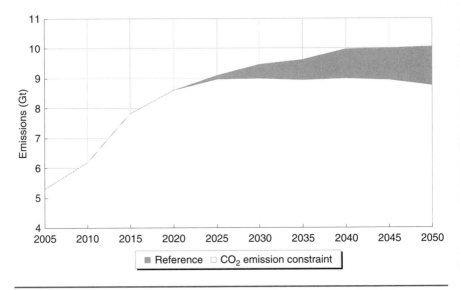

Figure 17.10 CO_2 emissions scenario comparison, 2005–2050.
Source: Tsinghua University MESSAGE model output.

levels in 2030 and by 13.3 percent below reference scenario levels in 2050 (see Figure 17.10).

This modeling exercise has illustrated one possible low-carbon scenario for China, which would require significant scaling of wind, advance coal, gas, and nuclear power technologies as well as significant improvements in energy efficiency nationwide. Such a shift will not come easily, but there are many policy options that could help to promote low-carbon technologies and an efficient use of energy in the Chinese context, as discussed in the final section of the chapter.

17.4.8 TOWARD A LOW-CARBON POWER SECTOR IN CHINA

China's power sector must be at the center of any global climate change solution. The world is watching to see how China will respond to the challenge of GHG mitigation, both in its domestic policy actions and in the international climate change negotiations that are working toward a new framework to build on the Kyoto Protocol after 2012.

National political leaders in China have shown increased attention to climate change in recent months. In 2007, China released its first national climate change plan composed of measures being taken across the economy that could help slow China's GHG emissions growth. Yet China's climate strategy remains centered on its energy development strategy, as driven by its overall economic development goals, and climate change has not yet surpassed economic development as a policy priority [16]. Given the lack of mandatory climate change policy in the United States and several other major emitting nations, this lack of policy urgency in China is not surprising. As a result, there is currently little incentive for China to pursue low-carbon power options that bring added costs and have little benefit other than their low-carbon characteristic.

One example of such a technological option is capturing the carbon dioxide from coal power plants and storing it geologically. CCS would reduce carbon emissions but would require a substantial up-front cost to build and would reduce the power generated over the lifetime of the plant by as much as one third. For this reason, the deployment of CCS only makes sense in a carbon-constrained economy or in the expectation of one. In a country such as China, with rapidly growing electricity demand, a technology that creates an energy penalty is a tough sell.

In addition to GHG mitigation benefits, many low-carbon technologies bring additional benefits to China's economic development in the form of reduced conventional pollution. Since China's fossil resources are increasingly imported from abroad, domestic low-carbon resources also can bring significant energy security benefits. Energy-efficiency measures can help promote efficient economic growth and can help reduce the energy intensity of the economy. Such measures have the dual benefit of reducing energy and emissions while also saving money and contributing to a more efficient economy. In addition, many renewable energy technologies are already providing substantial economic development benefits to China as new manufacturing industries fueling lucrative domestic and export markets are created. As a result, it would be judicious for China to focus on promoting these domestic policy strategies that can create economic benefit and reduce pollution while having the cobenefit of climate change mitigation.

The good news is that these are all areas that have already been targeted for domestic policy attention in China. For any of these options to drastically alter China's coal-dependent power sector, however, they will need to be dramatically expanded. Additional incentives or targets can be placed on low-carbon technologies to facilitate their utilization, or market-based carbon constraints such as taxes or targets can be used to achieve similar goals. In addition, increased research, development, and demonstration will be crucial to expand the technological options available in the coming years. Widespread adoption of CCS will require a targeted government support program that addresses crucial barriers such as high costs, energy losses, lack of experience, lack of a regulatory framework for sequestration, legal uncertainties related to liability and property rights, technical coordination among utilities and pipeline operators, and the need to increase public understanding and acceptance.

Because China's power sector must be at the center of any global climate-change solution, shaping its development is likely to be at the center of any international climate-change policy regime. China has consistently resisted taking on economywide emissions caps, but at the 2007 climate negotiations in Bali it agreed to take measurable, reportable, and verifiable mitigation actions, supported by finance and technology from the developed world [28]. There are several ways in which China's internationally agreed mitigation actions could focus on the power sector. For example, China could agree to take on policy-based commitments targeting the power sector in the form of an expanded renewable energy target, energy-intensity target, or technology mandate such as requiring new coal plants to be capture ready or begin capturing CO_2 by a certain date [17]. Alternatively, China could agree to a sectoral target in the power sector and earn international carbon credits for any reductions it achieves below this target [25]. In either case, there may be an increased role for the international community to play in helping China finance the incremental cost of a low-carbon transition.

17.5 Conclusion

The power sector is at the core of China's climate change challenge. A complex multidimensional universe comprising many actors and competing forces, China's power sector is expanding at a massive scale, with momentum driving it toward a coal-based future. Redirecting this momentum toward a low-carbon future is a monumental challenge. Yet in this challenge lies vast opportunity in the form of technology innovations, increased national security, and significant avoided damage to the environment, to public health, and to the global climate system.

REFERENCES

[1] Andrews-Speed P, Dow S, Wang A, Mao J, Wei B. Do the Power Sector Reforms in China Reflect the Interests of Consumers? The China Quarterly 1999;159:430–46.
[2] Bradsher K. Westinghouse Wins China Nuclear Reactor Bid. International Herald Tribune. Available at www.iht.com/articles/2006/12/17/business/nuke.php; December 1, 2006.
[3] BTM Consult ApS. International Wind Energy Development, World Market Update 2007. March 2008.
[4] Carbon Dioxide Information Analysis Center (CDIAC). Oak Ridge National Laboratory. CO_2 Emissions from Fossil Fuels. Available at http://cdiac.ornl.gov/trends/emis/meth_reg.html; 2008.

[5] China's Ministry of Land and Resources. 2007. 2006 China Land and Resources Communiqué. Beijing.

[6] Energy Information Administration (EIA). International Energy Outlook 2007. U.S. Department of Energy. Report #:DOE/EIA-0484(2007). Available at www.eia.doe.gov/oiaf/ieo/; 2007.

[7] Energy Information Administration (EIA). International Energy Annual, Table 8.2: World Estimated Recoverable Coal. U.S. Department of Energy. Available at www.eia.doe.gov/emeu/iea/table82.html; 2008.

[8] Houser T. China's Energy Consumption and Opportunities for U.S.-China Cooperation to Address the Effects of China's Energy Use. testimony before the U.S. China Economic and Security Review Commission. www.uscc.gov/hearings/2007hearings/written_testimonies/07_06_14_15wrts/07_06_14_houser_statement.php; June 14, 2007.

[9] Houser T, Bradley R, Childes B, Werksman J, Keilmayr R. Leveling the Carbon Playing Field: International Competition and US Climate Policy Design. Washington, DC: Peterson Institute for International Economics and World Resources Institute; May 2008.

[10] Intergovernmental Panel on Climate Change (IPCC). Working Group 1: The Physical Science Basis of Climate Change. Available: http://ipcc-wg1.ucar.edu/wg1/wg1-report.html; 2007.

[11] International Energy Agency (IEA). A Strategy for Cleaner Coal in China. First review draft, December 29, 2007. Paris: IEA/OECD; 2007a.

[12] International Energy Agency (IEA). World Energy Outlook, 2007. Paris: IEA/OECD; 2007b.

[13] International Energy Agency (IEA). CO_2 from Fossil Fuel Emissions, 1971–2005. Paris: IEA/OECD; 2007c.

[14] International Energy Agency (IEA). Energy Technology Perspectives, 2008. Paris: IEA/OECD; 2008.

[15] Lewis JI. 2005. From Technology Transfer to Local Manufacturing: China's Emergence in the Global Wind Power Industry. Ph.D. dissertation in energy and resources. Berkeley: University of California; Fall 2005.

[16] Lewis JI. China's Strategic Priorities in International Climate Negotiations. The Washington Quarterly Center for Strategic and International Studies and Massachusetts Institute of Technology 2007; 31(1):155–74.

[17] Lewis J, Diringer E. Policy-Based Commitments in a Post-2012 Climate Framework, Pew Center on Global Climate Change. Available at www.pewclimate.org/docUploads/Policy-Based%20Commitments%20in%20a%20Post-2012%20Climate%20Framework.pdf; 2007.

[18] Lu Y. Fueling One Billion: An Insider's Story of Chinese Energy Policy Development. Washington Institute Press; 1993.

[19] Netherlands Environmental Assessment Agency (MNP). China Now No. 1 in CO_2 Emissions; USA in Second Position. www.mnp.nl/en/dossiers/Climatechange/moreinfo/Chinanowno1-inCO2emissionsUSAinsecondposition.html; 2008.

[20] National Bureau of Statistics (NBS) and National Development and Reform Commission (NDRC). China Energy Statistical Yearbook. Beijing: China Statistics Press; 2006.

[21] National People's Congress. Renewable Energy Law of The People's Republic of China. Adopted at the 14th meeting of the Standing Committee of the 10th National People's Congress on February 28, 2005. English translation available at www.resource-solutions.org/lib/librarypdfs/RE_law_english_version.doc; 2005.

[22] Price L, Wang X. Constraining Energy Consumption of China's Largest Industrial Enterprises Through Top-1000 Energy-Consuming Enterprise Program. Lawrence Berkeley National Laboratory, http://ies.lbl.gov/iespubs/LBNL-62874.pdf; June 2007.

[23] Regulatory Assistance Project (RAP). China's Power Sector: A Backgrounder for International Regulators and Policy Advisors. Prepared for the Energy Foundation China Sustainable Energy Program; February 2008.

[24] Reuters. China Seen World Leader in Clean Energy. Available at www.chinadaily.com.cn/english/doc/2005-09/29/content_481754.htm; September 29, 2005.

[25] Schmidt J, Helme N, Lee J, Houdashelt M. Sector-based Approach to the Post-2012 Climate Change Policy Architecture. Washington, DC. Center for Clean Air Policy. Available at http://ccap.org/docs/resources/68/Sector_Straw_Proposal-FINAL_for_FAD_Working_Paper.pdf; 2006.

[26] Sinton JE, Fridley DG, Lewis JI, Chen Y, Lin J, Zhou N. China Energy Databook. 6th revised ed. (CD-ROM). Lawrence Berkeley National Laboratory, China Energy Group. LBNL-53856; 2004.

[27] UNFCCC. Investment and Financial Flows to Address Climate Change. Available at http://unfccc.int/files/cooperation_and_support/financial_mechanism/application/pdf/background_paper.pdf; 2007a.

[28] UNFCCC. The Bali Action Plan. (Decision 1/CP.13). FCCC/CP/2007/6/Add.1*. Available at http://unfccc.int/resource/docs/2007/cop13/eng/06a01.pdf#page = 3; 2007b.

[29] Wallace WL, Li J, Gao S. The Use of Photovoltaics for Rural Electrification in Northwestern China. Paper presented at the 2nd World Conference and Exhibition on Photovoltaic Solar Energy Conversion, hosted by the NREL, Vienna, Austria, July 6–10; 1981. p. 1.

[30] Windpower Monthly. Global Wind Market Status 2008. Denmark; March 2008.

[31] World Bank. Clean Energy and Development: Towards an Investment Framework. Available at http://siteresources.worldbank.org/DEVCOMMINT/Documentation/20890696/DC2006-0002 (E)-CleanEnergy.pdf; April 5, 2006. Draft.

[32] World Bank and China State Environmental Protection Administration. Cost of Pollution in China. Washington, DC: World Bank; 2007.

[33] Wu Y. Manufacturing Technology of Wind Turbines (Grid Connected) in China. Proceedings of the World Wind Energy Congress, Beijing, China, October 31–November 4, 2004.

[34] Yeh ET, Lewis JI. State Power and the Logic of Reform in China's Electricity Sector. Pacific Affairs 2004;77(3).

[35] Wang Y. Tariffs to Reduce Energy Consumption. China Daily. Available at www.chinadaily.com.cn/china/2006-10/31/content_720485.htm; October 31, 2006.

[36] Yi Y, Wang J. The Influence of Wind Turbine Price to the Budgetary Estimate of Wind Farm. Proceedings of the World Wind Energy Congress, Beijing, China, October 31–November 4, 2004.

California Dreaming: The Economics, Politics, and Mechanics of Meeting California's Carbon Mandate

Frank Harris and Gary Stern

Southern California Edison Company

Abstract

In 2006, the California legislature passed a landmark greenhouse gas emission reduction law. In doing so, it has set a standard for other states to follow. Southern California Edison Company (SCE) is the state's largest electric utility and as such is committing considerable resources to aid in the implementation of this law in a manner that minimizes the economic burden to its customers. This chapter outlines the challenges facing California in general and SCE in particular and describes the policy approaches that the company believes will produce the best results.

18.1 Introduction

18.1.1 CALIFORNIA GHG LAW

In January 2007, California Governor Arnold Schwarzenegger signed Assembly Bill 32 (AB-32), the California Global Warming Solutions Act. This law establishes that California must reduce its greenhouse gas (GHG) output to 1990 levels by 2020.[1] This means that California must reduce its GHG emissions by an estimated 169 million metric tons (MMT), or 30 percent, from the 596 MMT business-as-usual emissions forecast for 2020.[2] Specific limits and adoption of measures to achieve these limits must be established by the California Air Resources Board (CARB) by January 2011, to be implemented beginning January 1, 2012. CARB will be establishing these rules with input and recommendations from the California Energy Commission (CEC) and the California Public Utilities Commission (CPUC) as well as other stakeholders. CARB has completed the first two steps of this process, namely by:

- Establishing the reporting process and determining the 1990 baseline against which the future GHG emissions will be compared
- Developing the Scoping Plan; CARB released the Proposed Scoping Plan (PSP) in October 2008

The final regulatory structure established in the PSP aims to capture approximately 80 percent of mandated emission reductions via expanded direct regulation through so-called command-and-control measures. Additionally, the PSP recommends partnering with the Western Climate Initiative (WCI) to develop a cap-and-trade program for emissions. Although CARB has indicated that a carbon tax *could* be implemented as a means to reduce emissions, the PSP is focused on overlaying a cap-and-trade market-based approach on top of an expanded menu of command-and-control rules. Although the PSP provides insights into the process and substance behind the CARB regulatory design, at the time of this writing the final regulatory structure is far from certain. Significant questions remain, including:

---

[1] It is important to note that AB-32 specifically includes emissions from imported electricity under the state limit.

[2] CARB Proposed Scoping Plan. Emissions are measured in metric tons of CO_2 equivalent (CO_2e).

■ Will the WCI cap-and-trade system prevail?

• How would allowances be allocated?

• What offset projects would be approved?

■ Will the additional command-and-control regulations provide the emission reduction indicated in the PSP?

■ How will the state achieve its mandated reductions if the additional command-and-control measures fail to provide the expected emission reductions?

■ In the event that federal GHG legislation is passed, how will California integrate its actions with the federal program?

These uncertainties regarding the precise regulatory approach that CARB will take in implementing AB-32 create a critical challenge for Southern California Edison (SCE). Clearly, California's success or failure in implementing the law will be followed with interest by other states as well as other countries and jurisdictions around the world. In this chapter the authors evaluate the current state of the debate in California and the impact that various forms of regulation could have on the utility and its ratepayers and describe actions SCE is currently taking to best position itself in an emissions-constrained market.

18.1.2 AB-32 PRESENTS A CRITICAL CHALLENGE FOR SOUTHERN CALIFORNIA EDISON

SCE is California's largest electric utility, serving over 4 million ratepayers over a region encompassing south and central California.[3] SCE's service territory (see Figure 18.1) is characterized by extremely hot, summer peaking regions along with coastal regions that tend to peak in the winter. AB-32 challenges utilities such as SCE to reduce the emissions from an already low-carbon energy portfolio. To understand the magnitude of the challenge AB-32 poses for SCE, it is important to understand the current status of SCE's fuel mix, the existing regulatory environment in California, the nature of the proposed cap-and-trade scheme, and the interaction between cap-and-trade and command-and-control regulations.

---

[3]California's other two major investor-owned utilities, both electric and gas utilities, are Pacific Gas & Electric Company and San Diego Gas & Electric Company. The service area of each is depicted in Figure 18.1.

The California ISO
Control Area

(PG&E)

(SCE)

(SDG&E)

Figure 18.1 Map of California showing service areas of three major investor-owned utilities.
Source: California Independent System Operator.

18.1.2.1 Existing renewable and environmental regulations in the California electric sector

California's electric power sector is characterized by already aggressive renewable energy and environmental regulations. These regulations address issues such as energy efficiency, renewable energy procurement, and procurement of fossil-fuel generation. The PSP calls for the state to increase the depth and scope of existing regulations as well as to introduce a cap-and-trade program within the WCI to achieve AB-32 compliance.

18.1.2.2 California's renewable portfolio standard

A key regulation facing SCE is the state's Renewable Portfolio Standard (RPS), which requires that California retail sellers procure 20 percent of their retail load using certified renewable resources, including small hydroelectric power (30 MW nameplate capacity), geothermal, solar, wind, and biomass, by 2010. An Executive Order issued by the Governor in November 2008 established a new 33 percent target by 2020.[4]

Using renewable resources provides valuable environmental benefits and SCE is a leader among utilities in procurement of these resources. However, renewable procurement often increases the cost of electricity. Since renewable resources are generally not dispatchable, expanded renewable procurement presents a unique challenge to running the electricity grid. Intermittent resources need to be complemented with dispatchable resources for electricity generation to match customer demand in real time.[5] Solar and wind resources present a particular challenge to the grid due to their intermittent and unpredictable output patterns. However, SCE has made substantial steps toward achieving the 20 percent RPS goal with our current RPS value in the 17 percent range as of late 2008. Additional renewable procurement is currently constrained largely by transmission limitations. The bulk of renewable resources are not near existing transmission capacity.

As Table 18.1 indicates, California currently supplies over 6000 MW of renewable generation. However, the state needs nearly 8000 MW of additional renewable generation to meet the 20 percent RPS goal by 2010. An additional 11,000 MW of renewable generation will be needed to reach a 33 percent renewable procurement level by 2020.[6]

California retail energy sellers are required to procure 20 percent of their total retail energy sales with renewable resources by 2010. The state average in 2007 was just below 13 percent. On November 17, Governor Schwarzenegger signed an executive order mandating a 33 percent renewable procurement by 2020.

18.1.2.3 California's energy efficiency standards

The California Energy Commission has developed a resource loading order, essentially a preferred resource list, which ranks electricity-generating resources in order of procurement preference. SCE and other IOUs are required to reference this loading order in all procurement planning activities. Leading this list is energy efficiency (EE). Although EE does not directly generate electricity, it reduces the amount of energy needed and as such is modeled as a capacity

---

[4]Executive Order # S-14-08, November 2008.

[5]A dispatchable generation resource is one that can be dispatched, or turned off or on in response to electricity grid demand. Some baseload generation, such as coal and nuclear resources, takes time to ramp up or down, whereas intermittent resources, such as wind and solar, cannot be controlled by grid operators.

[6]Source: "CAISO's Plan for Integration of Renewable Resources," Presentation to the Energy Commission's 33% Renewable Workshop on July 21, 2008.

Table 18.1 Total renewable generation capacity in California must increase more than 400 percent to meet 33 percent RPS by 2020

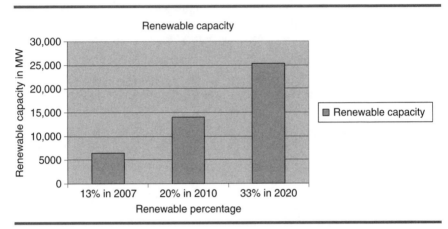

resource.[7] SCE and the other IOUs in California are instructed to implement essentially all available cost-effective EE programs before they pursue other supply-side options.[8] This has resulted in the offset of nearly 30,000 MW of generating capacity statewide since 1972. As Figure 18.2 shows, per capita electricity sales in California have been fairly constant since 1972, whereas the U.S. average has increased more than 50 percent over the same time period.[9] California regulators are proud of this record and attribute its success to the state's strong commitment to energy efficiency.

However, California continues to strive for more aggressive EE targets. Between 2012 and 2020, the CPUC expects to save more than 16,000 gigawatt-hours (gWh) of electricity through EE programs that are forecast to reduce capacity needs by more than 4500 MW. By 2020, cumulative electricity savings from EE programs in California are forecast to reach as high as 31,930 gWh.[10]

---

[7]For an expanded discussion of EE, refer to Chapter 8, by Prindle et al., in this volume.

[8]Due to California's above-average electricity rates, a greater quantity of EE will be cost effective than would be anticipated in regions with lower electricity prices.

[9]Source: Efficient Use of Energy in California Power Electronics Conference, Long Beach, CA, October 25, 2006. Presentation Arthur H. Rosenfeld, Commissioner CEC. The values from 2005 to 2008 are forecast.

[10]Source: "Assistance in Updating the Energy Efficiency Savings Goals for 2012 and Beyond Task A4.1 Final Report: Scenario Analysis to Support Updates to the CPUC Savings Goals," CPUC Report, March 24, 2007.

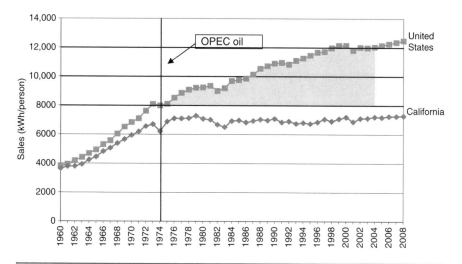

Figure 18.2   Annual per capita electricity sales in California vs. United States, 1960–2008 (kWh/person).
Source: "Assistance in Updating the Energy Efficiency Savings Goals for 2012 and Beyond Task A4.1 Final Report: Scenario Analysis to Support Updates to the CPUC Savings Goals," CPUC Report, March 24, 2007.

18.1.2.4 California's emissions performance standard

In 2006, the California legislature passed Senate Bill 1368, or SB-1368, the Greenhouse Gas Emission Performance Standard, which restricts California utilities from signing new procurement contracts longer than 5 years with any generating resource with an emissions factor greater than a combined cycle gas turbine (CCGT). SB-1368 is a direct strike at reducing the contracting between California utilities and coal generators beyond California's borders. Essentially, this law restricts long-term procurement with any resource with GHG emissions greater than 1100 pounds per MWH. Although SCE does not hold any long-term contracts with coal generators, it does own a share of the Four Corners coal-fired generating station. Additionally, many publicly owned utilities have long-term contracts with out-of-state coal generators. The practical impact of SB-1368 is to remove long-term contracting with out-of-state coal resources from the potential procurement choices available to California retail sellers. This is highly significant since California imports roughly 25 percent of its energy needs from other states, some from hydro resources in the Pacific Northwest, with the balance from coal-fired generation to the East.

18.2 Southern California Edison's fuel mix

As a result of relatively low use of coal-fired resources, existing renewable procurement activities, and EE initiatives, most California utilities deliver electricity that has a relatively low emissions rate. California's fuel mix is characterized by a heavy reliance on gas-fired generation as well as carbon-free nuclear, hydroelectric, and renewable generation. The relatively small amount of coal that is procured is imported from out of state.[11] As indicated in Figure 18.3, California's fuel mix is less coal intensive than the national average, and SCE's fuel mix is less coal intensive than the California average.

As can also be seen in Figure 18.3, California makes greater use of renewable resources and lower use of coal resources for electricity generation than the national average. This means that California's emission rate from

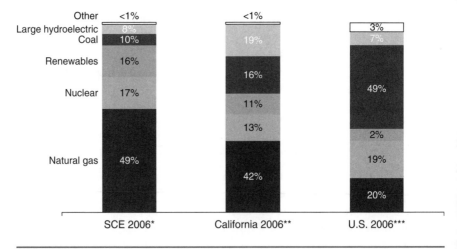

Figure 18.3 Power mix comparison: SCE, California, and United States, 2006.
Sources: *SCE's 2006 Annual Power Content Label, computed according to CEC methodology and included in SCE customer bills. **CEC's "2006 Net System Power Report," April 2007. ***Energy Information Administration's Electric Power Annual, November 2007, Figure ES-1.

---

[11]Although there are a handful of coal-fired cogeneration resources in California, there are no central station coal-fired resources in the state. SCE owns a share of the Four Corners coal-fired generating station, and PG&E, SDG&E, and SCE all procure market power that includes some coal. Beyond this, the bulk of the coal-fired power sold in California is procured on behalf of the Southern California municipal utilities.

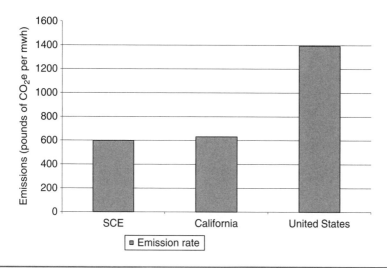

Figure 18.4 Emission rates in the Electric Sector, SCE, California, and U.S. average, in pounds of CO_2e per MWh, 2006.
Source: SCE Emissions as reported in the California Climate Action Registry Public Database at www.climateregistry.org. California and U.S. emission rates from the U.S. Department of Energy's Energy Information Agency at www.eia.doe.gov.

generating electricity is lower than the national average. As shown in Figure 18.4, SCE's portfolio emission rate compares favorably to the statewide emission rate for California, which itself is lower than the national emission rate.

However, the higher use of renewable resources and the lower emission rate of electricity generation in California have come at a price. At an average cost of 12.36 cents per kWh, California electricity rates are significantly higher than the national average of 8.72 cents.[12] Forecasts of the costs of meeting the AB-32 emission reduction goals vary widely. CARB continues to assume that meeting the AB-32 emission reduction goals will provide a *net* benefit to the California economy, but modeling performed by an energy and environmental consulting firm retained by the CPUC indicates that there will be an increase in electricity rates as California increases its use of renewable energy resources and further reduces its emissions rate.[13] This modeling presented various policy cases, from moderate to aggressive. The aggressive policy case included increased renewable procurement, energy efficiency, and other demand-side

[12]Energy Information Agency for 2007.
[13]Draft Scoping Plan at ES-4.

solutions within the electric sector. Additionally, the model included enforce-ment of the low-carbon fuel standard and lower vehicle miles traveled. But these policy instruments come at a cost. Specific estimates show that average retail rates are expected to rise from 14.9 cents per kWh in the reference case to 16.9 cents per kWh in the aggressive policy. Though the aggressive pol-icy case reduces IOU emissions by 29.6 million tons, it does so at a cost of $131 per ton.[14]

18.3 A brief description of cap and trade for California

A cap-and-trade mechanism is expected to be part of the regulatory solution developed by CARB in addition to expanded command-and-control rules. Section 1.2 presented a review of the three main command-and-control mechanisms to reduce emissions in the electric sector. A further challenge for SCE has been to consider how a state or regional cap-and-trade program would integrate with expanded state-level command-and-control mechanisms and how all this might integrate with a federal program. Although cap-and-trade has been used in the electricity sector before, applying a cap-and-trade system to GHG emissions poses a unique set of opportunities and challenges.[15] The remainder of this section includes a broad description of cap and trade and describes some of the activities SCE has undertaken to inform and advise stakeholders and regulators on issues relevant to the development of a cap-and-trade program.[16]

A cap-and-trade mechanism for limiting GHG emissions would be imple-mented by establishing an annual cap on the level of emissions for which allowances, permission slips to emit, would be issued. By limiting the pool of allowances and requiring regulated entities to submit allowances associated with their measured GHG responsibility subject to a substantial penalty, total emissions within the regulated sector can be capped. The entities subject to these regulatory requirements would need to acquire allowances associated with their level of emissions. The scarce supply of allowances will lead to a rising market price for emitting carbon as organized trading mechanisms evolve for trading allowances.

---

[14]Source: Energy and Environmental Economics, forecast provided to the California Public Utilities Commission on May 13, 2008; www.ethree.com. Data are in 2007 dollars.

[15]The U.S. Environmental Protection Agency currently administers a cap-and-trade program. The EPA's Clean Air Markets Division (CAMD) includes the Acid Rain Program and the NO_x Programs, which reduce emissions of sulfur dioxide (SO_2) and nitrogen oxides (NO_x).

[16]This section provides a high-level summary of cap and trade. For a more detailed treatment of the topic, see Adib and Musier, Chapter 3 in this volume.

Entities with potential GHG emissions and associated allowance responsibility are thus faced with two main compliance options. The most direct option is to reduce emissions internally. Firms have an incentive to seek out direct emission reduction alternatives as long as the reduction alternatives have lower costs than the cost of allowances. Alternatively, firms can fund the emission reductions of another entity or sector by purchasing allowances. The benefit of a broad-based cap-and-trade approach is demonstrated by this choice. As all regulated entities evaluate their direct and indirect abatement options, the abatement options will be selected in the most cost-effective order. Lower-cost abatement options will occur initially, with higher-cost reductions occurring until the regulated sector has achieved the cap. A well-designed cap-and-trade mechanism will facilitate efficient abatement choices,[17] as shown in Figure 18.5.

The upward-sloping line represents the expected business-as-usual (BAU) level of emissions that would occur in the absence of the imposition of GHG-reducing rules. Below this line is the capped level of emissions to be achieved by the program. By equating the number of allowances created to this capped level and requiring that allowances must be retired by all regulated entities according to their emissions, the emission reduction goal is achieved.[18] The value of these allowances is determined by this limited supply and the

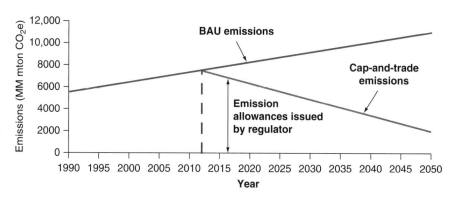

Figure 18.5 Emissions under cap and trade vs. BAU, 1990–2050.

[17]A well-designed cap-and-trade mechanism will result in a least-cost solution due to equi-marginality, the condition by which the marginal cost of abatement is equal across entities and sectors.

[18]An emission allowance is considered "retired" when it is presented to the regulatory agency for compliance purposes.

demand by regulated entities for allowances. The degree to which regulated entities will demand allowances depends on their GHG abatement options and costs.

18.3.1 THE POINT OF REGULATION IN A CALIFORNIA CAP-AND-TRADE PROGRAM

During 2007 and early 2008, much of the debate relating to a cap-and-trade market structure in California centered on the determination of the appropriate point of regulation (POR). California began its consideration of a cap-and-trade market for GHG with the recognition that much of the GHG emissions associated with serving the California electricity load emanate from power generated out of state and imported into the state—power that cannot be regulated by state law. In March 2007, the CPUC and CEC issued a joint decision[19] supporting the first deliverer as the recommended point of regulation for a California cap-and-trade market. A great deal of attention on the POR debate has focused on the electricity sector. An appropriate POR should:

- Allow the lowest possible cost of compliance
- Provide appropriate incentives to reduce emissions
- Be consistent with the normal operations of the electricity market
- Be easily adapted to a national cap and trade regulation
- Be simple to understand and implement

California has considered both a load-based and a source-based POR. Although neither is perfect at a state or regional level, it is useful to review the high-level structure of each.

18.3.2 SOURCE-BASED CAP AND TRADE

Traditional cap-and-trade programs such as the U.S. EPA's SO_2 trading program identify the emission source as the POR. A source-based POR would measure the emissions from all capped California sources and require each source to retire allowances sufficient to cover its emissions. However, a source-based POR would not of itself satisfy the directive in AB-32 to include

---

[19]D.08-030-18, *Interim Opinion on Greenhouse Gas Regulatory Policies*, issued by the CPUC on March 13, 2008.

emissions from imported generation in the California cap. The problem is not limited to the state's inability to include emissions from imported generation in its cap. Although a source-based cap-and-trade program would work well with the California electricity market, emissions leakage can become a significant challenge to the integrity of the program.[20] Additionally, over the long term, it creates a disincentive to invest in California generation and would provide a cost advantage to imported electricity.

18.3.3 LOAD-BASED CAP AND TRADE

In an attempt to avoid the challenges of the source-based POR and in an effort to capture the emissions from imported generation, the CPUC initially focused on a load-based POR. Under a load-based POR, load-serving entities (LSEs) in California would be the regulated entities, and each LSE would need to retire sufficient allowances to cover the emissions associated with the generation used to serve its retail load. Support for a load-based cap centered around the expectation that such a scheme would enable California to cap the emissions from all generation used to serve retail load in the state. However, a load-based cap also brings with it significant challenges, including incentives for regulated entities to "hide" emissions by way of contract shuffling.[21] A load-based cap also does not afford any mechanism by which the wholesale electricity markets can incorporate the cost of emissions in the dispatch decision.[22] This seriously handicaps the ability of the market to facilitate regulatory compliance.[23] An additional problem with a load-based cap becomes evident when one considers the potential for federal action on GHG emissions. All major federal GHG emissions proposals thus far have included a source-based POR in a cap-and-trade mechanism. Were California to adopt a load-based cap, migrating to a federal source-based cap would present a significant challenge for California.

A final and likely fatal flaw in a load-based POR is the challenge of reporting and tracking emissions from source to sink. There are thousands of daily energy trades in California, and it is virtually impossible to effectively track the emissions from the generating unit to the LSE unless both parties want

---

[20]Emissions leakage is the process by which some generators would shift their production across state lines. This would not result in any reduction of GHG emissions and would thereby reduce the effectiveness of the regulatory program.

[21]Contract shuffling is the process by which generators will merely allocate on paper the power from the cleanest plants in their portfolios to clients in California.

[22]Under MRTU, the California ISO will serve load via economic dispatch of generation resources.

[23]Chapter 19 describes the evolving role of ISOs and RTOs in meeting GHG emission targets.

to demonstrate a clear connection. Table 18.2 evaluates both first deliverer and load-based regulations based on a variety of factors.

18.3.4 FIRST DELIVERER

The first deliverer regulatory structure places the compliance obligation on the first deliverer of the energy within the state of California. In simple terms, the deliverer of power is the generation source for power generated within California, or, in the case of power imported into California, the entity that first puts the power onto the grid within the state. Thus, if an LSE inside California arranges for an import in which it takes delivery outside the state and brings

Table 18.2   First deliverer vs. load-based cap

Issue	First Deliverer (FD)	Load-Based (LB)	Evaluation
Contract Shuffling & Leakage	No in-state concems Imports still a concern	Both in-state and imports are concerns	FS minimizes leakage
Federal Legal Compliance			Both approaches raise similar legal questions, LB questions continue under regional market
Regional Integration	Consistent with regional and national proposals	Difficult to integrate and expand	FS significantly mitigates integration problems
Direct Access Compatibility	Compatible— compliance burden on sellers	Imposes significant burden on LSEs	FS is more compatible with DA

(Continued)

Table 18.2 *(Continued)*

Issue	First Deliverer (FD)	Load-Based (LB)	Evaluation
Emission Mapping and Accuracy	Assignment of emissions is more accurate and direct	Assignment very difficult under MRTU. Emissions above default emissions factor may be lost	FS is preferred under MRTU
EE and Renewables			Existing laws/ policies remain in both cases
Customer Costs			Costs are the same. LB may be more costly under a regional market

Source: SCE analysis.

the power into the state's grid, that LSE would be the deliverer and would have regulatory responsibility for the emissions associated with this import. On the other hand, if the out-of-state seller of the power engages in a transaction in which it imports the power into California and delivers the power to the state's grid, that seller would bear the regulatory responsibility associated with the GHG emissions from the import.

An important advantage of the first deliverer (or deliverer) POR is that it allows the wholesale market to incorporate emissions costs in the dispatch calculus. Additionally, a deliverer POR does not pervert the incentives of independent generators to bid into the market. Further, a deliverer-based POR can be adapted to a national, source-based mechanism with far less interruption of existing mechanisms than would a load-based approach. However, even under a deliverer POR, the challenges of emission leakage and contract shuffling remain.

18.3.4.1 Application of a first deliverer POR to a regional cap-and-trade program

The WCI released its design recommendations for a regional cap-and-trade program in September 2008.[24] In this document, the WCI recommended the application of a first jurisdictional deliverer point of regulation (FJD). The FJD point of regulation is most closely compared to a first deliverer POR. The distinction is the geographic footprint of the regulated region. The first deliverer approach assigns the compliance burden to the first deliverer of energy within the state of California. In the case of the WCI, the FJD approach assigns the compliance responsibility to the first deliverer of the electricity within the WCI-capped jurisdiction. Since the WCI is a multistate regulatory compact, this means that imported electricity would be that generated outside a WCI member region and imported into the WCI region. However, significant challenges remain to apply a regional cap-and-trade program within the WCI footprint. At this point, California remains the only WCI participant that has reinforced its commitment with legislative approval.

18.3.5 ALLOCATING ALLOWANCES

Once the specific regulatory approach has been designed to include a cap-and-trade program and the specific point of regulation has been identified, there remains one important design element to be addressed: the allowance allocation method. The cap-and-trade market mechanism works because the supply of allowances—permission slips to produce GHG—are limited by the goal of the regulatory mechanism. If the determination is that there can only be X tons of GHG allowed by the regulated sector in year Y, then only X allowances will be issued in that year.

Although flexible compliance mechanisms could alter the timing or even the quantity from that established by the regulatory goal, the basic premise remains—a limited supply of allowances needed by the set of entities identified by the determination of the point of regulation as those needing to produce allowances associated with their GHG responsibility. So, how do those entities gain access to the allowances they need?

There are two basic approaches to the problem of distribution of allowances within the regulated sector.

---

[24]Design Recommendations for the WCI Regional Cap-and-Trade Program, Western Climate Initiative, September 23, 2008.

■ *Allocate the allowances based on some predetermined criteria.* This process presents a distributional challenge. Those receiving the allocation might not be the same or even close to the same as those needing to acquire allowances. Thus a distribution process to allow for the transfer of allowances from those allocated to those who need them must exist. This could take the form of an unstructured bilateral negotiation process or some other form of secondary market, such as an electronic bulletin board.[25]

■ *Run an auction or a series of auctions in which those that demand allowances bid to acquire them and the highest bidders receive allowances.* This process transfers allowances to the regulated entities and dollars are accumulated from the auction process. Such a mechanism could be designed so that the auction proceeds would be distributed according to a predetermined allocation criterion. Although the precise disposition of the auction revenues would need to be determined, the auction solves the basic distribution problem in what is clearly a transparent and nondiscriminatory fashion.

18.3.6 THE DISTINCTION BETWEEN ALLOCATING ECONOMIC VALUE AND DISTRIBUTING ALLOWANCES

As the discussion of cap-and-trade turns to allowance allocation mechanisms, the question becomes one of auction or allocate. However, this question presents a false dichotomy. Policymakers do not need to choose between allocating and auctioning allowances. The only real difference between an initial auction of allowances and an allocation of allowances that is followed by a nondiscriminatory auction is that the distribution of the revenues associated with the auction in the latter case have been defined, whereas the distribution of the revenues from the auction in the former case might or might not be predetermined. It is certainly possible, and has been described specifically in the joint CPUC/CEC white paper on allowance allocation,[26] to define, *ex ante,* the recipients of the auction revenue rights from such an auction. In doing so, the economic implications of the allocation of allowances and the allocation of auction revenue rights (ARRs) can be identical, and the use of an auction with auction revenue rights can simply and directly resolve any distributional issues associated with allocating allowances.

---

[25]In California, concerns have been raised about potential market problems if one set of entities controls the allowances that a second set of entities must acquire in order to produce GHG emissions.

[26]Joint CPUC and CEC Staff Paper on Options for Allocation of GHG Allowances in the Electricity Sector, R.06-04-009 and D.07-OIIP-01, issued April 16, 2008.

Although this simple solution exists to the problem of the distribution of allowances under a cap-and-trade market system, no simple solution exists to the determination of who should receive the allocation, whether it's an allocation of allowances or an allocation of auction revenue rights.

18.3.7 ALLOCATION OBJECTIVES

Since there is no "correct" allocation mechanism, it is helpful to consider the objective(s) we are trying to achieve in an allocation mechanism. In previous cap-and-trade programs such as the U.S. SO_2 cap-and-trade system, the EU ETS GHG cap-and-trade system, and the South Coast Air Quality District's RECLAIM cap-and-trade market for NO_x, allowance allocations have gone almost exclusively to emitters based on historical emissions.[27] This might have simply seemed like the logical choice, since these were the entities that would need to acquire allowances in order to emit, or it might have been the result of debate and analysis. In the NO_x and SO_2 markets, this approach appears to have been appropriate and successful.

In the most recent of these, the EU ETS market (Europe's cap-and-trade system for GHG), several flaws were identified relating to the allocation of allowances. One of these, that too many allowances were originally allocated, should not be relevant to the debate in California or the United States as long as reporting and measurement are handled properly. Another identified flaw from Europe has received substantial review and debate, and policymakers are loathe to repeat it: allocating allowances in such a manner that some entities are able to achieve windfall profits.

What are windfall profits, and how did this happen in Europe? Simply defined in this context, the term *windfall profits* refers to profits that occur as a result of the imposition of the GHG reduction program that would not have occurred in its absence. It is recognized that some windfall profits are unavoidable, such as the profits of an independent generating company that is nonemitting and sells its output into the wholesale market. As wholesale prices rise to reflect the emissions cost of the marginal generating unit, which is GHG emitting and must recover its variable operating costs through its market clearing bid in this market, the nonemitting generator's revenues increase without any increase in its costs due to GHG regulation (since it is nonemitting).

---

[27]Source: Acid Rain Program SO_2 Allowances Fact Sheet at www.epa.gov/airmarket/trading/factsheet.html#how

Although these windfall profits have consequences, as described later, they are not the main cause of concern. Allocating allowances in such a manner that creates windfall profits, as occurred in the EU ETS for some sellers, has been the source of much criticism in the allowance allocation design. If a generator has an emissions rate that is higher than that of the marginal unit setting market prices, the cost to produce from this generator will be greater than the increased revenue it will receive from sales into the market under a GHG program.

Providing some allowances may mitigate this reduction in profits, but as long as the allocation of allowances is not greater than the profit lost from the GHG regulation, this allocation will not result in windfall profits. If a generator has emissions but at a rate that is less than or equal to the marginal generator, the costs of operating under a GHG program for this generator will be less than the additional revenue received from sales into the market. If an allowance allocation mechanism based solely on historical emissions provided an allowance for such a generator, that allowance would further increase the profits of that generator beyond its profits absent a GHG reduction program. Thus the windfall profits would have been either created or expanded directly as a result of the allocation of allowances.

This situation occurred for many generators in Europe, where gas and coal units are primarily the generators on the margin. U.S. policymakers in the current development of GHG market rules are loathe to repeat this mistake, and some are taking excessive measures to avoid the potential for any windfall profit creation from allocation of allowances. In the extreme, this is being used as justification for a full auction of allowances with revenues that do not flow back to the affected entities. One possible component of an objective may therefore be to avoid creating windfall profits from the allocation of allowances.

Is the allocation of allowances a question of efficiency or one of equity? Clearly under the prior models in which allocations were based solely on historical emissions, the only relevant issue may have been equity. This is so because such an allocation approach, predicated on historical fact and not altered by current or future behavior, thereby cannot incent any efficiency gains. Efficiencies may be achieved through the market solution of a cap-and-trade program resulting in a price for GHG emissions equating the demand and supply of allowances under the cap-and-trade program. These efficiencies are unaffected by who gets allowances, however, since that determination was based on historical data—data that cannot be altered by current or future behavior.

Some allocation approaches have equity implications but no efficiency considerations. This need not necessarily be the case, however. One could design an allowance allocation proposal that does depend on current and future behavior such that there were efficiency implications. The appropriate efficiencies to be achieved from a market structure are those derived from the clearing of the market through the intersection of supply and demand at a market-clearing price. Undoubtedly there are a set of underlying assumptions regarding the functioning of such a market that would be needed to prove that this is the most efficient outcome, but such analysis is beyond the scope of the chapter. Nevertheless, the maximum efficiency from the market had already been achieved through a well-functioning market outcome, and one cannot find a more efficient outcome than the maximum. Any allocation mechanism that attempts to achieve "additional" efficiencies through an allocation mechanism that alters the market solution would thus be suboptimal, or at least not better than equal to the efficiency of the solution from an allocation that does not alter behavior.

The prior argument implies that we should seek allowance allocation solutions that do not result in behavioral changes from the market solution. What then is left as the objective of such an allocation approach? If we assume that the cap in the cap-and-trade program is a binding constraint (otherwise there would be a zero price for GHG emissions and the allocation question would be moot), there will be a net cost of imposing the constraint. That cost will be borne by the full set of market participants. How these net costs should be spread among those entities then becomes a question of equity. So, what is the most equitable way to distribute the allowances and thus the shortfall?

Several opinions have been offered on what is equitable within the context of a GHG allowance allocation. These include:

■ Recognize past actions to acquire low GHG-emitting resources that have resulted in cost implications borne today. In such a circumstance, it could be argued that it would be equitable to allocate to those entities that made such sacrifices in years past.

■ Allocate based on current economic burden. It could be argued that prior to recent legislation there was no basis for choosing any but the lowest cost options available, and those who had access to coal and used it satisfy the needs of their customers will bear the brunt of the impacts of the change in rules and should get the bulk of the allowances.

■ Offer high-emitting resources a financial means to reduce their emissions. Some parties argue that those entities with high-emitting resources in their portfolios are those that will have to make investments to either clean those resources or replace them, and thus they should receive allowances.

A corollary argument is that those that have invested in high-emitting generation will potentially lose the financial capability to clean up or replace these resources due to the imposition of the GHG reduction program and the carbon prices that result from it.

■ Allocate according to retail sales or generation. An alternative definition of equity would suggest that providing for a fixed number of allowances for each kWh of sales by LSEs is fair. Others would define equity as a fixed number of allowances for each kWh produced, though this particular solution falls within the category of proposals that has efficiency implications and, by the aforementioned logic, adverse efficiency implications.

■ Minimize any wealth redistribution. Some define equity as minimizing the wealth redistribution that would result from the imposition of the GHG reduction program.

Many stakeholders argue that any allocation proposal that mutes the retail price signal associated with GHG in the energy sold would have adverse efficiency implications and should be dismissed. Note that an allocation to an LSE based on any number of allocation approaches could but need not result in changes to retail rates, as rate structure can be impacted independently of total average ratepayer costs. (For example, allocations to LSEs could be sold to convert them into dollars and provided to customers as a rebate check unassociated with any going-forward usage.)

There is no single and unique answer to the question: What is most equitable? However, there is a reasonable case to be made that the imposition of a GHG reduction program will impose substantial economic harm on some set of entities while either not impacting or positively impacting others. Achieving a desired level of GHG reduction while mitigating the economic harm caused in total appears to be a generally accepted goal of any GHG reduction program. So, if mitigating total economic harm is a desired goal, allocating to those experiencing economic harm seems to be a desirable property of an allocation approach. One benefit of allocating allowances only to those experiencing economic harm is that this is also consistent with not allocating allowances to any entity that is not suffering economic harm. This latter condition ensures that the creation of windfall profits from the allocation approach is avoided.[28]

---

[28]Although windfall profits may still exist from the implementation of a cap and trade (i.e., generators selling to a wholesale market).

18.3.8 Southern California Edison's Harm-Based Allocation Proposal

Although there is no "right" allocation method, in its regulatory filings SCE has suggested that California evaluate any allocation protocol against two general principles:

■ Equitably mitigating economic burden

■ Economic efficiency

Allocating allowances (or ARRs) is a key point of disagreement within the electricity sector in California. Although various proposals for allocating allowances have been offered over the years, no single proposal has enjoyed broad support. Proposals that allocate to generators based on historical emissions fail to recognize the structure of energy markets and often create economic windfalls for generators. Alternatively, allocating directly to LSEs based on load served creates windfalls for utility ratepayers. Other suggestions include allocating to generators based on electricity output or allocating to LSEs based on the emissions of utilities owned or procured resources. Both of these allocation mechanisms create unfavorable economic incentives.

SCE's approach recognizes the economic burden of a cap-and-trade but avoids the economic windfalls that occurred in the EU-ETS.

18.3.9 Economic Impact of Allocation

So, what is economic harm, and how can it be measured? Economic harm can be defined as the difference in economic outcomes for an entity between the business as usual (no imposition of a carbon-reducing regime) counterfactual and the actual economic result on that entity of imposing rules to reduce GHG. In the case of the electricity sector, there are at least three or four major sources of economic harm:

■ *Independent generator harm.* One source of economic harm occurs when an independent generator has an emissions rate that is higher than the emission rate of the marginal unit setting the market-clearing price in the market in which the generator transacts. In such a circumstance the generator will be incurring emissions costs to a greater degree than it will be receiving additional market revenue as a result of the imposition of the GHG regulation. The dollar value of economic harm it will suffer will be determined by the difference in emission rates between the unit's emissions and the market price-setting unit's emissions, times the price of emissions in the market, times the volume of power sold.

■ *LSE economic harm from its portfolio.* Another source of economic harm occurs when a load-serving entity owns the generation that has GHG emissions (or is responsible for the emissions cost of generation it has purchased by contract). In such a circumstance, the generation is not receiving any market revenues, since it is being used to serve load directly. Thus all the emissions costs associated with operating this generator, emissions costs that did not exist but for the GHG reduction program, constitute economic harm.

■ *LSE economic harm from its procurement.* A third major source of economic harm is suffered by an LSE (or its customers, typically, since the LSE passes on its costs to its customers in the form of higher rates) when the LSE is purchasing power from the market to meet its customers' needs but the market price has increased as a result of GHG reduction regulation. A competitive electricity market will yield prices equal to the marginal cost of the last unit needed to clear the market. Absent GHG regulation, these marginal costs are typically determined by the operating efficiency of the generating unit, the cost of the fuel used by this unit (typically natural gas for California, though in some other parts of the country coal generation is on the margin a reasonable fraction of the time), plus any variable O&M incurred from operating the unit. In the presence of GHG regulation, if the generating unit is responsible for acquiring allowances associated with its GHG output, an additional marginal cost component will be the product of the marginal unit's GHG emission rate and the market price of GHG allowances. Thus this higher marginal cost is expected to be observed in the market-clearing prices under a GHG reduction regulation regime. The economic harm suffered by an LSE purchasing from the market is this new increased cost component that is reflected in market-clearing prices.

18.3.10 COST CONTAINMENT AND FLEXIBLE COMPLIANCE OPTIONS

A well-designed market-based solution such as a cap-and-trade program will facilitate the least cost compliance opportunities. But SCE remains concerned that the allowance market and the electricity market may interact in unpredictable ways, particularly in the short run. As a result, SCE has consistently advised CARB to implement a cap-and-trade system with flexible compliance options to mitigate any unexpected price spikes. Flexible compliance options can include allowance banking and borrowing, multiyear compliance periods, and a potential alternative compliance payment. A key flexible compliance tool, and one that is often debated, is the emission offset.

It is anticipated that "offsets," qualified GHG emission reduction measures outside the regulated sector, could also be procured as an alternative that reduces one's emission profile for the purposes of meeting allowance requirements. In other words, if a deliverer under a California cap-and-trade program can acquire a qualifying credit from the reduction of emissions from some action unrelated to electricity production, the emissions for which it would be responsible to produce allowances would be reduced by the magnitude of the credits acquired. The eligibility and use of offsets is among several market details yet to be decided for a California cap-and-trade market. Under a cap-and-trade structure, the use of offsets can be an important mechanism to minimize the cost impact of imposing GHG caps.

Other potential cost mitigation elements to be considered in a cap-and-trade structure include the borrowing and banking of allowances across the time periods for regulatory compliance and the potential for a safety valve or other emissions price-limiting tool. California's AB-32 legislation contemplates the possible need for cost-control measures to ensure that the impact on California's economy is not too great for the state to bear.

18.4 What this means for SCE ratepayers

Going forward, California utilities such as SCE are facing the potential for an expansion of the California RPS program, expanding EE programs and possibly forced divestiture of remaining coal resources in addition to a cap-and-trade program being applied across multiple economic sectors. Figure 18.6 shows an approximate breakdown of the reductions outlined in the PSP.

The PSP calls for direct regulations such as expanding the California RPS regulation to 33 percent and expanding EE initiatives by as much as 32,000 gWh. The PSP also lays over these direct measures a multisector cap-and-trade program.[29] Of course, these numbers are statewide, and the particular abatement attributed to SCE is uncertain, particularly under the cap-and-trade program. In this regard, it is important to remember that AB-32 imposes a statewide emissions cap. AB-32 does not impose a cap on specific sectors or entities. Increasing the RPS requirements will increase the procurement costs for SCE.

---

[29]Source: CARB's Climate Change Proposed Scoping Plan, a Framework for Change, June 2008.

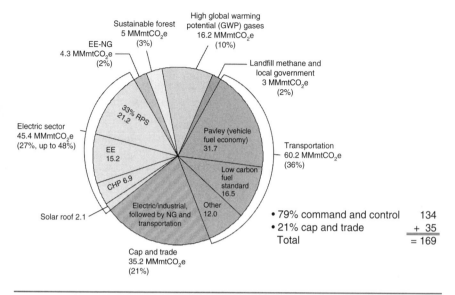

Figure 18.6   CARB reduction measures. The electric sector faces CHG reductions ranging from 27 to 48 percent, depending on the impact of the cap-and-trade.

Source: CARB's Climate Change Proposed Scoping Plan, a Framework for Change, June 2008.

However, expanding EE programs requires additional ratepayer incentives to make EE investments in their homes and businesses. Additionally, some of these EE savings may require government action in the form of building and operating standards. Such policies will directly impact SCE ratepayers.

18.4.1 SCE Is Looking for Real, Permanent Reductions at the Lowest Possible Cost

SCE currently serves retail load at an emissions rate that is lower than both the California and the national averages. The basic economic concept of increasing marginal costs indicates that the marginal cost of emissions abatement increases as the total quantity of abatement increases. Due to the high ratio of direct, prescriptive reduction measures outlined by CARB, the structure presented in the PSP will not allow California to achieve its AB-32 emission reduction goals at the lowest possible cost. Even within a cap-and-trade program, it is likely that a large share of the least costly emission reduction opportunities lies beyond the California border.

Given SCE's history of energy efficiency and procurement of low-emission resources, the most cost-effective way to achieve additional emission reductions will require that SCE look outside California and perhaps outside of the United States. A key element to a successful abatement policy must allow SCE and other California IOUs to utilize an offset program to fund abatement at the lowest possible cost. As a result, utilities such as SCE must be allowed to reduce emissions outside California and possibly outside the United States. Such an offset program is critical to enabling SCE ratepayers to fund the greatest amount of GHG abatement at a given cost.[30]

18.5 Conclusion

At this stage, it is clear that CARB intends to implement additional command-and-control–type regulations, coupled with the introduction of a multisector cap-and-trade program. Although AB-32 does not impose a sector- or entity-specific cap that drives the actions of specific entities, the types of regulations that would be imposed on SCE will likely include increased mandatory renewable procurement and an expansion of existing EE programs. The CARB PSP outlines expanded direct regulation to create approximately 80 percent of mandated reductions with the remaining 20 percent to come from a regional cap-and-trade program. While a cap-and-trade program would likely offer the type of flexibility and cost containment needed to minimize the economic burden of reducing emissions and complying with AB-32, the challenge remains for SCE to advocate for the most efficient reduction opportunities possible.

---

[30]While the final rules for the WCI have not yet been adopted, the WCI has recommended that offsets be limited to 49 percent of the mandatory emission reductions under the cap.

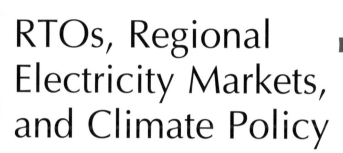

RTOs, Regional Electricity Markets, and Climate Policy

Udi Helman

California ISO

Harry Singh

RBS Sempra

Paul Sotkiewicz

PJM Interconnection, LLC

Abstract

This chapter examines how policies to mitigate greenhouse gas (GHG) emissions from the electric power sector could affect the core functions of the Regional Transmission Organizations (RTOs) that encompass approximately two thirds of the U.S. power system. Following a review of policy options for emissions abatement and their general implications for wholesale power markets, the chapter examines how GHG policies affect four functions central to RTOs: reliable power system operations; organized markets for spot energy and ancillary services; long-term resource adequacy, including capacity markets; and regional grid planning and interconnection procedures. Additionally, the chapter considers how features of these RTO functions could facilitate the implementation and effectiveness of GHG policies.

19.1 Introduction

The U.S. electric power sector comprises about 34 percent of total U.S. CO_2 emissions, with significant variation by state or region, depending on the fuel mix for power generation [22]. The prevalence of large point sources of GHG emissions in the sector also makes it likely to be regulated more intensively than other sectors of the economy for purposes of emission mitigation, since reductions will be easier to measure and control. Moreover, other sectors may attempt to reduce their own emissions through electrification, most notably automotive transportation. Climate policy could thus result in one of the largest power infrastructure development and technology transformation initiatives in U.S. history in the span of a few decades.

Developments in climate policy come at a time when the U.S. electric power sector is completing roughly the first decade of a sector-wide regulatory reform and "restructuring." This process started with the establishment of a transmission open-access regime and formation of Independent System Operators (ISOs) under Federal Energy Regulatory Commission (FERC) Orders 888 and 889 [23, 24], along with the evolution of most ISOs into Regional Transmission Organizations (RTOs) under FERC Order 2000 [25]. For purposes of this chapter, both ISOs and RTOs have sufficiently similar functions to be considered under the abbreviation RTO. As shown in Figure 19.1 and Table 19.1, approximately two thirds of the U.S. power system is currently under the control of RTOs (as are several Canadian provinces that will not be considered here).

To ensure independence, RTOs are only operators of the transmission system and do not own any infrastructure assets (other than those used to perform their functions).[1] In addition, to improve the efficient usage of system resources and support unbundling of transmission and generation, RTOs

---

[1]In the United States, transmission assets are still largely owned by the utilities that previously provided access and scheduling and dispatch functions. However, under open access, transmission ownership in RTOs does not confer any transmission usage priority.

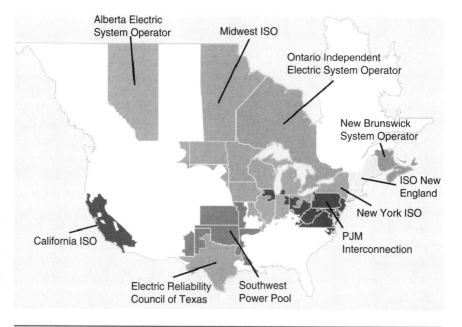

Figure 19.1 Geographic area of RTOs in North America.
Source: ISO/RTO Council (www.isorto.org).

Table 19.1 Selected measures of the size of U.S. ISOs and RTOs

RTO	Installed Capacity (MW)	Miles of Transmission	Population (Millions)
CAISO	54,000	25,526	30
ERCOT	71,812	38,000	20
ISO-NE	32,000	8,000	14
MISO	156,000	93,600	40
NYISO	44,851	12,000	19
PJM	164,634	56,250	51
SPP	50,392	40,364	4.5

operate organized day-ahead and real-time wholesale spot auction markets, with locational marginal pricing (LMP) of energy and, typically, zonal procurement and pricing of ancillary services—regulation and operating reserves.[2]

---

[2]Other ancillary services, such as voltage support and black start, are procured through tariff-based formulas or rates.

The spot markets are integrated with system operational functions, reflecting all needed generation and transmission constraints when calculating hourly and subhourly schedules and prices. The recent FERC Orders 890 and 719 [28, 30] have further required the participation of "nongeneration resources," such as demand-side and storage resources, in the energy and ancillary service markets. RTOs also allocate and operate markets for short-term (annual or less) and long-term (multi-year) financial transmission property rights that can be used to hedge congestion charges due to LMP. Almost all the RTO regions also have resource adequacy requirements on load-serving entities (LSEs), and most operate forward markets for capacity (although with different designs). RTOs have also, for the most part, integrated grid planning functions, including generation interconnection procedures (but not siting), over their territories. They have also made progress in interregional coordination to reduce RTO market seams and jointly plan transmission and allocate costs.

The purpose of this chapter is to examine how the regional scope of RTOs and the transparency of their core functions, with some needed adaptations, will facilitate the implementation of U.S. climate policy as it evolves at the federal, regional, and state levels. As of the writing of this chapter, the level and timeline of U.S. GHG emissions reductions remains uncertain. If federal policy unfolds roughly along the lines of recent proposed congressional legislation, perhaps with more aggressive measures in some states or regions, then current short-term and long-term technology and economic assessments make the following predictions.[3] The bulk of near-term, that is, 5–10 years, GHG emissions reductions will be achieved through switching from coal to gas (driven by the impact of carbon pricing on relative gas and coal fuel prices), increased renewable penetration, particularly wind generation, and end-use energy efficiency improvements. These reductions will be augmented to some degree by demand response and additional storage that emerges in this timeframe.

In the medium-term horizon (10 –20 years), additional nuclear power plants and carbon capture and storage (CCS) will become plausible from a technological perspective and possibly from an economic and political perspective (depending on the region). Additional distributed energy resources, including electrification of transportation, supported by "smart grid" features, could substantially reshape load profiles to support emissions reductions [17]. Some forecasting studies look out beyond 2030, a time horizon that obviously incorporates a high degree of technological uncertainty, to envision a radically

---

[3]See EPRI (2005, 2007), CEC (2007), ISO-NE (2007), IEA (2008), DOE (2008b), and PJM (2009b). See also chapters of this volume on particular fuel types.

transformed power sector, possibly with dramatic penetration of clean distributed energy and storage resources.

Given the uncertainties, RTOs need to be flexible and innovative to adapt, as needed, system operations, wholesale market designs and participation criteria, resource adequacy mechanisms, and grid planning procedures to the evolving aspects of climate policy. And they must do so while continuing to meet their traditional mandates of maintaining short-term and long-term reliability and supporting efficient, least-cost procurement of wholesale spot power.

The remainder of the chapter is organized as follows: Section 19.2 briefly reviews alternative GHG policy instruments for the electricity sector, including direct regulatory measures and market-based approaches. Section 19.3 examines some issues for RTO system operations raised by renewables policies that could accelerate under climate policy. Section 19.4 reviews how the RTO markets will support a dispatch that minimizes costs and GHG emissions and how the markets may be transformed by nongeneration resources that are likely to become more cost-effective under GHG policy. Section 19.5 considers how to evaluate resource adequacy with substantial renewables and nongeneration resources. Section 19.6 examines grid planning and generation interconnection. Section 19.7 concludes.

19.2 Alternative GHG policy instruments and implications for RTO functions

The trajectory of future technological and infrastructural changes, and implications for RTO functions and wholesale power prices, will clearly be heavily influenced by the choice of federal and state policy and regulatory instruments for GHG emissions abatement. The key instruments include (1) direct regulation measures, such as energy efficiency standards and procurement mandates, particularly renewable portfolio standards (RPSs), and (2) market-based mechanisms such as subsidies, fees, taxes, and cap-and-trade systems. Table 19.2 provides a summary. Current state and federal policy initiatives suggest that both types of approaches will be employed concurrently under GHG policy, along with regulatory initiatives to expand transmission so as to improve access to renewables.[4] The implications of this regulatory model

---

[4] For example, under state Assembly Bill (AB) 32, California is developing a multisector GHG policy that includes significant direct regulatory measures, including energy efficiency and a 33 percent RPS in the power sector, along with a GHG cap-and-trade program for more efficient abatement of any residual emissions not subject to the direct measures (CARB, 2008); on this policy, see also the discussion in Chapter 18, by Stern and Harris, in this volume.

Table 19.2 Types of GHG regulatory instruments and selected implications for RTO functions

Type of GHG Regulatory Instrument	Description	Implications for RTO Functions
Energy-efficiency requirements	These include standards for end-use appliances, building codes, and programs to encourage production efficiency (e.g., through combined heat and power, or CHP). In some states, up to 50% of planned GHG emissions reductions by 2020 are planned through such measures [7].	From an operational and market perspective, energy efficiency appears to the RTO as load reduction or can be incorporated into market design as a supply resource in capacity markets and for planning purposes (see Section 19.5).
Renewable portfolio standards (RPSs)	An RPS requires that a mandatory percentage of retail electricity sales come from renewable generation resources. Most implemented RPS policies have gradually increasing targets for renewable energy sales over a number of years, allowing time for renewable resources to be developed.	At higher levels of renewable penetration, RTOs have to consider significant operational and grid planning requirements (see Section 19.3).

| Production tax credits (PTCs) and subsidies | Production tax credits (PTCs) and subsidies targeted at preferred resources are designed to reduce the effective cost of those resources to developers so that they are more cost competitive with traditional resources. Currently in the United States there are PTCs for most renewable resources and for nuclear power. | In the absence of GHG pricing, PTCs or subsidies attempt to encourage development of less GHG-intensive resources. However, as with RPSs, they do not expose other generating resources or electricity consumers to price signals that include the cost of GHG abatement. Depending on the resources being targeted, these instruments may encourage rapid penetration of resources that create operational and planning issues, such as types of variable generation renewables. PTCs for wind may create barriers to achieving more dispatchability through price signals. |
| Emissions taxes | Emissions taxes directly place a price on emissions, or on the fuel inputs based on expected emissions from using that fuel, as though those emissions are an input to production. For example, emissions taxes designed to reduce CO_2 emissions could be placed directly on post-combustion emissions or placed on the carbon content of fuel that would lead to CO_2 emissions. An emission tax provides price certainty to emitters about the cost of emissions, but how much emissions will be abated is uncertain. | Emissions taxes, in contrast to an RPS or PTC, provide a direct, transparent price signal for the price of GHG emissions to both producers and consumers of electric power. In response to an emissions tax, polluting resources will reduce GHG emissions up to the point where the marginal cost of abatement is equal to the level of the tax. Entities that can reduce GHG emissions at low cost will reduce emissions in larger amounts than those with higher abatement costs. |

(Continued)

Table 19.2 (*Continued*)

Type of GHG Regulatory Instrument	Description	Implications for RTO Functions
Cap-and-trade systems	In contrast to an emissions tax, cap and trade provides certainty about the level of emissions allowed (the cap) and allows the price to be placed on emissions to be endogenously determined. The cap-and-trade mechanism allows resources that emit GHGs to trade allowances so that low abatement cost emitters can reduce emissions by larger amounts, which may allow them to sell any excess allowances to high abatement cost emitters who might need to purchase allowances. However, similar to an emissions tax regime, emitters will only reduce emissions up to the point at which an emitter's marginal cost of abatement is equal to the price of an emissions allowance.	A cap-and-trade system can have various impacts on RTO markets and possibly also affect reliability if compliance is not sufficiently flexible. See discussion in this section and in Stern and Harris, Chapter 18 in this volume.

for the structure of the power sector—and the interaction of markets and types of integrated resource planning that are being developed by some state regulators to guide regulated LSEs in their procurement decisions under RPS and GHG targets—will be clarified over time.

Because several other chapters of this book[5] have covered most aspects of such policy instruments, this section only briefly highlights some particular interactions of RPS and GHG cap-and-trade mechanisms with RTO functions, whereas subsequent sections provide further details. It is worth noting that RTOs, as organizations that do not own the generation or transmission assets in the power system, are in general agnostic with respect to these policy instruments. The exception is instances in which policy, regulatory or market design might unnecessarily interfere with aspects of power system reliability or with the efficient functioning of RTO markets.

RPS mandates have preceded implementation of GHG policy in the United States [53], and these procurement requirements are likely to be continued and expanded, even with explicit pricing of carbon. This is because although they are likely to be a less economically efficient method for GHG emissions abatement than purely market-based mechanisms, they are seen as a more certain driver of GHG reductions in the early years of climate policy as well as stimulating job creation and incubating new technologies [7].

As targets increase through mandates, RPSs also present operational and planning challenges for RTOs, which have increasingly become more proactive in the analysis of renewable integration and regional planning requirements [11, 12, 13, 38, 39]. Given likely initiatives to further promote regional development of renewables, it is expected that RTOs will take a more central collaborative role with state and federal governments in facilitating state and regional RPS implementation. Although less understood at present, increases in renewable generation and other changes on the grid that accompany them will also change the scheduling, dispatch, and pricing in RTO energy and ancillary service markets over time.[6] These issues are discussed further in subsequent sections of this chapter.

---

[5]See in particular chapters by Dijkema et al.; Musier and Adib; Stern and Harris; Wiser and Hand.
[6]Since they have minimal short-term variable costs and are not dispatched (at present), these resources, such as wind or solar, are scheduled as price takers; as such they will lower spot market prices for energy (PJM 2009b, Potomac Economics 2005), although increased variability of output will likely lead to increased procurement and costs of ancillary services (see Section 19.4). The impact of renewables on capacity markets may be to lower capacity prices in the areas where they locate, but because renewables typically cannot locate where capacity is needed for resource adequacy, overall capacity prices might not decrease. Other fixed costs of renewables will be recovered through bilateral contracts.

As of this writing, cap-and-trade systems are likely to be the central market-based policy instrument for specific U.S. GHG reductions. From the perspective of RTO functions, a cap-and-trade system assigning allowance requirements to generation sources (the "point of regulation") should be designed so as to ensure the continued efficiency and transparency of the RTO spot markets as well as to maintain operational reliability by ensuring that there is sufficient compliance flexibility. The market for GHG allowances would operate separately from the RTO wholesale markets, but with allowance prices being factored into bilateral contracts and spot energy and ancillary service market offers and hence into transparent RTO market prices.[7] The RTO markets should thus become a major tool in coordinating efficient GHG reductions.[8] This topic is discussed further in Section 19.4.

Other key design decisions in GHG cap-and-trade systems that can have implications for the wholesale markets and possibly system reliability include the allocation and/or auction of allowances and mechanisms for compliance flexibility.[9] In general, under the current RTO wholesale market designs, the price of allowances will be reflected in supplier offers and hence in market prices, regardless of the method of allocation. However, in an auction model, the use of auction proceeds could have a bearing on wholesale price signals and possibly distort the incentives facing end-use customers. For example, in the Regional Greenhouse Gas Initiative (RGGI), auction proceeds are being used to invest in energy efficiency and demand-response technologies. Another idea under consideration is to rebate the auction proceeds directly back to consumers to protect them from significant price increases, which would clearly diminish the incentive to alter behavior. The cap-and-trade systems being designed provide compliance flexibility via the ability to employ any available

---

[7]For RTOs that provide rules for cost-based market offers for purposes of market power mitigation and any other functions requiring evaluation of costs, allowance costs are now included. See, for example, PJM (2009a), pp. 11-15, which allows suppliers to include the costs of sulfur dioxide, nitrogen oxide and CO_2 allowances as part of their fuel-related costs.

[8]As discussed by Burtraw (2008), Stern and Harris (Chapter 18 in this volume), and CPUC (2008b), there have been aspects of cap-and-trade design that could create adverse bidding and scheduling incentives in the RTO markets. In particular, such impacts would have been likely if LSEs had become the point of regulation for GHG emissions—that is, the entity required to retire allowances—rather than the sources of the emissions (generation). This "load-based" point of regulation was considered under some early California design proposals (CPUC 2008b).

[9]There are various choices for allocation of allowances: free allocation to emitting sources based on some historical measure of emissions, inputs, or outputs, in perpetuity; free allocation of allowances initially based on a historical measure of emissions, inputs, or outputs but updated to account for more recent information; and an auction of allowances. See also Stern and Harris, Chapter 18 in this volume.

carbon abatement technology or method, the opportunity to procure additional allowances, multiyear periods for compliance (e.g., the 3-year period under RGGI), the availability of offsets, and possibly multisector trading that could bring other cost-effective reductions into the system. However, even with such flexibility, there could be constraints on the ability of generators needed for system reliability to obtain allowances regardless of price. For example, in considering the situation in New England, [34] points to the potential for lack of liquidity in the allowance market, higher-than-forecast energy demand, or poor operation of carbon-free resources that could lead to a shortage of allowances or offsets. This is an issue that has not yet been examined sufficiently in any area of the country [43].

19.2.1 STATE AND REGIONAL GHG REGULATION

As noted, a number of U.S. states and regional entities have begun to design cap-and-trade systems ahead of federal legislation. The two major regional initiatives that are most developed are the RGGI in the Northeast and the Western Climate Initiative (WCI); more recently, there has been some development of a Midwestern Regional Greenhouse Gas Reduction Accord. The RGGI is a cap-and-trade program that targets the electric power sector alone; it takes effect in 2009 and includes 10 Northeastern and Mid-Atlantic states. During the 2009–2014 period the goal is to stabilize emissions at the approximate average 2000–2004 levels, and then from 2015–2018 to reduce emissions 2.5 percent per year so that emissions in 2018 will be 10 percent below the 2009 allowed emissions. Most states within RGGI have chosen to allocate their allowances via auctions.

In contrast, the WCI is developing a more ambitious multisector cap-and-trade system that also includes transportation and industrial, commercial, and residential fuel combustion sources [52]. The WCI is scheduled to take effect in 2012. California, which is a member of WCI, is also subject to state law AB-32, which requires emissions reductions in the state to begin in 2012.[10] Table 19.3 summarizes the various regional initiatives.

Most RTOs cross multiple states or operate within single-state power systems that engage in substantial power trade with other states (such as California). GHG reduction regulations, whether through an emissions tax or cap-and-trade, that differ across states or regions in an electrically interconnected region can lead to emissions leakage and also contract "shuffling" that can defeat the objective of reducing emissions [6, 50].

---

[10]See discussion by Stern and Harris, Chapter 18 in this volume.

Table 19.3 U.S. regional GHG cap-and-trade policies

Initiative	Jurisdiction/(RTO Coverage)	Implementation	Key Features
Regional Greenhouse Gas Initiative (RGGI)	Members: Connecticut, Delaware, Maine, Maryland, Massachusetts, New Hampshire, New Jersey, New York, Rhode Island, and Vermont (ISO New England, New York ISO, PJM)	Compliance period started on January 1, 2009. Auctions for allowances began in September 2008.	Cap-and-trade for power sector to reduce CO_2 levels by 10% by 2018. Does not directly address leakage. Allows offsets from reductions in other sectors.
Western Climate Initiative (WCI)	Partners: Arizona, British Columbia, California, Manitoba, Montana, New Mexico, Ontario, Oregon, Quebec, Utah, Washington Observers: Alaska, Colorado, Idaho, Kansas, Nevada, Wyoming, Saskatchewan, and six Mexican states (California ISO)	Initiated in February 2007; implementation target in 2012.	Multisector cap-and-trade program to reduce GHG levels 15% below 2005 levels by 2020; reductions begin in 2012. Has rules to minimize leakage by requiring imports not linked to specific resources to retire allowances.

Midwest Regional Greenhouse Gas Reduction Accord	Members: Iowa, Illinois, Kansas, Manitoba, Michigan, Minnesota, Wisconsin Observers: Indiana, Ohio, Ontario, South Dakota (Midwest ISO, Southwest Power Pool)	Signed November 15, 2007; goal to complete basic cap-and-trade agreement within 12 months and other details within 30 months.	Establish multisector market-based cap-and-trade approach for GHG reduction. Minimize leakage and address interaction with future national program.
California AB-32	State of California (California ISO, which does not completely cover California)	Signed into law in 2006. California Air Resources Board (CARB) responsible for implementation, with input for power sector from the California Public Utilities Commission (CPUC) and California Energy Commission (CEC).	Requires GHG reduction to 1990 levels by 2020; reductions begin in 2012.

In the short term, leakage would take place if generation that is subject to GHG regulation, and is thus relatively more expensive, is displaced by generation from unregulated states. Over the long term, leakage takes place through the migration of end-use consumption to the unregulated region with cheaper power. Contract shuffling allows entities outside the regulated region to "shuffle" bilateral sales into the region from higher-emitting to lower-emitting resources while retaining the output of the higher-emitting resources. Hence, if overall energy needs remain fixed, emissions reductions can be largely undone by an increase in generation not subject to emissions constraints. RGGI [50] examined ways to prevent leakage, but did not implement any direct mechanisms to do so.[11] The WCI is attempting to reduce leakage by requiring imports into the WCI to retire allowances. However, there could still be contract shuffling (unless prevented by further restrictions on trade). Ultimately, a federal system is needed to preserve the integrity of regional emissions reductions in the power sector under cap and trade.

19.3 Implications of GHG policy for power system operations

RTOs operate the power system over large regions, allowing for improved reliability and efficient use of system resources. At the core of system operations is the real-time balancing of scheduled and spot energy, sometimes called imbalance energy;[12] correction of area control error (ACE); maintaining frequency; maintaining sufficient operating reserves to meet potential system contingencies; and maintaining voltage levels within prescribed limits. RTOs undertake these functions through a sequence of integrated market and operational procedures, beginning with a day-ahead market with security-constrained unit commitment that establishes an initial hourly schedule for the operating day (along with any nonmarket-based commitments needed for local reliability, such as for "reliability must-run" resources). The sequence continues with a post-day-ahead reliability unit commitment (to establish physical supply adequacy against the RTO's next-day load forecast) along with subsequent

---

[11]Under RGGI, direct mechanisms to prevent leakage were not favored as the growing prospects for a national cap-and-trade program would eliminate the need for such a mechanism. However, indirect actions such as energy efficiency were recommended.

[12]The term *imbalance* implies that there is a pre-real-time physical schedule that is being deviated from. In the RTO markets, this term is no longer accurate, since real-time energy, though in principle the deviation from a day-ahead financial schedule, is the only physical market for financial settlement.

Regulation/AGC	Load following Operating reserves Balancing markets	Unit commitment Day-ahead markets	Resource adequacy Capacity markets Transmission planning	Function
Seconds	——— Minutes	——— Days	——— Years	——→ Timeframe

Figure 19.2   Timeframe of reliability functions.

adjustments of schedules until approximately 1 hour before the real-time market, in which operators will dispatch already committed resources to follow load on 5- to 10-minute timeframes, with regulation to cover frequency deviations within those time steps (see RTO technical manuals for details; [31]). Figure 19.2 shows the timeframe of reliability functions performed by an RTO.

The operational challenges encountered under GHG policy, including renewables policy, are likely to be driven by the following main factors:

■ Operational constraints on and early retirement of existing coal and gas thermal resources with high GHG emissions

■ Operational requirements created by substantial penetration of variable generation renewable resources

■ Increasing participation of nongeneration resources in providing ancillary services

These issues have only recently begun to be analyzed by RTOs and other reliability entities in depth [11, 35, 36, 40, 42, 43]. Operational reliability requirements are not necessarily barriers over time to GHG emissions abatement but should not be understated: If changes to the grid and the resource mix are not evaluated carefully, the result could be reduced reliability and increased cost of meeting GHG emissions targets.

GHG restrictions, increasing renewable penetration, and other environmental policies could be forcing thermal resources that can provide ancillary services and support local reliability off the system ahead of planned retirements, although, as discussed further in Section 19.4, the revenue pressures driving early retirements may be offset in part by additional cash flows from forward capacity markets [4]. The generation mix resulting from early retirements may also change the inertial response of the power system in some regions for frequency control. For example, the state of California has regulatory objectives to retire or repower a large number of thermal generation plants (up to 22,000 MW of gas and nuclear generation) by 2021 that use once-through cooling (which causes loss of marine and estuary biodiversity), many of which can provide regulation and ramp capability and are located in load pockets.[13] Notably, although planners can suppose that older, less efficient plants can be kept available as reserves or to support local reliability until the low-carbon power system of the future is considered reliable, such plants also typically emit more GHGs per MWh at low operating levels than at higher output [40]. Hence, there may be problematic trade-offs for a number of years in seeking to reduce emissions while maintaining reliability.

Variable-generation renewable resources are likely to be the primary operational challenge of the next decade. The major types of renewable resources roughly divide into those with predictable output on daily operational time frames, such as hydro, geothermal, and biomass, and those with inherently variable, or intermittent, output, such as wind and solar. Variability creates potentially substantial operational requirements to maintain system balancing and frequency control at higher levels of penetration [11, 42].

These requirements will vary by region. For example, in California, an assessment of operational needs found that approximately tripling the wind capacity on the system to meet a 20 percent RPS would at times require up to roughly a doubling of the regulation that the ISO would need to procure, with variations by season and time of day [11]. When the variability of wind output was added to load variability, there is the potential for substantial increases in load-following requirements and ramp duration.[14] NERC [42] is also considering whether to add wind variability to the definition of contingency used in defining operating reserve requirements; depending on the RTO, this would increase the procurement of contingency reserves to the extent that wind capacity exceeds the largest single contingency on the system.

These new ancillary service requirements will need to be met by RTOs procuring additional regulation and operating reserves and by providing market signals for bringing onto the system additional capabilities for fast ramping, quick start, lower operating minimums, and other operational flexibility [11]. With increased variable generation, and as carbon constraints become more binding, additional types of low- or no-carbon resources will be needed to

---

[13]In addition, in California, limited availability of NO_x permits in Southern California is already restricting the ability to build new gas plants (Allen et al., 2009).

[14]In California, where wind tends to produce more output in off-peak hours, the wind can drop substantially just as the morning ramp begins; with sufficient wind on the system simply to meet the 20 percent RPS, this could more than double the ramping capability needed in some hours. Although the existing generation fleet in the CAISO system is considered capable of meeting these requirements at 20 percent RPS, higher levels of RPS could require much greater additional operational flexibility (CAISO, 2007).

provide ancillary services. These will include what are now called non-generation resources [28], such as the demand response and storage discussed further in Section 19.4.[15] RTOs are also taking measures to facilitate wind resources themselves becoming more dispatchable, such that they can ramp down in response to price (and possibly also ramp up, perhaps to provide frequency response). RTOs will also seek to address variability and reduce potential ancillary service requirements through better wind and solar forecasts integrated into the day-ahead and real-time market and operational procedures [11, 43].

A further operational issue that could become more frequent at higher levels of renewable integration is overgeneration. This is a situation typically currently experienced in off-peak, light-load hours in which the existing baseloaded generation resources on the system are already backed down to their minimum operating levels, but there remains excess supply on the system. Overgeneration will have different impacts and possibly different solutions in different systems. In California, with high hydro levels in some seasons and where wind may be largely an off-peak supplier, off-peak overgeneration will become a more frequent occurrence with high wind penetration [11]. Solar power projections for California under a 33 percent RPS also show substantial on-peak overgeneration (relative to projected demand). At some point in the future, demand response and storage, including electric vehicles, may provide the load-shifting needed to absorb overgeneration consistent with GHG policy goals.

Finally, climate policy is stimulating the already growing interest in "smart grids" and much greater penetration of distributed energy resources, including electrification of transportation [17]. As such, many of the fundamental procedures of system (and market) operations might have to be revised and new operational technologies developed on a large scale, such as real-time interfaces with automated price- or frequency-responsive distributed systems. This chapter does not address those topics in depth, but they may be integral to climate policy over time.

---

[15]Although reliability standards allow demand response to provide products such as contingency reserves, this is subject to technical feasibility. Demand response provides near-instantaneous interruption, but it differs from the natural frequency response available from a generator in response to frequency changes in the system. Consequently, RTO markets currently vary in the participation of demand response for provision of spinning reserves. Some RTOs already allow for up to 25 percent of their reserve requirements to be met by demand-side resources (although this limit has not been reached). The experience to date—for example, in PJM—has been that demand-side resources have performed as well or better than generation resources in responding to contingencies. Moreover, RTOs have modified rules or initiated pilots to further test capabilities and economic feasibility of demand response and storage technologies to provide regulation and operating reserves.

19.4 Implications of GHG policy for design of RTO spot markets

One of the benefits of the RTO structure is the capability to integrate system operations and wholesale spot markets over large regions, allowing for locational pricing of spot energy and bid-based ancillary services under both normal operating conditions and shortage or scarcity conditions. As such, a key topic on the RTO agenda for the coming decade will be the evaluation of how the RTO markets can facilitate investments in a coordinated fashion over wide areas, so as to achieve carbon emissions abatement reliably and at least cost. With the operational needs created by renewables under expanding RPSs, supplemented over time by higher GHG allowance prices and/or emissions taxes, the spot markets will provide the transparent and decentralized price signals for specific attributes and capabilities needed in both new technologies and any repowering of existing generation resources (while the forward RTO markets for capacity discussed in the next section will provide an additional source of revenues to cover the fixed cost requirements of new and existing resources that are not otherwise revenue sufficient). The impact of the spot prices on improving the efficiency of investments in carbon abatement will depend, as noted earlier, on the mix of policy instruments used to achieve emissions abatement and the technologies that they favor.

This section provides a brief description of each RTO spot market product; the market design for that product, including complementary or substitution relationships to other products; and some implications of GHG policy options for the design.[16] Key issues that will be discussed and evaluated more comprehensively over the coming years include:

- How the existing wholesale market designs support investments in the system operational requirements created by GHG policy
- Whether the market designs need to be modified or supplemented— for example, through provision of additional pricing features or new products
- How nongeneration resources, such as demand response and small-scale storage technologies, are reliably integrated into the markets

---

[16]More extensive surveys of RTO market design can be found in O'Neill et al. (2006) and Helman et al. (2008).

19.4.1 Spot Energy Markets

Wholesale spot energy is energy (MWh) bought or sold on a day-ahead or real-time basis in the RTO markets in the sequence described in the prior section. Although these markets can have high volumes of transacted energy, buyers and sellers will also have forward contracts that hedge them when they engage in most spot purchases.[17] The locational marginal price (LMP) of energy is the cost of withdrawing an incremental MWh at a bus on the transmission system and is set either by the combination of the offer prices ($/MWh) of the marginal generators that are delivering power to that location or the bids ($/MWh) of price-sensitive demand (although, as discussed in a moment, spot prices are also be set in combination with administrative actions in "scarcity" conditions). LMP incorporates the effect of marginal transmission congestion and marginal losses.

As noted, whether reflecting a carbon tax or the cost of GHG allowances under a cap-and-trade system, suppliers will factor the cost of carbon abatement into their spot energy offers, and LMPs will thus provide a signal of the carbon content of the marginal suppliers. Both the temporal (day-ahead, real-time) and the spatial aspects of LMP will thus be of value to induce both behavioral changes on the part of consumers (dispatchable demand response or load-shifting stimulated by retail rate reforms) and provide information on the best locations and types of investments to reduce emissions. Inframarginal nonGHG emitting resources (such as renewables) will essentially earn "clean generation" rents in hours when resources that require allowances are on the margin [14].

As already observed, the effect of carbon pricing on wholesale power prices will differ across the country, depending on the resource mix. In California, where wind production is on average highest in off-peak hours, such hours are likely to remain low LMP hours, even with a carbon price, thus encouraging load shifting to the off-peak hours [11]. In general, higher penetration of renewables, whether driven by RPS or GHG policy, lowers spot energy prices compared to what they would be otherwise [46, 49]. This is because renewables are typically price-takers in the spot markets. The impact of carbon pricing on wholesale prices will depend on the cost of allowances for the marginal carbon-emitting unit, which will vary by region, but is likely to be more significant in regions with substantial base-load coal plants. In simulations of

---

[17]For example, in PJM, less than 5 percent of all energy is scheduled through the real-time spot market, approximately 35 percent is scheduled through the day-ahead spot market, and the remainder of energy is accounted for by bilateral transactions. See PJM MMU (2008), Table 2-83, p. 91.

its market with carbon pricing, a PJM study [46] finds that, all else equal, the increase in load-weighted average LMP in its area is 75–80 percent of the CO_2 price expressed in dollars per short ton of CO_2. Although PJM did not investigate how LMP increases due to CO_2 prices would change investment, the study did find a growing displacement of coal resources by combined cycle gas as CO_2 prices increased.[18]

Energy pricing during the overgeneration periods that are likely to become more frequent with more variable-generation renewable resources is another topic of increased interest among RTOs. In principle, locational prices for energy become zero or negative during low-load or overgeneration periods, signaling generation to ramp down to minimum operating levels or decommit. A negative price indicates that generation is paying to inject power into the grid and, concomitantly, load and storage are being paid to withdraw power. Baseload thermal generation may have startup costs and startup times that make it uneconomical and operationally undesirable to decommit during overgeneration periods. Hence, for operational and market reasons, RTOs are examining alternative measures to address overgeneration, including examining options for wind resources to back down on a price basis and encouraging price-based load shifting, including additional storage capacity, to the overgeneration hours when negative pricing would pay to consume in those hours.

19.4.2 ANCILLARY SERVICE MARKETS

Ancillary service markets are another mechanism for eliciting investment in the operational capabilities needed for renewable integration and efficient GHG emissions reductions. Ancillary services are procured by RTOs on behalf of LSEs, although self-provision is also allowed. Although they currently typically account for between 5 and 10 percent of total wholesale expenditures, increased procurement associated with renewable integration could raise this amount over time.

There are two categories of ancillary services currently offered through bid-based auction markets in RTOs: regulation and operating reserves. Both types of ancillary services can be provided by traditional generation resources as well as, more recently, by nongeneration resources. Table 19.4 shows that RTOs do not offer these products uniformly, and pricing rules for particular products

---

[18]It is worth noting that analyzing the way that carbon pricing will affect LMPs is a separate question from that of how to design allocation of GHG allowances so as to minimize "windfall" profits. RTOs generally do not offer such policy advice, which is the domain of regulatory authorities.

Table 19.4 RTO bid-based markets for ancillary services

Ancillary Service	Description	ISO/RTO
Regulation (or automatic generation control, AGC)	The ability to increase or decrease energy output on a second-by-second basis for energy balancing	NYISO, ISO-NE, PJM, CAISO, MISO
Ten-minute spinning (or synchronous) reserve	Reserves available (MW) within 10 minutes from generators synchronized with the grid (or demand response)	NYISO, ISO-NE, PJM, CAISO, MISO
Ten-minute nonspinning (or nonsynchronous) reserve	Reserves available (MW) within 10 minutes from generators not synchronized with the grid or demand response	NYISO, ISO-NE, CAISO, MISO
Thirty-minute (or supplemental) reserves	Reserves available (MW) within 30 minutes or more from generators either synchronized or not synchronized with the grid or demand response	NYISO, ISO-NE, PJM

Source: RTO websites.

also differ between them. However, the requirements of renewable integration are prompting interest in establishing markets for particular ancillary services where they do not already exist and, where they do, possibly modifying pricing rules or adding additional products [11].

The increased demand for regulation that may accompany renewable integration has prompted additional interest in eliciting storage and demand response into the regulation market as well as possibly requiring wind resources to respond to regulation signals. Although conventional thermal resources are currently the main providers of regulation, the addition of a carbon price may eventually provide a competitive advantage to noncarbon-emitting resources, such as storage resources, that can provide the service, or at least ones with lower net carbon impact. Preliminary research suggests that operating such plants primarily for quick-response ancillary services results in higher emissions per MWh due to a loss of efficiency [40].

As variable-generation penetration increases, a key issue will be the allocation of the costs of ancillary services caused by these resources: If allocated to the cause of the increased costs, they will encourage variable resources to firm their output through investment in on-site storage. Alternatively, if they are averaged to load, there could be less incentive to find the least cost technological solution.

19.4.3 Scarcity Pricing

In both energy and ancillary service pricing, RTOs have been introducing administrative scarcity pricing, which raises market prices above market offer caps [30]. Typically, such scarcity pricing is triggered by a shortage of operating reserves and, in some cases, regulation; alternatively, it can be triggered by the implementation of emergency procedures. In the scenario of substantial penetration of variable renewable generation, at least some current studies foresee significant increases in regulation requirements [11], possibly leading to increased frequency of regulation shortages. Similarly, depending on how the operating reserve requirements are modified, there could be increased frequency of shortages in reserves. Scarcity pricing triggered by ancillary service shortages can thus provide incentives for enhanced capabilities to provide renewable integration capabilities, whether through providing additional revenues for investments in repowering of existing plants or stimulating entry of storage and demand response.

19.4.4 Market Participation of Demand Response and Storage Technology

As noted, a further transformation of the energy and ancillary services markets that could play a role in supporting efficient GHG emissions reductions, is the participation of demand response resources and storage technologies, recently termed nongeneration resources [28]. Efforts to reduce regulatory barriers and promote dispatchable demand response figure prominently in recent market reforms within and outside RTOs [28, 30, 54]. In principle, such demand response can provide energy, most importantly when supply is short (i.e., peak shaving), as well as regulation, operating reserves, and capacity for resource adequacy (as discussed in the next section).

Similarly, there are nascent efforts in all the RTOs to develop rules and pricing methods to bring smaller scale storage devices—such as flywheels, batteries, and compressed air storage that could scale up in size in coming years— into the energy and ancillary services markets for which they can qualify. These wholesale market-based efforts may be complemented in coming years by penetration of electrified transportation and other distribution-level storage-and-demand response technologies that, via the "smart grid" [17] and with appropriate retail rates, can provide large-scale load shifts in response to hourly and locational spot market prices and also frequency response.

Although demand response and storage might not heretofore have been central components of GHG policy, when GHG emissions are explicitly priced the market will provide these resources with the right price signals for location and for shifting consumption to lower emissions hours. For example, PJM [46] estimates the added value of energy efficiency and demand response in the presence of carbon pricing [46]. Hence, the programs and incentives for demand response and storage that are developing in the RTO markets (and elsewhere) could become important components of an efficient GHG policy.

19.5 Implications of GHG policy for resource adequacy

Vertically integrated utilities relied on multiyear integrated resource plans that included generation, transmission, and demand side investments to meet load growth over time and provide a sufficient reserve margin. In principle liberalized energy markets could, with much looser regulation of market offers and prices and with appropriate forward contracting, maintain resource adequacy, but many states and regional reliability entities in the United States have retained a resource adequacy requirement. This entails that LSEs procure installed capacity to ensure a planning reserve margin that is a fixed percentage, typically at least 15 percent, above their forecast peak loads. These requirements are specified to provide both overall system reliability and the reliability of load pockets. Failure by LSEs to meet these targets results in deficiency penalties (whereas failure to perform by capacity resources also encounters physical deratings or financial penalties).

The question of how to maintain resource adequacy under GHG policy is complex. As noted, a general concern in the electric power industry, unrelated to the design of particular resource adequacy mechanisms, is that the regulatory uncertainty over GHG policy will cause delays in generation investment, causing a reduction in supply adequacy margins [32, 43]. Compounding this trend is the likelihood of substantial fuel switching from coal to gas in anticipation of carbon regulation, causing reliability and fuel security concerns [42]. There is also concern about shortages of GHG allowances in key compliance periods that might affect particular plants needed for local reliability. And, as noted, in some states other environmental regulations, including for air and water quality, may force early retirements of existing plants and further create constraints on the types of new investments in generation. The western U.S. states also rely more substantially on hydropower, which adds uncertainty to resource adequacy and may be even more variable with forecast climatic changes that could shift snowpack accumulation and rainfall patterns.

With this backdrop, there are several alternative market and regulatory designs to achieve resource adequacy requirements in RTOs as well as continuing evaluation of their performance in setting market clearing prices, which this section will not review [2, 3, 4]. Rather, it discusses some of the implications of the changes in the resource mix under GHG policy for resource adequacy eligibility and capacity market outcomes.

In general, resource adequacy requirements and capacity markets are not intended to address the question of fuel diversity or GHG emissions abatement through technical specification of resource characteristics. Rather, all eligible capacity MW—determined by measures of availability, especially during peak periods—are interchangeable and tradable, regardless of fuel source.[19] Indeed, the current market design objective for regions with multiyear forward markets for capacity is to establish a framework in which all types of generation, demand response, and energy efficiency investments can compete to provide capacity (with future transmission assets represented in the auction to ensure accurate locational representation of capacity). The particular capabilities of the resources that enter into bilateral capacity contracts or earn capacity market revenues will be shaped by such resources' value in energy and ancillary services markets as well as by their relative costs of GHG allowances.[20]

Annual RTO assessments of summer peak loads have progressively incorporated demand response and variable-generation renewables. However, there is still little experience with these technologies as providing a substantial percentage of peak load capacity. The qualification of wind and solar capacity as resource adequacy resources will thus be the subject of considerable examination in coming years [42]. Apart from the general issue of variability, in some parts of the country, wind resources often operate at low-capacity factors during peak hours, when generation is most needed for resource adequacy. Table 19.5 shows the current RTO rules for counting nameplate wind capacity toward resource adequacy requirements, but methodologies are in evolution. There is also interest in coupling storage with wind or solar resources to improve the capacity value of such resources.

---

[19]In some states with resource adequacy requirements but without centralized capacity markets, resource adequacy MW are procured through bilateral contracts.

[20]In principle, the capacity payment covers the fixed costs that are not recovered through expected spot or forward market revenues.

Table 19.5 Calculation of wind capacity for resource adequacy in selected regions

ISO/RTO/Regulatory Agency	Methodology
PJM	Initial year of operation: 13% of nameplate capacity based on class average capacity factor during summer peak hours. Subsequent years based on rolling 3-year average of output during the hours of 2:00 and 6:00 pm from June 1 to August 31.
New York ISO	*Summer capacity credit:* Initial year of operation: 10% of nameplate capacity. Subsequent years use an average of output during the hours of 2:00 and 6:00 pm from June to August of the prior year. *Winter capacity credit:* Initial year of operation: 30% of nameplate capacity. Subsequent years use an average of output during the hours of 4:00 and 8:00 pm from December to February of the prior year.
ISO-New England	*Summer capacity credit:* Rolling 5-year average of output during the hours of 2:00 and 6:00 pm from June to September. *Winter capacity credit:* Rolling 5-year average of output during the hours of 6:00 and 7:00 pm from October to May.
California Public Utilities Commission (CPUC)	Resource adequacy requirement is monthly. Hence, for each month, a rolling 3-year average of monthly output during the hours of 12:00 and 6:00 pm on weekdays. Initial years of operation can use average for existing wind resources in zone.

The centralized capacity markets may take on additional value as a revenue stream for existing GHG-emitting thermal resources as variable-generation renewables, demand response, and storage technologies are loaded onto the grid, initially displacing the older, more inefficient coal and gas thermal plants from energy markets but then over time also displacing baseload thermal coal and gas resources unless carbon capture is developed. System operators are studying how to maintain reliability under such scenarios, and there will likely be a multiyear learning curve during which thermal resources may be needed for reliability, even as they lose market revenues. For that potential transitional period, market-based capacity contracts or cost-based reliability must-run (RMR) contracts could be needed to maintain the availability of such plants. Over the past few years, RTOs have been aiming to reduce RMR contracts; however, unless market-based revenues are sufficient, the reliability requirements caused by GHG policy may lead to an increase in such contracts.

From an RTO market perspective, energy efficiency investments are not energy or ancillary service resources (except indirectly, as they affect load forecasts), but they can be valued as peaking capacity offsets. Although regulatory proceedings will often assign energy efficiency investments a peaking capacity value when calculating avoided costs, the availability of a capacity market can provide a more accurate temporal and spatial value for such investments as well as a revenue stream for nonutility investments (if they meet the minimum MW to qualify). This extension of the capacity market design has only recently begun. Both the ISO-NE Forward Capacity Market (FCM) and the PJM Reliability Pricing Model (RPM) capacity markets allow both demand response and energy efficiency investments to participate as capacity resources.[21]

Demand response and energy efficiency resources are often mentioned together, but they are different. A demand response resource such as an interruptible load is comparable to generation when serving as a capacity resource. On the other hand, energy efficiency resources are different in that once the investment is made, the demand reduction stays in place over multiple years. This raises the question of whether the energy efficiency investment should be compensated over multiple years and also whether allocation of capacity charges to the reduced demand could overcompensate the investment. Additionally, effective measurement and verification of the investment are critical.

There will be significant challenges in revising the resource adequacy paradigm under GHG policy, due both to the expected penetration of variable generation along with newer nongeneration resources and energy efficiency. The RTO capacity market framework, although still under evaluation and subject to further refinement, does have the advantage of providing a transparent

---

[21]In ISO-NE, the first FCM auction, held in February 2008 for meeting capacity obligations in 2010–11, selected approximately 2500 MW of demand-side resources that included new and existing demand response and energy-efficiency capability. In PJM, in the base residual RPM auction for 2012–2013 delivery, 7047 MW of demand response (out of 9847 MW offered) along with 569 MW of energy efficiency (out of 653 MW offered) cleared the auction. Consequently, demand resources account for just over 5 percent of PJM's all time system peak of just under 145 GW.

approach to setting capacity prices and a basis for evaluating the performance of alternative resources over time.

19.6 Implications of GHG policy for transmission policy and planning

Climate policy is already affecting transmission policy, primarily through renewable energy initiatives, and in turn requiring adaptation of RTO transmission planning and generation interconnection procedures.[22] This section first outlines some U.S. policy goals and then turns to planning issues.

Prior to the current attention on climate policy, there has been the perception in the United States that parts of the transmission grid have been underbuilt, especially following the August 14, 2003, electrical blackout in the U.S. Midwest and Northeast. As such, the federal government has taken a role in providing incentives for rate-based transmission investment as well as providing the authority to site transmission in locations considered vital to the reliability of the grid. Under the Energy Policy Act of 2005 (EPAct, 2005), these include FERC Order 679 [27], which offers higher incentive rates for certain transmission projects, the repeal of the Public Utilities Holding Company Act of 1935 (PUHCA),[23] designation of national interest electricity transmission corridors for congestion (NIETC), and federal backstop siting authority.[24] At the time of this publication, DOE has designated NIETCs, but FERC has yet to assert federal siting authority to ensure that projects are built. There are also several proposed pieces of legislation that would reinforce federal authority for transmission siting but they vary in details on whether this is limited to renewables or is broader to include transmission upgrades for reliability and economics.

---

[22]Although RTOs conduct planning, they do not, except in limited circumstances, have the ability to enforce transmission investment. Unlike in some countries in Europe, U.S. RTOs are not transcos, that is, owners and operators of the transmission grid subject to performance-based incentives (although there are transcos in the United States within the boundaries of RTOs). Instead, the transmission infrastructure is largely owned by utilities that receive a fixed regulated rate of return on their investment but allow access to their transmission facilities to be controlled by the RTO. Further details on transmission structures, rate incentives, and pricing in the United States can be found in Singh (2008).

[23]PUHCA is a 1935 law that limited the merger of electric utilities and established stringent financial regulation holding companies.

[24]As noted, transmission siting remains a state-level decision except in the case of areas designated as NIETC.

This recent interest in transmission investment at the federal level could dramatically accelerate under climate policy, since some changes in the generation mix to reduce GHG emissions could require major investments in transmission.[25] This includes, most notably, greatly increased access to renewables. Heretofore, planning for access to renewables has taken place under state jurisdictions, but it could also become a new focus for federal transmission policy. Most renewables with substantial near-term expansion potential, including wind, solar, and, in some areas, geothermal, that are forecast to provide a growing percentage of low- or no-carbon generation are typically sited in locations remote from major loads.[26] There would need to be a substantial investment in high-voltage transmission to access such locations [12, 13, 39], which will need to be justified on a cost-effectiveness basis and consider the barriers to transmission in some regions that might delay renewables development beyond anticipated GHG reduction targets (see Wiser and Hand, Chapter 9 in this volume; [8, 9, 18]. Any future revival of nuclear power could also require significant investment in new or upgraded transmission, unless new nuclear plants are sited largely to displace existing coal or gas base-load plants already on the grid. To assist efficient investment for carbon abatement, all these considerations will need to be evaluated in the context of RTO planning functions.

19.6.1 RTO PLANNING FUNCTIONS

RTOs are now the planners of the U.S. transmission network across large regions of the country (individually and cooperatively with neighboring RTOs and non-RTO areas), and they manage generation and transmission interconnection once the siting of those assets has been approved by state regulators (with the exception of the federal siting jurisdiction discussed previously).

The regulatory framework for RTO transmission planning and interregional coordination is developed within FERC Order 2000 [25], which delineates the essential functions of RTOs, with additional requirements under FERC Order 890 [28], such as improved transparency and participation of demand response. Most RTOs now develop and publish a 10- to 15-year transmission plan for their regions [12, 13, 34, 41, 47]. The primary driver

---

[25]Note that transmission investment targeted for enabling remote renewables interconnection will not lend itself to every initiative under EPAct 2005. For example, the designation of NIETC requires a determination of the presence of significant levels of existing congestion. This is unlikely in the case of yet-to-be-tapped generation capacity from wind and solar resources.

[26]For a review of conceptual transmission plans and issues on both the Western and Eastern RTO regions, see CAISO 2008a,b, CEC 2007, 2008, DOE 2008b, and JCSP (2009).

in the transmission plans is an assessment of reliability criteria subject to load growth, generation additions and deactivations, and other anticipated changes during the time horizon. Transmission plans have also increasingly attempted to quantify economic benefits of alternative transmission upgrades—for example, through evaluation of reduced congestion and the corresponding net change in social welfare. They also consider alternative nontransmission investments, such as generation and demand-side management, including energy efficiency and demand response.[27]

The forward-looking nature of RTO transmission plans, encompassing both reliability and economic considerations, will increasingly require assumptions regarding factors such as carbon prices, availability of carbon allowances, operational requirements created by variability of renewable resources, and costs of renewable integration, in addition to other assumptions, such as fuel prices, that are already used in developing the plans [12, 13, 34, 39]. Ignoring the impact of carbon costs and restrictions that are expected within the time horizon covered by the plans could result in suboptimal decisions—yet accounting for carbon costs with any precision can be challenging. ISO-New England, which includes states in the RGGI, has been among the first to incorporate some measure of carbon pricing in its initial forecasts of supply impacts and to raise issues about availability of allowances [34]. In other regions, such as California, the plans already include a focus on transmission needed for a substantial penetration of renewables under the California RPS [11, 12], while the state has assumed the carbon displacement of the RPS in meeting its GHG abatement target [7, 15]. However, the two environmental policy goals (RPS and GHG) and their implications for future transmission and resource decisions have not yet been formally evaluated jointly through the planning and resource adequacy process.

19.6.2 REDUCING TRANSMISSION AND INTERCONNECTION BARRIERS TO REMOTE RENEWABLE RESOURCES

Because of the jurisdictional issues in transmission and generation siting between federal and state agencies, there is a growing focus on new types of cooperative planning, incorporating RTOs to develop substantial new renewable resources at remote locations. These planning approaches have been in

---

[27]Different methodologies for evaluating the costs and benefits of alternative investments have been developed in different RTOs (CAISO, 2004; PJM, 2008). See also Singh (2008) for a discussion of congestion metrics.

development for several years at the state and regional level and are likely to be at the forefront of the next phase of major U.S. transmission expansion in support of renewables and climate policy.

Elements of this new planning framework include:

■ Identification and evaluation of appropriate sites for renewable development

■ Evaluation of cost-effective transmission expansion options, including costs of transmission facilities and other requirements of renewable integration

■ Reduction of barriers to interconnection by renewable developers (i.e., generation interconnection)

■ Reduction of barriers to transmission siting

A first step to effectively planning for access to remote renewables is the identification of promising renewable energy zones. State agencies and RTOs then evaluate alternative transmission projects to reach those zones. There are several such initiatives that have been largely completed in the United States, notably in California and Texas, with several more under way on a regional basis. In Texas, the legislature created a new paradigm for transmission development by giving the Public Utility Commission of Texas (PUCT) statutory authority to designate five Competitive Renewable Energy Zones (CREZs) in West Texas and the Texas Panhandle and authorize the construction of transmission to enable wind generation in the CREZs to be deliverable to load centers in Texas. ERCOT (the RTO in Texas) initiated a study to develop an optimal transmission plan to achieve this objective and outlined scenarios for PUCT consideration and eventual selection (intended to support development of 18,000 MW of wind generation) [21]. Transmission project construction will not be limited to incumbent utilities.

A roughly parallel approach has also been developed in California with the identification of CREZs and a state Renewable Energy Transmission Initiative (RETI) to identify and rank needed additional or upgraded transmission facilities. In California, recent estimates by CAISO project at least seven transmission projects at a cost of $6.4 billion required in California and to interconnect with and upgrade the Western grid under a 33 percent RPS [13, 14]. However, barriers to new transmission in California remain high; every project that proposes to interconnect renewables must currently go through a detailed and lengthy state regulatory approval process [16].

The CREZ approach is now being considered nationally in proposed legislation under which areas across the country would be declared as National

Renewable Energy Zones (NREZ) if they could generate more than 1000 MW of renewable generation. There is also a Western multistate initiative, the Western Renewable Energy Zones (WREZ), being developed for the states outside California. An interregional planning effort supported by most of the major transmission operators in the Eastern Interconnection has identified conceptual transmission plans to support wind integration that cross the territories of Midwest ISO, Southwest Power Pool, PJM, and other Midwestern and Southern utilities [39]. In this analysis, meeting a 20 percent RPS requirement with wind resources across the Eastern Interconnection would require an investment of $80 billion in transmission, along with a $1 trillion investment in capital costs of new generation [39].

Even with identification of CREZs and the appropriate transmission facilities, there are further barriers to transmission development and generator interconnection. One of these barriers is that with each line supporting multiple developers, upgrades needed to interconnect a specific generator or generators on a radial segment are paid by the generator. However, planning such upgrades for each wind or solar generator likely to interconnect at a remote location would not be efficient, leading to a barrier to investment as both transmission and generation wait for the other to appear first. A regulatory approach to reducing this barrier was developed in California, where Southern California Edison (SCE) worked with the CAISO to obtain authorization from state and federal regulators to build the rate-based transmission in advance of the generation, with assured cost recovery. As renewable providers do interconnect, they then pay for an increasing portion of the line through transmission access charges. The transmission projects that would qualify would need to satisfy a set of conditions, such as access of significant locational constrained resources, a minimum level of interest, and operational control of the project by the CAISO [29].

19.6.3 Merchant Transmission

Within RTO regions, there are a small number of "merchant" transmission operators, owners of transmission facilities that are not under cost-based regulated rates but rather are financed on the basis of forward contracts for energy price differences (congestion rents).[28] A recent example of a merchant line that was subscribed primarily by renewables is the Montana/Alberta link, with a

---

[28]Merchant facilities are also eligible for awards of financial transmission rights (whereas builders of cost-based regulated facilities are typically eligible for awards of such rights only if they also are load-serving entities using the transmission facilities). However, it is also noteworthy that most merchant projects have been across regions and without the award of financial transmission rights.

300 MW interconnection between Montana and the Canadian province of Alberta. Merchant facilities typically rely on "open seasons," when transmission customers line up in advance to support the project. However, one challenge in such approaches may be to not fully exploit the economies of scale in transmission by being limited in the size of the project. There have been references to "supersizing" transmission projects but without a clear policy directive on how this might be supported. Recent decisions by FERC have indicated some flexibility in cases where policies (e.g., the requirement to conduct an open-season) formulated to promote open-access seemed to be too restrictive for specific projects promoting renewable or clean energy.[29]

19.6.4 GENERATION INTERCONNECTION

Finally, there has been some effort by federal and state regulators, with RTO involvement, to standardize rules and streamline queues for generation interconnection, motivated in part by the large number of applications for renewable and gas plants. FERC has required transmission providers, including RTOs, to incorporate approved procedures for new generator interconnection into their transmission tariffs [26]. These have been separated into Large Generator Interconnection Procedures (LGIPs) for generators of 20 MW or greater and Small Generator Interconnection Procedures (SGIPs) for smaller units. Besides specifying standards that a new generator must meet, a significant focus of interconnection procedures has been to study the impact a new generator will have on the transmission grid and determine the transmission upgrades that will be necessary. The interconnecting generator typically pays for the upgrades and receives a rebate in transmission access charges after it is in operation. There are regional differences regarding precisely what share of upgrade costs a new generator must pick up on the front end.

The "first-come, first-served" approach of the procedures has come under significant stress as the applications for interconnections have increased dramatically in recent years, as shown in Figure 19.3. More than half the new requests have come from renewable sources. Clearly, not every request is likely to correspond to an actual new build, but they must all be studied. Further,

---

[29]For example, the Zephyr and Chinook projects that seek to connect wind resources in the Northern Rockies to load centers in the Southwest were approved by FERC to allocate half their capacity to anchor customers representing wind generation. In another decision involving participant funding, the structure for a proposed 1200 MW HVDC project from Quebec to New Hampshire was approved where the entire capacity would be used by an anchor customer to sell hydro power to utilities in New England under long-term Power Purchase Agreements (PPAs).

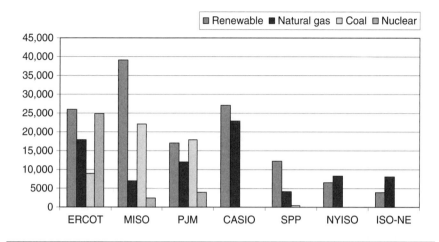

Figure 19.3   Proposed new generation by fuel type based on generation interconnection queues in RTOs.
Source: ISO/RTO Council [37].

each time a request that has already been studied drops out, it impacts others farther down the queue, leading to a restudy. To improve the process, recent reforms have focused on the concept of clustering similar requests and studying them together as well as increasing the size of deposits for studies.

19.7 Conclusion

After more than a decade of U.S. electric power sector restructuring, RTOs are now major institutions in much of the country, with the responsibility to ensure short-term and long-term reliable system operations, efficient and non-discriminatory allocation of transmission capacity, and efficient auction markets for spot energy and ancillary services as well as certain forward products, such as capacity and financial transmission rights. The RTO planning process has evolved such that it can play a leading role in integrating reliability and economic assessments of proposed infrastructure. As such, RTOs will play a prominent role in aiding the power sector in meeting challenges associated with climate-change policies.

Because of their regional scope, the RTOs provide a significant capability to coordinate the many changes on the grid that are likely to take place under climate policy. RTOs already engage in the long-term planning and operational assessments that must precede efficient investment in generation, transmission, and demand-side management as well as provide, with FERC guidance, standardized and increasingly streamlined generator interconnection procedures. Such long-term planning capability, combined with the price signals provided by the wholesale markets, will be brought to the fore as the sector pursues alternative approaches to GHG emissions abatement. The next few years could see a more coordinated national approach to expanding and upgrading the high-voltage transmission grid in much the same way as the interstate highway system was built (as opposed to the more piecemeal approaches driven by incremental upgrades that have comprised the bulk of transmission investment over the past decade). At the same time, there will be increased interest in exploiting the capabilities of "smart grids" and distributed energy resources.

However, there is much further assessment and development to be done before RTOs and other power system operators can fully evaluate the choices and explain all trade-offs associated with various strategies for meeting GHG goals. For example, the analysis of how to integrate substantial quantities of variable-generation renewable resources reliably and at reasonable cost requires much further investigation. Nevertheless, it is clear that regional diversity of renewable resources could be used to reduce variability of generation, supported by a central regional operator with improved weather-forecasting capabilities and the ability to call on a least-cost mix of regional resources to balance that variability and provide frequency control.

From a wholesale market perspective, the scheduling, dispatch, and pricing functions of the RTO markets already allow for a least-cost regional market solution that will transparently reflect the cost of GHG abatement in energy and ancillary service market prices (as soon as it is priced explicitly) as well as the effect of congestion and losses on the cost of delivering power to specific locations. Moreover, spot market prices will also help guide the use of clean distributed energy, storage resources, demand response, and energy efficiency in providing energy services, ancillary services, and resource adequacy. RTOs will continue to evaluate how to adapt existing market designs and operational procedures to assist penetration of viable technologies that mitigate GHG emissions. Finally, as independent regional energy institutions in their own right, RTOs will increasingly be called on to assist their stakeholders and policymakers in understanding reliability and power market considerations when setting policy goals for the electric power sector.

ACKNOWLEDGMENT

The authors would like to thank the editor, Fereidoon Sioshansi for helpful comments and great patience and Jorge Riveros for research assistance.

REFERENCES

[1] Allen E, Brown D, Jaske M, Loyer J, Ratliff R. Potential Impacts of the South Coast Air Quality Management District Air Credit Limitations and Once-Through Cooling Mitigation on Southern California's Electricity System. California Energy Commission Staff Paper (February); 2009.

[2] Adib P, Schubert E, Oren S. Competitive Electricity Markets. In: Sioshansi F, editor. Resource Adequacy: Alternate Perspectives and Divergent Paths. Amsterdam: Elsevier; 2008.

[3] Bowring JE. Competitive Electricity Markets. In: Siohansi F, editor. The Evolution of PJM's Capacity Market. Amsterdam: Elsevier; 2008.

[4] Brattle Group. Review of PJM's Reliability Pricing Model (30 June), www.pjm.com; 2008.

[5] Burtraw D. State Efforts to Cap the Commons: Regulating Sources or Consumers? Resources for the Future. RFF DP 07-49; November 2008.

[6] Bushnell J. The Implementation of California AB 32 and Its Impact on Wholesale Electricity Markets. CSEM WP 170; August 2007.

[7] California Air Resources Board (CARB). Climate Change Proposed Scoping Plan: A Framework for Change (October), available at www.arb.ca.gov; 2008.

[8] California Energy Commission (CEC). 2007 Integrated Energy Policy Report, Committee Final Report. CEC-100-2007-008-CTF. Available at www.energy.ca.gov; November 2007.

[9] California Energy Commission (CEC). 2008 Integrated Energy Policy Report Update, CEC-100-2008-008-CMF. Available at www.energy.ca.gov; 2008.

[10] California ISO (CAISO). Transmission Economic Assessment Methodology (TEAM), available at www.caiso.com; June 2004.

[11] California ISO (CAISO). Integration of Renewable Resources: Transmission and Operating Issues and Recommendations for Integrating Renewable Resources on the California ISO-Controlled Grid, available at www.caiso.com; November 2007.

[12] California ISO (CAISO). 2008 CAISO Transmission Plan: A Long-Term Assessment of the California ISO's Controlled Grid (2008 – 2017), Planning and Infrastructure Development Department, www.caiso.com; January 2008.

[13] California ISO (CAISO). Report on Preliminary Renewable Transmission Plans, Final Version, available at www.caiso.com; August 6, 2008b.

[14] California Public Utility Commission (CPUC). Renewables Portfolio Standard. Quarterly Report; October 2008.

[15] California Public Utility Commission (CPUC). Final Opinion on Greenhouse Gas Regulatory Strategies. Rulemaking 06-04-009 (Filed April 13, 2006), Date of Issuance October 22, 2008. 2008b.

[16] California Public Utility Commission (CPUC). Alternate Proposed Decision Granting a Certificate of Public Convenience and Necessity for the Sunrise Powerlink Transmission Project. November 18, 2008c.

[17] Department of Energy (DOE). Smart Grid: Enabler of the New Energy Economy, Electricity Advisory Committee. Available at www.oe.energy.gov/eac.htm; December 2008a.

[18] Department of Energy (DOE). 20% Wind Energy by 2030: Increasing Wind Energy's Contribution to U.S. Electricity Supply. DOE/GO-102008-2567; May 2008b.

[19] Electric Power Research Institute (EPRI). The Power to Reduce CO_2 Emissions: The Full Portfolio. Discussion Paper. The EPRI Energy Technology Assessment Center; August 2007.

[20] Electric Power Research Institute (EPRI). Program on Technology Innovation: Electric Technology in a Carbon-Constrained World. Final Report, 1013041. 2005.

[21] Electricity Reliability Council of Texas (ERCOT). Competitive Renewable Energy Zones (CREZ) Transmission Optimization Study, ERCOT System Planning. Available at www .ercot.com; April 2, 2008.

[22] Environmental Protection Agency (EPA). Inventory of U.S. Greenhouse Gas Emissions and Sinks: 1990–2006, available at www.epa.gov; April 2008.

[23] Federal Energy Regulatory Commission (FERC). Promoting Wholesale Competition Through Open Access Non-Discriminatory Transmission Services by Public Utilities; Recovery of Stranded Costs by Public Utilities and Transmitting Utilities. Order No. 888, FERC Stats. & Regs. 31,036, April 24, 1996a.

[24] Federal Energy Regulatory Commission (FERC). Open Access Same-Time Information System (formerly Real-Time Information Networks) and Standards of Conduct. Order No. 889, 75 FERC 61,078, April 24, 1996b.

[25] Federal Energy Regulatory Commission (FERC). Regional Transmission Organizations. Order No. 2000, 89 FERC 61,285, December 20, 1999c.

[26] Federal Energy Regulatory Commission (FERC). Standardization of Generator Interconnection Agreements and Procedures. Order No. 2003, 68 FR 49845, FERC Stats. & Regs. 31,146, August 19, 2003.

[27] Federal Energy Regulatory Commission (FERC). Promoting Investment Through Pricing Reform. Order No. 679, Docket No. RM06-4-001, FERC 31,222, July 31, 2006.

[28] Federal Energy Regulatory Commission (FERC). Preventing Undue Discrimination and Preference in Transmission Service. Order No. 890, 18 CFR Parts 35 and 37, February 16, 2007a.

[29] Federal Energy Regulatory Commission (FERC). Order Granting Petition for Declaratory Order. Docket No. EL07-33-000, April 19, 2007b.

[30] Federal Energy Regulatory Commission (FERC). Wholesale Competition in Regions with Organized Electric Markets. Order No. 719, 18 CFR Part 35, October 17, 2008.

[31] Helman U, Hobbs BF, O'Neill R. The Design of U.S. Wholesale Energy and Ancillary Service Auction Markets: Theory and Practice. In: Sioshansi F, editor. Competitive Electricity Markets. Amsterdam: Elsevier; 2008.

[32] Holt L, Sotkiewicz P, Berg S. (When) to Build or Not to Build?: The Role of Uncertainty in Nuclear Power Expansion. Texas Journal of Oil, Gas, and Energy Law 2008;3(2):174–214.

[33] International Energy Agency (IEA). Energy Technology Perspectives 2008: Scenarios & Strategies to 2050. Paris: International Energy Agency; 2008.

[34] ISO New England Inc. 2008 Regional System Plan, available at www.iso-ne.com; October 16, 2008.

[35] ISO New England Inc. New England Electricity Scenario Analysis: Exploring the Economic, Reliability, and Environmental Impacts of Various Resource Outcomes for Meeting the Region's Future Electricity Needs, available at www.iso-ne.com; August 2, 2007.

[36] ISO New England Inc. Evaluation of Impact of Regional Greenhouse Gas Initiative CO_2 Cap on the New England Power System, Jim Platts, Hang Wang, System Planning Dept, ISO New England Inc. Available at www.iso-ne.com; October 26, 2006.

[37] ISO/RTO Council (IRC). White Paper on Interconnection Queue Management Process, available at www.isorto.org; January 10, 2008.

[38] ISO/RTO Council (IRC). Increasing Renewable Resources: How ISOs and RTOs Are Helping Meet This Public Policy Objective, available at www.isorto.org; October 16, 2007.

[39] Joint Coordinated System Planning (JCSP). Joint Coordinated System Plan 2008, available at www.jcspstudy.org; February 2009.

[40] Kirby B. Frequency Regulation Basics and Trends. Oak Ridge: Tennessee; December 2004.

[41] Midwest ISO. Transmission Expansion Plan 2008. December 2008.

[42] North American Electric Reliability Corporation (NERC). Reliability Issues White Paper: Accommodating High Levels of Variable Generation, available at www.nerc.com; draft November 2008b.

[43] North American Electric Reliability Corporation (NERC). Special Report: Electric Industry Concerns on the Reliability Impacts of Climate Change Initiatives, available at www.nerc .com; November 2008a.

[44] O'Neill R, Helman U, Hobbs BF, Baldick R. Independent System Operators in the USA: History, Lessons Learned, and Prospects. In: Sioshansi F, Pfaffenberger W, editors. Electricity Market Reform: An International Perspective. Amsterdam: Elsevier; 2006.

[45] PJM. PJM Manual 15: Cost Development Guidelines, Revision 10, available at www.pjm .com; June 1, 2009a.

[46] PJM. Potential Effects of Proposed Climate Change Policies on PJM's Energy Market, available at www.pjm.com; January 27, 2009b.

[47] PJM. 2007 Regional Transmission Expansion Plan (RTEP), Report, available at www.pjm .com; February 27, 2008.

[48] PJM Market Monitoring Unit (PJM MMU). 2007 State of the Market Report: Volume 2 Detailed Analysis, available at www.pjm.com; March 11, 2008.

[49] Potomac Economics, Ltd. Estimated Market Effects of the New York Renewable Portfolio Standard, By Potomac Economics, Ltd. Independent Market Advisor to the New York Independent System Operator; June 2005.

[50] Regional Greenhouse Gas Initiative (RGGI). Potential Emissions Leakage and the Regional Greenhouse Gas Initiative (RGGI): Final Report of the RGGI Emissions Leakage Multi-State Staff Working Group to the RGGI Agency Heads, available at www.rggi.org; March 2008.

[51] Singh H. Transmission Markets, Congestion Management and Investment. In: Sioshansi F, editor. Competitive Electricity Markets. Amsterdam: Elsevier; 2008.

[52] Western Climate Initiative (WCI). Design Recommendations for the WCI Regional Cap-and-Trade Program, available at www.westernclimateinitiative.org; September 23, 2008.

[53] Wiser R, Barbose G. Renewable Portfolio Standards in the United States: A Status Report with Data through 2007. Lawrence Berkeley National Laboratory; LBNL-154E. April 2008.

[54] Zarnikau J. Demand Participation in Restructured Markets. In: Sioshansi F, editor. Competitive Electricity Markets. Amsterdam: Elsevier; 2008.

Two Surprises and One Insight

Fereidoon P. Sioshansi

Menlo Energy Economics

When I started working on this project, my genuine hope and expectation were that by capturing the inherent synergies in the options and solutions examined by the various contributing experts to this book, one would be able to claim that the problem of carbon emissions associated with electricity generation could be successfully addressed. Moreover, I was convinced that although the electricity sector faces a challenging transition to a lower-carbon future, there were opportunities to become a part of the solution rather than remain part of the problem.

Along the way, I was confronted by two surprises and one insight, perhaps trivial to everyone else. The first surprise was that as I learned more about the scale of the problem, I became alarmed about the sheer immensity of the challenges ahead. The second surprise was that as I studied the various chapters describing the obstacles and limitations of each technology or solution, it began to dawn on me that the task at hand is more daunting than I had originally imagined.

My personal insight—and I must emphasize that this view is *not* necessarily shared by the contributors to the book and should *not* be attributed to anyone other than me—is that although a lot can, and should, be done on the electricity *generation* side, in all likelihood these efforts will *not* be sufficient. If the goal is to limit concentration of carbon in the atmosphere to, say, 450–550 parts per million by 2050, as many scientists have suggested, I am now convinced that we must work equally hard on the *demand* side, and that effort should not be limited to improving the efficiency with which electricity—or energy—is used but on altering the nature of the way we *work* and ultimately examining our *values* and *lifestyles*.

The following excerpt from the preface to the book by Professor Wolfgang Pfaffenberger states the same idea but more eloquently:

As the chapters of this book make clear, we cannot rely entirely on changes in the supply side of the equation to reduce the industry's carbon footprint. Changes in the demand side as well as changes in energy consumption habits—and perhaps more profoundly—lifestyles changes may ultimately be needed to address the carbon problem.

Index